f

μ

ξ

ν

g

ε

λ

γ

η

i

α ν

A

π

ρ ζ

ο

b

ω

θ

Q

ς

s

ERAU-PRESCOTT LIBRARY
WITHDRAWN

Bb

ALSO BY MARCIA BARTUSIAK

Thursday's Universe

Through a Universe Darkly

A Positron Named Priscilla (coauthor)

Einstein's Unfinished Symphony

ARCHIVES
OF THE
UNIVERSE

1/07

ARCHIVES OF THE UNIVERSE

A TREASURY OF ASTRONOMY'S
HISTORIC WORKS OF DISCOVERY

EDITED AND WITH INTRODUCTIONS BY

Marcia Bartusiak

PANTHEON BOOKS, NEW YORK

Copyright © 2004 by Marcia Bartusiak

All rights reserved under International and Pan-American Copyright
Conventions. Published in the United States by Pantheon Books,
a division of Random House, Inc., New York, and simultaneously in
Canada by Random House of Canada Limited, Toronto.

Pantheon Books and colophon are registered trademarks of
Random House, Inc.

Owing to limitations of space, permissions acknowledgments follow
the index, constituting an addition to this copyright page.

Library of Congress Cataloging-in-Publication Data

Archives of the universe: a treasury of astronomy's historic works of
discovery / edited and with introductions by Marcia Bartusiak.
p.   cm.
Includes index.
ISBN 0-375-42170-X
1. Astronomy—History. I. Bartusiak, Marcia, [date]

QB15.A75 2004
520'.9—dc22   2004040057

www.pantheonbooks.com

Printed in the United States of America

First Edition

2 4 6 8 9 7 5 3 1

In memory of my father
Czeslaw A. Bartusiak
My first guide to the stars

Contents

Contents

Preface

> I do not doubt that in the course of time this new science will be improved by further observations, and still more by true and conclusive proofs. But this need not diminish the glory of the first observer. My regard for the inventor of the harp is not made less by knowing that his instrument was very crudely constructed and still more crudely played. Rather, I admire him more than I do the hundreds of craftsmen who in ensuing centuries have brought this art to the highest perfection.
>
> —*Galileo, writing in praise of William Gilbert,*
> *investigator of magnetism*

Often missing from astronomy textbooks are the voices of the scientists themselves. When Nicolaus Copernicus bravely placed the Sun at the center of the solar system in his noted *De revolutionibus*, he confesses, "The scorn which I had to fear on account of the newness and absurdity of my opinion almost drove me to abandon a work already undertaken." Yet his friends convinced him to reconsider and announce to the world that "the sun, as if resting on a kingly throne, governs the family of stars which wheel around." The Danish astronomer Tycho Brahe, upon seeing a brilliant new star (now known to be a supernova) in the heavens in 1572, wrote, "I was so astonished at this sight that I was not ashamed to doubt the trustworthiness of my own eyes." Galileo informed the readers of his *Sidereus nuncius* that whatever region of the Milky Way to which "you direct your spyglass, an immense number of stars immediately offer themselves to view, of which very many appear rather large and very conspicuous but the multitude of small ones is truly unfathomable." More recently when Margaret Geller, John Huchra, and Valérie de Lapparent in 1986 recognized that galaxies congregated to form a bubblelike lattice through the universe, they playfully compared the distribution to "suds in the kitchen sink."

This book was compiled and written to reacquaint us with these words of discovery. Within these pages are excerpts from the seminal reports that first introduced both scientists and the public to a wondrous variety of celestial phenomena and in the process moved our understanding of the cosmos forward. The arc of astronomy's history and development over the centuries is traced through its key discovery papers, from the time of ancient Greece to the present day.

How does one define an astronomical discovery? It is essentially a new theory or cosmic feature that, once revealed, will have fundamental effects on how we view our place in the universe, and will lead to new and fruitful explorations. Pinpointing such a discovery can be an ever-evolving affair. The initial finding may be found over time to be faulty and incomplete, as further evidence accumulates and the original discovery is amended. But this does not alter its ultimate importance. Ptolemy's influential model of an Earth-centered universe proved erroneous in the end, but it was a valuable tool nonetheless in making initial predictions concerning the motions of the planets. It was a necessary first step in a continuing journey of further testing and improvement. In the twentieth century the steady-state theory of the universe was a potent presence, galvanizing astronomers to seek and weigh observational evidence that ultimately confirmed the Big Bang model of the universe's origin. And so these findings are included here. A full history of astronomy's evolution is incomplete without including such material.

The excerpts within this work were gleaned from a range of sources and reflect the evolution of science communication over the centuries. Books, such as Ptolemy's famous *Almagest*, were at first the primary mode of disseminating information about the heavens, often taking years to assemble and publish. Not until the mid-seventeenth century were special journals at last established, which accelerated the time between a discovery and its announcement to the scientific community at large. The Royal Society of London first established its *Philosophical Transactions* in 1665; similar journals followed in France, Germany, the United States, and elsewhere. Three centuries later Helmut Abt, for many years editor in chief of the *Astrophysical Journal,* noted in the journal's centennial issue in 1999 that "the current rate of publication in astronomy and astrophysics is so large that we rarely have the time to read the fundamental papers upon which our current research rests. In fact, some people, especially younger astronomers and students, are tempted to think that most of the important results were derived in the past decade or two." A cursory check of this book's table of contents vividly confirms this conjecture: the pace of discovery in astronomy has increased exponentially over the centuries, accompanied by a parallel increase in the papers announcing them. The first quarter of this book deals with discoveries made from around 400 B.C. to A.D. 1800, twenty-two centuries of time. It takes only one more century, though, from 1800 to 1900, to provide the next quarter's worth of discoveries. The remaining papers—fully half of this book—were published in the twentieth century.

This pattern is duplicated in the way astronomy has been conducted over time. Starting in antiquity and continuing into the modern era, astronomy was primarily an effort carried out by amateurs and scholars. Not until the last few centuries did astronomy start becoming a full-fledged profession—initially at universities and observatories in Europe, then in the United States. The first important observatories were established with the support of monarchies and republics: Tycho Brahe's Uraniborg and the great observatories of Paris and Greenwich, as well as the U.S. Naval Observatory. With both the rise of astrophysics at the turn of the twentieth century and the wealth accumulated as a result of the industrial revolution, private and corporate benefactors came to sponsor the construction of major observatories devoted to astrophysical research. Then after World War II, new technologies and the lure of space-based research once again enhanced government support of the astronomical enterprise.

The number of professional astronomers was initially small. In 1899, for example, the Astronomical and Astrophysical Society of America held its first annual meeting at the Yerkes Observatory in Wisconsin. Fifty astronomers attended, presenting a total of thirty-one papers. A century later, the American Astronomical Society, as it is now known, registered some twelve hundred astronomers for its annual meeting in Chicago, where about eight hundred papers were submitted. Worldwide, professional astronomers presently number in the thousands, working in not only academia but also government institutions, industry, and the military. And they are no longer just skilled observers but often trained physicists applying their knowledge to problems of astrophysics or geophysicists interpreting the wide-ranging data arriving from the many spacecraft probing our solar system. It is now an era of astronomical teams, akin to the large scientific collaborations long found in particle physics.

Advances in astronomy are intimately linked with advances in instrumentation. Each new development in technology—the telescope, the spectroscope, and later the various detectors for gathering wavelengths of light beyond the visible spectrum—immediately led to new celestial discoveries. Crucial as well has been the development of clever electronic instruments such as photoelectric devices for precise measurement of stellar luminosities and spectral properties. After World War II, there was a bloom of new technologies: radio telescopes, rockets, satellites, and computers for handling theoretical calculations once too intricate to carry out manually. This book emphasizes the new findings rather than advances in instrumentation, though recognizing that the two often go hand in hand. (The paper on x-ray astronomy is the perfect example—a discovery made simultaneously as new instrumentation is tested for the first time.) Surveys and catalogs, so vital to the astronomical enterprise, are also left out for lack of space.

The essays introducing each moment of discovery were written to stand alone, inviting the reader to peruse his or her interests in no particular order. Yet it can be profitable to read the sections in sequence to perceive how the questions that observers asked of the heavens evolved over time. The advances were sometimes subtle, at other times wrenching and abrupt. At first astronomers asked: What is out there? Where is it located? How is it moving? Are there patterns to the motion? The first accurate measurements were made of such celestial phenomena as the lunar cycles, planetary motions, and eclipses. Later came the more sophisticated interpretations of those celestial measurements. Copernicus, with his doctrine of an Earth revolving about the Sun, removed us from the center of the cosmos. Galileo, one of the first to scrutinize the heavens with a telescope, introduced us to a universe of unexpected change and intricacy. Newton, from a theoretical perspective never before undertaken, ushered in an age of reason, convincing us that motions everywhere, both in the heavens and on Earth, are described by the same set of physical laws.

All the while, observers continued their surveys, enlarging their catalogs of stellar positions and magnitudes. Well into the nineteenth century, astronomers seemed fated to be no more than celestial librarians, spending their nights painstakingly monitoring the positions of the planets and stars, whose motions were guided by Newton's laws. But then there was a decided shift in the basic job of an astronomer. With the introduction of spectroscopes, devices that could break a star's light into its array of colors, astronomers were at last able to discern and interpret the chemistry of the heavens. Astronomy was revolutionized once again, shortly after World War II, when new technologies allowed astronomers to expand their searches into other regions of the electromagnetic spectrum, beyond the narrow band visible to the human eye. With radio and x-ray telescopes came the discovery of such celestial objects as quasars and pulsars. As a consequence, the calm and dignified universe of the ancients was transformed into a cosmos of titanic energies, full of explosive and violent behaviors. At the same time, the space age offered a new means of exploring the solar system and heavens, beyond the Earth's obscuring atmosphere.

Astronomers' curiosity, once confined to the Milky Way, extended outward to encompass an entire universe of galaxies, as if carried by the cosmic expansion itself. By the beginning of the twenty-first century, astronomy was pushing the borders of the visible universe further and further back in time, almost to the very moment of creation. Today the physics of the microcosm—the quantum mechanics that describes the nature of elementary particles—is joining the physics of the macrocosm to interpret what is being found in that primordial era.

With such a wealth of choices before me, I decided to feature those discoveries that evoke a particular resonance within both popular culture and the history of astronomy. These are the discoveries that came to define the universe as we now know it: its composition, its various members, its structure, its evolution. Left out are the many papers that established the theoretical basis for analyzing the data that led to those discoveries—breakthroughs in understanding such phenomena as radiative transfer, ionization, and polarization. This was not intended to diminish their importance, for without such physical understanding astronomers would have had no means of interpreting the radiation that falls upon their detectors. Satellite explorations of the planets, as well as surveys of the celestial sky from space at various wavelengths, have also offered breathtaking new vistas to explore, but it is difficult to include all of these studies in one book. Consequently the focus is on astronomy's great moments — insightful predictions or discoveries that revealed entirely new paradigms, rather than the observations that provided more detail on phenomena already known (although references to such additional works are included in the notes).

Some discoveries are obvious and immediate: the discovery of gamma-ray bursts, for instance, or the sudden realization of a quasar's true distance. Here the choice of the discovery paper is clear. Other findings emerge slowly, such as the dark-matter mystery or the source of stellar power. In such cases, one definitive paper is often difficult to choose. The discovery is progressively unveiled in a series of papers from a variety of contributors. In those cases, a particular paper is chosen to represent a more extended set.

Many of the papers, particularly those from the modern era, are hardly narrative literature. The mathematics can often be unwieldy and the issues fairly complex. The more technical details have been omitted to make the papers more accessible to a general audience. Yet it is still informative to view the excerpts from a step back and notice the increasing complexity of the scientific issues addressed and the manner in which they are conveyed. For uniformity and readability of this book, the punctuation and spelling of the British papers from the seventeenth and eighteenth centuries have been adjusted, but sparingly, to retain as much as possible the literary flavor of that time. It is quite fascinating to observe the evolution of the scientific enterprise through a comparison of the style and substance in the papers over the centuries. The principal origin of publication also evolves, from Europe to the United States, as the world order shifts from empire to superpower.

Galileo cannot contain his emotions when he introduces his telescopic discoveries. "In this short treatise," he writes, "I propose great things for inspection and contemplation by every explorer of Nature. Great, I say, because of the excellence of the things themselves, because of their newness, unheard of through the ages, and also because of the instrument with the benefit of which they make themselves manifest to our sight." The style today is to keep such subjective comments out of the scientific literature, relegating them to statements at a news conference or in a memoir. Astronomical papers today are terse, highly mathematical, and far less introspective. And almost no original findings are deferred for publication in a book, as centuries ago.

Ideas as well can go in and out of fashion. In Harlow Shapley's *Source Book in Astronomy,* published in 1929, Laplace's nebular hypothesis for the formation of the solar system is declared extinct, though noted as a "valuable and powerful stimulant to scientific thought throughout the nineteenth century." By then it had been replaced by the tidal evolution theory, the Sun's reputed close encounter with a passing star, which now seems quaint as astronomers have once again returned to Laplace's basic scheme. Likely there will be papers in this volume that also will lose their luster in the face of additional data, better modeling, and more astute reasoning. What is captured here is a snapshot of astronomical thinking at the start of the twenty-first century.

I

THE ANCIENT SKY

The fortuitous alliance of two agents led to the birth of astronomy: curiosity and necessity. From savannas, mountaintops,
and forest clearings, the first celestial observers looked up at
the nighttime sky and beheld a vast, pitch-black bowl covered with
sparkling pinpoints of light. While likely awed at first by this jewel-
like canopy, imagining it as a vaulted roof through which the fires of
the gods flickered, prehistoric peoples eventually learned there were
practical benefits to studying the sky's incessant motions and cycles.

Tracing out patterns of stars—constellations—became a useful
procedure for establishing a coordinate system across the heavens,
and the leisurely parade of these stellar figures over the seasons
served as valuable markers for navigation, agriculture, and time-
keeping. As the Greek poet Hesiod advised in the eighth century
B.C., "When the Pleiades, daughters of Atlas, are rising, begin the
harvest, the plowing when they set."[1] Here the farmer was instructed
to reap winter wheat in the spring, when the Pleiades rise with the
Sun, and to plant seeds in the fall, when the notable constellation
sets in the west before sunrise. In ancient Egypt observers noticed
that the brilliant star Sirius rose in the east right before dawn, at the
very time that the Nile river experienced its annual flooding.

In the high northern latitudes it was the Sun's recurrent passage
that held particular significance. As winter approaches there, the
Sun's path moves steadily southward, just as the days and nights get

colder. Primitive megaliths were built to mark the pivotal moment—winter solstice—when the Sun would (to much thanksgiving) turn back and once again rise higher in the sky.

Relics from the first days of civilization showcase the ancients' intense intellectual curiosity about the nighttime sky. Inscriptions on Chinese oracle bones recorded the appearance of bright comets and "guest stars"; Mayan hieroglyphic books documented the movements of Venus with remarkable precision; clay tablets in Babylonia, dating back nearly four thousand years, chronicled the cyclic movements of the Moon and the "wanderers"—the planets—among the fixed stars. With Alexander the Great conquering Persia in 331 B.C., Babylonia's tradition of keen skywatching merged with Greece's focus on geometric models of the universe's workings.

It was the ancient Greeks who were most influential in moving contemplation of the cosmos from pure mythology to a more reasoned cosmology. They began to wonder about the essential nature of heavenly bodies: how they moved, what they were made of. The first challenge was explaining why that small, elite group of wanderers—the Sun, Moon, Mercury, Venus, Mars, Jupiter, and Saturn—moved at differing speeds and in some cases even stopped and moved backwards in the sky. The Pythagoreans, so enamored of numbers and harmonic relationships, influenced Greek astronomers to solve this problem by thinking of the heavens as a geometric system. With beauty and harmony requiring uniform motion, imaginative models were devised to have the planets move via a set of nested spheres. It was the first attempt at a grand unified theory: explaining celestial motion with a single, all-encompassing mechanism. At the same time, these early astronomers came to understand the source of the Moon's light, the cause of eclipses, and the true shape of the Earth. They also used their knowledge of geometry to tackle such other questions as the size of our planet and the distances to the Sun and Moon.

There were some prescient speculations on the nature of the solar system in this ancient era. In the fourth century B.C. Heraclides of Pontus suggested that night and day were due to the rotation of the Earth. Aristarchus of Samos later put the Sun at the center of his model of the universe. But these ideas never flourished, as they were overshadowed by the authoritative cosmology espoused by the noted philosopher Aristotle. At the center of his cosmos was the Earth, composed of one of the four basic elements. Surrounding this were water and air. The last element, fire, extended outward to the Moon. In this realm, life was mortal and imperfect. The heavenly bodies, on the other hand, inhabited a domain that was flawless and eternal—the celestial spheres in perpetual circular motion. This model held sway for nearly twenty centuries, and astronomy progressed only when observers such as Hipparchus and Ptolemy dared to tinker with its precepts. Hipparchus discovered the precession of the equinoxes, and Ptolemy cleverly amended Aristotle's standard model to make it agree better with observation. From these early creative attempts to understand the star-studded sky, a science was born.

1 / Mayan Venus Tables

Some 3,500 years ago the Maya came to occupy a large territory in Central America that now covers southern Mexico, Guatemala, and northern Belize. By A.D. 200 (or even earlier) these native Mesoamericans had advanced from a simple Stone Age existence cultivating maize and squash to a sophisticated civilization whose cities contained impressive stone temples, palaces, and pyramids.

Along with hieroglyphic writing, the Maya developed refined astronomical methods that were representative of astronomical techniques carried out by early societies in other parts of the world—for example, in ancient Egypt and Babylonia. Like the observations made by those other ancient cultures, Mayan stargazing focused on cycles. They viewed the cosmos as a repetitive machine whose operation could offer their society advance knowledge of its fate if the celestial movements could be accurately tracked. Their meticulous observations of the nighttime sky were closely linked with their ritualistic needs.

Of particular importance to the Maya was the planet Venus, whose appearance in the sky follows a distinct pattern. When Venus passes between the Earth and the Sun (a configuration known as inferior conjunction), it cannot be seen for eight days. Eventually Venus is spotted in the morning sky, after it proceeds in its orbit and rises just before the Sun. For 263 days on average it remains visible

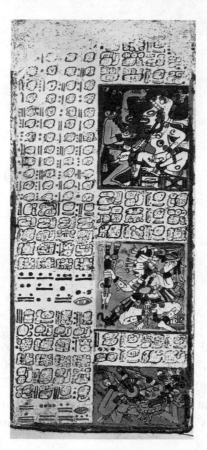

Figure 1.1: A representative page from the Venus section of the Dresden Codex. On the right, three pictures are aligned from top to bottom. A celestial observer, often seated on a throne covered with planetary symbols, resides in the top box. On this page, a fierce god looks at the god of maize holding a vase or cup. In the middle is a depiction of the Venus god, menacing his victims with spears at heliacal rising—when Venus appears with the Sun in the east just after inferior conjunction. This was the time associated with the most dire omens. The victim is at the bottom, knocked down by the dart that has pierced his shield.

in the morning, until it passes behind the Sun (superior conjunction) and again disappears. Fifty days later it comes back into view but this time as the evening star, remaining in the night sky for another 263 days until it reaches inferior conjunction once again. The period from one inferior conjunction to the next totals 584 days.

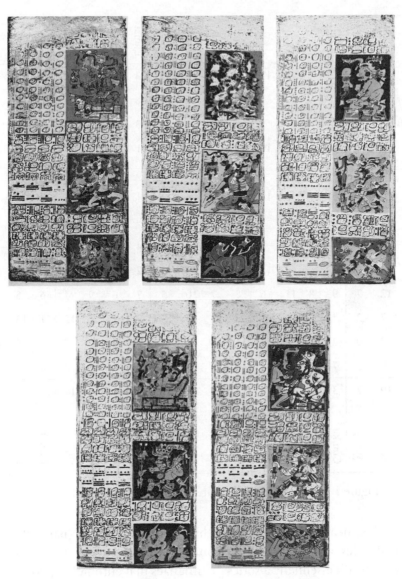

Figure 1.2: Pages 46–50 of the Dresden Codex, which display the Mayan Venus tables.

The Maya followed this cycle and recorded their knowledge of its predictability in the Dresden Codex, one of three surviving Mayan hieroglyphic books transported to Europe as spoils of the Spanish conquest. In each book, intricate glyphs are displayed on a single sheet of paper, pounded from the inner bark of a wild ficus tree and folded into separate leaves like a screen. The Dresden

Cib	Cimi	Cib	Kan	Ahau	Oc	Ahau	Lamat	Kan	Ix	Kan	Eb	Lamat	Etz'nab	Lamat	Cib	Eb	Ik	Eb	Ahau
3	2	5	13	2	1	4	12	1	13	3	11	13	12	2	10	12	11	1	9
11	10	13	8	10	9	12	7	9	8	11	6	8	7	10	5	7	6	9	4
6	5	8	3	5	4	7	2	4	3	6	1	3	2	5	13	2	1	4	12
1	13	3	11	13	12	2	10	12	11	1	9	11	10	13	8	10	9	12	7
9	8	11	6	8	7	10	5	7	6	9	4	6	5	8	3	5	4	7	2
4	3	6	1	3	2	5	13	2	1	4	12	1	13	3	11	13	12	2	10
12	11	1	9	11	10	13	8	10	9	12	7	9	8	11	6	8	7	10	5
7	6	9	4	6	5	8	3	5	4	7	2	4	3	6	1	3	2	5	13
2	1	4	12	1	13	3	11	13	12	2	10	12	11	1	9	11	10	13	8
10	9	12	7	9	8	11	6	8	7	10	5	7	6	9	4	6	5	8	3
5	4	7	2	4	3	6	1	3	2	5	13	2	1	4	12	1	13	3	11
13	12	2	10	12	11	1	9	11	10	13	8	10	9	12	7	9	8	11	6
8	7	10	5	7	6	9	4	6	5	8	3	5	4	7	2	4	3	6	1
4	14	19	7	3	8	18	6	17	7	12	0	11	1	6	14	10	0	5	13
Yaxkin	Zac	Zec	Xul	Cumku	Zotz'	Pax	Kayab	Yax	Muan	Ch'en	Yax	Zip	Mol	Uo	Uo	Kankin	Uayeb	Mac	Mac

Figure 1.3: Translation of the dates on the Mayan Venus tables.

Codex, nearly four yards long, has thirty-nine leaves (painted on both sides) and takes its name from the German city in which it now resides. It's essentially a series of almanacs that chronicle upcoming astronomical events, including lunar and solar eclipses. The glyphs depict a number of gods—some benevolent, others auguring bad tidings. They include the rain gods, the god of maize, a merchant god, a sun god, and a moon goddess, as well as several deities associated with death. Astronomy in this case was being used for divine forecasting, to help farmers predict times of drought, fearsome storms, or an abundant crop.

    The Venus tables are found on six pages of the Dresden Codex and tell the reader when Venus will appear and disappear in the morning and evening sky over time. One of the Maya's greatest achievements in their tracking of Venus was recognizing that the planet's cycle was not a full 584 days but slightly less (583.92 days). They adjusted their calendar for this difference with astounding accuracy. Concern for such precision is essentially what transformed an astrological endeavor into a science.

The Maya had names for units of time comparable to days, months, decades, and centuries, although on a far different counting system. The *uinal* (or *winal*), for example, consisted of 20 days, a sort of month. At times 5 extra days were added. A *tun*, close to a year, was 360 days. Twenty *tuns* made up a *katun*, while 20 *katuns* was a *baktun*. A listing of the number of these "centuries," "decades," "years," "months," and days since some day zero was one way that the Maya generated a calendar.[2] The Mayan Venus tables, though, use another system, where each day is represented by a set of numbers (a dot is one; a bar is five) and names. These dates are listed on the upper left of a page. Notice in Figure 1.1 that every line in this section has four symbol groups. Each specifies an important date in one complete cycle of the Venus period: first the day when Venus will disappear at superior conjunction; next when it reappears as the evening star; then when it disappears at inferior conjunction; and finally when it becomes visible once again as the morning star. Continuing along a selected line across five of the tables (see Figures 1.2 and 1.3) covers a unique period of 2,920 days, over which five Venus cycles equal eight Earth years. At the end, the user of the table moves on to the next line of the five-table chart, where the cycle begins again.

2 / Proof That the Earth Is a Sphere

In 342 B.C. King Philip II of Macedonia brought the learned philosopher Aristotle to his court to tutor his son, who as a man would become Alexander the Great. Soon after Alexander assumed the throne, Aristotle established a school in Athens where he continued his wide-ranging studies in philosophy, logic, politics, physics, and biology.

Aristotle's writings on astronomy were compiled in a four-volume text entitled *De caelo*, "On the Heavens." The cosmology that he established within this work wielded a powerful influence on astronomers for nearly twenty centuries. Aristotle reasoned that the Earth was an arena of change and imperfection. Its basic elements, earth and water, moved downward, because they sought their natural place. The other essential elements, air and fire, moved upward. To Aristotle, though, the region inhabited by the planets and stars was far different. That was because celestial bodies did not move up or down but rather traveled in circles, an eternal path of perfection and uniformity. Given that difference, he concluded that the heavens had to be composed of another substance altogether, the *aether*.

There were irregularities in the heavenly movements that required clarification. To explain retrograde, the appearance of a planet's traveling backward on the nighttime sky, Aristotle adopted a model of planetary motion devised by his contemporary Eudoxus of

Cnidus.* Eudoxus, a geometer, introduced the idea of the planets and stars being moved by heavenly spheres rotating about the Earth. Each planet was attached to several spheres. The orderly motion of these spheres, once combined, produced a planet's deceivingly irregular movement. Aristotle modified this system and advocated it so commandingly that it was difficult for celestial observers even to consider models that didn't incorporate his vision of circular perfection. During the rise of Christianity, it was transformed into God's chosen design. Modern astronomy would not emerge until astronomers were willing to break away from such Aristotelian notions and consider other possibilities to explain their observations (see Part II, "Revolutions").

De caelo is more philosophy than true astronomy, but there is one section in which Aristotle does rally decent evidence in behalf of his conclusion: the sphericity of the Earth. That the Earth is round was likely recognized by Greek thinkers, such as the Pythagoreans, two centuries beforehand. For that matter, ancient seafarers saw far-off landmarks or ships dipping below sea level and probably realized that the Earth was curved. The earliest surviving proof, however, can be traced to Aristotle. The major part of his argument comes from his physics: a spherical shape, he theorizes, is naturally generated as the terrestrial elements fall downward to seek the center. But he doesn't ignore observational data (including, as we shall see, the range of elephants). Absorbing the wisdom of others before him, he notes that the Earth's shadow, as it passes over the Moon during an eclipse, is always circular. He adds to this by noting that travelers will see different stars come into view as they travel north and south, a change that would not occur if the Earth were flat. Such reasoning was a tremendous advance over earlier guesses on the Earth's shape, such as Anaximander's suggestion in the sixth century B.C. that it was a cylinder freely suspended in space. As described by Hippolytus in his *Refutation of All Heresies*, "Its form is

* Over the course of an evening, a planet moves east to west with the celestial sphere, like any other celestial object. However, from one night to the next, a planet will be seen to move from west to east against the background of stars because of its orbital motion around the Sun. Occasionally, though, the planet's long-term movement reverses direction, and the planet will move, for a short time, from east to west against the background constellations. This reversal is known as retrograde motion. It occurs when the Earth overtakes the planet in its orbit.

rounded, circular, like a stone pillar; of its plane surfaces one is that on which we stand, the other is opposite."[3]

Aristotle's astronomical commentary also includes the earliest recorded mention of the Earth's circumference, 400,000 stades, although he doesn't provide his source or the method of the calculation. Originally a stade was the length of a traditional Greek racetrack, but eventually different types of stades came into use. Values can vary from roughly 8 to 10 stades per mile. So Aristotle's declared circumference is between 40,000 to 50,000 miles, not outrageously larger than the true measurement of nearly 25,000.

From *De caelo*
by Aristotle

Translated by J. L. Stocks

The shape of the heaven is of necessity spherical; for that is the shape most appropriate to its substance and also by nature primary.

First, let us consider generally which shape is primary among planes and solids alike. Every plane figure must be either rectilinear or curvilinear. Now the rectilinear is bounded by more than one line, the curvilinear by one only. But since in any kind the one is naturally prior to the many and the simple to the complex, the circle will be the first of plane figures. Again, if by complete, as previously defined, we mean a thing outside which no part of itself can be found, and if addition is always possible to the straight line but never to the circular, clearly the line which embraces the circle is complete. If then the complete is prior to the incomplete, it follows on this ground also that the circle is primary among figures. And the sphere holds the same position among solids. For it alone is embraced by a single surface, while rectilinear solids have several. The sphere is among solids what the circle is among plane figures. Further, those who divide bodies into planes and generate them out of planes seem to bear witness to the truth of this. Alone among solids they leave the sphere undivided, as not possessing more than one surface: for the division into surfaces is not just dividing a whole by cutting it into its parts, but division of another fashion into parts different in form. It is clear, then, that the sphere is first of solid figures. . . .

With regard to the shape of each star, the most reasonable view is that they are spherical. It has been shown that it is not in their nature to move themselves, and, since nature is no wanton or random creator, clearly she will have given things which possess no movement a shape particularly unadapted to movement. Such a shape is the sphere, since it possesses no instrument of movement. Clearly then their mass will have the form of a sphere. Again, what holds of one holds of all, and the evidence of our eyes shows us that the moon is spherical. For how else should the moon as it waxes and wanes show for the most part a crescent-shaped or gibbous figure, and only at one moment a half-moon? And astronomical arguments give further confirmation; for no other hypothesis accounts for the crescent shape of the sun's eclipses. One, then, of the heavenly bodies being spherical, clearly the rest will be spherical also. . . .

There are similar disputes about the *shape* of the earth. Some think it is spherical, others that it is flat and drum-shaped. For evidence they bring the fact that, as the sun rises and sets, the part concealed by the earth shows a straight and not a curved edge, whereas if the earth were spherical the line of section would have to be circular. In this they leave out of account the great distance of the sun from the earth and the great size of the circumference, which, seen from a distance on these apparently small circles appears straight. Such an appearance ought not to make them doubt the circular shape of the earth. But they have another argument. They say that because it is at rest, the earth must necessarily have this shape. For there are many different ways in which the movement or rest of the earth has been conceived. . . .

Its shape must necessarily be spherical. For every portion of earth has weight until it reaches the center, and the jostling of parts greater and smaller would bring about not a waved surface, but rather compression and convergence of part and part until the center is reached. The process should be conceived by supposing the earth to come into being in the way that some of the natural philosophers describe. Only they attribute the downward movement to constraint, and it is better to keep to the truth and say that the reason of this motion is that a thing which possesses weight is naturally endowed with a centripetal movement. When the mixture, then, was merely potential, the things that were separated off moved similarly from every side towards the center. Whether the parts which came together at the center were distributed at the extremities evenly, or in some other way, makes no difference. If, on the one hand, there were a similar movement from each quarter of the extremity to the single center, it is obvious that the resulting mass would be similar on every side. For if an equal amount is added on every side the extremity of the mass will be everywhere equidistant from its center, i.e. the figure will be spherical. But neither will it in any way affect the argument if there is not a similar accession of concurrent fragments from every side. For the greater quantity, finding a lesser in front of it, must necessarily drive it on, both having an impulse whose goal is the center, and the greater weight driving the lesser forward till this goal is reached. In this we have also the solution of a possible difficulty. The earth, it might be argued, is at the center and spherical in shape: if, then, a weight many times that of the earth were added to one hemisphere, the center of the earth and of the whole will no longer be coincident. So that either the earth will not stay still at the center, or if it does, it will be at rest without having its center at the place to which it is still its nature to move. Such is the difficulty. A short consideration will give us an easy answer, if we first give precision to our postulate that any body endowed with weight, of whatever size, moves towards the center. Clearly it will not stop when its edge touches the center. The greater quantity must prevail until the body's center occupies the center. For that is the goal of its impulse. Now it makes no difference whether we apply this to a clod or common fragment of earth or to the earth as a whole. The fact indicated does not depend upon degrees of size but applies universally to everything that has the centripetal impulse. Therefore earth in motion, whether in a mass or in fragments, necessarily continues to move until it occupies the center equally every way, the less being forced to equalize itself by the greater owing to the forward drive of the impulse.

If the earth was generated, then, it must have been formed in this way, and so clearly its generation was spherical; and if it is ungenerated and has remained so always, its character must be that which the initial generation, if it had occurred, would have given it. But the spherical shape, necessitated by this argument, follows also from the fact that the motions of heavy bodies always make equal angles, and are not parallel. This would be the natural form of movement towards what is naturally spherical. Either then the earth is spherical or it is at least naturally spherical. And it is right to call anything that which nature intends it to be, and which belongs to it, rather than that which it is by constraint and contrary to nature. The evidence of the senses further corroborates this. How else would eclipses of the moon show segments shaped as we see them? As it is, the shapes which the moon itself each month shows are of every kind—straight, gibbous, and concave—but in eclipses the outline is always curved: and, since it is the interposition of the earth that makes the eclipse, the form of this line will be caused by the form of the earth's surface, which is therefore spherical. Again, our observations of the stars make it evident, not only that the earth is circular, but also that it is a circle of no great size. For quite a small change of position to south or north causes a manifest alteration of the horizon. There is much change, I mean, in the stars which are overhead, and the stars seen are different, as one moves northward or southward. Indeed there are some stars seen in Egypt and in the neighborhood of Cyprus which are not seen in the northerly regions; and stars, which in the north are never beyond the range of observation, in those regions rise and set. All of which goes to show not only that the earth is circular in shape, but also that it is a sphere of no great size: for otherwise the effect of so slight a change of place would not be so quickly apparent. Hence one should not be too sure of the incredibility of the view of those who conceive that there is continuity between the parts about the pillars of Hercules and the parts about India, and that in this way the ocean is one. As further evidence in favor of this they quote the case of elephants, a species occurring in each of these extreme regions, suggesting that the common characteristic of these extremes is explained by their continuity. Also, those mathematicians who try to calculate the size of the earth's circumference arrive at the figure 400,000 stades. This indicates not only that the earth's mass is spherical in shape, but also that as compared with the stars it is not of great size.

3 / Celestial Surveying

The ancient Greeks were very interested in estimating the sizes of the heavenly bodies and their distances from the Earth. The Greek mathematician Aristarchus was the first investigator to shift this discussion from sheer philosophical speculation to a rigorous scientific examination. A native of Samos, a large island in the Aegean Sea, Aristarchus lived from around 310 to 230 B.C.

Skilled in geometry and anticipating the methods of trigonometry to come, Aristarchus conceived a means for calculating relative celestial distances. His report, *On the Sizes and Distances of the Sun and Moon*, is the only one of his works that still exists, and it was the first astronomical treatise of its kind put forth in ancient Greece. Just as he would create a proof in abstract geometry, he initially established a set of axioms and then proceeded through a series of deductions to arrive at his astronomical conclusions.

The complete work involves eighteen propositions in all, the most cited being the seventh, which deals with the distance of the Sun from the Earth (in terms of the Earth-Moon distance). He determined this by envisioning the moment when the Moon is half full and the angle between the Earth, Moon, and Sun is a right angle. By then estimating the second angle—made between the Moon, Earth, and Sun—he recognized that he could determine the ratio of the Earth-Moon distance to the Earth-Sun distance. Aristarchus's technique was a valid one and ingenious for its time;

he failed to find today's accepted values only because one of his measurements was in serious error. Aristarchus underestimated the Moon-Earth-Sun angle by nearly three degrees, and this led him to incorrectly derive that the Sun was some nineteen times farther out from the Earth than the Moon (a distance twenty times too small). Later in his treatise, extending his geometric logic and taking the size of the Earth's shadow during an eclipse into account, he went on to conclude that the size of the Sun was only seven times that of the Earth. (It's more than a hundred times wider.)

Yet this attempt at sizing up the universe, however flawed, did lead to some intriguing repercussions, for which Aristarchus is probably best known. Though his celestial sizes and distances were vastly off, he did come to realize that the Sun was bigger than the Earth. It is suspected that this impelled Aristarchus to develop the first heliocentric model of the universe. He deemed it more reasonable that the smaller body, the Earth, would be in orbit around the larger one, the Sun. Aristarchus's direct words on this idea do not survive, but his theory was mentioned by others in letters and books, including Archimedes in the *Sand Reckoner*. "His hypotheses," wrote Archimedes, "are that the fixed stars and the Sun remain unmoved, that the Earth revolves about the Sun in the circumference of a circle, the Sun lying in the middle of the orbit, and that the sphere of the fixed stars, situated about the same center as the Sun, is so great that the circle in which he supposes the Earth to revolve bears such a proportion to the distance of the fixed stars as the center of the sphere bears to its surface."[4] But in that ancient era, such speculation created more problems in physics than it answered. Why, for example, wouldn't the clouds in the sky be ripped away as the Earth traveled about? In addition, the fixed stars should appear to move as the Earth traveled in its orbit, an effect not observed. Aristarchus's fellow philosophers couldn't imagine the stars being so immensely distant from the Earth that such stellar parallax would be imperceptible. Aristarchus was declared impious "for having disturbed the peace of Hestia," by removing the Earth from its proper place as the central hub of the cosmos.[5] An Earth-centered universe held firm for nineteen more centuries.

From *On the Sizes and Distances of the Sun and Moon*
by Aristarchus

Translated by Thomas L. Heath

(Hypotheses)

1. *That the moon receives its light from the sun.*

2. *That the earth is in the relation of a point and center to the sphere in which the moon moves.*

3. *That, when the moon appears to us halved, the great circle which divides the dark and the bright portions of the moon is in the direction of our eye.*

4. *That, when the moon appears to us halved, its distance from the sun is then less than a quadrant by one thirtieth of a quadrant.*

5. *That the breadth of the* [earth's] *shadow is* [that] *of two moons.*

6. *That the moon subtends one fifteenth part of a sign of the zodiac.*

We are now in a position to prove the following propositions:

1. *The distance of the sun from the earth is greater than eighteen times, but less than twenty times, the distance of the moon* [from the earth]; this follows from the hypothesis about the halved moon.

2. *The diameter of the sun has the same ratio* [as aforesaid] *to the diameter of the moon.*

3. *The diameter of the sun has to the diameter of the earth a ratio greater than that which 19 has to 3, but less than that which 43 has to 6;* this follows from the ratio thus discovered between the distances, the hypothesis about the shadow, and the hypothesis that the moon subtends one fifteenth part of a sign of the zodiac . . .

Proposition 7

The distance of the sun from the earth is greater than eighteen times, but less than twenty times, the distance of the moon from the earth.

For let *A* be the center of the sun, *B* that of the earth [see Figure 3.1].

Let *AB* be joined and produced.

Let *C* be the center of the moon when halved;
let a plane be carried through *AB* and *C*, and let the section made by it in the sphere on which the center of the sun moves be the great circle *ADE*.

Let *AC, CB* be joined, and let *BC* be produced to *D*.

Then, because the point *C* is the center of the moon when halved, the angle *ACB* will be right.

Let *BE* be drawn from *B* at right angles to *BA*;
then the circumference *ED* will be one thirtieth of the circumference *EDA*; for, by hypothesis, when the moon appears to us halved, its distance from the sun is less than a quadrant by one thirtieth of a quadrant [Hypothesis 4].

Thus the angle *EBC* is also one thirtieth of a right angle.

Let the parallelogram *AE* be completed, and let *BF* be joined.

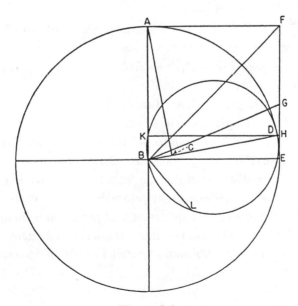

Figure 3.1

Then the angle *FBE* will be half a right angle.

Let the angle *FBE* be bisected by the straight line *BG;* therefore the angle *GBE* is one fourth part of a right angle.

But the angle *DBE* is also one thirtieth part of a right angle; therefore the ratio of the angle *GBE* to the angle *DBE* is that which 15 has to 2:
for, if a right angle be regarded as divided into 60 equal parts, the angle *GBE* contains 15 of such parts, and the angle *DBE* contains 2.

Now, since *GE* has to *EH* a ratio greater than that which the angle *GBE* has to the angle *DBE,*
therefore *GE* has to *EH* a ratio greater than that which 15 has to 2.

Next, since *BE* is equal to *EF,* and the angle at *E* is right, therefore the square on *FB* is double of the square on *BE.*

But, as the square on *FB* is to the square on *BE,* so is the square on *FG* to the square on *GE;*
therefore the square on *FG* is double of the square on *GE.*

Now 49 is less than double of 25,
so that the square on *FG* has to the square on *GE* a ratio greater than that which 49 has to 25;
therefore *FG* also has to *GE* a ratio greater than that which 7 has to 5.*

Therefore, *componendo, FE* has to *EG* a ratio greater than that which 12 has to 5, that is, than that which 36 has to 15.

But it was also proved that *GE* has to *EH* a ratio greater than that which 15 has to 2;
therefore *ex aequali, FE* has to *EH* a ratio greater than that which 36 has to 2, that is, than that which 18 has to 1;
therefore *FE* is greater than 18 times *EH.*

And *FE* is equal to *BE;*
therefore *BE* is also greater than 18 times *EH;*
therefore *BH* is much greater than 18 times *HE.*

But, as *BH* is to *HE,* so is *AB* to *BC,* because of the similarity of the triangles; therefore *AB* is also greater than 18 times *BC.*

And *AB* is the distance of the sun from the earth, while *CB* is the distance of the moon from the earth; therefore the distance of the sun from the earth is greater than 18 times the distance of the moon from the earth.

Again, I say that it is also less than 20 times that distance.

For let *DK* be drawn through *D* parallel to *EB,* and about the triangle

* The Pythagorean approximation to √2, namely, 7/5.

DKB let the circle *DKB* be described; then *DB* will be its diameter, because the angle at *K* is right.

Let *BL,* the side of a hexagon, be fitted into the circle.

Then, since the angle *DBE* is one thirtieth of a right angle, the angle *BDK* is also one thirtieth of a right angle;
therefore the circumference *BK* is one sixtieth of the whole circle.

But *BL* is also one sixth part of the whole circle.

Therefore the circumference *BL* is 10 times the circumference *BK*.

And the circumference *BL* has to the circumference *BK* a ratio greater than that which the straight line *BL* has to the straight line *BK;*
therefore the straight line *BL* is less than 10 times the straight line *BK.*

And *BD* is double of *BL;*
therefore *BD* is less than 20 times *BK.*

But, as *BD* is to *BK,* so is *AB* to *BC;*
therefore *AB* is also less than 20 times *BC.*

And *AB* is the distance of the sun from the earth,
while *BC* is the distance of the moon from the earth;
therefore the distance of the sun from the earth is less than 20 times the distance of the moon from the earth.

And it was before proved that it is greater than 18 times that distance.

4 / Measuring the Earth's Circumference

Aristotle was the first to mention a specific girth for the Earth, a circumference of 400,000 stades (roughly 45,000 miles). Archimedes in his noted work *Sand Reckoner* later gave a figure of 300,000 stades (around 34,000 miles). But no detailed calculation emerges until the work of Eratosthenes, a contemporary of Archimedes in the third century B.C.

Born around 276 B.C. in Cyrene, a Greek town on the North African coast in what is now Libya, Eratosthenes studied in Athens and became the tutor to a pharaoh of Egypt. He rose to become head of the great library in Alexandria. Eclectic in his interests, he was nicknamed "Beta" for his failure to be preeminent in any one field. He was a literary scholar, philosopher, and mathematician, as well as a specialist in geography. Indeed, he was the first to divide the world into a series of temperate, tropical, and frigid zones, work that Julius Caesar consulted two centuries later.

An interest in applying geometric techniques to geography likely led Eratosthenes to develop his famous method for measuring the Earth. His own words do not survive, but his general argument was summarized by a number of writers over the following centuries. Most notable was Cleomedes, author of an introductory astronomy book entitled *On the Orbits of the Heavenly Bodies*, written around the first century A.D. Eratosthenes had noticed that the Sun's rays fully reached the bottom of a well at noon on a midsummer day in the town of Syene along the upper Nile (now the site of the city of Aswan). There, near the Tropic of Cancer, a pole pointing straight upward at the summer solstice casts no shadow. Later, Eratosthenes measured the shadow cast by a vertical rod at noon in Alexandria, situated farther north. Applying some basic geometry, as explained by Cleomedes, Eratosthenes concluded that the two cities were separated by one-fiftieth of a circle. Figuring that the distance between Alexandria and Syene was 5,000 stades, fifty times that distance (250,000 stades) thus encompasses the full circumference of the Earth.

Eratosthenes introduced some errors into his calculation: Syene is not directly on the tropic, it does not share the same longitude as Alexandria (it's actually a bit farther east), and the two towns are somewhat closer than he assumed. But it was still an admirable achievement for its time and one not improved upon for many centuries. As mentioned in Chapter 2, the actual number of stades per modern mile is hard to assess, as there were different stades in use over that era. On average, 250,000 stades converts to 28,000 or 29,000 miles, fairly close to the real value of 24,900 miles for the Earth's circumference.

From *On the Orbits of the Heavenly Bodies* by Cleomedes

Translated by Thomas L. Heath

About the size of the earth the physicists, or natural philosophers, have held different views, but those of Posidonius and Eratosthenes are preferable to the rest. . . .

The method of Eratosthenes depends on a geometrical argument and gives the impression of being slightly more difficult to follow. But his statement will be made clear if we premise the following. Let us suppose . . . that Syene and Alexandria lie under the same meridian [longitude] circle; secondly, that the distance between the two cities is 5,000 stades; and thirdly, that the rays sent down from different parts of the sun on different parts of the earth are parallel; for this is the hypothesis on

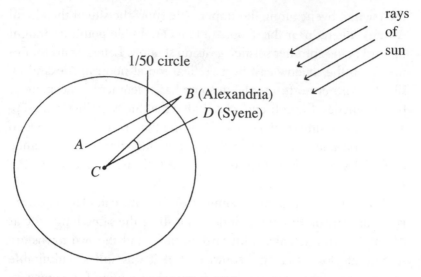

Figure 4.1

which geometers proceed. Fourthly, let us assume that, as proved by the geometers, straight lines falling on parallel straight lines make the alternate angles equal, and fifthly, that the arcs standing on (i.e., subtended by) equal angles are similar, that is, have the same proportion and the same ratio to their proper circles—this, too, being a fact proved by the geometers. Whenever, therefore, arcs of circles stand on equal angles, if any one of these is (say) one-tenth of its proper circle, all the other arcs will be tenth parts of their proper circles.

Any one who has grasped these facts will have no difficulty in understanding the method of Eratosthenes, which is this. Syene and Alexandria lie, he says, under the same meridian circle. Since meridian circles are great circles in the universe, the circles of the earth which lie under them are necessarily also great circles. Thus, of whatever size this method shows the circle on the earth passing through Syene and Alexandria to be, this will be the size of the great circle of the earth. Now Eratosthenes asserts, and it is the fact, that Syene lies under the summer tropic. Whenever, therefore, the sun, being in the Crab [constellation Cancer] at the summer solstice, is exactly in the middle of the heaven, the gnomons (pointers) of sundials necessarily throw no shadows, the position of the sun above them being exactly vertical; and it is said that this is true throughout a space three hundred stades in diameter. But in Alexandria, at the same hour, the pointers of sundials throw shadows, because Alexandria lies further to the north than Syene. The two cities lying under the same meridian great circle, if we draw an arc from the extremity of the shadow to the base of the pointer of the sundial in Alexandria, the arc will be a segment of a great circle in the (hemispherical) bowl of the sundial, since the bowl of the sundial lies under the great circle (of the meridian). If now we conceive straight lines produced from each of the pointers through the earth, they will meet at the center of the earth [see Figure 4.1]. Since then the sundial at Syene is vertically under the sun, if we conceive a straight line coming from the sun to the top of the pointer of the sundial, the line reaching from the sun to the center of the earth will be one straight line. If now we conceive another straight line drawn upwards from the extremity of the shadow of the pointer of the sundial in Alexandria, through the top of the pointer to the sun, this straight line and the aforesaid straight line will be parallel, since they are straight lines coming through from different parts of the sun to different parts of the earth. On these straight lines, therefore, which are parallel, there falls the straight line drawn from the center of the earth to the pointer at Alexandria, so that the alternate angles which it makes are equal. One of these angles is that formed at the center of the earth, at the intersection of the straight lines which were drawn from the sundials to the center of the earth; the other is at the point of intersection of the top of the pointer at Alexandria and the straight line drawn from the extremity of its shadow to the sun through the point (the top) where it meets the pointer. Now on this latter angle stands the arc carried round from the extremity of the shadow of the pointer to its base, while on the angle at the center of the earth stands the arc reaching from Syene to Alexandria. But the arcs are similar, since they stand on equal angles. Whatever ratio, therefore, the arc in the bowl of the sundial has to its proper circle, the arc reaching from Syene to Alexandria has that ratio to *its* proper circle. But the arc in the bowl is found to be one-fiftieth of its proper circle. Therefore the distance from Syene to Alexandria must necessarily be one-fiftieth part of the great circle of the earth. And the said distance is 5,000 stades; therefore the complete great circle measures 250,000 stades. Such is Eratosthenes' method.

5 / Precession of the Equinoxes

The Greek astronomer Hipparchus, the most accomplished observer of his time, was reportedly inspired by a *nova stella*, a new star, that he saw in the heavens in 134 B.C. This extraordinary event encouraged him to compile a highly accurate star catalog, the first of its kind ever completed. A catalog, he surmised, would come in handy in determining whether a star had altered either its position or its magnitude in later years. This work, which pegged the positions of some 850 stars, led to one of his greatest discoveries, an observation that presented a serious challenge to Aristotle's vision of an unchanging universe.

As with other notable personages of that far-off era, little is known of Hipparchus's background. The only firm fact is that he was born in Nicaea (now Iznik, Turkey) in the second century B.C. and died sometime after 127 B.C. At some point in his career he moved to Rhodes, where he made great advances in tracking the motions of the Sun and Moon, allowing him to make improved predictions of lunar and solar eclipses.

Hipparchus's work is primarily known through Ptolemy, the second-century astronomer who incorporated Hipparchus's findings into his major tome on astronomy, *Almagest*. Within its pages we learn that Hipparchus used the data in his catalog to make his most startling finding, the precession of the equinoxes. Hipparchus noticed that the entire celestial sphere slowly and regularly shifts over time. As Ptolemy put it, the stars may be "fixed," but not the sphere. This means that a constellation's position on the sky on a particular date (say, the spring or autumnal equinox) slightly changes from year to year. Hipparchus estimated the shift at one degree every hundred years. (Today we know it is one degree every seventy-two years.) In astrology, for example, a person's zodiac sign denotes the position of the Sun at the time of his or her birth. Since those signs were initially established, though, the constellations have moved considerably, by one entire sign (and a fraction more) eastward. A person said to be born under the sign of Aquarius today is actually born when the Sun is in the constellation Capricorn. This is why calendars that were dependent on the appearance of a certain constellation or star on a particular day drifted out of date.

According to Ptolemy, Hipparchus first discerned this effect as he was comparing the positions of certain stars in his catalog with the positions recorded by the astronomers Aristyllus and Timocharis of Alexandria 150 years earlier. As we see in the excerpt below, Hipparchus found that the star Spica was situated six degrees west of the autumnal equinox point. Timocharis, however, had measured it to be eight degrees west. When Hipparchus discovered other stars experiencing the same shift, he came to suspect that *all* the stars were involved in this movement. Ptolemy, a couple of centuries later, confirmed this discovery.

The actual reason for this apparent shift lies with the Earth. The Earth not only spins, it gyrates. Its axis continually traces out a circle, much the way a spinning top does, though the Earth is far more sluggish. It takes the Earth twenty-six thousand years to complete just one revolution in its precessional dance with respect to the "fixed" stars. As the Earth's axis slowly gyrates westward, the celestial sphere appears to move eastward (or as Ptolemy describes it, "rearward," since the movement is opposite to the daily motion of the stars from east to west). Astronomers had to wait for Isaac Newton's law of gravity to explain this peculiar motion. As the Earth rapidly spins, it bulges at its equator. With both the Sun and Moon gravitationally tugging on this bulge, the Earth ends up precessing like a lazy whirligig. And as the centuries progress, the constellations methodically shift their appearance from season to season.

From *Almagest*, Book VII
by Ptolemy

Translated by G. J. Toomer

1. {*That the fixed stars always maintain the same position relative to each other*}

First of all we must make the following introductory point. Concerning the terminology we use, in as much as the stars themselves patently maintain the formations [of their constellations] unchanged and their distances from each other the same, we are quite right to call them "fixed"; but in as much as their sphere, taken as a whole, to which they are attached, as it were, as they are carried around, also [like the other spheres] has a regular motion of its own towards the rear and the east with respect to the first [daily] motion, it would not be appropriate to call this [sphere] too "fixed." For we find that both these statements are true, at least on the [observational] basis afforded by the amount of time [preceding us]: even before this Hipparchus conceived of both these notions on the basis of the phenomena available to him, but under conditions which forced him, as far as concerns the effect over a long period, to conjecture rather than to predict, since he had found very few observations of fixed stars before his own time, in fact practically none besides those recorded by Aristyllos and Timocharis, and even these were neither free from uncertainty nor carefully worked out; but we too come to the same conclusions by comparing present phenomena with those of that time, but with more assurance, both because our examination is conducted [with material taken] from a longer time-interval, and because the fixed-star observations recorded by Hipparchus, which are our chief source for comparisons, have been handed down to us in a thoroughly satisfactory form.

First, then, no change has taken place in the relative positions of the stars even up to the present time. On the contrary, the configurations observed in Hipparchus' time are seen to be absolutely identical now too. . . . This can easily be seen by anyone who is willing to make an inspection of the matter and examine, in the spirit of love of truth, whether present phenomena agree with those recorded for Hipparchus' time. . . .

[Omitted here is a long list of stars surveyed by Hipparchus, which Ptolemy includes to demonstrate that no change in their relative positions to each other has occurred since Hipparchus's time.]

2. {*That the sphere of the fixed stars, too, performs a rearward motion along the ecliptic*}

From these considerations, and others like these, we can be assured that absolutely all the so-called fixed stars maintain one and the same position relative [to each other], and share one and the same motion. But the sphere of the fixed stars also performs a motion of its own in the opposite direction to the revolution of the universe, that is, [the motion of] the great circle through both poles, that of the equator and that of the ecliptic. We can see this mainly from the fact that the same stars do not maintain the same distances with respect to the solsticial and equinoctial points in our times as they had in former times: rather, the distance [of a given star] towards the rear with respect to [one of] those same points is found to be greater in proportion as the time [of observation] is later.

For Hipparchus too, in his work "On the displacement of the solsticial and equinoctial points," adducing lunar eclipses from among those accurately observed by himself, and from those observed earlier by Timocharis, computes that the distance by which Spica is in advance of the autumnal [equinoctial] point is about 6° in his own time, but was about 8° in Timocharis' time. For his final conclusion is expressed as follows: "If, then, Spica, for example, was formerly 8°, in zodiacal longitude, in advance of the autumnal [equinoctial] point, but is now 6° in advance," and so forth. Furthermore he shows that in the case of almost all the other fixed stars for which he carried out the comparison, the rearward motion was of the same amount. And we also, comparing the distances of fixed stars from the solsticial and equinoctial points as they appear in our time with those observed and recorded by Hipparchus, find that their motion towards the rear with respect to the ecliptic is, proportionally, similar to the above amount. . . .

. . . From this we find that 1° rearward motion takes place in approximately 100 years, as Hipparchus too seems to have suspected, according to the following quotation from his work "On the length of the year": "For if the solstices and equinoxes were moving, from that cause, not less than $\frac{1}{100}$th of a degree in advance [i.e. in the reverse order] of the signs, in the 300 years they should have moved not less than 3°."

6 / Ptolemy's *Almagest*

Claudius Ptolemaeus, Greek mathematician and scholar, lived from about 100 to 175, under the reigns of four Roman emperors, from Trajan to Marcus Aurelius. Ptolemy (as he is best known) worked in Alexandria, Egypt's vibrant intellectual center and site of the ancient world's finest library. A prodigious researcher, he wrote treatises on geography, optics, and astrology. But his greatest influence came in astronomy with the creation of his *Almagest*.

In the original Greek, Ptolemy's work was entitled *Mathematike syntaxis* or "Mathematical Treatise." When translated into Arabic, this august compilation gained the title *Al-Majisti*, "the greatest," and finally in medieval Latin *Almagestum* (hence its most noted name). Ptolemy made some original contributions to astronomy with his *Almagest*, but he also incorporated the technical work of many of his predecessors. It was the culmination of hundreds of years of thought and refinement, becoming the authoritative source in astronomy for fourteen centuries and pushing all previous works into the shadows. Ptolemy's choices set the standards for generations of astronomers to come. There had long been disagreements, for example, on how the planets lined up around the Earth. Ptolemy decided the order would be the Moon, Mercury, Venus, Sun, Mars, Jupiter, Saturn, and finally the fixed stars. Balance was maintained: there were just as many planets on one side of the Sun, a special body because of its radiance, as on the other. He also arrived at a measurement of the universe, the distance out to the celestial sphere of the fixed stars. He deduced it was at least 19,865 times the radius of the Earth, or some 75 million miles. Though now known to be wildly wrong, it was still an astounding figure for its time, a size difficult to comprehend when most people over their lifetimes never ventured more than a few miles from their homes. Moreover, Ptolemy refined the system of stellar magnitudes that had preceded him (and which is often speculatively credited to Hipparchus). The brightest stars in the sky were designated as magnitude 1, while those just barely visible to the naked eye were magnitude 6. Though awkward, this system is still in use today, modified and extended beyond magnitude 6 to accommodate the telescopic era.

Almagest is a formidable tome, its thirteen sections densely packed with diagrams, tables, and mathematical instructions for calculating planetary positions into both the past and the future. Books I and II set the stage by describing the overall structure of the universe. Ptolemy agrees with Aristotle that the Earth was the central celestial body and at rest with the universe. That the Earth would not be moving or spinning seemed rational. "Neither clouds nor other flying or thrown objects would ever be seen moving towards the east," he wrote, "since the earth's motion towards the east would always outrun and overtake them, so that all other objects would seem to move in the direction of the west and the rear."[6] This argument could not be adequately refuted until the seventeenth century, when natural philosophers began to understand the principle of inertia (that an object thrown into the air still shares the Earth's rotation and moves with it).

Book III goes on to establish the length of the year and promotes a resourceful innovation to account for the differing lengths of the seasons. The Moon is the subject of Books IV and V, while Book VI involves eclipse theory. Books VII and VIII deal with the fixed stars and include his catalog of some one thousand stars and forty-eight constellations. This catalog would not be surpassed until the fifteenth century with the work of Ulugh Beg at his great observatory in central Asia and of Tycho Brahe a hundred years later. The last five books of *Almagest* lay out Ptolemy's planetary theory, which more fully developed the idea of "epicycles," first discussed by such scholars as the Greek mathematician Apollonius and others a few centuries earlier. Here a planet does not just revolve in a perfect circular path around the Earth but also revolves in a circle around a point along that path. Such wheels within wheels in that pre-Copernican era helped explain retrograde (when a planet appears to be moving backward on the sky as the Earth in its orbit overtakes it).

Although Ptolemy's cosmic model would inevitably crumble with the advent of modern astronomy, it was a profoundly brave and brilliant step away from the perfect universe established by Aristotle. Championing and extending an idea used by Hipparchus three centuries earlier, Ptolemy shifted the planetary orbits with regard to the Earth, the universe's center. The orbits became *eccentric*. (See Figure 6.1.) It was a radical decision, given the prevailing wisdom of the time. Ptolemy recognized that from a central Earth observers watching the planets moving along their off-center orbits would see the celestial bodies vary their orbital speeds—fastest when nearest the Earth (perigee) and slowest when farthest away (apogee). By adopting this Hipparchian alteration he could explain why the Sun took more time traveling from the spring equinox to the fall equinox than from the fall to the spring. For the planets, Ptolemy also introduced into his model the *equant point*, the mirror point to the Earth on the opposite side of the center of the planet's orbit. If the planet, moving at its variable speed, could be viewed from that off-kilter vantage point, it would *appear* to be moving uniformly.

Ptolemy's insistence on matching theory with observation was a bold and modern move. (The fact that it would also help his astrological forecasting was a pleasant fringe benefit.) Yet his modification greatly disturbed scholars, both Islamic and Western medieval academics, because uniformity of motion was now centered on an imaginary point (the equant) rather than the Earth. Ptolemy's decision to introduce the equant violated a cardinal Aristotelian rule: the celestial spheres were supposed to be steadfastly revolving about a central axis, but the equant was not the center of the universe. Questions and philosophical doubts about Ptolemy's choices kept an industry flourishing throughout the Middle Ages, an industry that ultimately led to the Copernican revolution.

This excerpt from *Almagest* highlights the basic ideas of the Ptolemaic model of a geocentric universe. Also included are sections that describe eccentricity, epicycles, and Ptolemy's justification for his divergence from uniform heavenly motion.

From *Almagest*
by Ptolemy

Translated by G. J. Toomer

Book I, 2. {*On the order of the theorems*}

. . . The general preliminary discussion covers the following topics: the heaven is spherical in shape, and moves as a sphere; the earth too is sensibly spherical in shape, when taken as a whole; in position it lies in the middle of the heavens very much like its center; in size and distance it has the ratio of a point to the sphere of the fixed stars; and it has no motion from place to place. We shall briefly discuss each of these points for the sake of reminder.

Book I, 3. {*That the heavens move like a sphere*}

It is plausible to suppose that the ancients got their first notions on these topics from the following kind of observations. They saw that the sun, moon and other stars were carried from east to west along circles which were always parallel to each other, that they began to rise up from below the earth itself, as it were, gradually got up high, then kept on going round in similar fashion and getting lower, until, falling to earth, so to speak, they vanished completely, then, after remaining invisible for some time, again rose afresh and set; and [they saw] that the periods of these [motions], and also the places of rising and setting, were, on the whole, fixed and the same.

What chiefly led them to the concept of a sphere was the revolution of the ever-visible stars, which was observed to be circular, and always taking place about one center, the same [for all]. For by necessity that point became [for them] the pole of the heavenly sphere: those stars which were closer to it revolved on smaller circles, those that were farther away described circles ever greater in proportion to their distance, until one reaches the distance of the stars which become invisible. . . .

For if one were to suppose that the stars' motion takes place in a straight line towards infinity, as some people have thought, what device could one conceive of which would cause each of them to appear to begin their motion from the same starting-point every day? How could the stars turn back if their motion is towards infinity? Or, if they did turn back, how could this not be obvious? [On such a hypothesis], they must gradually diminish in size until they disappear, whereas, on the contrary, they are seen to be greater at the very moment of their disappearance,* at which time they are gradually obstructed and cut off, as it were, by the earth's surface.

But to suppose that they are kindled as they rise out of the earth and are extinguished again as they fall to earth is a completely absurd hypothesis. For even if we were to concede that the strict order in their size and number, their intervals, positions and periods could be restored by such a random and chance process; that one whole area of the earth has a kindling nature, and another an extinguishing one, or rather that the same part [of the earth] kindles for one set of observers and extinguishes for another set; and that the same stars are already kindled or extinguished for some observers while they are not yet for others: even if, I say, we were to concede all these ridiculous consequences, what could we say about the ever-visible stars, which neither rise nor set? Those stars which are kindled and extinguished ought to rise and set for observers everywhere, while those which are not kindled and extinguished ought always to be visible for observers everywhere. What cause could we assign for the fact that this is not so? We will surely not say that stars which are kindled and extinguished for some observers never undergo this process for other observers. Yet it is utterly obvious that the same stars rise and set in certain regions [of the earth] and do neither at others. . . .

The following considerations also lead us to the concept of the sphericity of the heavens. No other hypothesis but this can explain how sundial constructions produce correct results; furthermore, the motion of the heavenly bodies is the most unhampered and free of all motions, and freest motion belongs among plane figures to the circle and among solid shapes to the sphere; similarly, since of different shapes having an equal boundary those with more angles are greater [in area or volume], the circle is greater than [all other] surfaces, and the sphere greater than [all other] solids; [likewise] the heavens are greater than all other bodies.

Furthermore, one can reach this kind of notion from certain physical considerations. E.g., the ether is, of all bodies, the one with constituent

---

* A reference to the psychological illusion that the Sun and Moon appear larger when close to the horizon.

parts which are finest and most like each other; now bodies with parts like each other have surfaces with parts like each other; but the only surfaces with parts like each other are the circular, among planes, and the spherical, among three-dimensional surfaces. And since the ether is not plane, but three-dimensional, it follows that it is spherical in shape. Similarly, nature formed all earthly and corruptible bodies out of shapes which are round but of unlike parts, but all ethereal and divine bodies out of shapes which are of like parts and spherical. For if they were flat or shaped like a discus they would not always display a circular shape to all those observing them simultaneously from different places on earth. For this reason it is plausible that the ether surrounding them, too, being of the same nature, is spherical, and because of the likeness of its parts moves in a circular and uniform fashion. . . .

Book I, 5. {*That the earth is in the middle of the heavens*}

. . . If one next considers the position of the earth, one will find that the phenomena associated with it could take place only if we assume that it is in the middle of the heavens, like the center of a sphere. For if this were not the case, the earth would have to be either

[a] not on the axis [of the universe] but equidistant from both
     poles, or
[b] on the axis but removed towards one of the poles, or
[c] neither on the axis nor equidistant from both poles . . .

. . . [I]f the earth did not lie in the middle [of the universe], the whole order of things which we observe in the increase and decrease of the length of daylight would be fundamentally upset. Furthermore, eclipses of the moon would not be restricted to situations where the moon is diametrically opposite the sun (whatever part of the heaven [the luminaries are in]), since the earth would often come between them when they were not diametrically opposite, but at intervals of less than a semi-circle. . . .

Book I, 7. {*That the earth does not have any motion from place to place, either*}

One can show by the same arguments as the preceding that the earth cannot have any motion in the aforementioned directions, or indeed ever move at all from its position at the center. For the same phenomena would result as would if it had any position other than the central one. Hence I think it is idle to seek for causes for the motion of objects towards the center, once it has been so clearly established from the actual phenomena that the earth occupies the middle place in the universe, and that all heavy objects are carried towards the earth. The following fact alone would most readily lead one to this notion [that all objects fall towards the center]. In absolutely all parts of the earth, which, as we said, has been shown to be spherical and in the middle of the universe, the direction and path of the motion (I mean the proper, [natural] motion) of all bodies possessing weight is always and everywhere at right angles to the rigid plane drawn tangent to the point of impact. It is clear from this fact that, if [these falling objects] were not arrested by the surface of the earth, they would certainly reach the center of the earth itself, since the straight line to the center is also always at right angles to the plane tangent to the sphere at the point of intersection [of that radius] and the tangent.

Those who think it paradoxical that the earth, having such a great weight, is not supported by anything and yet does not move, seem to me to be making the mistake of judging on the basis of their own experience instead of taking into account the peculiar nature of the universe. They would not, I think, consider such a thing strange once they realized that this great bulk of the earth, when compared with the whole surrounding mass [of the universe], has the ratio of a point to it. For when one looks at it in that way, it will seem quite possible that that which is relatively smallest should be overpowered and pressed in equally from all directions to a position of equilibrium by that which is the greatest of all and of uniform nature. For there is no up and down in the universe with respect to itself, any more than one could imagine such a thing in a sphere: instead the proper and natural motion of the compound bodies in it is as follows: light and rarefied bodies drift outwards towards the circumference, but seem to move in the direction which is "up" for each observer, since the overhead direction for all of us, which is also called "up," points towards the surrounding surface; heavy and dense bodies, on the other hand, are carried towards the middle and the center, but seem to fall downwards, because, again, the direction which is for all us towards our feet, called "down," also points towards the center of the earth. These heavy bodies, as one would expect, settle about the center because of their mutual pressure and resistance, which is equal and uniform from all directions. Hence, too, one can see that it is plausible that the earth, since its total mass is so great compared with the bodies which fall towards it, can remain motionless under the impact of these very small weights (for they strike it from all sides), and receive, as it were, the objects falling on it. If the earth had a single motion in common with other heavy objects, it is obvious that it would be carried down faster than all of them because of its much greater size: living things and individual heavy objects would be left behind, riding on the air, and the earth itself would very soon have fallen completely out of the heavens. But such things are utterly ridiculous merely to think of.

But certain people, [propounding] what they consider a more persuasive view, agree with the above, since they have no argument to bring against it, but think that there could be no evidence to oppose their view if, for instance, they supposed the heavens to remain motionless, and the earth to revolve from west to east about the same axis [as the heavens], making approximately one revolution each day; or if they made both heaven and earth move by any amount whatever, provided, as we said, it is about the same axis, and in such a way as to preserve the overtaking of one by the other.* However, they do not realize that, although there is perhaps nothing in the celestial phenomena which would count against that hypothesis, at least from simpler considerations, nevertheless from what would occur here on earth and in the air, one can see that such a notion is quite ridiculous. Let us concede to them [for the sake of argument] that such an unnatural thing could happen as that the most rare and light of matter should either not move at all or should move in a way no different from that of matter with the opposite nature (although things in the air, which are less rare [than the heavens] so obviously move with a more rapid motion than any earthy object); [let us concede that] the densest and heaviest objects have a proper motion of the quick and uniform kind which they suppose (although, again, as all agree, earthy objects are sometimes not readily moved even by an external force). Nevertheless, they would have to admit that the revolving motion of the earth must be the most violent of all motions associated with it, seeing that it makes one revolution in such a short time; the result would be that all objects not actually standing on the earth would appear to have the same motion, opposite to that of the earth: neither clouds nor other flying or thrown objects would ever be seen moving towards the east, since the earth's motion towards the east would always outrun and overtake them, so that all other objects would seem to move in the direction of the west and the rear. But if they said that the air is

* In the fourth century B.C., both Heraclides of Pontus and Aristarchus considered the idea that the Earth rotated on its axis.

carried around in the same direction and with the same speed as the earth, the compound objects in the air would none the less always seem to be left behind by the motion of both [earth and air]; or if those objects too were carried around, fused, as it were, to the air, then they would never appear to have any motion either in advance or rearwards: they would always appear still, neither wandering about nor changing position, whether they were flying or thrown objects. Yet we quite plainly see that they do undergo all these kinds of motion, in such a way that they are not even slowed down or speeded up at all by any motion of the earth. . . .

Book III, 3. {*On the hypotheses for uniform circular motion*}

Our next task is to demonstrate the apparent anomaly of the sun. . . . The reason for the appearance of irregularity can be explained by two hypotheses, which are the most basic and simple. When their motion is viewed with respect to a circle imagined to be in the plane of the ecliptic, the center of which coincides with the center of the universe (thus its center can be considered to coincide with our point of view), then we can suppose, either that the uniform motion of each [body] takes place on a circle which is not concentric with the universe, or that they have such a concentric circle, but their uniform motion takes place, not actually on that circle, but on another circle, which is carried by the first circle, and [hence] is known as the "epicycle." It will be shown that either of these hypotheses will enable [the planets] to appear, to our eyes, to traverse unequal arcs of the ecliptic (which is concentric to the universe) in equal times.

In the eccentric hypothesis: we imagine the eccentric circle [see Figure 6.1], on which the body travels with uniform motion, to be ABGD on center E, with diameter AED, on which point Z represents the observer [our view from Earth]. Thus A is the apogee, and D the perigee. We cut off equal arcs AB and DG, and join BE, BZ, GE and GZ. Then it is immediately obvious that the body will traverse the arcs AB and GD in equal times, but will [in doing so] appear to have traversed unequal arcs of a circle drawn on center Z. . . .

In the epicyclic hypothesis: we imagine the circle concentric with the ecliptic [see Figure 6.2] as ABGD on center E, with diameter AEG, and the epicycle carried by it, on which the body moves, as ZHΘK on center A.

Then here too it is immediately obvious that, as the epicycle traverses circle ABGD with uniform motion, say from A towards B, and as the body traverses the epicycle with uniform motion, then when the body is at points

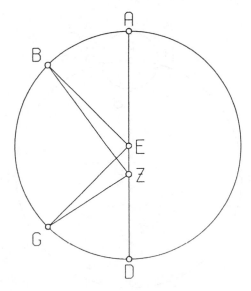

Figure 6.1

Z and Θ, it will appear to coincide with A, the center of the epicycle, but when it is at other points it will not. Thus when it is, e.g., at H, its motion will appear greater than the uniform motion [of the epicycle] by arc AH, and similarly when it is at K its motion will appear less than the uniform by arc AK.

Now in this kind of eccentric hypothesis the least speed always occurs at the apogee and the greatest at the perigee, since ∠ AZB [in Figure 6.1] is always less than ∠ DZG. But in the epicyclic hypothesis both this and the reverse are possible. For the motion of the epicycle is towards the rear with respect to the heavens, say from A towards B [in Figure 6.2]. Now if the motion of the body on the epicycle is such that it too moves rearwards from the apogee, that is from Z towards H, the greatest speed will occur at the apogee, since at that point both epicycle and body are moving in the same direction. But if the motion of the body from the apogee is in advance on the epicycle, that is from Z towards K, then the reverse will occur: the least speed will occur at the apogee, since at that point the body is moving in the opposite direction to the epicycle.

Having established that, we must next make the additional preliminary point that for bodies which exhibit a double anomaly both the above hypotheses may be combined, as we shall prove in our discussions of such bodies. . . .

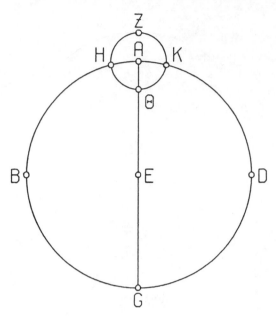

Figure 6.2

Book IX, 2. {*On our purpose in the hypotheses of the planets*}

. . . Now it is our purpose to demonstrate for the five planets, just as we did for the sun and moon, that all their apparent anomalies can be represented by uniform circular motions, since these are proper to the nature of divine beings, while disorder and nonuniformity are alien [to such beings]. Then it is right that we should think success in such a purpose a great thing, and truly the proper end of the mathematical part of theoretical philosophy. But, on many grounds, we must think that it is difficult, and that there is good reason why no-one before us has yet succeeded in it. . . . In general, observations [of planets] with respect to one of the fixed stars, when taken over a comparatively great distance, involve difficult computations and an element of guesswork in the quantity measured, unless one carries them out in a manner which is thoroughly competent and knowledgeable. . . .

. . . If we are at any point compelled by the nature of our subject to use a procedure not in strict accordance with theory (for instance, when we carry out proofs using without further qualification the circles described in the planetary spheres by the movement [of the body, i.e.] assuming that these circles lie in the plane of the ecliptic, to simplify the course of the proof); or [if we are compelled] to make some basic assumptions which we arrived at not from some readily apparent principle, but from a long period of trial and application,* or to assume a type of motion or inclination of the circles which is not the same and unchanged for all planets; we may [be allowed to] accede [to this compulsion], since we know that this kind of inexact procedure will not affect the end desired, provided that it is not going to result in any noticeable error; and we know too that assumptions made without proof, provided only that they are found to be in agreement with the phenomena, could not have been found without some careful methodological procedure, even if it is difficult to explain how one came to conceive them (for, in general, the cause of first principles is, by nature, either non-existent or hard to describe); we know, finally, that some variety in the type of hypotheses associated with the circles [of the planets] cannot plausibly be considered strange or contrary to reason (especially since the phenomena exhibited by the actual planets are not alike [for all]); for, when uniform circular motion is preserved for all without exception, the individual phenomena are demonstrated in accordance with a principle which is more basic and more generally applicable than that of similarity of the hypotheses [for all planets].

* An example would be Ptolemy's introduction of the equant point.

II

REVOLUTIONS

For fourteen centuries, from the time of Ptolemy to that of Copernicus, Western astronomy was essentially in hibernation. Although some Greek works on astronomy survived in musty libraries after the decline of Greece and later Rome, continuation of the field largely took place in the great Islamic civilization that arose in the seventh and eighth centuries. In this golden age, Islamic leaders actively sought out ancient texts from Greece, India, and China and had them translated into Arabic. Spreading throughout the Middle East and over northern Africa into Spain, these manuscripts eventually fell into Christian hands and were translated into Hebrew and Latin, stimulating the Renaissance and scientific revolutions to come. Without the Islamic intervention, many of the early findings might have been lost forever.

Islamic religious requirements (sighting the new Moon to begin a new month, timing the start of prayers throughout the day, properly orienting mosques toward Mecca) led to advances in celestial and terrestrial mapping. Observatories were built and instruments such as sextants and astrolabes improved for determining more precise positions of celestial bodies (oftentimes funded by rich patrons interested in astrological forecasts). The notable Islamic influence on astronomy is preserved in the names of many stars, such as Aldebaran, Altair, and Betelgeuse, as well the astronomical terms *zenith* and *nadir*. The culmination of this period was the construction of the great observatory at Samarkand in central Asia (what is now Uzbekistan), built in 1420 by the Persian prince Ulugh Beg, grandson of the conqueror Tamerlane. The observatory's great accomplishment was a catalog of over one thousand stars, one of the best surveys of its kind in the Middle Ages.

By the end of the medieval era scholars, reacquainted with the ancient texts, began to question the Ptolemaic model of the universe, as well as Aristotle's physics. Revolution was in the air: politically, theologically, and scientifically. Nicolaus Copernicus, Tycho Brahe, and Johannes Kepler were like dominoes, each contributing a push that toppled Ptolemy's hoary ethereal spheres and ultimately led to the most significant achievement of the age: Newton's law of gravitation. Coupled with Galileo's radical new use of the telescope to unveil a universe previously unseen—a spotted Sun, rugged lunar landscapes, moons around Jupiter, and a star-filled Milky Way—astronomy was completely redefined by the end of the seventeenth century.

Copernicus initiated this chain of events in 1543 by daring to assert that the Earth, for so long singled out as the center of the universe, was orbiting the Sun with all the other planets. He thrust the Earth into motion, which inspired others to examine new rules of mechanics. Tycho Brahe, the last and greatest naked-eye astronomer, did not fully agree with Copernicus, but he delivered his own blows to the long-held vision of the universe. He proved that a *nova stella*—new star—that appeared in 1572 resided among the fixed stars, challenging the Aristotelian edict that the heavens never changed. He went on to prove that comets were not an atmospheric phenomenon but objects that easily passed through the transparent spheres allegedly transporting the planets.

After Tycho's death in 1601 Johannes Kepler, who had begun working under Tycho in early 1600, used the decades of meticulous observations recorded by Tycho to demonstrate that planets do not orbit in circles but rather follow paths more elliptical in shape. In some ways, Kepler's hard-earned revelation (it took years of bone-grinding calculations) was as wrenching a jolt as Copernicus's alteration of the solar system. The circle, long considered the perfect geometric form by philosophers and theologians, was at last cast aside. The scientific value of Kepler's work was immense. His "planetary laws" became a foundation upon which Isaac Newton could later forge his triumphant law of gravitation.

With his monumental *Principia,* published in 1687, Newton established that motions everywhere, in the heavens and on the Earth, are described by the same physical principles. As a result, any specialness of the heavens was banished. Newton was the first scientist to demonstrate that terrestrial and celestial phenomena behave according to a set of *mathematical* laws. That this was true was proven most effectively when Edmond Halley pored over historic records to match the highly elliptical path of a comet he had observed in 1682 with comet sightings of the past. Using Newton's laws as his guide, he confidently predicted the celestial visitor would return in 1758. When it arrived virtually on schedule, Newton's laws were confirmed and astronomers gradually came to think of the cosmos as utterly predictable—a clockwork universe ticking on like a gigantic watchspring.

7 / Copernicus and the
Sun-Centered Universe

Nicolaus Copernicus was born in Toruń, Poland (then Royal Prussia), in 1473 and raised by his uncle, the bishop of Warmia. As a young man he was educated in medicine and canon law at a moment in the Renaissance when Plato's works and their stress on the mathematical nature of the universe were being actively revived. During his academic studies, first in Cracow and then in Italy, he began an independent pursuit of astronomy and became particularly dissatisfied with the digressions from uniformity within Ptolemy's model of the universe, with its eccentric orbits and equant points (see Chapter 6). "A theory of this kind seemed neither perfect enough nor sufficiently in accord with reason," he later concluded.[1]

In 1503 Copernicus returned to his homeland, eventually settling near the Baltic Sea to serve the remaining years of his life as an administrative canon at the cathedral of Frauenburg (now Frombork in northeastern Poland). He continued to work privately on his astronomy and sometime before 1514 wrote the *Commentariolus* (Little Commentary), his preliminary outline postulating a sun-centered or heliocentric universe. In this early version, discreetly circulated, he boldly placed the center of the universe near the Sun, which allowed him to get rid of Ptolemy's pesky violations to uniform circular motion. "We revolve around the Sun just as any other planet," he wrote. "The retrograde and direct motion that appears in the planets belongs not to them but to the [motion] of the earth. Thus the motion of the Earth by itself accounts for a considerable number of apparently irregular motions in the heavens."[2] Copernicus was not disturbed at all by a moving Earth; he was more troubled by a rotating sky. The farther out one moves from a stationary Earth, the faster and faster the sky must move to stay in place. To Copernicus, a heaven at rest was a more noble condition, with the Sun at last in its proper place, "as if resting on a kingly throne, [governing] the family of stars which wheel around," he would later write.[3]

Copernicus spent the rest of his life developing mathematical proofs that showed how planetary movements could be calculated from his new perspective. His completed work, *De revolutionibus orbium coelestium* (On the Revolutions of the Heavenly Spheres), was not published until 1543, the year of his death. A well-intentioned theologian, Andreas Osiander, hired to proofread the publication and wanting to protect Copernicus from ecclesiastical controversy, inserted an unsigned introduction maintaining that placing the Sun at the center was merely a mathematical convenience that allowed more precise predictions of planetary motions. Do not take it literally, he warned, "lest, if anyone take as true that which has been constructed for another use, he go away from this discipline a bigger fool than when he came to it."[4] But Copernicus, then on his deathbed and unaware of Osiander's addition, took his new cosmic order quite seriously. In the preface to his great work, he rails against those ignorant of astronomy and mathematics, who would attack his model based on some twist of scripture.

Osiander's meddling, though, was probably quite useful in the end. Copernicus's book was not banned for some seventy years, allowing it to be discussed within the intellectual community. The lack of immediate uproar over its controversial ideas was likely aided by its complex presentation; consisting of six books in all, *De revolutionibus* was written in such a way that it was accessible only to scholars with considerable mathematical knowledge. The first book is largely devoted to sketching out Copernicus's general system, followed by some mathematical propositions and tables. The second book dwells on the motions of the celestial sphere, while the third discusses the precession of the equinoxes and its origin in a slow motion of the Earth's axis. The fourth book focuses on the Moon, and the last two books deal with the motion of the planets. Throughout the work, one can sense the excitement and pride Copernicus has in his discovery, coupled with the discomfiting knowledge that he was upsetting longheld, sacred beliefs.

Postulates in Copernicus's *Commentariolus*[5]

1. There is no one center of all the celestial spheres (*orbium*) or spheres (*sphaerarum*).
2. The center of the earth is not the center of the universe, but only the center towards which heavy things move and the center of the lunar sphere.
3. All spheres surround the sun as though it were in the middle of all of them, and therefore the center of the universe is near the sun.
4. The ratio of the distance between the sun and the earth to the height of the sphere of the fixed stars is so much smaller than the ratio of the semidiameter of the earth to the distance of the sun that the distance between the sun and the earth is imperceptible compared to the great height of the sphere of the fixed stars.
5. Whatever motion appears in the sphere of the fixed stars belongs not to it but to the earth. Thus the entire earth along with the nearby elements rotates with its daily motion on its fixed poles while the sphere of the fixed stars remains immovable and the outermost heaven.
6. Whatever motions appear to us to belong to the sun are not due to [motion] of the sun but [to the motion] of the earth and our sphere with which we revolve around the sun just as any other planet. And thus the earth is carried by more than one motion.
7. The retrograde and direct motion that appears in the planets belongs not to them but to the [motion] of the earth. Thus the motion of the earth by itself accounts for a considerable number of apparently irregular motions in the heavens.

It should be stressed that new astronomical observations were not forcing Copernicus to introduce his heliocentric model, an explanation often mistakenly made. Given the level of their technology at the time, astronomers were still able to calculate reasonably accurate planet positions using Ptolemy's geocentric system.

When it came to predicting planetary motions, the Ptolemaic and Copernican systems were fairly equivalent. Instead, Copernicus was driven by the coherency of his new outlook and its return to the Aristotelian aesthetic of uniformity. So in some ways it was a throwback to earlier times, the culmination of the centuries-long quest to establish a single harmonious model of the universe's structure. Copernicus's one bold move—a revolutionary one at that—was putting the Earth into motion.

With this new model, planetary movements such as retrograde now had a natural and simplified explanation: when the Earth in its orbit overtakes another planet, that planet will briefly appear to move backward in the sky. No need for all those big, awkward epicycles (though Copernicus did retain some minor epicycles in his model to reproduce more accurately a planet's nonuniform motion on the sky). Copernicus's new scheme also clarified why Venus and Mercury, being inside the Earth's orbit, always stayed by the Sun. And like Aristarchus centuries earlier, Copernicus reasoned that the stars did not shift their positions during the Earth's annual orbital journey simply because they were so far away. "... [T]he heavens are immense in comparison with the Earth," he wrote, "and present the aspect of an infinite magnitude."[6]

For many of Copernicus's contemporaries, a Sun-centered universe was hardly a simplification. They had to start worrying about the universe's dynamics, such as why clouds (or us) weren't ripped off the face of the planet as the Earth spun on its axis and moved in its orbit. With the revisions suggested within his masterpiece, Copernicus opened the door to contemplating a universal law of gravitation and to establishing new rules of mechanics. Moreover, *De revolutionibus*, dedicated to the pope, introduced a new humility into astronomy. It initiated the gradual displacement of humanity from the hub of cosmic affairs.

From *De revolutionibus orbium coelestium* by Nicolaus Copernicus

Translated by Charles Glenn Wallis

Preface and Dedication to Pope Paul III

I can reckon easily enough, Most Holy Father, that as soon as certain people learn that in these books of mine which I have written about the revolutions of the spheres of the world I attribute certain motions to the terrestrial globe, they will immediately shout to have me and my opinion hooted off the stage. . . . [F]or a long time I was in great difficulty as to whether I should bring to light my commentaries written to demonstrate the Earth's movement, or whether it would not be better to follow the example of the Pythagoreans and certain others who used to hand down the mysteries of their philosophy not in writing but by word of mouth and only to their relatives and friends. . . . [W]hen I weighed these things in my mind, the scorn which I had to fear on account of the newness and absurdity of my opinion almost drove me to abandon a work already undertaken.

But my friends made me change my course in spite of my long-continued hesitation and even resistance . . . letting come to light a work which I had kept hidden among my things for not merely nine years, but for almost four times nine years. . . .

. . . If perchance there are certain "idle talkers" who take it upon themselves to pronounce judgment, although wholly ignorant of mathematics, and if by shamelessly distorting the sense of some passage in Holy Writ to suit their purpose, they dare to reprehend and to attack my work; they worry me so little that I shall even scorn their judgments as foolhardy. . . .

Book One

Among the many and varied literary and artistic studies upon which the natural talents of man are nourished, I think that those above all should be embraced and pursued with the most loving care which have to do with things that are very beautiful and very worthy of knowledge. Such studies are those which deal with the godlike circular movements of the world and the course of the stars, their magnitudes, distances, risings and settings, and the causes of the other appearances in the heavens; and which finally explicate the whole form. For what could be more beautiful than the heavens which contain all beautiful things? . . .

5. *Does the Earth Have a Circular Movement? And of Its Place*

Now that it has been shown that the Earth too has the form of a globe, I think we must see whether or not a movement follows upon its form and what the place of the Earth is in the universe. For without doing that it will not be possible to find a sure reason for the movements appearing in the heavens. Although there are so many authorities for saying that the Earth rests in the center of the world that people think the contrary supposition inopinable and even ridiculous; if however we consider the thing attentively, we will see that the question has not yet been decided and accordingly is by no means to be scorned. For every apparent change in place occurs on account of the movement either of the thing seen or of the spectator, or on account of the necessarily unequal movement of both. For no movement is perceptible relatively to things moved equally in the same directions—I mean relatively to the thing seen and the spectator. Now it is from the Earth that the celestial circuit is beheld and presented to our sight. Therefore, if some movement should belong to the Earth it will appear, in the parts of the universe which are outside, as the same movement but in the opposite direction, as though the things outside were passing over. And the daily revolution in especial is such a movement. For the daily revolution appears to carry the whole universe along, with the exception of the Earth and the things around it. And if you admit that the heavens possess none of this movement but that the Earth turns from west to east, you will find—if you make a serious examination—that as regards the apparent rising and setting of the sun, moon, and stars the case is so. . . .

6. *On the Immensity of the Heavens in Relation to the Magnitude of the Earth*

. . . [T]he heavens are immense in comparison with the Earth and present the aspect of an infinite magnitude, and that in the judgment of sense-perception the Earth is to the heavens as a point to a body and as a finite to an infinite magnitude. But we see that nothing more than that has been shown, and it does not follow that the Earth must rest at the center of the world. And we should be even more surprised if such a vast world should

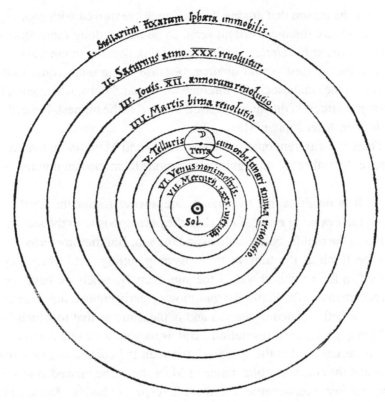

Figure 7.1

wheel completely around during the space of twenty-four hours rather than that its least part, the Earth, should. . . .

. . . But it is not at all clear how far this immensity stretches out. On the contrary, since the minimal and indivisible corpuscles, which are called atoms, are not perceptible to sense, they do not, when taken in twos or in some small number, constitute a visible body; but they can be taken in such a large quantity that there will at last be enough to form a visible magnitude. So it is as regards the place of the earth; for although it is not at the center of the world, nevertheless the distance is as nothing, particularly in comparison with the sphere of the fixed stars. . . .

10. *On the Order of the Celestial Orbital Circles*

I know of no one who doubts that the heavens of the fixed stars is the highest up of all visible things. We see that the ancient philosophers wished to take the order of the planets according to the magnitude of their revolutions, for the reason that among things which are moved with equal speed those which are the more distant seem to be borne along more slowly, as Euclid proves in his *Optics*. And so they think that the moon traverses its circle in the shortest period of time, because being next to the Earth, it revolves in the smallest circle. But they think that Saturn, which completes the longest circuit in the longest period of time, is the highest. Beneath Saturn, Jupiter. After Jupiter, Mars.

There are different opinions about Venus and Mercury, in that they do not have the full range of angular elongations from the sun that the others do. . . .

. . . It is manifest that the planets are always nearer the Earth at the time of their evening rising, *i.e.,* when they are opposite to the sun and the Earth is in the middle between them and the sun. But they are farthest away from the Earth at the time of their evening setting, *i.e.,* when they are occulted in the neighborhood of the sun, namely, when we have the sun between them and the Earth. All that shows clearly enough that their center is more directly related to the sun and is the same as that to which Venus and Mercury refer their revolutions. But as they all have one common center, it is necessary that the space left between the convex orbital circle of Venus and the concave orbital circle of Mars should be viewed as an orbital circle or sphere homocentric with them in respect to both surfaces, and that it should receive the Earth and its satellite the moon and whatever is contained beneath the lunar globe. For we can by no means separate the moon from the Earth, as the moon is incontestably very near to the Earth—especially since we find in this expanse a place for the moon which is proper enough and sufficiently large. Therefore we are not ashamed to maintain that this totality—which the moon embraces—and the center of the Earth too traverse that great orbital circle among the other wandering stars in an annual revolution around the sun; and that the center of the world is around the sun. I also say that the sun remains forever immobile and that whatever apparent movement belongs to it can be verified of the mobility of the Earth; that the magnitude of the world is such that, although the distance from the sun to the Earth in relation to whatsoever planetary sphere you please possesses magnitude which is sufficiently manifest in proportion to these dimensions, this distance, as compared with the sphere of the fixed stars, is imperceptible. I find it much more easy to grant that than to unhinge the understanding by an almost infinite multitude of spheres—as those who keep the earth at the center of the world are forced to do. But we should rather follow the wisdom of nature, which, as it takes very great care not to have produced anything superfluous or useless, often prefers to endow one thing with many effects. And though all these things are difficult, almost inconceivable, and quite contrary to the opinion of the multitude, nevertheless in what follows we will with God's help make them clearer than day—at least for those who are not ignorant of the art of mathematics.

Therefore if the first law is still safe—for no one will bring forward a better one than that the magnitude of the orbital circles should be measured by the magnitude of time—then the order of the spheres will follow in this way—beginning with the highest: the first and highest of all is the sphere of the fixed stars, which comprehends itself and all things, and is accordingly immovable. In fact it is the place of the universe, *i.e.,* it is that to which the movement and position of all the other stars are referred. . . . Saturn, the first of the wandering stars, follows; it completes its circuit in 30 years. After it comes Jupiter moving in a 12-year period of revolution. Then Mars, which completes a revolution every 2 years. The place fourth in order is occupied by the annual revolution in which we said the Earth together with the orbital circle of the moon as an epicycle is comprehended. In the fifth place, Venus, which completes its revolution in [9] months. The sixth and final place is occupied by Mercury, which completes its revolution in a period of [80] days.* In the center of all rests the sun. For who would place this lamp of a very beautiful temple in another or better place than this wherefrom it can illuminate everything at the same time? As a matter of fact, not unhappily do some call it the lantern; others, the mind and still others, the pilot of the world. Trismegistus calls it a "visible god"; Sophocles' Electra, "that which gazes upon all things." And so the sun, as if resting on a kingly throne, governs the family of stars which wheel around. . . .

Therefore in this ordering we find that the world has a wonderful commensurability and that there is a sure bond of harmony for the movement and magnitude of the orbital circles such as cannot be found in any other way. For now the careful observer can note why progression and retrogradation appear greater in Jupiter than in Saturn and smaller than in Mars; and in turn greater in Venus than in Mercury. And why these reciprocal events appear more often in Saturn than in Jupiter, and even less often in

---

* In his translation, Wallis had here inserted 7½ months for Venus's orbital period and 88 days for Mercury. Copernicus did calculate those more accurate values, but they first appear later in his manuscript. At this point Copernicus wrote 9 months and 80 days.

Mars and Venus than in Mercury. In addition, why when Saturn, Jupiter, and Mars are in opposition [to the mean position of the sun] they are nearer to the Earth than at the time of their occultation and their reappearance. And especially why at the times when Mars is in opposition to the sun, it seems to equal Jupiter in magnitude and to be distinguished from Jupiter only by a reddish color, but [otherwise he is scarce equal to a star of the second magnitude, and can be recognized only when his movements are carefully followed]?[7] All these things proceed from the same cause, which resides in the movement of the Earth.

But that there are no such appearances among the fixed stars argues that they are at an immense height away, which makes the circle of annual movement or its image disappear from before our eyes since every visible thing has a certain distance beyond which it is no longer seen, as is shown in optics. For the brilliance of their lights shows that there is a very great distance between Saturn the highest of the planets and the sphere of the fixed stars. It is by this mark in particular that they are distinguished from the planets, as it is proper to have the greatest difference between the moved and the unmoved. How exceedingly fine is the godlike work of the Best and Greatest Artist!

8 / Tycho Brahe and the Changing Heavens

Tycho Brahe was the last (and best) of the celestial observers of the pretelescopic era. Born in 1546 to a noble Danish family, he became one of astronomy's most colorful characters. Hot-tempered and fractious, Tycho (in Danish, Tyge) lost the bridge of his nose during a duel as a young man and wore a metal replacement, a mix of gold, silver, and copper to resemble the color of flesh. With his large handlebar mustache and close-cropped reddish hair, this aristocrat assumed a regal air and lived accordingly.

Tycho was drawn to astronomy as a teenager and was particularly influenced by a conjunction (close passing) of Jupiter and Saturn that took place in 1563. He noticed that old planetary tables from the thirteenth century, based on Ptolemy's model, were days off in predicting the event. This spurred his interest in building instruments and making more accurate measurements. By 1572 his skills were tested when a bright, starlike object suddenly appeared in the sky within the constellation Cassiopeia. Rivaling the planet Venus, it gradually faded to yellow and then red, until it disappeared altogether sixteen months later. We now know that it was a supernova, the explosion of a star, a rare sight to the unaided eye; written records list fewer than a dozen bright supernovae in the last two thousand years. Throughout the appearance of this new star, Tycho periodically pegged its position to see whether it was moving. No parallax was sighted, leading Tycho to conclude that the stellar interloper was located farther out than the Moon (beyond the "region of the Element," as he put it) and likely resided among the fixed stars.[8] Here was a direct challenge to the long-standing view that the celestial sphere never changed. Being an astrologer as well, Tycho predicted that the new star was foretelling the end of old religious regimes, a claim borne out for some by the rise of Protestantism. Tycho wrote of his observations in *De nova et nullius aevi memoria prius visa stella* (often succinctly translated as On the New and Never Previously Seen Star), despite the fact that it was considered improper for a nobleman to publish a book.

Within five years, a comet appeared, and again Tycho studied the visitor meticulously. He determined it was not an atmospheric phenomenon, the result of friction as the realm of fire rubbed up against the ether, as Aristotle taught, but an object effortlessly passing through the invisible spheres then thought to transport the planets around the Earth, putting the mechanism into doubt. He determined the comet moved around the Sun, just outside the orbit of Venus. Plotting out a circular orbit, though, Tycho had to conclude that the comet moved irregularly. Uniformity was restored if the orbit was "not exactly circular but somewhat oblong, like a figure commonly called oval," he wrote in his thick tome on the comet, *De mundi aetheri recentioribus phaenomenis* (Concerning the New Phenomena in the Ethereal World).[9] This may have been the first time an astronomer suggested that a celestial body moved in an orbit differing from a circle. Comets did not have the same status as planets, but such an observation serves in a way as a harbinger of Johannes Kepler's later insights on orbital motion (see Chapter 9).

In 1580 Tycho completed construction of a grand observatory, the first of its kind in Christian Europe, on the island of Hven in the sound between Denmark and Sweden. The island was granted to him by King Frederick II of Denmark, and he presided over its affairs like a dominating feudal lord (he even had his own prison). Tycho called his observatory Uraniborg, "heavenly castle," and by 1584 expanded its operations to include a satellite observatory called Stjerneborg, or "castle of the stars." His sextants and quadrants with their specially constructed sights were stably mounted on observing decks beneath movable roofs. This enabled Tycho and his staff of assistants to carry out with the naked eye the most accurate measurements of planetary motions to date, to within one-sixtieth of a degree.

He also measured the timing of the year to within one second, determined how starlight was refracted by the Earth's atmosphere, and obtained the first accurate precession rate. He repeated observations to reduce accidental sources of error. Uraniborg established a model for the university and national observatories to come. Losing his royal patronage in 1597, he settled in Prague for the last few years of his life.

In the history of astronomy, Tycho is a transitional figure between the medieval and Renaissance eras. Although his findings on both the new star of 1572 and the comet were revolutionary, helping astronomy break away from Aristotelian physics, he remained quite conservative when it came to cosmic models. He stubbornly stayed in the Earth-centered-universe camp, but with a twist. He established his own model of the cosmos in which the Earth was at rest but all the other planets circled the Sun. The Sun, in turn, revolved around the Earth. On his deathbed, a victim of uremia, Tycho pleaded with Kepler, his chief assistant, to carry on his research within the Tychonic system of the universe, not the Copernican. "Let me not be seen to have lived in vain," he cried out in delirium.[10] He did not. The keen accuracy of his observations and his cometary evidence that the legendary celestial spheres were a sham at last allowed his successors to determine how the planets truly moved within the solar system.

From *On a New Star, Not Previously Seen Within the Memory of Any Age Since the Beginning of the World* by Tycho Brahe

Translated by John H. Walden

Its First Appearance in 1572

Last year [1572], in the month of November, on the eleventh day of that month, in the evening, after sunset, when, according to my habit, I was contemplating the stars in a clear sky, I noticed that a new and unusual star, surpassing the other stars in brilliancy, was shining almost directly above my head; and since I had, almost from boyhood, known all the stars of the heavens perfectly (there is no great difficulty in attaining that knowledge), it was quite evident to me that there had never before been any star in that place in the sky, even the smallest, to say nothing of a star so conspicuously bright as this. I was so astonished at this sight that I was not ashamed to doubt the trustworthiness of my own eyes. But when I observed that others, too, on having the place pointed out to them, could see that there was really a star there, I had no further doubts. A miracle indeed, either the greatest of all that have occurred in the whole range of nature since the beginning of the world, or one certainly that is to be classed with those attested by the Holy Oracles, the staying of the Sun in its course in answer to the prayers of Joshua, and the darkening of the Sun's face at the time of the Crucifixion. For all philosophers agree, and facts clearly prove it to be the case, that in the ethereal region of the celestial world no change, in the way either of generation or of corruption, takes place; but that the heavens and the celestial bodies in the heavens are without increase or diminution, and that they undergo no alteration, either in number or in size or in light or in any other respect; that they always remain the same, like unto themselves in all respects, no years wearing them away. Furthermore, the observations of all the founders of the science, made some thousands of years ago, testify that all the stars have always retained the same number, position, order, motion, and size as they are found, by careful observation on the part of those who take delight in heavenly phenomena, to preserve even in our own day. Nor do we read that it was ever before noted by any one of the founders that a new star had appeared in the celestial world, except only by Hipparchus, if we are to believe Pliny. For Hipparchus, according to Pliny (Book II of his Natural History), noticed a star different from all others previously seen, one born in his own age. . . .

Its Position with Reference to the Diameter of the World and Its Distance from the Earth, the Center of the Universe

It is a difficult matter, and one that requires a subtle mind, to try to determine the distances of the stars from us, because they are so incredibly far removed from the earth; nor can it be done in any way more conveniently and with greater certainty than by the measure of the parallax [diurnal], if a star have one. For if a star that is near the horizon is seen in a different place than when it is at its highest point and near the vertex, it is necessarily found in some orbit with respect to which the Earth has a sensible size. How far distant the said orbit is, the size of the parallax compared with the semi-diameter of the Earth will make clear. If, however, a [circumpolar] star, that is as near to the horizon [at lower culmination] as to the vertex [at upper culmination], is seen at the same point of the Primum Mobile, there is no doubt that it is situated either in the eighth sphere or not far below it, in an orbit with respect to which the whole Earth is as a point.

In order, therefore, that I might find out in this way whether this star was in the region of the Element or among the celestial orbits, and what its distance was from the Earth itself, I tried to determine whether it had a parallax, and, if so, how great a one; and this I did in the following way: I observed the distance between this star and Schedir of Cassiopeia (for the latter and the new star were both nearly on the meridian), when the star was at its nearest point to the vertex, being only 6 degrees removed from the zenith itself (and for that reason, though it were near the Earth, would produce no parallax in that place, the visual position of the star and the real position then uniting in one point, since the line from the center of the Earth and that from the surface nearly coincide). I made the same observation when the star was farthest from the zenith and at its nearest point to the horizon, and in each case I found that the distance from the above-mentioned fixed star was exactly the same, without the variation of a minute: namely 7 degrees and 55 minutes. Then I went through the same process, making numerous observations with other stars. Whence I conclude that this new star has no diversity of aspect, even when it is near the horizon. For otherwise in its least altitude it would have been farther away from the above-mentioned star in the breast of Cassiopeia than when in its greatest altitude. Therefore, we shall find it necessary to place this star, not in the region of the Element, below the Moon, but far above, in an orbit with respect to which the Earth has no sensible size. For if it were in the highest region of the air, below the hollow region of the Lunar sphere, it would, when nearest the horizon, have produced on the circle a sensible variation of altitude from that which it held when near the vertex.

[Omitted here is Tycho's geometric proof that the new star was located beyond the Moon, likely in the celestial sphere of fixed stars itself or just below it.]

. . . That it is not in the orbit of Saturn, however, or in that of Jupiter, or in that of Mars, or in that of any one of the other planets, is clear from this fact: after the lapse of six months it had not advanced by its own motion a single minute from that place in which I first saw it; and this it must have done if it were in some planetary orbit. For, unlike the Primum Mobile, it would be moved by the peculiar motion of the orbit itself, unless it were at rest at one or the other pole of the orbits of the Secundum Mobile; from which, however, as I have shown above, it is removed 28 degrees. For the entire orbits, revolving on their own poles, carry along their own stars, or (as I see Pliny and some others hold) are carried along by them; unless, indeed, one would deny the belief accepted by philosophers and mathematicians, and assert (what is absurd) that the stars alone revolve, while the orbits are fixed. Therefore, if this star were placed in some one of the orbits of the seven wandering stars, it would necessarily be carried around with the orbit itself to which it were affixed, in the opposite direction to the daily revolution. And, furthermore, this motion, even in the case of the orbit which moves the slowest, that of Saturn, would, after such a length of time, be noticed, though one were to make his observation without any instrument at all.

Therefore, this new star is neither in the region of the Element, below the Moon, nor among the orbits of the seven wandering stars, but it is in the eighth sphere, among the other fixed stars, which was what we had to prove. Hence it follows that it is not some peculiar kind of comet or some other kind of fiery meteor become visible. For none of these are generated in the heavens themselves, but they are below the Moon, in the upper region of the air, as all philosophers testify; unless one would believe with Albategnius that comets are produced, not in the air, but in the heavens.* For he believes that he has observed a comet above the Moon, in the sphere of Venus. That this can be the case, is not yet clear to me. But, please God, sometime, if a comet shows itself in our age, I will investigate the truth of the matter. Even should we assume that it can happen (which I, in company with other philosophers, can hardly admit), still it does not follow that this star is a kind of comet; first, by reason of its very form, which is the same as the form of the real stars and different from the form of all the comets hitherto seen, and then because, in such a length of time, it advances neither latitudinally nor longitudinally by any motion of its own, as comets have been observed to do. For, although these sometimes seem to remain in one place several days, still, when the observation is made carefully by exact instruments, they are seen not to keep the same position for so very long or so very exactly. I conclude, therefore, that this star is not some kind of comet or a fiery meteor, whether these be generated beneath the Moon or above the Moon, but that it is a star shining in the firmament itself—one that has never previously been seen before our time, in any age since the beginning of the world.

* Al-Battānī, known in the West as Albategnius, was a noted Islamic astronomer and mathematician of the ninth century.

9 / Johannes Kepler and Planetary Motion

Johannes Kepler was a temblor, a precursor to the seismic Newtonian revolution to come. A confirmed Copernican, he spent his professional life as a pioneer in determining the dynamics of the solar system's behavior—"to demonstrate from observations," he noted.[11] The mathematical tools were not yet available for him to fully succeed, but in his pursuit he discovered the true shape of a planetary orbit and established two more planetary "laws" that would later aid Isaac Newton in developing his triumphant theory of gravitation.

Kepler was born in 1571, within a tumultuous age in both politics and religion when Protestants and Catholics in the duchies and principalities that then made up Austria and Germany were in constant conflict. He underwent banishment several times. Scholarly and unassuming, a man of few social graces, Kepler entered Tübingen University to train for the Lutheran clergy. There a revered teacher of mathematics, Michael Mästlin, introduced his particularly bright student to Copernicus's heliocentric theory. Kepler's theological studies were cut short by a teaching assignment, at which time he became obsessed with discovering the geometric rules of God's grand design—the reason for an orbit's size and a planet's speed. Kepler published his grand and fanciful unifying theory in *Mysterium cosmographicum* (Cosmographic Mystery) in 1596. He imagined that the five Platonic solids (cube, tetrahedron, dodecahedron, icosahedron, and octahedron), when nested inside each other, could explain the relative sizes of the planetary orbits. The Danish astronomer extraordinaire Tycho Brahe, impressed by Kepler's accomplishment, allowed the young man to join him in Prague as an assistant in 1600. Within a year Tycho died, but Kepler inherited the post of imperial mathematician and at last had access to Tycho's rich trove of observational data stretching back some thirty-five years.[12]

Kepler focused on discerning the orbit of Mars, a task that Tycho had assigned to him. He was sure he could explain its behavior within a week. He ended up grappling with its orbital complexities for five years. He published his findings in 1609 in *Astronomia nova* (full title: "New Aetiological Astronomy or Celestial Physics Treated by Means of Commentaries on the Movements of the Planet Mars"), possibly astronomy's most unusual (and frank) account of a discovery. A terribly difficult read—seventy chapters dense with charts, computations, and diagrams—it was written as he progressed through his tortuous calculations. His dead ends and blind alleys are included side by side with his successes.

Much of his work was motivated by the prescient assumption that the Sun was somehow responsible for moving the planets. Inspired by William Gilbert's recent claim that the Earth was a giant magnet, he figured that the Sun flung out rays of magnetic power that pushed the planets around. Though this guess was wrong, he was moving astronomy from sheer geometric concerns to issues of physics, forces, and dynamics. His interest in deriving the motion of Mars from the physics of its movements led him to the discovery of a unique relationship: that a planet, joined to the Sun by an imaginary line, sweeps out equal areas in equal times as it orbits the Sun (a fact that would later be explained with Newton's mechanics).

Kepler had first calculated the distance from the Sun to Mars at each and every degree along its orbit—assuming that the orbit was circular but that the Sun was located off center to account for the planet's variable speed. The result, however, did not match Tycho's observations. He gradually recognized that the area rule worked better if the orbit was squeezed in to form an oval (a courageous move, as circles had dominated astronomers' thoughts since the time of Aristotle). He considered some twenty different configurations. Chapter by chapter he plugs numbers into his models and with a seasoned wit shares his gripes with the readers. "If this wearisome method has filled you with loathing," he writes in Chapter 16, "it should more properly fill you with compassion for me, as I have gone through it at least seventy times at the expense of a great deal of time, and you will cease to wonder that the fifth year has now gone by since I took up Mars."[13] He's still moving through his myriad computations into Chapter 50: "How small a heap of grain we have gathered from this threshing! But you also see what a huge cloud of husks there is now."[14] The struggle became his personal "war with Mars":

> While I am thus celebrating a triumph over the motions of Mars, and fetter him in the prison of tables and the leg-irons of eccentric equations, considering him utterly defeated, it is announced in various places that the victory is futile, and war is breaking out again with full force. For while the enemy was in the house as a captive, and hence lightly esteemed, he burst all the chains of the equations and broke out of the prison of the tables. . . . And now there is not much to prevent the fugitive enemy's joining forces with his fellow rebels and reducing me to desperation, unless I send new reinforcements of physical reasoning in a hurry to the scattered troops and old stragglers, and, informed with all diligence, stick to the trail without delay in the direction whither the captive has fled.[15]

Kepler at last saw theory match observation when his orbit took the shape of an ellipse, with the Sun at one of the two foci. The realization came, he said, "as if I were roused from a dream and saw a new light."[16] It was Tycho's data, accurate to within one to two minutes of arc (four times better than previous measurements), that allowed Kepler to detect the subtle difference in shape.* Astronomers in the late seventeenth century would later call this finding Kepler's "first planetary law." Kepler's area rule, though it came first, became his "second law."

Toward the end of his life, Kepler published *Epitome astronomiae Copernicanae* (Epitome of Copernican Astronomy), a text-

---

* With the exceptions of Mercury and Pluto, planetary orbits are very close to circular. But even Mercury's minor axis is only 2 percent shorter than its major axis.

book summarizing Copernican/Keplerian science, and *Harmonice mundi* (The Harmony of the World),[*] an attempt to find musical harmony in the spacing of the planets. In carrying out this work, Kepler chanced upon what came to be known as his third planetary law: that the square of a planet's period, the time it takes to round the Sun, is related to the cube of the orbit's radius. This law was a treasure to seventeenth-century astronomers; since they knew the planets' orbital periods very well, they could use the law to calculate orbital widths more easily.

With the telescope introduced in his lifetime, Kepler received reports from Galileo on the instrument's astounding findings. In 1627, three years before his death, he published the Rudolphine Tables, a far more accurate resource for calculating the positions of the planets based upon his own theories. This endeavor, more than all his other works, secured Kepler's fame in his day. Today we know that his conviction to derive the motions of the planets from physical considerations, leading to the introduction of the elliptical orbit, was his greatest act. Gone for good were the divine circles and an immovable Earth. Like Moses, Kepler brought astronomy to the edge of the promised land. It would take others, the Galileos and the Newtons, to complete the journey.

Kepler's Three Planetary "Laws"

1. The orbital paths of the planets are elliptical, with the Sun at one focus.
2. A line drawn to connect the Sun to any planet sweeps out equal areas of the ellipse in equal intervals of time.
3. The cube of a planet's average distance from the Sun is proportional to the square of its orbital period about the Sun. Or,

$$\frac{\text{Period}^2 \text{ (planet)}}{\text{Period}^2 \text{ (Earth)}} = \frac{\text{Orbital radius}^3 \text{ (planet)}}{\text{Orbital radius}^3 \text{ (Earth)}}$$

---

[*] Often mistranslated as "The Harmonies of the World."

From *Epitome of Copernican Astronomy* by Johannes Kepler

Translated by Charles Glenn Wallis

To the Reader

It has been ten years since I published my *Commentaries on the Movements of the Planet Mars*. As only a few copies of the book were printed, and as it had so to speak hidden the teaching about celestial causes in thickets of calculations and the rest of the astronomical apparatus, and since the more delicate readers were frightened away by the price of the book too; it seemed to my friends that I should be doing right and fulfilling my responsibilities, if I should write an epitome, wherein a summary of both the physical and astronomical teaching concerning the heavens would be set forth in plain and simple speech and with the boredom of the demonstrations alleviated. . . .

Book 4, I, 1. *Are there solid spheres* [orbs] *whereon the planets are carried? And are there empty spaces between the spheres?*

Tycho Brahe disproved the solidity of the spheres by three reasons: the first from the movement of comets; the second from the fact that light is not refracted; the third from the ratio of the spheres.

For if spheres were solid, the comets would not be seen to cross from one sphere into another, for they would be prevented by the solidity; but they cross from one sphere into another, as Brahe shows.

From light thus: since the spheres are eccentric, and since the Earth and its surface—where the eye is—are not situated at the center of each sphere; therefore if the spheres were solid, that is to say far more dense than that very limpid ether, then the rays of the stars would be refracted before they reached our air, as optics teaches; and so the planet would appear irregularly and in places far different from those which could be predicted by the astronomer.

The third reason comes from the principles of Brahe himself; for they bear witness, as do the Copernican, that Mars is sometimes nearer the Earth than the sun is. But Brahe could not believe this interchange to be possible if the spheres were solid, since the sphere of Mars would have to intersect the sphere of the sun.

Then what is there in the planetary regions besides the planets?

Nothing except the ether which is common to the spheres and to the intervals: it is very limpid and yields to the movable bodies no less readily than it yields to the lights of the sun and stars, so that the lights can come down to us. . . .

Book 4, II. On the Movement of the Bodies of the World

. . . Now according as each of the primary bodies is nearer the sun, so it is borne around the sun in a shorter period, under the same common circle of the zodiac, and all in the same direction in which the parts of the solar body precede them—Mercury in the space of three months, Venus in seven and one-half months, the Earth with the lunar heaven in twelve months, Mars in twenty-two and one-half months or less than two years, Jupiter in twelve years, Saturn in thirty years. But for Copernicus the sphere of the fixed stars is utterly immobile.

The Earth meanwhile revolves around its own axis too, and the moon around the Earth—still in the same direction (if you look towards the outer parts of the world) as all the primary bodies.

Now for Copernicus all these movements are direct and continuous, and there are absolutely no stations or retrogradations in the truth of the matter. . . .

How is the ratio of the periodic times, which you have assigned to the mobile bodies, related to the aforesaid ratio of the spheres wherein those bodies are borne?

[Kepler's third planetary law]. The ratio of the times is not equal to the ratio of the spheres, but greater than it, and in the primary planets exactly the ratio of the 3/2th powers. That is to say, if you take the cube roots of the 30 years of Saturn and the 12 years of Jupiter and square them, the true ratio of the spheres of Saturn and Jupiter will exist in these squares. This is the case even if you compare spheres which are not next to one another. For example, Saturn takes 30 years; the Earth takes one year. The cube root of 30 is approximately 3.11. But the cube root of 1 is 1. The squares of these roots are 9.672 and 1. Therefore the sphere of Saturn is to the sphere of the Earth as 9.672 is to 1,000. And a more accurate number will be produced, if you take the times more accurately.

What is gathered from this?

Not all the planets are borne with the same speed, as Aristotle wished, otherwise their times would be as their spheres, and as their diameters; but according as each planet is higher and farther away from the sun, so it traverses less space in one hour by its mean movement: Saturn—according to the magnitude of the solar sphere believed in by the ancients—traverses 240 German miles (in one hour), Jupiter 320 German miles, Mars 600, the center of the Earth 740, Venus 800, and Mercury 1,200.* And if this is to be according to the solar interval proved by me in the above, the number of miles must everywhere be tripled. . . .

Book 4, II, 2. Concerning the Causes of the Movement of the Planets

. . . We are convinced by the astronomical observations which have been taken correctly that the route of a planet is approximately circular and as a matter of fact eccentric—that is, the center [of the circle] is not at the center of the world or of some body; and furthermore that during the succession of ages the planet crosses from place to place [in other words, precesses]. Now as many arguments can be drawn up against the discovery of such an orbit as there are parts of it already described.

For firstly, the orbit of the planet is not a perfect circle. But if mind caused the orbit, it would lay out the orbit in a perfect circle, which has beauty and perfection to the mind. On the contrary, the elliptic figure of the route of the planet and the laws of the movements whereby such a figure is caused smell of the nature of the balance or of material necessity rather than of the conception and determination of the mind . . .

Finally, in order that we may grant that a different idea from that of a circle shines in the mind of the mover: it is asked by what means the mind can apply this or that [idea] to the regions of the world. Now the circle is described around some one fixed center, but the ellipse, which is the figure of the planetary orbits, is described around two centers [Kepler's first planetary law]. . . .

Book 4, II, 3. By what reasons are you led to make the sun the moving cause or the source of movement for the planets?

1. Because it is apparent that in so far as any planet is more distant from the sun than the rest, it moves the more slowly—so that the ratio of

* 1 German mile = 7.5 kilometers = 4.7 English miles.

the periodic times is the ratio of the 3/2th powers of the distances from the sun. Therefore we reason from this that the sun is the source of movement.

2. Below we shall hear the same thing come into use in the case of the single planets—so that the closer any one planet approaches the sun during any time, it is borne with an increase of velocity in exactly the ratio of the square.

3. Nor is the dignity or the fitness of the solar body opposed to this, because it is very beautiful and of a perfect roundness and is very great and is the source of light and heat, whence all life flows out into the vegetables: to such an extent that heat and light can be judged to be as it were certain instruments fitted to the sun for causing movement in the planets.

4. But in especial, all the estimates of probability are fulfilled by the sun's rotation in its own space around its immobile axis, in the same direction in which all the planets proceed: and in a shorter period than Mercury, the nearest to the sun and fastest of all the planets. . . .

Book 4, III. *Then what is that true movement of the planets through their surroundings?*

It is constant with respect to the whole periods; and proceeds around the sun, the center of the world, always eastward towards the signs which follow. It never sticks in one place, as though stationary, and much less does it ever retrograde. But nevertheless it is of irregular speed in its parts; and it makes the planet in one fixed part of its circuit digress rather far from the sun, and in the opposite part come very near to the sun: and so the farther it digresses, the slower it is; and the nearer it approaches, the faster it is. . . .

Book 4, III, 2. *What are the laws and instances of this speed and slowness?*

There is a genuine instance in the lever. For there, when the arms are in equilibrium, the ratio of the weights hanging from each arm is the inverse of the ratio of the arms. For a greater weight hung from the shorter arm makes a moment equal to the moment of the lesser weight which is hung from the longer arm. And so, as the short arm is to the long, so the weight on the longer arm is to the weight on the shorter arm. And if in our mind we remove the other arm, and if instead of the weight on it we conceive at the fulcrum an equal power to lift up the remaining arm with its weight; then it is apparent that this power at the fulcrum does not have so much might over a weight which is distant as it does over the same weight when near. So too astronomy bears witness concerning the planet that the sun does not have as much power to move it and to make it revolve when the planet is farther away from the sun in a straight line, as it does when the interval is decreased. And, in brief, if on the orbit of the planet you take two arcs which are equally distant, the ratio between the distances of each arc from the sun is the same as the ratio of the times which the planet spends in those arcs [Kepler's second planetary law]. . . .

Book 5, I. *If you set up no solid spheres in the heavens and if all the movements of the planets are regulated by natural faculties, which are implanted in the bodies of the planets: then I ask what will the theory* [ratio] *of astronomy be? For it seems that the theory cannot do without the imagining of circles and spheres*

It can easily do without the useless furniture of fictitious circles and spheres. But there is such great need of imagining the true figures, in which the routes of the planets are arranged, that we are impoverishing Astronomy and that the big job to be worked on by the true astronomer is to demonstrate from observations what figures the planetary orbits possess; and to devise such hypotheses, or physical principles, as can be used to demonstrate the figures which are in accord with the deductions made from observations. Therefore when once the figure of the planetary orbit has been established, then will come the second and more popular exercise of the astronomer: to formulate, and to give the rules of, an astronomical calculus in accordance with this true figure, or even to make use of the figure as expressed in material instruments not otherwise than the solid spheres of the ancients were used, and through these figures to lay the movements of the planets before the eyes.

10 / Galileo Initiates the Telescopic Era

Galileo Galilei was born in 1564, the son of an accomplished musician and composer. It was his father's wish for him to train as a physician, but Galileo ignored his tedious medical studies to focus on mathematics. Renowned for his skill in this arena, he was appointed in 1589 to the chair of mathematics at the University of Pisa. There he made his first forays into the fundamental laws of motion, making enemies by discrediting Aristotle and promoting the testing of mathematical laws with physical observations. Moving to the University of Padua in 1592, Galileo dramatically transformed the field of astronomy when he heard of a new invention: a tube with two pieces of glass—one concave, the other convex—that made far-off objects appear up to three times closer.

In 1608 the Dutch lens grinder Hans Lipperhey had applied for a patent on the novel device. By the next year, news of the wondrous instrument reached Italy, and Galileo, immediately grasping the principle, built one of his own that he claimed "exceeds in fame the one of Flanders."[17] Possessing a showman's fondness for fame and fortune, he acquainted Venetian officials with the military and economic advantages of spotting ships from afar through the "optical tubes."*

* The word *telescope* was not coined until 1611.

Before moving to Florence, Galileo built more-powerful telescopes. Turning a twenty-power spyglass to the nighttime sky, he revealed a universe filled with more richness and complexity than any previous astronomer had dared to imagine. Others were beginning to look as well, but Galileo was the first to publish, and he made history with the keen analysis of his observations. He compiled the notes and letters he wrote during the first months of his observations into a book entitled *Sidereus nuncius* ("The Sidereal Messenger" or "The Starry Messenger"), printed in the spring of 1610. The sixty-page pamphlet was a best-seller in its day and generated excitement throughout Europe. Given advance word of the discoveries before publication, the grand duke of Tuscany offered Galileo a lifetime appointment in Florence as his resident mathematician and philosopher.

Since the days of the ancient Greeks, it had been commonly assumed that the heavenly bodies were perfectly smooth and uniform. Through his telescope, though, Galileo discovered a lunar landscape filled with mountains and craters. He watched as the craggy mountaintops on the Moon caught the light of the rising or setting Sun, while the plains remained in shadow. From these sightings, he estimated that the lunar mountains were up to four miles high. The Moon's darker regions, he surmised, might be water, which led others to call them "seas." He saw that the Milky Way, which Tycho Brahe and Johannes Kepler had supposed was some kind of nebulous substance, was actually composed of a multitude of stars, whose light blended together to form a luminous band of milky white. The stars themselves, upon closer telescopic inspection, still appeared pointlike, which boosted the Copernican viewpoint that they were indeed very far off.

For many, Galileo's most astounding astronomical discovery involved the planet Jupiter. On the night of January 7, 1610, he spied three points of light lined up on either side of Jupiter. Six days later, he saw a fourth. At first he thought they were simply stars, but, watching over the course of a week, he concluded they were really moons, regularly circling around the giant planet like a mini solar system. The pivotal moment of his realization is captured in his notebook: at first writing in Italian, he switches to Latin, the language of serious academic concerns, at the instant he recognizes the true identity of the tiny Jovian dots (which he attempted unsuccessfully to name after his patron's family, the Medici).[18] Here was proof that the Earth could indeed travel around the Sun without losing the Moon, a major objection to the Copernican theory. Kepler, the first scholar to publicly support Galileo's findings, was elated by the news.

Galileo's astronomical discoveries did not end with the *Sidereus nuncius*. On July 30, 1610, he notified a colleague about "a most extraordinary marvel . . . that the planet Saturn is not one alone, but is composed of three, which almost touch one another."[19] He couldn't yet recognize the edges of Saturn's ring. With his limited telescopic resolution, it looked as if Saturn were accompanied by two close companions. The Dutch researcher Christiaan Huygens explicitly discerned Saturn's impressive ring five decades later. Galileo also observed Venus go through crescent and gibbous phases like the Moon, which convinced him "that Venus must necessarily revolve around the sun, just like Mercury and all the other planets, a fact believed by the Pythagoreans, Copernicus, Kepler, and me, but not actually proved."[20]

The following year Galileo was studying sunspots. The Chinese as early as 29 B.C. had noticed the dark blemishes by looking at the Sun through thin slices of jade. Western astronomers did not officially note the phenomenon until the fourteenth century. The German astronomer Johannes Fabricius, who made regular observations of sunspots with his father, David, wrote a tract in 1611 proposing that sunspots were on the surface of the Sun. Moreover, their regular movement suggested the Sun was rotating from west to east. Others were saying that the spots were celestial bodies separately orbiting the Sun. From his studies, though, Galileo agreed with Fabricius: the Sun was marred. While Tycho Brahe cast the first stone against the long-standing view of a perfect, unchanging heavens, Galileo completed the assault.

Galileo lived in a time of theological tension, given the relative newness of the Reformation. The astronomical evidence he was gathering on behalf of a Sun-centered universe was a challenge to scripture. In the Bible, Joshua had commanded the moving Sun to stand still. By 1616, the Church in Rome was taking the position that it was the astronomer's job to make accurate predictions of celestial movements, not to propose ultimate truths about the nature of the universe.

With the election in 1623 of a personal friend and patron of science, Maffeo Barberini, as Pope Urban VIII, Galileo believed it was safe once again to air new viewpoints on the structure of the cosmos. Combative in nature, he thrived on sardonically putting down his opponents. In 1632 he published *Dialogue Concerning the Two Chief World Systems,* his discussion of the Copernican universe versus the Ptolemaic. But he seriously misjudged the temper of the times. Within a year, he was brought before the Inquisition and eventually put under house arrest for going too far in his support of a moving Earth. His final work was *Discourses and Mathematical Demonstrations Concerning Two New Sciences,* his pioneering treatise on the science of kinematics, which served as a foundation for modern mechanics. The manuscript was smuggled out of the country for publication in 1638. Galileo died four years later at the age of seventy-seven.

From *Sidereus nuncius*
by Galileo

Translated by Albert Van Helden

SIDEREAL MESSENGER
unfolding great and very wonderful sights
and displaying to the gaze of everyone,
but especially philosophers and astronomers,
the things that were observed by
GALILEO GALILEI,
Florentine patrician
and public mathematician of the University of Padua,
with the help of a spyglass lately devised by him,
about the face of the Moon, countless fixed stars,
the Milky Way, nebulous stars,
but especially about
four planets
flying around the star of Jupiter at unequal intervals
and periods with wonderful swiftness;
which, unknown by anyone until this day,

the first author detected recently
and decided to name
MEDICEAN STARS*

Venice
1610

In this short treatise I propose great things for inspection and contemplation by every explorer of Nature. Great, I say, because of the excellence of the things themselves, because of their newness, unheard of through the ages, and also because of the instrument with the benefit of which they make themselves manifest to our sight.

Certainly it is a great thing to add to the countless multitude of fixed stars visible hitherto by natural means and expose to our eyes innumerable others never seen before, which exceed tenfold the number of old and known ones.

It is most beautiful and pleasing to the eye to look upon the lunar body, distant from us about sixty terrestrial diameters, from so near as if it were distant by only two of these measures, so that the diameter of the same Moon appears as if it were thirty times, the surface nine hundred times, and the solid body about twenty-seven thousand times larger than when observed only with the naked eye.† Anyone will then understand with the certainty of the senses that the Moon is by no means endowed with a smooth and polished surface, but is rough and uneven and, just as the face of the Earth itself, crowded everywhere with vast prominences, deep chasms, and convolutions.

Moreover, it seems of no small importance to have put an end to the debate about the Galaxy or Milky Way and to have made manifest its essence to the senses as well as the intellect; and it will be pleasing and most glorious to demonstrate clearly that the substance of those stars called nebulous up to now by all astronomers is very different from what has hitherto been thought.

But what greatly exceeds all admiration, and what especially impelled us to give notice to all astronomers and philosophers, is this, that we have discovered four wandering stars, known or observed by no one before us.

---

* In ancient cosmology, all heavenly bodies were routinely called stars. The planets were the wandering stars.

† The translator notes that Galileo mistakenly uses *diameters* here and elsewhere, when the Moon was commonly known to be about sixty terrestrial *radii* distant.

These, like Venus and Mercury around the Sun, have their periods around a certain star notable among the number of known ones, and now precede, now follow, him, never digressing from him beyond certain limits. All these things were discovered and observed a few days ago by means of a glass contrived by me after I had been inspired by divine grace.

Perhaps more excellent things will be discovered in time, either by me or by others, with the help of a similar instrument, the form and construction of which, and the occasion of whose invention, I shall first mention briefly, and then I shall review the history of the observations made by me.

[*The Telescope*] About 10 months ago a rumor came to our ears that a spyglass had been made by a certain Dutchman [Hans Lipperhey] by means of which visible objects, although far removed from the eye of the observer, were distinctly perceived as though nearby. About this truly wonderful effect some accounts were spread abroad, to which some gave credence while others denied them. The rumor was confirmed to me a few days later by a letter from Paris from the noble Frenchman Jacques Badovere. This finally caused me to apply myself totally to investigating the principles and figuring out the means by which I might arrive at the invention of a similar instrument, which I achieved shortly afterward on the basis of the science of refraction. And first I prepared a lead tube in whose ends I fitted two glasses, both plane on one side while the other side of one was spherically convex and of the other concave. Then, applying my eye to the concave glass, I saw objects satisfactorily large and close. Indeed, they appeared three times closer and nine times larger than when observed with natural vision only. Afterward I made another more perfect one for myself that showed objects more than sixty times larger. Finally, sparing no labor or expense, I progressed so far that I constructed for myself an instrument so excellent that things seen through it appear about a thousand times larger and more than thirty times closer than when observed with the natural faculty only. It would be entirely superfluous to enumerate how many and how great the advantages of this instrument are on land and at sea. But having dismissed earthly things, I applied myself to explorations of the heavens. And first I looked at the Moon from so close that it was scarcely two terrestrial diameters distant. Next, with incredible delight I frequently observed the stars, fixed as well as wandering, and as I saw their huge number I began to think of, and at last discovered, a method whereby I could measure the distances between them. In this matter, it behooves all those who wish to make such observations to be forewarned. For it is necessary first that they prepare a most accurate glass that shows objects brightly, distinctly, and not veiled by any obscurity, and second that it multiply them at least four hundred times and show them twenty times closer. For if it is not an instrument such as that, one will try in vain to see all the things observed in the heavens by us and enumerated below. . . .

[*Observations of the Moon*] Let us speak first about the face of the Moon that is turned toward our sight, which, for the sake of easy understanding, I divide into two parts, namely a brighter one and a darker one. The brighter part appears to surround and pervade the entire hemisphere, but the darker part, like some cloud, stains its very face and renders it spotted. Indeed, these darkish and rather large spots are obvious to everyone, and every age has seen them. For this reason we shall call them the large or ancient spots, in contrast with other spots, smaller in size and occurring with such frequency that they besprinkle the entire lunar surface, but especially the brighter part. These were, in fact, observed by no one before us. By oft-repeated observations of them we have been led to the conclusion that we certainly see the surface of the Moon to be not smooth, even, and perfectly spherical, as the great crowd of philosophers have believed about this and other heavenly bodies, but, on the contrary, to be uneven, rough, and crowded with depressions and bulges. And it is like the face of the Earth itself, which is marked here and there with chains of mountains and depths of valleys. The observations from which this is inferred are as follows.

On the fourth or fifth day after conjunction [new moon], when the Moon displays herself to us with brilliant horns, the boundary dividing the bright from the dark part does not form a uniformly oval line, as would happen in a perfectly spherical solid, but is marked by an uneven, rough, and very sinuous line, as the figure shows. For several, as it were, bright excrescences extend beyond the border between light and darkness into the dark part, and on the other hand little dark parts enter into the light. Indeed,

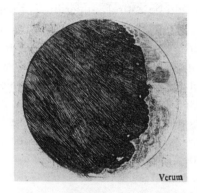

Verum

a great number of small darkish spots, entirely separated from the dark part, are distributed everywhere over almost the entire region already bathed by the light of the Sun, except, at any rate, for that part affected by the large and ancient spots. We noticed, moreover, that all these small spots just mentioned always agree in this, that they have a dark part on the side toward the Sun while on the side opposite the Sun they are crowned with brighter borders like shining ridges. And we have an almost entirely similar sight on Earth, around sunrise, when the valleys are not yet bathed in light but the surrounding mountains facing the Sun are already seen shining with light. And just as the shadows of the earthly valleys are diminished as the Sun climbs higher, so those lunar spots lose their darkness as the luminous part grows.

Not only are the boundaries between light and dark on the Moon perceived to be uneven and sinuous, but, what causes even greater wonder, is that very many bright points appear within the dark part of the Moon, entirely separated and removed from the illuminated region and located no small distance from it. Gradually, after a small period of time, these are increased in size and brightness. Indeed, after 2 or 3 hours they are joined with the rest of the bright part, which has now become larger. In the meantime, more and more bright points light up, as if they are sprouting, in the dark part, grow, and are connected at length with that bright surface as it extends farther in this direction. An example of this is shown in the same figure. Now, on Earth, before sunrise, aren't the peaks of the highest mountains illuminated by the Sun's rays while shadows still cover the plain? Doesn't light grow, after a little while, until the middle and larger parts of the same mountains are illuminated, and finally, when the Sun has risen, aren't the illuminations of plains and hills joined together? These differences between prominences and depressions in the Moon, however, seem to exceed the terrestrial roughness greatly. . . .

[*The Stars and the Milky Way*] Up to this point we have discussed the observations made of the lunar body. We will now report briefly on what has been observed by us thus far concerning the fixed stars. And first, it is worthy of notice that when they are observed by means of the spyglass, stars, fixed as well as wandering, are seen not to be magnified in size in the same proportion in which other objects, and also the Moon herself, are increased. In the stars, the increase appears much smaller so that you may believe that a glass capable of multiplying other objects, for example, by a ratio of 100 hardly multiplies stars by a ratio of 4 or 5. . . .

The difference between the appearance of planets and fixed stars also seems worthy of notice. For the planets present entirely smooth and exactly circular globes that appear as little moons, entirely covered with light, while the fixed stars are not seen bounded by circular outlines but rather as pulsating all around with certain bright rays. With the glass they appear in the same shape as when they are observed with natural vision, but so much larger that a little star of the fifth or sixth magnitude appears to equal the Dog Star [Sirius], which is the largest of all fixed stars. Indeed, with the glass you will detect below stars of the sixth magnitude such a crowd of others that escape natural sight that it is hardly believable. For you may see more than six further gradations of magnitude. The largest of these, which we may designate as of the seventh magnitude, or the first magnitude of the invisible ones, appear larger and brighter with the help of the glass than stars of the second magnitude seen with natural vision. But in order that you may see one or two illustrations of the almost inconceivable crowd of them, and from their example form a judgment about the rest of them, I decided to reproduce two star groups. In the first I had decided to depict the entire constellation of Orion, but overwhelmed by the enormous multitude of stars and a lack of time, I put off this assault until another occasion. For there are more than five hundred new stars around the old ones, spread over a space of 1 or 2 degrees. For this reason, to the three in Orion's belt and the six in his sword that were observed long ago, I have added eighty others seen recently, and I have retained their separations as accurately as possible. For the sake of distinction, we have depicted the known or ancient ones larger and outlined by double lines, and the other inconspicuous ones smaller and outlined by single lines. We have also preserved the distinction in size as much as possible. In the second example we have depicted the six

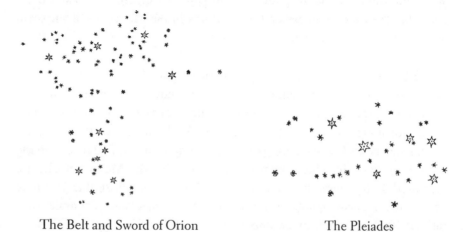

The Belt and Sword of Orion   The Pleiades

stars of the Bull [Taurus] called the Pleiades (I say six since the seventh almost never appears) contained within very narrow limits in the heavens. Near these lie more than forty other invisible stars, none of which is farther removed from the aforementioned six than scarcely half a degree. We have marked down only thirty-six of these, preserving their mutual distances, sizes, and the distinction between old and new ones, as in the case of Orion.

What was observed by us in the third place is the nature or matter of the Milky Way itself, which, with the aid of the spyglass, may be observed so well that all the disputes that for so many generations have vexed philosophers are destroyed by visible certainty, and we are liberated from wordy arguments. For the Galaxy is nothing else than a congeries of innumerable stars distributed in clusters. To whatever region of it you direct your spyglass, an immense number of stars immediately offer themselves to view, of which very many appear rather large and very conspicuous but the multitude of small ones is truly unfathomable.

And since that milky luster, like whitish clouds, is seen not only in the Milky Way, but dispersed through the ether, many similarly colored patches shine weakly; if you direct a glass to any of them, you will meet with a dense crowd of stars. Moreover—and what is even more remarkable—the stars that have been called "nebulous" by every single astronomer up to this day are swarms of small stars placed exceedingly closely together. While each individual one escapes our sight because of its smallness or its very great distance from us, from the commingling of their rays arises that brightness ascribed up to now to a denser part of the heavens capable of reflecting the rays of the stars or Sun. . . .

[*Discovery of Jupiter's Moons*] We have briefly explained our observations thus far about the Moon, the fixed stars, and the Milky Way. It remains for us to reveal and make known what appears to be most important in the present matter: four planets never seen from the beginning of the world right up to our day, the occasion of their discovery and observation, their positions, and the observations made over the past 2 months concerning their behavior and changes. And I call on all astronomers to devote themselves to investigating and determining their periods. Because of the shortness of time, it has not been possible for us to achieve this so far. We advise them again, however, that they will need a very accurate glass like the one we have described at the beginning of this account, lest they undertake such an investigation in vain.

Accordingly, on the seventh day of January of the present year 1610, at the first hour of the night, when I inspected the celestial constellations through a spyglass, Jupiter presented himself. And since I had prepared for myself a superlative instrument, I saw (which earlier had not happened because of the weakness of the other instruments) that three little stars were positioned near him—small but yet very bright. Although I believed them to be among the number of fixed stars, they nevertheless intrigued me because they appeared to be arranged exactly along a straight line and parallel to the ecliptic, and to be brighter than others of equal size. And their disposition among themselves and with respect to Jupiter was as follows:

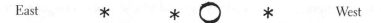

East   *   * ◯   *   West

That is, two stars were near him on the east and one on the west; the more eastern one and the western one appeared a bit larger than the remaining one. I was not in the least concerned with their distances from Jupiter, for, as we said above, at first I believed them to be fixed stars. But when, on the eighth, I returned to the same observation, guided by I know not what fate, I found a very different arrangement. For all three little stars were to the west of Jupiter and closer to each other than the previous night, and separated by equal intervals, as shown in the adjoining sketch.

East   ◯ * * *   West

Even though at this point I had by no means turned my thought to the mutual motions of these stars, yet I was aroused by the question of how Jupiter could be to the east of all the said fixed stars when the day before he had been to the west of two of them. I was afraid, therefore, that perhaps, contrary to the astronomical computations, his motion was direct and that, by his proper motion, he had bypassed those stars. For this reason I waited eagerly for the next night. But I was disappointed in my hope, for the sky was everywhere covered with clouds.

Then, on the tenth, the stars appeared in this position with regard to Jupiter. Only two stars were near him, both to the east. The third, as I thought, was hidden behind Jupiter. As before, they were in the same straight line with Jupiter and exactly aligned along the zodiac.

East   *   * ◯   West

When I saw this, and since I knew that such changes could in no way be assigned to Jupiter, and since I knew, moreover, that the observed stars were always the same ones (for no others, either preceding or following Jupiter, were present along the zodiac for a great distance), now, moving from doubt to astonishment, I found that the observed change was not in Jupiter but in the said stars. And therefore I decided that henceforth they should be observed more accurately and diligently.

And so, on the eleventh, I saw the following arrangement:

East   **✳ ✳**   **◯**   West

There were only two stars on the east, of which the middle one was three times as far from Jupiter than from the more eastern one, and the more eastern one was about twice as large as the other, although the previous night they had appeared about equal. I therefore arrived at the conclusion, entirely beyond doubt, that in the heavens there are three stars wandering around Jupiter like Venus and Mercury around the Sun. This was at length seen clear as day in many subsequent observations, and also that there are not only three, but four wandering stars making their revolutions about Jupiter. . . .

These are the observations of the four Medicean planets recently, and for the first time, discovered by me. From them, although it is not yet possible to calculate their periods, something worthy of notice may at least be said. And first, since they sometimes follow and at other times precede Jupiter by similar intervals, and are removed from him toward the east as well as the west by only very narrow limits, and accompany him equally in retrograde and direct motion, no one can doubt that they complete their revolutions about him while, in the meantime, all together they complete a 12-year period about the center of the world. Moreover, they whirl around in unequal circles, which is clearly deduced from the fact that at the greatest separations from Jupiter two planets could never be seen united while, on the other hand, near Jupiter two, three, and occasionally all four planets are found crowded together at the same time. It is further seen that the revolutions of the planets describing smaller circles around Jupiter are faster. For the stars closer to Jupiter are often seen to the east when the previous day they appeared to the west, and vice versa, while from a careful examination of its previously accurately noted returns, the planet traversing the largest orb appears to have a semimonthly period. We have moreover an excellent and splendid argument for taking away the scruples of those who,

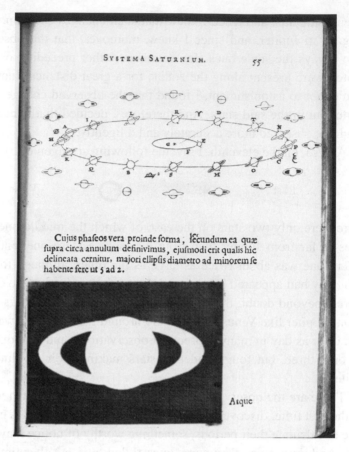

Figure 10.1: The Dutch astronomer and physicist Christiaan Huygens improved the quality of lenses, which led to the construction of vastly superior telescopes. Using one in 1656, he discovered the first moon of Saturn, which he named Titan. He was also able to discern that Saturn was accompanied by, not separate bodies (as Galileo thought), but a ring. Complete details were given in his *Systema saturnium* (1659). In a diagram from that book (top), he shows how Saturn's appearance changes due to the changing positions of both the Earth (E) and Saturn as they orbit the Sun (G). At the bottom, Huygens displays his observation of Saturn's ring at its greatest inclination to us. Starting in 1671, Giovanni Cassini at the Paris Observatory discovered four more moons of Saturn and also discerned a gap in the ring, now known as Cassini's division. This made him suspect that the ring was not solid but rather composed of small particles, a notion spurned by most astronomers until James Clerk Maxwell proved Cassini was right nearly two hundred years later.

while tolerating with equanimity the revolution of the planets around the Sun in the Copernican system, are so disturbed by the attendance of one Moon around the Earth while the two together complete the annual orb around the Sun that they conclude that this constitution of the universe must be overthrown as impossible. For here we have only one planet revolving around another while both run through a great circle around the Sun: but our vision offers us four stars wandering around Jupiter like the Moon around the Earth while all together with Jupiter traverse a great circle around the Sun in the space of 12 years.

11 / Newton's Universal Law of Gravity

Isaac Newton made public his important discoveries at the age of forty-four, when he published *Philosophiae naturalis principia mathematica* (Mathematical Principles of Natural Philosophy), his monumental work on gravitation. It's best known as simply the *Principia*. With its publication in 1687, Newton offered an entirely new approach to physics and astronomy; he was the first to show that nature's actions, from the path of a cannon ball to the movement of a planet, could be described and predicted by a few universal physical laws. The heavens and the Earth shared a mother tongue: it was the language of mathematics.

In casual conversations with acquaintances, Newton claimed that the seeds of his major achievements—the theory of colors, the calculus, and the law of gravity—were planted in the years 1665 and 1666, when he "was in the prime of my age for invention, and minded mathematics and philosophy more than at any time since."[21] He had temporarily left his undergraduate studies at Cambridge University and returned to his childhood home in the rural hamlet of Woolsthorpe to wait out a plague epidemic. Possibly by watching that fabled apple fall in his country garden, he came to reflect on the tendency of bodies to fall toward the Earth with a set acceleration. Did this same force extend to the Moon? he asked himself. He computed that the Moon did seem to be continually "falling" toward the Earth—its path becoming curved—by an earthly pull that diminished outward by about the square of the distance, the sign that a force is spreading its influence equally in all directions. But since the fit wasn't perfect (his size for the Earth was not accurate, which threw off the calculation), Newton put the problem aside for many years.

By the seventeenth century the greatest minds in Europe were wondering about the causes of planetary motion. Johannes Kepler suggested magnetic forces emanating from the Sun; the French philosopher René Descartes imagined the planets were carried around like leaves trapped within a swirling whirlpool by vortices of ether. By 1674 Robert Hooke in England, curator of experiments for the Royal Society, developed an intriguing set of suppositions for explaining celestial motion: that all celestial bodies have a gravitating power directed toward their centers, that they can attract other bodies, and that the attraction is more powerful the closer you are. Triggered by an exchange of letters with Hooke in the winter of 1679–80, Newton returned to the problem of his youth.

Newton was an intensely private man, wary of his jealous rival Hooke, and often fearful of exposing his work to criticism. At first he let his revolutionary findings on gravitation go unpublished, possibly not yet satisfied they were of importance.[22] That he wrote the *Principia* at all is largely due to Edmond Halley (of comet fame). While visiting in 1684, Halley asked Newton how a planet would move under an inverse square law. Newton confidently replied, "An ellipsis."[23] Only with Halley's persistent prodding and financial backing was Newton convinced to write his masterpiece. He abandoned his work on classical mathematics, chemistry, and alchemy, his most recent fascinations, and at last applied his legendary power of concentration to completing his work on gravity. Armed with better measurements of the Earth, he was at last able to prove that the Moon was indeed attracted to the Earth by an inverse square law and that such a force directly leads to objects moving in elliptical orbits, just as Kepler had observed.* It took him nearly two years to complete the *Principia*.

His historic tome consists of three books, which lay out his argu-

---

* In other words, the strength of the force between two objects falls off as the square of the distance ($1/r^2$). Triple the distance, for example, and the force between them is reduced to $1/3^2$, or one-ninth of its original strength.

ments over hundreds of pages in a well-organized progression of lemmas, propositions, geometric proofs, and diagrams. Book 1 opens by defining basic quantities, such as mass, momentum (what he called "quantity of motion"), and force, and proving that keeping a body in orbit must involve a particular force—a centripetal force—directed toward the center of the orbit. Newton's famous second law is stated only verbally. The more familiar $F = ma$ (force = mass × acceleration) derives from Leonhard Euler in the 1740s. Book 2 discusses forces at work in various media, such as fluids and air, and is a section usually neglected. But it was crucial in proving that Descartes's vortices were a fiction. Newton showed that the orbit of an object moving through a resisting medium would decay, which was not happening to the planets. The universe suddenly emptied. Gone were the Cartesian vortices of endless matter.

Not until his final book does Newton fully establish his law of gravitation, and it led naturally to the Copernican system, given the Sun as the solar system's center of mass. Long-standing problems in astronomy seemed to melt away as Newton considered more and more cases with his new rules. It was what we now call conservation of angular momentum that made a planet speed up as it approaches the Sun and slow down as it recedes, as Kepler noted observationally with his second law (see Chapter 9). Gravity could also explain the tides, the Earth's precession (due to the Moon's and Sun's tugging on the Earth's bulge), and the trajectory of a comet. In a grand conjectural leap, Newton was declaring that gravity was a fundamental and universal force of nature. What draws an apple to the ground also keeps the Moon in orbit about the Earth. "For nature is simple," Newton writes in Book 3, "and does not indulge in the luxury of superfluous causes."[24] The cosmos and terra firma were no longer separate realms, as Aristotle had reasoned; the heavens and the earth were now blissfully wedded under one set of physical laws.

But Newton was also aware that he was presenting the ideal case: that if a planet or object deviates from his law, it was a sure sign that other forces were at work. He was already aware that the planets themselves gravitationally affected each other in subtle ways. "To consider simultaneously all these causes of motion and to define these motions by exact laws admitting easy calculation exceeds, if I am not mistaken, the force of any human mind," he said.[25] It would lead to an astronomical industry, with astronomers in the following century devoted to endless calculations of planetary perturbations.

There was one problem with Newton's law of gravity, though. It implied that imperceptible ribbons of attraction somehow emanated over distances, both short and long, to keep moon to planet and boulder to Earth. This feat appeared more resonant with the occult than science. Critics demanded a mechanism, which led to Newton's famous statement: "I have not as yet been able to deduce from phenomena the reason for these properties of gravity, and I do not feign hypotheses."[26] It was enough for him (and eventually the entire physics community) that his laws allowed successful predictions to be made. Newton's law of gravity held firm for more than two hundred years. Not until 1915 was a new understanding of gravity revealed, when Albert Einstein amended Newton's laws with his general theory of relativity (see Chapter 36).

Excerpted here are Newton's three laws of motion; his rules for the study of natural philosophy, which argue for the universality of gravitation; his "moon test" showing that Earth's gravity extends out to its satellite; and finally the answer to his critics who were demanding a reason for gravity's properties.

From *The Principia*
by Isaac Newton

Translated by I. Bernard Cohen and Anne Whitman, assisted by Julia Budenz

Axioms, Or the Laws of Motion

Law 1: *Every body perseveres in its state of being at rest or of moving uniformly straight forward, except insofar as it is compelled to change its state by forces impressed.*

Projectiles persevere in their motions, except insofar as they are retarded by the resistance of the air and are impelled downward by the force of gravity. A spinning hoop, which has parts that by their cohesion continually draw one another back from rectilinear motions, does not cease to rotate, except insofar as it is retarded by the air. And larger bodies— planets and comets—preserve for a longer time both their progressive and their circular motions, which take place in spaces having less resistance.

Law 2: *A change in motion is proportional to the motive force impressed and takes place along the straight line in which that force is impressed.*

If some force generates any motion, twice the force will generate twice the motion, and three times the force will generate three times the motion, whether the force is impressed all at once or successively by degrees. And if the body was previously moving, the new motion (since motion is always in the same direction as the generative force) is added to the original motion if that motion was in the same direction or is subtracted from the original motion if it was in the opposite direction or, if it was in an oblique direction, is combined obliquely and compounded with it according to the directions of both motions.

Law 3: *To any action there is always an opposite and equal reaction; in other words, the actions of two bodies upon each other are always equal and always opposite in direction.*

Whatever presses or draws something else is pressed or drawn just as much by it. If anyone presses a stone with a finger, the finger is also pressed by the stone. If a horse draws a stone tied to a rope, the horse will (so to speak) also be drawn back equally toward the stone, for the rope, stretched out at both ends, will urge the horse toward the stone and the stone toward the horse by one and the same endeavor to go slack and will impede the forward motion of the one as much as it promotes the forward motion of the other. If some body impinging upon another body changes the motion of that body in any way by its own force, then, by the force of the other body (because of the equality of their mutual pressure), it also will in turn undergo the same change in its own motion in the opposite direction. By means of these actions, equal changes occur in the motions, not in the velocities—that is, of course, if the bodies are not impeded by anything else. For the changes in velocities that likewise occur in opposite directions are inversely proportional to the bodies because the motions are changed equally. This law is valid also for attractions. . . .

Book 3, The System of the World
Rules for the Study of Natural Philosophy

Rule 1: *No more causes of natural things should be admitted than are both true and sufficient to explain their phenomena.*

As the philosophers say: Nature does nothing in vain, and more causes are in vain when fewer suffice. For nature is simple and does not indulge in the luxury of superfluous causes.

Rule 2: *Therefore, the causes assigned to natural effects of the same kind must be, so far as possible, the same.*

Examples are the cause of respiration in man and beast, or of the falling of stones in Europe and America, or of the light of a kitchen fire and the sun, or of the reflection of light on our earth and the planets.

Rule 3: *Those qualities of bodies that cannot be intended and remitted [i.e., qualities that cannot be increased and diminished] and that belong to all bodies on which experiments can be made should be taken as qualities of all bodies universally.*[27]

For the qualities of bodies can be known only through experiments; and therefore qualities that square with experiments universally are to be regarded as universal qualities; and qualities that cannot be diminished cannot be taken away from bodies. Certainly idle fancies ought not to be fabricated recklessly against the evidence of experiments, nor should we depart from the analogy of nature, since nature is always simple and ever consonant with itself. The extension of bodies is known to us only through our senses, and yet there are bodies beyond the range of these senses; but because extension is found in all sensible bodies, it is ascribed to all bodies universally. We know by experience that some bodies are hard. Moreover, because the hardness of the whole arises from the hardness of its parts, we justly infer from this not only the hardness of the undivided particles of bodies that are accessible to our senses, but also of all other bodies. That all bodies are impenetrable we gather not by reason but by our senses. We find those bodies that we handle to be impenetrable, and hence we conclude that impenetrability is a property of all bodies universally. That all bodies are movable and persevere in motion or in rest by means of certain forces (which we call forces of inertia) we infer from finding these properties in the bodies that we have seen. The extension, hardness, impenetrability, mobility, and force of inertia of the whole arise from the extension, hardness, impenetrability, mobility, and force of inertia of each of the parts; and thus we conclude that every one of the least parts of all bodies is extended, hard, impenetrable, movable, and endowed with a force of inertia. And this is the foundation of all natural philosophy. Further, from phenomena we know that the divided, contiguous parts of bodies can be separated from one another, and from mathematics it is certain that the undivided parts can be distinguished into smaller parts by our reason. But it is uncertain whether those parts which have been distinguished in this way and not yet divided can actually be divided and separated from one another by the forces of nature. But if it were established by even a single experiment that in the breaking of a hard and solid body, any undivided particle underwent division, we should conclude by the force of this third rule not only that divided parts are separable but also that undivided parts can be divided indefinitely.

Finally, if it is universally established by experiments and astronomical observations that all bodies on or near the earth gravitate [*lit.* are heavy] toward the earth, and do so in proportion to the quantity of matter in each body, and that the moon gravitates [is heavy] toward the earth in proportion to the quantity of its matter, and that our sea in turn gravitates [is heavy] toward the moon, and that all planets gravitate [are heavy] toward one another, and that there is a similar gravity [heaviness] of comets toward the sun, it will have to be concluded by this third rule that all bodies gravitate toward one another. Indeed, the argument from phenomena will be even stronger for universal gravity than for the impenetrability of bodies, for which, of course, we have not a single experiment, and not even an observation, in the case of the heavenly bodies. Yet I am by no means affirming that gravity is essential to bodies. By inherent force I mean only the force of inertia. This is immutable. Gravity is diminished as bodies recede from the earth.

Rule 4: *In experimental philosophy, propositions gathered from phenomena by induction should be considered either exactly or very nearly true notwithstanding any contrary hypotheses, until yet other phenomena make such propositions either more exact or liable to exceptions.*[28]

This rule should be followed so that arguments based on induction may not be nullified by hypotheses. . . .

Moon Test

[Here Newton makes his revolutionary claim for universality—that the same force that allows a heavy body to fall near the surface of the Earth is the same force that keeps the Moon in orbit.]

Book 3, Proposition 4: The moon gravitates toward the earth and by the force of gravity is always drawn back from rectilinear motion and kept in its orbit.

... Let us assume a mean distance of 60 semidiameters in the syzygies;* and also let us assume that a revolution of the moon with respect to the fixed stars is completed in 27 days, 7 hours, 43 minutes, as has been established by astronomers; and that the circumference of the earth is 123,249,600 Paris feet, according to the measurements made by the French.† If now the moon is imagined to be deprived of all its motion and to be let fall so that it will descend to the earth with all that force urging it by which (by prop. 3, corol.) it is [normally] kept in its orbit, then in the space of one minute, it will by falling describe $15\frac{1}{12}$ Paris feet. This is determined by a calculation carried out either by using prop. 36 of book 1 or (which comes to the same thing) by using corol. 9 to prop. 4 of book 1. For the versed sine of the arc which the moon would describe in one minute of time by its mean motion at a distance of 60 semidiameters of the earth is roughly $15\frac{1}{12}$ Paris feet, or more exactly 15 feet, 1 inch, and $1\frac{4}{9}$ lines [or twelfths of an inch]. Accordingly, since in approaching the earth that force is increased as the inverse square of the distance, and so at the surface of the earth is 60×60 times greater than at the moon, it follows that a body falling with that force, in our regions, ought in the space of one minute to describe $60 \times 60 \times 15\frac{1}{12}$ Paris feet, and in the space of one second $15\frac{1}{12}$ feet, or more exactly 15 feet, 1 inch, and $1\frac{4}{9}$ lines. And heavy bodies do actually descend to the earth with this very force. For a pendulum beating seconds in the latitude of Paris is 3 Paris feet and $8\frac{1}{2}$ lines in length, as Huygens observed. And the height that a heavy body describes by falling in the time of one second is to half the length of this pendulum as the square of the ratio of the circumference of a circle to its diameter (as Huygens also showed), and so is 15 Paris feet, 1 inch, $1\frac{7}{9}$ lines. And therefore that force by which the moon is kept in its orbit, in descending from the moon's orbit to the surface of the earth, comes out equal to the force of gravity here on earth, and so (by rules 1 and 2) is that very force which we generally call gravity. For if gravity were different from this force, then bodies making for the earth by both forces acting together would descend twice as fast, and in the space of one second would by falling describe $30\frac{1}{6}$ Paris feet, entirely contrary to experience. ...

---

* A syzygy is the straight-line configuration of three celestial bodies in a gravitational system (such as the Sun, Moon, and Earth during a solar or lunar eclipse). Here Newton assumes a Moon–Earth distance of 60 Earth radii at such a moment.

† 1 Paris foot = 1.07 English foot.

General Scholium

. . . Thus far I have explained the phenomena of the heavens and of our sea by the force of gravity, but I have not yet assigned a cause to gravity. Indeed, this force arises from some cause that penetrates as far as the centers of the sun and planets without any diminution of its power to act, and that acts not in proportion to the quantity of the *surfaces* of the particles on which it acts (as mechanical causes are wont to do) but in proportion to the quantity of *solid* matter, and whose action is extended everywhere to immense distances, always decreasing as the squares of the distances. Gravity toward the sun is compounded of the gravities toward the individual particles of the sun, and at increasing distances from the sun decreases exactly as the squares of the distances as far out as the orbit of Saturn, as is manifest from the fact that the aphelia of the planets are at rest, and even as far as the farthest aphelia of the comets, provided that those aphelia are at rest. I have not as yet been able to deduce from phenomena the reason for these properties of gravity, and I do not feign hypotheses. For whatever is not deduced from the phenomena must be called a hypothesis; and hypotheses, whether metaphysical or physical, or based on occult qualities, or mechanical, have no place in experimental philosophy. In this experimental philosophy, propositions are deduced from the phenomena and are made general by induction. The impenetrability, mobility, and impetus of bodies, and the laws of motion and the law of gravity have been found by this method. And it is enough that gravity really exists and acts according to the laws that we have set forth and is sufficient to explain all the motions of the heavenly bodies and of our sea.

12 / Halley's Comet

At the end of the *Principia*, his masterful treatise on gravitation, Isaac Newton laid out his mathematical theory on the motion of comets. It was his grand finale, an effort that he described to a colleague as "the most difficult of the whole book."[29] He concluded that "comets are a kind of planet and revolve in their orbits with a continual motion."[30] These paths could be very elongated ellipses or even hyperbolic orbits, in which case the comet would forever depart from the solar system. Newton was greatly inspired by the appearance of a comet in 1680, and the *Principia* goes into great detail tracing this particular comet's path. In fact, a diagram he included in Book 3 of the *Principia* is the first figure in astronomical history to show a comet completely swinging around the Sun due to gravity (see Figure 12.1). (Before that, observers were not sure that a comet approaching the Sun was actually the same object seen later to recede from it.) Comets were not omens of disaster, Newton was saying, but merely small planetoids.

Edmond Halley, Newton's Royal Society colleague, used his friend's mathematical laws to make the first prediction of a comet's return.* After poring over historic records, Halley figured that a comet sighted in 1682 had much in common with comets previously observed in 1531 and 1607. They shared the same orbital

---

* At times Halley's first name is also spelled "Edmund."

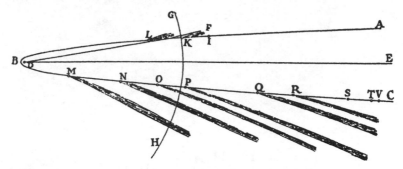

Figure 12.1: The comet of 1680, as illustrated in Newton's *Principia*.

characteristics (all went around the Sun in the opposite direction to the planets) and appeared every seventy-five to seventy-six years. Based on his calculations, he predicted the comet would return at the end of 1758, taking into account the additional tugs by Jupiter in the comet's journey through the solar system. He published this prediction in 1705, in various reports written in both Latin and English.

The comet did appear on schedule, just as Halley foretold. On Christmas Day in 1758, an amateur astronomer and gentleman farmer in Saxony named Johann Georg Palitzsch spotted "a nebulous star" in the nighttime sky. Noted observer Charles Messier, already on the lookout for the comet, saw the same fuzzy object four weeks later from Paris, and it was soon confirmed to be Halley's returnee. By March the comet was rounding the Sun. The public was bedazzled and Newton's critics were instantly silenced. It was at that moment that Newton's controversial law of gravity was at last triumphant among both scientists and the public, and Halley's name became forever linked to the periodic celestial visitor. Scientists came to view the universe as intrinsically knowable, ticking away predictably like a well-oiled timepiece.

Halley had been named astronomer royal in 1720 and made further contributions to astronomy, including advances in determining the distance to the Sun and detecting stellar motions (see Chapter 17). But, alas, he was not among those present to witness the return of his beloved comet. He died in 1742 at the age of eighty-five.

Into the nineteenth century, comets became a favored object for scrutiny and computation. The German astronomer Johann Encke calculated the orbit of a comet seen in 1819 and said it would return a scant three years later. It did return as predicted and is now known as Comet Encke. It was only the second time that a comet's return was accurately foretold and was also proof that a different type existed—the short-period comet that does not travel into the farthest reaches of the solar system but rather remains inside the orbit of Jupiter.

In the closing remarks of his prediction paper, knowing that comets at times made close approaches to the Earth, Halley couldn't help but wonder about the ultimate cosmic disaster: a collision with a comet. He noted this was "by no means impossible."[31] Astronomers at last saw for themselves when the separate components of Comet Shoemaker-Levy, which had broken into at least twenty-one fragments, slammed into Jupiter over a period of seven days in July 1994. It was the first collision of solar system bodies directly observed.

From *A Synopsis of the Astronomy of Comets* by Edmond Halley*

. . . All those that considered comets, until the time of Tycho Brahe (that great restorer of astronomy), believed them to be below the Moon and so took but little notice of them, reckoning them no other than vapors.

But in the year 1577 (Tycho seriously pursuing the study of the stars and having gotten large instruments for performing celestial mensurations, with far greater care and certainty than the ancients could ever hope for), there appeared a very remarkable comet; to the observation of which Tycho vigorously applied himself and found by many just and faithful trials that it had not a diurnal parallax that was at all perceptible. And consequently was not only no aireal vapor but also much higher than the Moon; nay, might be placed amongst the planets for anything that appeared to the contrary. The cavilling opposition made by some of the School-men, in the meantime, being to no purpose.

* The punctuation and spelling in this and other eighteenth-century papers have been updated in part to a more modern style.

Next to Tycho came the sagacious Kepler. He having the advantage of Tycho's labors and observations found out the true physical system of the World and vastly improved the astronomical science.

For he demonstrated that all the planets perform their revolutions in elliptic orbits, whose planes pass through the center of the Sun, observing this law: that the areas of the elliptic sectors taken at the center of the sun, which he proved to be in the common focus of these ellipses, are always proportional to the times in which the correspondent elliptical arches are described. He discovered also that the distances of the planets from the Sun are in the sesquialtera ratio of the periodical times, or (which is all one) that the cubes of the distances are as the squares of the times. This great astronomer had the opportunity of observing two comets, one of which was a very remarkable one. And from the observations of these (which afforded sufficient indications of an annual parallax) he concluded that the comets moved freely through the planetary orbs, with a motion not much different from a rectilinear one; but of what kind he could not then precisely determine. Next, Hevelius (a noble emulator of Tycho Brahe), following in Kepler's steps, embraced the same hypothesis of the rectilinear motion of comets, himself accurately observing many of them.* Yet, he complained that his calculations did not perfectly agree to the matter of fact in the heavens and was aware that the path of a comet was bent into a curved line towards the Sun. At length came that prodigious comet of the year 1680 which, descending (as it were) from an infinite distance perpendicularly towards the Sun, arose from him again with as great a velocity.

This comet (which was seen for four months continually) by the very remarkable and peculiar curvity of its orbit (above all others) gave the fittest occasion for investigating the theory of the motion. And the Royal Observatories at Paris and Greenwich having been for some time founded and committed to the care of most excellent astronomers, the apparent motion of this comet was most accurately (perhaps as far as human skill could go) observed by Messieurs Cassini and Flamsteed.†

Not long after that great geometrician, the illustrious Newton, writing his Mathematical Principles of Natural Philosophy, demonstrated not only that what Kepler had found did necessarily obtain in the planetary system;

---

* The seventeenth-century Polish astronomer Johannes Hevelius, noted for his accurate measurements of stellar positions.

† Giovanni Cassini, director of the Paris Observatory, and John Flamsteed, astronomer royal, based at Greenwich.

but also, that all the phenomena of comets would naturally follow from the same principles, which he abundantly illustrated by the example of the aforesaid comet of the year 1680. Showing, at the same time, a method of delineating the orbits of comets geometrically; wherein he (not without the highest admiration of all men) solved a problem, whose intricacy rendered it worthy of himself. This comet he proved to move round the Sun in a parabolic orb and to describe areas (taken at the center of the Sun) proportional to the times.

Wherefore (following the steps of so great a man), I have attempted to bring the same method to arithmetical calculation, and that with desired success. . . .

[Omitted here is a set of tables for calculating the motion of a comet in a parabolic orbit. Afterward, Halley gives detailed mathematical instructions on how to use the tables to determine a comet's path and includes two test cases to illustrate his methods.]

By comparing together the accounts of the motions of these comets, 'tis apparent their orbits are disposed in no manner of order; nor can they, as the planets are, be comprehended within a zodiac but move indifferently every way, as well retrograde as direct; from whence it is clear, they are not carried about or moved in vortices. Moreover, the distances in their periheliums are sometimes greater, sometimes less; which makes me suspect there may be a far greater number of them, which moving in regions more remote from the Sun, become very obscure and wanting tails, pass by us unseen.

Hitherto I have considered the orbits of comets as exactly parabolic; upon which supposition it would follow that comets being impelled towards the Sun by a centripetal force, descend as from spaces infinitely distant, and by their falls acquire such a velocity as that they may again run off into the remotest parts of the universe, moving upwards with such a perpetual tendency, as never to return again to the Sun. But since they appear frequently enough, and since none of them can be found to move with a hyperbolic motion, or a motion swifter than what a comet might acquire by its gravity to the Sun, 'tis highly probable they rather move in very eccentric orbits and make their returns after long periods of time. [Therefore] their number will be determinate, and, perhaps, not so very great. Besides, the space between the Sun and the fixed stars is so immense that there is room enough for a comet to revolve, though the period of its revolution be vastly long. . . . The principal use therefore of this table of the elements of their motions, and that which induced me to construct it, is that whenever a new comet shall appear, we may be able to know, by comparing together the elements, whether it be any of those which has appeared before, and consequently to determine its period and the axis of its orbit and to foretell its return. And, indeed, there are many things which make me believe that the comet which Apian* observed in the year 1531 was the same with that which Kepler and Longomontanus† took notice of and described in the year 1607 and which I myself have seen return and observed in the year 1682. All the elements [from the table] agree, and nothing seems to contradict this my opinion, besides the inequality of the periodic revolutions, which inequality is not so great neither, as that it may not be owing to physical causes. For the motion of Saturn is so disturbed by the rest of the planets, especially Jupiter, that the periodic time of that planet is uncertain for some whole days together. How much more therefore will a comet be subject to such like errors, which rises almost four times higher than Saturn and whose velocity, though increased but a very little, would be sufficient to change its orbit, from an elliptical to a parabolic one. This, moreover, confirms me in my opinion of its being the same; that in the year 1456 in the summer time, a comet was seen passing retrograde between the Earth and the Sun, much after the same manner. Which, though nobody made observations upon it, yet from its period and the manner of its transit, I cannot think different from those I have just now mentioned. Hence I dare venture to foretell that it will return again in the year 1758. And, if it should then return, we shall have no reason to doubt but the rest must return too. Therefore astronomers have a large field to exercise themselves in for many ages, before they will be able to know the number of these many and great bodies revolving about the common center of the Sun and reduce their motions to certain rules. I thought, indeed, that the comet which appeared in the year 1532 might be the same with that observed by Hevelius in the year 1661. But Apian's observations, which are the only ones we have concerning the first of these comets, are too crude and unskillful for any thing of certainty to be drawn from them in so nice a matter. I design to treat of all these things in a larger volume and contribute my utmost for the promotion of this part of astronomy, if it shall please God to continue my life and health. . . .

---

* The sixteenth-century German astronomer Peter Apian, who was also the first to note that the tails of comets invariably point away from the sun.

    † The seventeenth-century Danish astronomer Christian Longomontanus.

Table 12.1

Comet Of Year	Ascending Node			Inclin. of Orbit			Perihelion in Orbit			Perihel dist.	Log Per. di.	Time of Perihelion		
	deg '		"	deg '		"		deg '		"		day h		'
1337	Gem	24 21	0	32 11	0	Tau	7 59	0	0.40666	9.609236	Jun	2 6 25		
1472	Cap	11 46	20	5 20	0	Tau	15 33	30	0.54273	9.734584	Feb	28 22 23		
1531	**Tau**	**19 25**	**0**	**17 56**	**0**	**Aqr**	**1 39**	**0**	**0.56700**	**9.753583**	**Aug**	**24 21 18.5**		
1532	Gem	20 27	0	32 36	0	Cnc	21 7	0	0.5091	9.706803	Oct	19 22 12		
1556	Vir	25 42	0	32 6 30		Cap	8 50	0	0.46390	9.666424	Apr	21 20 3		
1577	Ari	25 52	0	74 32 45		Leo	9 22	0	0.18342	9.263447	Oct	26 18 45		
1580	Ari	18 57	20	64 40	0	Cnc	19 5 50		0.59628	9.775450	Nov	28 15 00		
1585	Tau	7 42	30	6 4	0	Ari	8 51	0	1.09538	0.038850	Sep	27 19 20		
1590	Vir	15 30	40	29 40 40		Sco	6 54 30		0.57661	9.760882	Jan	29 3 45		
1596	Aqr	12 12	30	55 12	0	Sco	18 16	0	0.51293	9.710058	Jul	31 19 55		
1607	**Tau**	**20 21**	**0**	**17 2**	**0**	**Aqr**	**2 16**	**0**	**0.58680**	**9.768490**	**Oct**	**16 3 50**		
1618	Gem	16 1	0	37 34	0	Ari	2 14	0	0.37975	9.579498	Oct	29 12 23		
1652	Gem	28 10	0	79 28	0	Ari	28 18 40		0.84750	9.928140	Nov	2 15 40		
1661	Gem	22 30	30	32 35 50		Cnc	25 58 40		0.44851	9.651772	Jan	16 23 41		
1664	Gem	21 14	0	21 18 30		Leo	10 41 25		1.025755	0.011044	Nov	24 11 52		
1665	Sco	18 02	0	76 05	0	Gem	11 54 30		0.10649	9.027309	Apr	14 5 15.5		
1672	Cap	27 30	30	83 22 10		Tau	16 59 30		0.69739	9.843476	Feb	20 8 37		
1677	Sco	26 49	10	79 03 15		Leo	17 37 5		0.28059	9.448072	Apr	26 00 37.5		
1680	Cap	2 2	0	60 56	0	Sgr	22 39 30		0.006125	7.787106	Dec	8 00 6		
1682	**Tau**	**21 16**	**30**	**17 56**	**0**	**Aqr**	**2 52 45**		**0.58328**	**9.765877**	**Sep**	**4 07 39**		
1683	Vir	23 23	0	83 11	0	Gem	25 29 30		0.56020	9.748343	Jul	3 2 50		
1684	Sgr	28 15	0	65 48 40		Sco	28 52	0	0.96015	9.982339	May	29 10 16		
1686	Psc	20 34	40	31 21 40		Gem	17 00 30		0.32500	9.511883	Sep	6 14 33		
1698	Sgr	27 44	15	11 46	0	Cap	00 51 15		0.69129	9.839660	Oct	8 16 57		

The information Halley gathered and computed from the historic record on twenty-four comets observed from 1337 to 1698. He noticed that three of the comets (1531, 1607, and 1682) shared similar orbital characteristics, making him suspect it was the same comet returning every seventy-fiv to seventy-six years.

One more thing perhaps it may not be improper or unpleasant to adver-
tise [to] the astronomical reader: that some of these comets have their
nodes so very near the annual orb [orbit] of the Earth, that if it shall so happen that the Earth be found in the parts of her orb next the node of such a
comet, whilst the comet passes by . . . what might be the consequences of
so near an appulse [conjunction of two heavenly bodies] or of a contact or,
lastly, of a shock of the celestial bodies (which is by no means impossible
to come to pass) I leave to be discussed by the studious of physical matters.

13 / Binary Stars

With the advance of telescopic instrumentation in the eighteenth century, astronomers examined the nighttime sky in finer and finer detail, enabling them to resolve new species of objects. By 1767 British geologist and astronomer John Michell, innovatively applying the techniques of probability to astronomy, argued that double stars were being seen in such abundance that some must be gravitationally bound together as pairs.

Michell was an original thinker who initially thought of earthquakes as seismic waves, invented a torsion balance for measuring the mass of the Earth, and first imagined what has come to be known as a "black hole" (see Chapter 42). His greatest contribution to astronomy was introducing statistics to the field's repertoire of mathematical tools in his analysis of double stars. Within that 1767 *Philosophical Transactions* paper, he also perceptively mentions (in a footnote, of all places) that observing a linked pair of stars would offer the means of determining the mass and luminosity of the stars. He even correctly surmises that a star's color might be a gauge of its absolute brightness.

In 1779 the great British astronomer William Herschel, a friend of Michell's, began collecting a catalog of stars that were positioned close together in the sky. His intent was to use these alignments to conduct distance measurements, based on an idea introduced by Galileo. Galileo reasoned that as the Earth traveled in its orbit from one end to the other, the nearer star would shift its position in relation to the farther star. This displacement (known as parallax) would enable an observer to geometrically peg the star's distance (see Chapter 19). Herschel's resolution wasn't keen enough to detect such a parallax, but he did end up confirming Michell's statistical conjecture. Herschel over the years cataloged some seven hundred double stars and, in reexamining these objects, saw that some were moving in such a way that left no doubt they were revolving around one another. He named them "binary stars."[32] The first such system he recognized was the star Castor in the constellation Gemini. By 1827 the French astronomer Félix Savary computed the actual motion of a binary star. He showed that the two yellowish stars of Xi Ursae Majoris, a system just 26 light-years from Earth, were moving in elliptical orbits in perfect agreement with Newton's laws.

The common wisdom had been that stars were closely aligned by chance alone. But Michell's statistics, coupled with Herschel's observations, proved otherwise. The majority of stars in the Milky Way, in fact, are members of binary or multiple star systems. Binaries provided evidence that Newton's law of gravity was truly universal, operating throughout the cosmos. Until Michell's landmark calculation and Herschel's keen measurements, there was no real evidence that gravity extended beyond the borders of our solar system. With stars of different luminosities sometimes paired together, binaries also provided proof that stars came in a variety of brightnesses and weren't all identical to the Sun, as Michell suspected all along.

From "An Inquiry into the Probable Parallax,
and Magnitude of the Fixed Stars, from the Quantity
of Light Which They Afford Us, and the Particular
Circumstances of Their Situation."
Philosophical Transactions, Volume 57 (1767)
by John Michell

. . . We have assumed the magnitude of the fixed stars, as well as their brightness, to be equal to those of the sun; it is however probable that there may be a very great difference amongst them in both these respects. . . . In other instances we may perhaps judge in some degree of the native brightness of different stars with respect to one another by their color; those, which afford the whitest light, being probably the most luminous. . . .*

It has always been usual with astronomers to dispose the fixed stars into constellations. This has been done for the sake of remembering and distinguishing them, and therefore it has in general been done merely arbitrarily and with this view only. Nature herself however seems to have distinguished them into groups. What I mean is, that from the apparent situation of the stars in the heavens, there is the highest probability that, either by the original act of the Creator or in consequence of some general law (such as perhaps gravity), they are collected together in great numbers in some parts of space, whilst in others there are either few or none.

The argument I intend to make use of, in order to prove this, is of that kind which infers either design or some general law from a general analogy and the greatness of the odds against things having been in the present situation, if it was not owing to some such cause.

Let us then examine what it is probable would have been the least apparent distance of any two or more stars, anywhere in the whole heavens, upon the supposition that they had been scattered by mere chance, as it might happen. Now it is manifest, upon this supposition that every star being as likely to be in any one situation as another, the probability that any

---

* ". . . We find . . . in general that those fires which produce the whitest light are much the brightest, and that the Sun, which produces a whiter light than any fires we commonly make, vastly exceeds them all in brightness. It is not therefore improbable from this general analogy that those stars which exceed the Sun in the whiteness of their light may also exceed [the Sun] in their native brightness. . . .

"If however it should hereafter be found that any of the stars have others revolving about them (for no satellites shining by a borrowed light could possibly be visible), we should then have the means of discovering the proportion between the light of the Sun and the light of those stars, relatively to their respective quantities of matter. For in this case, the times of the revolutions and the greatest apparent elongations of those stars that revolved about the others as satellites, being known, the relation between the apparent diameters and the densities of the central stars would be given, whatever was their distance from us. And the actual quantity of matter which they contained would be known, whenever their distance was known, being greater or less in the proportion of the cube of that distance. Hence, supposing them to be of the same density with the Sun, the proportion of the brightness of their surfaces, compared with that of the Sun, would be known from the comparison of the whole of the light which we receive from them with that which we receive from the Sun. . . ."

one particular star should happen to be within a certain distance (as for example one degree) of any other given star would be represented (according to the common way of computing chances) by a fraction, whose numerator would be to its denominator as a circle of one degree radius to a circle whose radius is the diameter of a great circle . . . (that is, about 1 in 13,131) and the complement of this to unity, viz. .999923846 or the fraction $13130/13131$, will represent the probability that it would not be so. But because there is the same chance for any one star to be within the distance of one degree from any given star as for every other, multiplying this fraction into itself as many times as shall be equivalent to the whole number of stars, of not less brightness than those in question, and putting n for this number, $.999923846^n$ . . . will represent the probability that no one of the whole number of stars n would be within one degree from the proposed given star, and the complement of this quantity to unity will represent the probability that there would be some one star or more, out of the whole number n, within the distance of one degree from the given star. And farther, because the same event is equally likely to happen to any one star as to any other, and therefore any one of the whole number of stars n might as well have been taken for the given star as any other, we must again repeat the last found chance n times, and consequently the number $(.999923846^n)^n$ . . . will represent the probability that, no where in the whole heavens, any two stars amongst those in question would be within the distance of one degree from each other, and the complement of this quantity to unity would represent the probability of the contrary.

[At this point Michell generalizes his argument. Omitted here is Michell's computation of the more general probability that no two stars would be a given distance—one at a distance x, the other at a distance z—from a particular star.]

If now we compute, according to the principles above laid down, what the probability is that no two stars in the whole heavens should have been within so small a distance from each other as the two stars β Capricorni, to which I suppose about 230 stars only to be equal in brightness, we shall find it to be about 80 to 1.

For an example, where more than two stars are concerned, we may take the six brightest of the Pleiades, and, supposing the whole number of those stars which are equal in splendor to the faintest of these to be about 1,500, we shall find the odds to be near 500,000 to 1 that no six stars out of that number, scattered at random in the whole heavens, would be within so small a distance from each other as the Pleiades are.

If, besides these examples that are obvious to the naked eye, we extend the same argument to the smaller stars, as well those that are collected together in clusters, such for example as the Praesepe Cancri [Beehive cluster in Cancer], the nebula in the hilt of Perseus's sword, etc. as to those stars which appear double, treble, etc. when seen through telescopes, we shall find it still infinitely more conclusive, both in the particular instances and in the general analogy, arising from the frequency of them.

We may from hence, therefore, with the highest probability conclude (the odds against the contrary opinion being many million millions to one) that the stars are really collected together in clusters in some places, where they form a kind of system, whilst in others there are either few or none of them, to whatever cause this may be owing, whether to their mutual gravitation or to some other law or appointment of the Creator. And the natural conclusion from hence is that it is highly probable in particular, and next to a certainty in general, that such double stars, etc. as appear to consist of two or more stars placed very near together, do really consist of stars placed near together and under the influence of some general law, whenever the probability is very great that there would not have been any such stars so near together, if all those that are not less bright than themselves had been scattered at random through the whole heavens. . . .

III

TAKING MEASURE

Through the eighteenth century and into the next the number of astronomical observatories grew around the world, in both the Northern and Southern Hemispheres. And with the rise of the industrial revolution, handcrafted astronomical instruments were supplanted by ones manufactured with superb exactness. Formerly installed on wood, telescopes were now mounted on solid metal for increased stability, and clockworks were made more precise. Astronomers who specialized in measuring the positions and magnitudes of stars reaped the benefits. These technological improvements led to both increased sensitivity for detecting dimmer and dimmer objects in space and more accuracy in astrometric measurements.

With his momentous law of gravitation and its codification within the renowned *Principia*, Isaac Newton stood like a colossus over astronomy for decades after his death. Astronomy became a science of patient computations, a domain of reflection and synthesis after the frenzy of new insights from Copernicus, Galileo, and Newton. The majority of astronomers in the eighteenth and nineteenth centuries were highly skilled mathematicians who primarily used Newton's revolutionary laws to predict the motions of the Moon, planets, and comets. These experts in celestial mechanics endlessly calculated the positions and orbital characteristics of the planetary bodies to discern their intricate patterns. They focused on such astronomical arcana as variations in orbital eccentricity, motion of the aphelion, obliquity of the ecliptic, nutation, and aberration. Such diligence led to the discovery of new planets (Neptune) and new, unexpected objects (asteroids) within our solar system. Timing the eclipses of Jupiter's moons provided the first fairly accurate measurement of the speed of light. Stars themselves were not yet as interesting or provocative to astronomers as determining a star's position on the celestial sphere with utmost precision and expanding stellar catalogs.

Newton's laws also led theorists to speculate on the nature and mechanisms of the cosmos, including how the solar system itself came into being. And since gravity was a universal force, it implied that stars could hardly be at rest but rather were moving due to their mutual attraction to one another. These motions were first successfully measured in the mid-1700s.

Observational astronomy in this era was synonymous with measurement. "The noblest problem in astronomy," according to George Biddell Airy, then Great Britain's astronomer royal, was determining the distance between the Earth and the Sun. From the age of Ptolemy to the time of Copernicus and Tycho Brahe, it was largely assumed that the Sun was at a distance of 1,200 Earth radii, about 4.8 million miles. The concern for precision in the eighteenth and nineteenth centuries, though, enabled astronomers to arrive within 10 percent of the true value of some 93 million miles. This was accomplished by carefully monitoring transits of the planet Venus— rare moments (pairs of them occur eight years apart every 105.5 or 121.5 years) when Venus can be seen crossing the surface of the Sun. Watched from different positions over the Earth, the timing of the passage leads to an angle that when coupled with the Earth's diameter gives the Sun's distance. Captain James Cook's first historic sea voyage in 1768–71 was arranged by the Royal Society of London so that astronomers could triangulate the distance to the Sun via a Venus transit. Such precision eventually extended to the stars. By the nineteenth century astronomers also made their first successful measurements of stellar distances.

Most astronomers in this era were primarily concerned with the solar system and its workings. The "fixed stars" were still largely a backdrop to these endeavors among sky observers—with one glaring exception. In Great Britain the musician and self-taught astronomer

William Herschel specifically constructed large reflecting tele-scopes in the late eighteenth century to explore beyond the solar sys-tem. Although he initially gained fame discovering Uranus, the first new planet revealed since the dawn of history, Herschel was funda-mentally interested in the stars and mysterious cloudlike nebulae ignored by others. A visionary ahead of his time, he undertook the first pioneering attempts at determining the universe's structure, its composition, and the extent of its vastness. He and others also began to contemplate that other disklike systems of stars—other "uni-verses"—might exist beyond the borders of the Milky Way.

14 / The Speed of Light

For most of history it was generally assumed that light was transmitted instantaneously. With this premise the light from a far-off star arrived at Earth as soon as it was emitted. By the seventeenth century, however, certain thinkers began to wonder whether light had a finite speed after all—like sound, only faster. In his treatise on mechanics in 1638, Galileo may have been the first to suggest an experiment to test this hypothesis directly. One man stands on a hill and uncovers a lantern, signaling a companion positioned on another hill less than a mile away. The second man, as soon as he sees the initial beacon, flashes a return light of his own. Performing this task on hills spaced farther and farther apart, Galileo figured the men would spot a successively longer delay between the dual flashes, which would reveal the speed of light. This very test was eventually carried out by members of the Florentine Academy. Of course, no delay was detected, given the crudeness of Galileo's experiment. Human reaction time is far too slow. The vast span of our solar system provided a far better test.

In the 1670s—Newton's day—the Danish mathematician and astronomer Ole Römer, then working in Paris for the French Royal Academy of Sciences, closely studied the movements of Jupiter's four largest moons, particularly the innermost one, Io. Specifically, he carefully monitored the moment when Io periodically moves behind Jupiter and gets eclipsed. In doing this, he noticed that the interval between successive eclipses (an event that occurred about every 42 hours) was not constant but regularly changed, depending on the position of the Earth in relation to Jupiter. When the Earth was moving away from Jupiter in its orbital motion around the Sun, the expected moment for Io to be eclipsed arrived later and later. This is because the light bringing that information to our eyes has to travel a bit more distance with each eclipse. By the time the Earth reached its farthest point from Jupiter, Römer's measured delay mounted up to 22 minutes (a better figure is 16.5 minutes). Others had noticed such changes before, but Römer shrewdly demonstrated that the delay was just the time needed for Io's light to traverse the extra width of the Earth's orbit. Dividing the Earth's orbital width (184 million miles) by the delay time, Römer's crude measurements pegged a light speed of around 140,000 miles per second. Though it was fast, Römer had shown that light did not travel instantaneously. This first measured estimate is reasonably close to the modern value of 186,282 miles per second.

It took about half a century, though, before the noninstantaneous transmission of light was generally accepted. Astronomers were more convinced once the British astronomer James Bradley in 1728 reported his discovery of aberration, an apparent change in stellar positions due to the Earth's continual motion around the Sun.[1] The amount of aberration is essentially the ratio of the Earth's orbital speed to the speed of light. Since Bradley could measure the angular displacement of a star due to aberration and since he also knew the Earth's velocity, he was able to determine a speed of light close to the modern value, confirming Römer's earlier work.

"A Demonstration Concerning the Motion of Light, Communicated from Paris, in the *Journal des Scavans*, and Here Made English." *Philosophical Transactions*, Volume 12 (1677), A Report on the Work of Ole Römer

Philosophers have been laboring for many years to decide by some experience whether the action of light be conveyed in an instance to distant places or whether it requires time. M. Römer of the Royal Academy of the

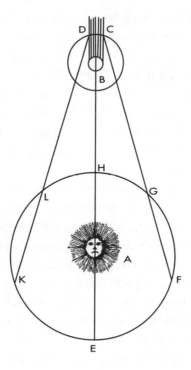

Figure 14.1

Sciences has devised a way, taken from the observations of the first satellite of Jupiter, by which he demonstrates that for the distance of about 3,000 leagues (such as is very near the bigness of the diameter of the Earth) light needs not one second of time.

Let (in Fig. [14.1]) A be the Sun, B Jupiter, C the first satellite of Jupiter, which enters into the shadow of Jupiter to come out of it at D. And let EFGHKL be the Earth placed at diverse distances from Jupiter.

Now, suppose the Earth, being in L towards the second quadrature of Jupiter, has seen the first satellite at the time of its emersion or issuing out of the shadow in D, and that about 42½ hours after (i.e. after one revolution of this satellite), the Earth being in K, do see it returned in D. It is manifest that if the light requires time to traverse the interval LK, the satellite will be seen returned later in D than it would have been if the Earth had remained in L, so that the revolution of this satellite being thus observed by the emersions will be retarded by so much time as the light shall have taken in passing from L to K. And that, on the contrary, in the other quadrature FG, where the Earth by approaching goes to meet the light, the revolutions of the immersions will appear to be shortened by so much, as those of the emersions had appeared to be lengthened. And because in 42½ hours, which this satellite very near takes to make one revolution, the distance between the Earth and Jupiter in both the quadratures varies at least 210 diameters of the Earth, it follows that if for the account of every diameter of the Earth there were required a second of time, the light would take 3½ minutes for each of the intervals of GF, KL; which would cause near half a quarter of an hour between two revolutions of the first satellite, one observed in FG and the other in KL, whereas there is not observed any sensible difference.

Yet [it does not] follow hence that light demands no time. For, after M. Römer had examined the thing more nearly, he found that what was not sensible in two revolutions became very considerable in many being taken together. And that, for example, forty revolutions observed on the side F might be sensibly shorter than forty [revolutions on the other side], and that in proportion of twenty-two [minutes] for the whole interval of HE, which is the double of the interval that is from hence to the Sun.*

The necessity of this new equation of the retardment of light is established by all the observations that have been made in the Royal Academy and in the Observatory for the space of eight years, and it has been lately confirmed by the emersion of the first satellite observed at Paris the 9th of November [1676] last at 5 [hours], 35 [minutes], 45 [seconds] at night, 10 minutes later than it was to be expected by deducing it from those that had been observed in the month of August, when the Earth was much nearer to Jupiter, which M. Römer had predicted to the said Academy from the beginning of September.

But to remove all doubt that this inequality is caused by the retardment of light, he demonstrates that it cannot come from an eccentricity or any other cause of those that are commonly alleged to explicate the irregularities of the Moon and the other planets, though he [is] well aware that the first satellite of Jupiter was eccentric, and that . . . his revolutions were advanced or retarded according as Jupiter did approach to or recede from the Sun, [and] also that the revolutions of the *primum mobile* were unequal. Yet [he has faith that] these three last causes of inequality do not hinder the first from being manifest.

---

* Here it was reported that it takes 22 minutes for light to cross the diameter of Earth's orbit, double the 11-minute interval Römer believed it took light to go from the Sun to the Earth. It's now known that better figures are a little more than 16 and 8 minutes, respectively.

15 / The Solar System's Origin

Pierre-Simon de Laplace was renowned in the late eighteenth century for his advances in mathematics and celestial mechanics (he coined the latter term). His work proving the gravitational stability of the solar system nurtured the idea of a clockwork universe. Emboldened by the successes of Newtonian mechanics, he once noted that if some demon had the capability of knowing the positions and speeds of all objects in the universe, he would then be able to know both the complete history of its past and the precise behavior of its future. "For such an intellect," he wrote, "nothing could be uncertain and the future just like the past would be present before its eyes."[2]

Laplace's work on celestial mechanics eventually led him to develop a hypothesis for the origin of the solar system, which he discussed at the end of *Système du monde* (System of the World), published in 1796, his nonmathematical summary of astronomy, mechanics, and gravity—essentially a handbook on the era's cosmology. He made additions and amendments to his theory in subsequent editions, up until his death in 1827. (The excerpt below is an early version.) Laplace originally imagined a ready-made Sun, whose hot atmosphere (then much larger) cooled and contracted, leaving the planets in its wake, all revolving in the same plane and direction. In later editions, he chose a colder nebulosity that came to organize over time through gravitational contraction. The idea of the planets originating from a rotating nebula was actually first mentioned by Immanuel Kant some forty years earlier in his *Universal Natural History and Theory of Heaven,* but Laplace's more widely circulated nebular theory became the definitive one discussed among scholars.

Slowly rotating and contracting, a ring of material at the edge of the nebula in this scheme separates off, the matter eventually coalescing into a planet. As the nebula continues to contract, a series of planets results, each a smaller distance from the center and each with a faster velocity than the one before. A scientist with exacting standards, Laplace may not have taken this idea as seriously as his other work. At the end of *Système du monde,* he cautions the reader that he is offering his conjecture "with that distrust which everything ought to inspire that is not the result of observation or calculation."[3] Yet, ironically, this one speculative section of the book continues to retain the most historical interest.

The nebular theory went in and out of repute as astronomers gathered more information on the solar system. By the turn of the twentieth century, the American geologist Thomas Chamberlin and his University of Chicago colleague astronomer Forest Ray Moulton were suggesting that a close encounter with another star may have thrown out enough material from the Sun to form the planets. For a while this theory was widely accepted because it helped explain why the Sun was not spinning fast, as it should have under Laplace's mechanism. With stellar collisions so rare, it also kept Earth special—an exceptional item in the universe, which some found attractive for philosophic and theological reasons. But the nebular theory resurged as soon as magnetic fields in the solar system were better understood. An early magnetic field, which links the Sun to the nebula, can slow down the Sun's rotation to its current levels. The collision theory, however, survives in a way. It serves to explain the origin of our Moon, whose matter probably spewed out when a planetesimal crashed into the hot, primordial Earth.

From *The System of the World* (1796)
by Pierre-Simon de Laplace

Translated by J. Pond

Considerations on the System of the Universe, and on the Future Progress of Astronomy

Let us now direct our attention to the arrangement of the solar system, and its relation with the stars. The immense globe of the Sun, the focus of these motions, revolves upon its axis in twenty-five days and a half. Its surface is covered with an ocean of luminous matter, whose active effervescence forms variable spots, often very numerous, and sometimes larger than the Earth. Above this ocean exists an immense atmosphere, in which the planets, with their satellites, move, in orbits nearly circular, and in planes little inclined to the ecliptic. Innumerable comets, after having approached the Sun, remove to distances, which evince that his empire extends beyond the known limits of the planetary system. This luminary not only acts by its attraction upon all these globes, and compels them to move around him, but imparts to them both light and heat; his benign influence gives birth to the animals and plants which cover the surface of the Earth, and analogy induces us to believe, that it produces similar effects on the planets; for, it is not natural to suppose that matter, of which we see the fecundity, develop itself in such various ways, should be sterile upon a planet so large as Jupiter, which, like the Earth, has its days, its nights, and its years, and on which observation discovers changes that indicate very active forces. Man, formed for the temperature which he enjoys upon the Earth, could not, according to all appearance, live upon the other planets; but ought there not to be a diversity of organization suited to the various temperatures of the globes of this universe? If the difference of elements and climates alone, causes such variety in the productions of the Earth, how infinitely diversified must be the productions of the planets and their satellites? The most active imagination cannot form any just idea of them, but still their existence is extremely probable.

However arbitrary the system of the planets may be, there exists between them some very remarkable relations, which may throw light on their origin; considering them with attention, we are astonished to see all the planets move round the Sun from west to east, and nearly in the same plane, all the satellites moving round their respective planets in the same direction, and nearly in the same plane with the planets. Lastly, the Sun, the planets, and those satellites in which a motion of rotation have been observed, turn on their own axis, in the same direction, and nearly in the same plane as their motion of projection.

A phenomenon so extraordinary is not the effect of chance, it indicates a universal cause, which has determined all these motions. To approximate somewhat to the probable explanation of this cause, we should observe that the planetary system, such as we now consider it, is composed of seven planets, and fourteen satellites. We have observed the rotation of the Sun, of five planets, of the Moon, of Saturn's ring, and of his farthest satellite; these motions with those of revolution, form together thirty direct movements, in the same direction. If we conceive the plane of any direct motion whatever, coinciding at first with that of the ecliptic, afterwards inclining itself towards this last plane, and passing over all the degrees of inclination, from zero to half the circumference; it is clear that the motion will be direct in all its inferior inclinations to a hundred degrees, and that it will be retrograde in its inclination beyond that; so that, by the change of inclination alone, the direct and retrograde motions of the solar system, can be represented. Beheld in this point of view, we may reckon twenty-nine motions, of which the planes are inclined to that of the Earth, at most $\frac{1}{4}$th of the circumference; but, supposing their inclinations have been the effect of chance, they would have extended to half the circumference, and the probability that one of them would have exceeded the quarter, would be $1-\frac{1}{2^{29}}$ or $\frac{536870911}{536870912}$. It is then extremely probable, that the direction of the planetary motion is not the effect of chance, and this becomes still more probable, if we consider that the inclination of the greatest number of these motions to the ecliptic, is very small, and much less than a quarter of the circumference.

Another phenomenon of the solar system equally remarkable, is the small eccentricity of the orbits of the planets and their satellites, while those of comets are much extended. The orbits of the system offer no intermediate shades between a great and small eccentricity. We are here again compelled to acknowledge the effect of a regular cause; chance alone could not have given a form nearly circular, to the orbits of all the planets. This cause then must also have influenced the great eccentricity of the orbits of comets, and what is very extraordinary, without having any influence on the direction of their motion; for, in observing the orbits of retrograde comets, as being inclined more than 100° to the ecliptic, we find that the mean inclination of the orbits of all the observed comets, approaches near to 100°, which would be the case if the bodies had been projected at random.

Thus, to investigate the cause of the primitive motions of the planets, we have given the five following phenomena: 1st, The motions of planets in the same direction, and nearly in the same plane. 2d, The motion of their satellites in the same direction, and nearly in the same plane with those of the planets. 3d, The motion of rotation of these different bodies, and of the Sun in the same direction as their motion of projection, and in planes but little different. 4th, The small eccentricity of the orbits of the planets, and of their satellites. 5th, The great eccentricity of the orbits of comets, although their inclinations may have been left to chance.

Buffon is the only one whom I have known, who, since the discovery of the true system of the world, has endeavored to investigate the origin of the planets, and of their satellites.* He supposes that a comet, in falling from the Sun, may have driven off a torrent of matter, which united itself at a distance, into various globes, greater or smaller, and more or less distant from this luminary. These globes are the planets and satellites, which, by their cooling, are become opaque and solid.

This hypothesis accounts for the first of the five preceding phenomena; for, it is clear that all bodies thus formed, must move nearly in the plane which passes through the center of the Sun, and in the direction of the torrent of matter which produces them. The four other phenomena appear to me inexplicable by his theory. In fact, the absolute motion of the particles of a planet would then be in the same direction of the motion of its center of gravity; but it does not follow that the rotation of the planet would be in the same direction. Thus, the Earth may turn from west to east, and yet the absolute direction of each of its particles may be from east to west. What I say of the rotatory motion of the planets, is equally applicable to the motion of their satellites in their orbits, of which the direction in the hypothesis he adopts, is not necessarily the same with the projectile motion of the planets.

The small eccentricity of the motion of the planetary orbits is not only very difficult to explain on this hypothesis, but the phenomenon contradicts it. We know by the theory of central forces, that if a body moving in an

---

* Georges-Louis Leclerc, comte de Buffon (1707–88), French aristocrat trained in law and medicine.

orbit around the Sun, touched the surface of this luminary, it would uniformly return to it at the completion of each revolution, from whence it follows, that if the planets had originally been detached from the Sun, they would have touched it at every revolution, and their orbits, far from being circular, would be very eccentric. It is true, that a torrent of matter, sent off from the Sun, cannot correctly be compared to a globe which touches its surface. The impulse which the particles of this torrent receive from one another, and the reciprocal attraction exercised among them, may change the direction of their motion, and increase their perihelion distances; but their orbits would uniformly become very eccentric, or at least it must be a very extraordinary chance that would give them eccentricities so small as those of the planets. In a word, we do not see, in this hypothesis of Buffon, why the orbits of about eighty comets, already observed, are all very elliptical. This hypothesis, then, is far from accounting for the preceding phenomena. Let us see if it is possible to arrive at their true cause.

Whatever be its nature, since it has produced or directed the motion of the planets and their satellites, it must have embraced all these bodies, and considering the prodigious distance which separates them, they can only be a fluid of immense extent. To have given in the same direction, a motion nearly circular round the Sun, this fluid must have surrounded the luminary like an atmosphere. This view, therefore, of planetary motion, leads us to think, that in consequence of excessive heat, the atmosphere of the Sun originally extended beyond the orbits of all the planets, and that it has gradually contracted itself to its present limits, which may have taken place from causes similar to those which caused the famous star that suddenly appeared in 1572, in the constellation Cassiopæa, to shine with the most brilliant splendor during many months.

The great eccentricity of the orbits of comets, leads to the same result; it evidently indicates the disappearance of a great number of orbits less eccentric, which indicates an atmosphere round the Sun, extending beyond the perihelion of observable comets, and which, in destroying the motion of these which they have traversed in a duration of such extent, have reunited themselves to the Sun. Thus, we see that there can at present only exist such comets as were beyond this limit at that period. And as we can observe only those which in their perihelion approach near the Sun, their orbits must be very eccentric; but, at the same time, it is evident that their inclinations must present the same inequalities as if the bodies had been sent off at random, since the solar atmosphere has no influence over their motions. Thus, the long period of the revolutions of comets, the great eccentricity of their orbits, and the variety of their inclinations, are very naturally explained by means of this atmosphere.

But how has it determined the motions of revolution and rotation of the planets? If these bodies had penetrated this fluid, its resistance would have caused them to fall into the Sun. We may then conjecture, that they have been formed at the successive bounds of this atmosphere, by the condensation of zones, which it must have abandoned in the plane of its equator, and in becoming cold have condensed themselves towards the surface of this luminary, as we have seen in the preceding Book. One may likewise conjecture, that the satellites have been formed in a similar way by the atmosphere of the planets. The five phenomena, explained above, naturally result from this hypothesis, to which the rings of Saturn add an additional degree of probability.

Whatever may have been the origin of this arrangement of the planetary system, which I offer with that distrust which every thing ought to inspire that is not the result of observation or calculation; it is certain that its elements are so arranged, that it must possess the greatest stability, if foreign observations do not disturb it. Through this cause alone, that the motions of planets and satellites are nearly circular, and impelled in the same direction, and in planes differing but little from each other, it arises that this system can only oscillate to a certain extent, from which its deviation must be extremely limited; the mean motions of rotation and revolution of these different bodies are uniform, and their mean distances to the foci of the principal forces which animate them, are uniform. It seems that nature has disposed every thing in the heavens to insure the duration of the system by views similar to those which she appears to us so admirably to follow upon Earth, to preserve the individual and insure the perpetuity of the species.

16 / Discovery of Uranus

F riedrich Wilhelm (William) Herschel, a musician and self-taught astronomer, was scanning the heavens on March 13, 1781, with a 7-foot-long home-built telescope set up in the back of his home in Bath, England. With the telescope's 6½-inch mirror he noticed a new object. "A curious either nebulous star or perhaps a comet," he wrote in his journal. Others had seen it before and dismissed it as an ordinary star, but Herschel, an exceptionally keen celestial surveyor, recognized that it was different. When he increased the magnification, its disk grew in size; a far-off star would have looked the same. Carefully tracking the object's movements over the ensuing weeks, he reported to the Royal Society on April 26 that he had discovered an approaching comet. Herschel, then a relative unknown, would be thrust into celebrity by his discovery.

A native of Hanover and a member of the Hanoverian guard band, Herschel had fled to England in 1757 at the age of nineteen during the Seven Years' War. By 1766 he moved to fashionable Bath to serve as a chapel organist and orchestra leader. More financially secure, he began to pursue astronomy as a serious hobby, inspired by a classic text on optics. He ground and polished his own mirrors and eyepieces, achieving high magnifications that professional astronomers for a while found dubious. Ahead of his time, Herschel devised a new approach to astronomy: he observed, counted, and classified all the various objects he surveyed. He didn't dwell on the familiar—the planets, Moon, and bright stars—but looked at the heavens as an integrated system, ably assisted by his younger sister Caroline (who discovered eight comets and three nebulae on her own). It was Herschel's intimate knowledge of the overall celestial landscape that enabled him to recognize, so quickly, that his new-found "star" was an unusual object.

Others, following up on Herschel's report of the comet, worked on its trajectory and concluded that it was not eccentric, like that of most comets, but rather circular. The British astronomer Nevil Maskelyne and the Swedish mathematician Anders Lexell suspected right off that it was a planet. Lexell was the first to compute an orbit and found that its radius extended far beyond Saturn (which is 9.5 AU from the Sun).* Over time, more accurate measurements showed the new object's orbital radius was 19.2 AU. In discovering the first planet since the dawn of history, Herschel had doubled the size of the solar system.

A name for the new planet became a cause célèbre. Among the suggested choices were Hypercronius ("above Saturn"), Cybele (the wife of Saturn), Minerva (the Roman goddess of wisdom), and even Herschel (extending the tradition of naming comets after their discoverers to planets). Herschel himself preferred Georgium Sidus ("the Georgian star") in honor of England's King George III, a fellow Hanoverian. Uranus, a suggestion by the Berlin astronomer Johann Bode and the choice of astronomers outside England, eventually won out. It was an apt selection; in classical mythology Uranus is the father of Saturn and grandfather of Jupiter. Astronomers wanted to continue the tradition of a nomenclature based on ancient mythologies.

Procuring a royal pension from the king after his great discovery, Herschel was at last able to devote himself to astronomy and build the ever-larger telescopes that were unsurpassed in his day for the study of faint objects. His largest had a 4-foot-wide mirror and a focal length of 40 feet. Set in the garden of his home in the town of Slough, west of London, where he settled for the rest of his life, the long tube was positioned up and down by a series of pulleys and turned back and forth on wheels. A platform, perched 50 feet high

* 1 AU (astronomical unit) is the mean distance of the Earth from the Sun, 93 million miles.

atop the large wooden structure, allowed Herschel to look through the eyepiece at the upper end of the telescope. Giuseppe Piazzi, the discoverer of the first asteroid, Ceres (see Chapter 18), once visited and broke an arm falling from a wooden ladder on the side of the great reflector. Herschel was most productive, though, with his easier-to-handle 20-foot telescope with its 18-inch mirror.

Herschel pioneered many of astronomy's current concerns: large-scale surveys, interest in big mirrors to look farther into space, and a curiosity about the universe's ultimate structure (see Chapter 21). He did not forget Uranus, though. In 1787 he discovered its first two moons, Titania and Oberon. (By the end of the twentieth century, astronomers had discovered twenty in all.) Herschel died in 1822, just three months shy of eighty-four, the number of years it takes Uranus to complete one orbit about the Sun. The inscription on his grave reads *Coelorum perrupit claustra*—"He broke through the barriers of the heavens."

"Account of a Comet." *Philosophical Transactions of the Royal Society of London*, Volume 71 (1781) by William Herschel

On Tuesday the 13th of March, between ten and eleven in the evening, while I was examining the small stars in the neighborhood of H Geminorum, I perceived one that appeared visibly larger than the rest. Being struck with its uncommon magnitude, I compared it to H Geminorum and the small star in the quartile between Auriga and Gemini, and finding it so much larger than either of them, suspected it to be a comet.

I was then engaged in a series of observations on the parallax of the fixed stars, which I hope soon to have the honor of laying before the Royal Society. . . .* The power I had on when I first saw the comet was 227. From experience I knew that the diameters of the fixed stars are not proportionally magnified with higher powers, as the planets are; therefore I now put

* See Chapter 13.

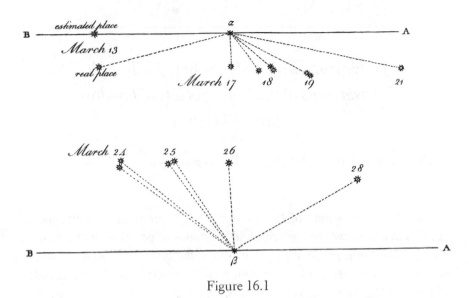

Figure 16.1

on the powers of 460 and 932 and found the diameter of the comet increased in proportion to the power, as it ought to be on a supposition of its not being a fixed star, while the diameters of the stars to which I compared it were not increased in the same ratio. Moreover the comet, being magnified much beyond what its light would admit of, appeared hazy and ill-defined with these great powers, while the stars preserved that luster and distinctness which from many thousand observations I knew they would retain. The sequel has shown that my surmises were well founded, this proving to be the comet we have lately observed. . . .

Miscellaneous Observations and Remarks

March 19. The comet's apparent motion is at present 2¼ seconds *per* hour [see Figure 16.1]. It moves according to the order of the signs, and its orbit declines but very little from the ecliptic.

March 25. The apparent motion of the comet is accelerating, and its apparent diameter seems to be increasing.

March 28. The diameter is certainly increased, from which we may conclude that the comet approaches to us. . . .

April 6. With a magnifying power of 278 times the comet appeared perfectly sharp upon the edges and extremely well defined, without the least appearance of any beard or tail. . . .

"A Letter from William Herschel." *Philosophical Transactions of the Royal Society of London*, Volume 73 (1783)

To Sir Joseph Banks [President, Royal Society],

Sir,

By the observations of the most eminent astronomers in Europe, it appears that the new star, which I had the honor of pointing out to them in March, 1781, is a primary planet of our solar system. A body so nearly related to us by its similar condition and situation, in the unbounded expanse of the starry heavens, must often be the subject of the conversation, not only of astronomers, but of every lover of science in general. This consideration then makes it necessary to give it a name, whereby it may be distinguished from the rest of the planets and fixed stars.

In the fabulous ages of ancient times the appellations of Mercury, Venus, Mars, Jupiter, and Saturn were given to the planets, as being the names of their principal heroes and divinities. In the present more philosophical era, it would hardly be allowable to have recourse to the same method and call on Juno, Pallas, Apollo, or Minerva for a name to our new heavenly body. The first consideration in any particular event, or remarkable incident, seems to be its chronology: if in any future age it should be asked, *when* this last-found planet was discovered, it would be a very satisfactory answer to say, "In the Reign of King George the Third." As a philosopher then, the name of GEORGIUM SIDUS presents itself to me, as an appellation which will conveniently convey the information of the time and country where and when it was brought to view. But as a subject of the best of Kings, who is the liberal protector of every art and science;—as a native of the country from whence this illustrious family was called to the British throne;—as a member of that Society, which flourishes by the distinguished liberality of its Royal Patron;—and, last of all, as a person now more immediately under the protection of this excellent Monarch, and owing everything to his unlimited bounty;—I cannot but wish to take this opportunity of expressing my sense of gratitude, by giving the name Georgium Sidus . . . to a star, which (with respect to us) first began to shine under his auspicious reign.

By addressing this letter to you, Sir, as President of the Royal Society, I take the most effectual method of communicating that name to the literati of Europe, which I hope they will receive with pleasure. I have the honor to be, with the greatest respect, Sir, your most humble and most obedient servant, W. Herschel.

17 / Stars Moving and Changing

Well into the nineteenth century, the solar system and the dynamics of the planets remained astronomy's prime areas of concern. Stars were of minor interest. But a few pioneers started to examine the realm of the "fixed stars," and their work set the stage for astronomy's greatest advances. Hipparchus in the second century B.C. had noticed that the celestial sphere as a whole, the entire panorama of stars, regularly shifts eastward over time or precesses (see Chapter 5). Centuries later Isaac Newton figured it was due to the Earth's axis slowly gyrating. But not until the eighteenth century did observers begin to seriously wonder whether each star in the heavens had its own individual motion.

Edmond Halley filed the first report on this possibility. In 1718, while investigating precession, he looked at some stellar positions that Ptolemy had listed in his *Almagest* some eighteen centuries earlier and compared them with contemporary observations. From this examination, he was surprised to find that such major stars as Sirius, Aldebaran, Betelgeuse, and Arcturus seemed to have moved (and not just due to precession). The stars had separately migrated either to the north or to the south. Was this a real effect? he asked his fellow colleagues of the Royal Society. "It is scarce credible," he writes, "that the ancients could be deceived in so plain a matter . . . these stars being the most conspicuous in heaven are in all probability the nearest to the Earth, and if they have any particular motion of their own, it is most likely to be perceived in them."[4]

Answers could not arrive until astronomers better understood the many variables that can affect locating a star's exact position on the sky. Not until 1728, for example, did astronomers come to recognize aberration, an apparent change in a star's location due to the motion of the Earth in its orbit about the Sun. A star's position also changes due to nutation, a subtle nodding of the Earth's axis due to its gravitational interaction with the Moon. Instrumentation improved enough by the 1750s for Johann Tobias Mayer of Göttingen to detect subtle displacements between his measurements of stellar positions and those made just fifty years earlier, confirming the "proper motion"—the change of a position on the celestial sphere—of a number of stars.

By 1760 Mayer suggested that the motion of the Sun itself through space should be detectable by looking for a specific effect. He compared it to someone walking in a forest. The trees ahead appear to move to your sides as you approach. So too, he said, as the solar system is moving toward a particular spot on the sky, the stars would appear to move aside. He looked for this stellar wake but reported he saw no evidence of it. By 1783, though, William Herschel, working with data then available on the motions of seven stars, disclosed that the Sun did seem to be traveling toward the constellation Hercules. He followed up with another study in 1805 that used thirty-six stars and came to the same conclusion. Some doubted his results (even his own astronomer son), but by 1837 Herschel's discovery was confirmed as more data came in involving the proper motions of some 390 stars.

That individual stars could alter their appearance had been recognized even earlier, when Tycho Brahe spotted a nova in 1572, soon followed by another in 1604 studied by Johannes Kepler. David Fabricius believed he had seen a nova in the constellation Cetus in 1596. On August 13 he noticed a third-magnitude star in the neck of the Whale that he hadn't noticed before. Two months later, it disappeared. Others later realized that this particular star repeatedly varied in brightness over a period of many months, making it the first bona fide variable star. It was labeled *mira stella*, for "wonderful star."* In 1667, the French priest-astronomer Ismaël

---

* Mira or Omicron Ceti is a cool red-giant star 400 light-years distant. It undergoes dramatic pulsations every 332 days that cause it to become hundreds of times brighter over the course of a year.

Boulliau (Latinized as Bullialdus) had a clever (though incorrect) explanation: as the Sun was known to rotate and have sunspots, perhaps this variable star was turning and varying its brightness due to extensive spotting on its surface.

Variable-star hunting became quite popular, but in the seventeenth century it was difficult to judge variability, because the standards were fuzzy in judging the brightness of a star. By the end of the eighteenth century, Herschel helped by establishing a catalog of stellar brightnesses, which compared neighboring stars to one another in any given constellation.

The Italian astronomer Geminiano Montanari in the 1660s had noticed that the star Algol varied in brightness. A century later, two close friends and amateur astronomers in the north of England, Edward Pigott and John Goodricke, came to recognize that Algol was varying in a very regular and relatively quick-paced manner.* The name Algol comes from the Arabic *al-ghoul,* the "demon star," which suggests it was long known as an unusual star. At its faintest Algol is just a quarter of its maximum brightness. Pigott and Goodricke determined the period of this cyclical change to be "two days and nearly twenty hours and three quarters." (Goodricke later pegged it at two days, twenty hours, and forty-nine minutes, essentially matching today's measurements.) They speculated (correctly) that Algol had an orbiting companion that periodically eclipses it, an idea that created a sensation in London scientific circles. They reported this to the Royal Society, although they later abandoned the eclipse model, possibly because it didn't fit other changing stars they found. (Firm proof arrived a century later; see Chapter 26.) Pigott and Goodricke also discovered the short-term variables δ Cephei and η Aquilae, now known to be Cepheid stars, which vary in brightness as they physically pulsate. Stars of this type would later play a vital role in one of cosmology's greatest discoveries (see Chapter 51).

---

* Goodricke, though deaf and mute since infancy, was an astronomy prodigy who won the Royal Society's prestigious Copley medal for his work on Algol at the age of nineteen. He died three years later.

"Considerations on the Change of the Latitudes of Some of the Principal Fixt Stars."
Philosophical Transactions, Volume 30 (1718)
by Edmond Halley

Having of late had occasion to examine the quantity of the precession of the equinoctial points, I took pains to compare the declinations of the fixed stars delivered by Ptolemy in the 3rd chapter of the 7th book of his *Almagest* as observed by Timocharis and Aristyllus* near 300 years before Christ and by Hipparchus about 170 years after them, that is about 130 years before Christ, with what we now find. And by the result of very many calculations, I concluded that the fixed stars in 1,800 years were advanced somewhat more than 25 degrees in longitude or that the precession is somewhat more than 50" [arc seconds] per year. But that with so much uncertainty, by reason of the imperfect observations of the ancients, that I have chosen in my tables to adhere to the even proportion of five minutes in six years, which from other principles we are assured is very near the truth. But while I was upon this inquiry, I was surprised to find the latitudes of three of the principal stars in [the] heaven directly to contradict the supposed greater obliquity of the ecliptic, which seems confirmed by the latitudes of most of the rest; they being set down in the old catalogue as if the plane of the Earth's orb [orbit] had changed its situation among the fixed stars about 20 minutes [of arc] since the time of Hipparchus. Particularly, all the stars in Gemini are put down: those to the northward of the ecliptic with so much less latitude than we find, and those to the southward with so much more southerly latitude. Yet the three stars Palilicium or the Bull's Eye [Aldebaran], Sirius and Arcturus do contradict this rule directly. For by it, Palilicium being in the days of Hipparchus in about 10 gr. of Taurus ought to be about 15 minutes more southerly than at present, and Sirius being then in about 15 of Gemini ought to be 20 minutes more southerly than now. Yet . . . Ptolemy places the first 20 minutes and the other 22 more northerly in latitude than we now find them. Nor are these errors of

* Two third-century B.C. astronomers based in Alexandria.

transcription but are proved to be right by the declinations of them set down by Ptolemy, and observed by Timocharis, Hipparchus and himself, which show that those latitudes are the same as those authors intended. As to Arcturus, he is too near the equinoctial colure to argue from him concerning the change of the obliquity of the ecliptic, but Ptolemy gives him 33' [arc minutes] more north latitude than he now has; and that greater latitude is likewise confirmed by the declinations delivered by the abovesaid observers.* So then all these three stars are found to be above half a degree more southerly at this time than the ancients reckoned them. When on the contrary at the same time, the bright shoulder of Orion [Betelgeuse] has in Ptolemy almost a degree more southerly latitude than at present.† What shall we say then? It is scarce credible that the ancients could be deceived in so plain a matter, three observers confirming each other. Again these stars being the most conspicuous in heaven are in all probability the nearest to the Earth, and if they have any particular motion of their own, it is most likely to be perceived in them, which in so long a time as 1,800 years may show itself by the alteration of their places, though it be utterly imperceptible in the space of a single century of years. Yet as to Sirius it may be observed that Tycho Brahe makes him 2 minutes more northerly than we now find him, whereas he ought to be above as much more southerly from his ecliptic. . . . One half of this difference may perhaps be excused, if refraction were not allowed in this case by Tycho; yet two minutes in such a star as Sirius is somewhat too much for him to be mistaken.

But a further and more evident proof of this change is drawn from the observation of the application of the Moon to Palilicium [in the year 509] when in the beginning of the night the Moon was seen to follow that star very near and seemed to have eclipsed it. . . . Now from the undoubted principles of astronomy, it was impossible for this to be true at Athens or near it, unless the latitude of Palilicium were much less than we at this time find it.

This argument seems not unworthy of the Royal Society's consideration, to whom I humbly offer the plain fact as I find it and would be glad to have their opinion.

But whether it were really true that the obliquity of the ecliptic was, in the time of Hipparchus and Ptolemy, really 22 minutes greater than now

---

* The colure is one of two great circles that intersect each other at right angles at the poles and divide the equinoctial and the ecliptic into four equal parts.

† Halley was correct about the motions of Sirius and Arcturus but mistaken about how Aldebaran and Betelgeuse were moving, due to errors in the ancient positions.

may well be questioned, since Pappas Alexandrinus, who lived but about 200 years after Ptolemy, makes it the very same that we do.

"On the Proper Motion of the Sun and Solar System; With an Account of Several Changes That Have Happened Among the Fixed Stars Since the Time of Mr. Flamsteed."
Philosophical Transactions of the Royal Society of London, Volume 73 (1783)
by William Herschel

The new lights that modern observations have thrown upon several interesting parts of astronomy begin to lead us now to a subject that cannot but claim the serious attention of everyone who wishes to cultivate this noble science. That several of the fixed stars have a proper motion is now already so well confirmed that it will admit of no further doubt. From the time this was first suspected by Dr. Halley we have had continued observations that show Arcturus, Sirius, Aldebaran, Procyon, Castor, Rigel, Altair, and many more to be actually in motion; and considering the shortness of the time we have had observations accurate enough for the purpose, we may rather wonder that we have already been able to find the motions of so many, than that we have not discovered the like alterations in all the rest. Besides, we are well prepared to find numbers of them apparently at rest, as, on account of their immense distance, a change of place cannot be expected to become visible to us till after many ages of careful attention and close observation, though every one of them should have a motion of the same importance with Arcturus. This consideration alone would lead us strongly to suspect that there is not, in strictness of speaking, one *fixed* star in the heavens; but many other reasons, which I shall presently adduce, will render this so obvious that there can hardly remain a doubt of the general motion of all the starry systems, and consequently of the solar one among the rest.

I might begin with principles drawn from the theory of attraction, which evidently oppose every idea of absolute rest in any one of the stars, when once it is known that some of them are in motion: for the change that must arise by such motion, in the value of a power which acts inversely as the squares of the distances, must be felt in all the neighboring stars; and if these be influenced by the motion of the former, they will again affect those that are next to them, and so on till all are in motion. Now as we know several stars, in diverse parts of the heavens, do actually change their place, it will follow that the motion of our solar system is not a mere hypothesis; and what will give additional weight to this consideration is that we have the greatest reason to suppose most of those very stars, which have been observed to move, to be such as are nearest to us; and, therefore, their influence on our situation would alone prove a powerful argument in favor of the proper motion of the sun had it actually been originally at rest. But I shall waive every view of this subject which is not chiefly derived from experience.

To begin with my own, I will give a short but general account of the most striking changes I have found to have happened in the heavens since Flamsteed's* time. I have now almost finished my third review [sky survey]. . . . This review extended to all the stars in Flamsteed's catalogue, together with every small star about them as far as the tenth, eleventh, or twelfth magnitudes, and occasionally much farther, to the amount of a great many thousands of stars. . . .

[Omitted here are several sections in which Herschel describes the many stars that have undergone a change, can no longer be seen, or altered their magnitude since Flamsteed's day.]

To return to the principal subject of this paper, which is the proper motion of the sun and solar system: does it not seem very natural that so many changes among the stars—many increasing their magnitude, while numbers seem gradually to vanish; several of them strongly suspected to be newcomers, while we are sure that others are lost out of our sight; the distance of many actually changing, while many more are suspected to have a considerable motion—I say, does it not seem natural that these observations should cause a strong suspicion that most probably every star in the heaven is more or less in motion? . . . Now, if the proper motion of the stars in general be once admitted, who can refuse to allow that our sun, with all its planets and comets (that is, the solar system), is no less liable to such a general agitation as we find to obtain among all the rest of the celestial bodies.

* The seventeenth-century English astronomer John Flamsteed, the first astronomer royal based at the Greenwich Observatory, and a contemporary of Newton and Halley.

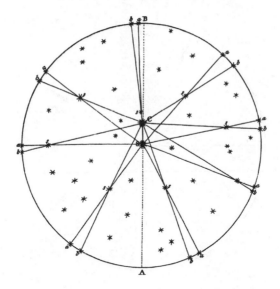

Figure 17.1

Admitting this for granted, the greatest difficulty will be how to discern the proper motion of the sun between so many other (and variously compounded) motions of the stars. This is an arduous task indeed. . . . I shall therefore now point out the method of detecting the direction and quantity of the supposed proper motion of the sun by a few geometrical deductions, and at the same time show by an application of them to some known facts that we have already some reasons to guess which way the solar system is probably tending its course.

Suppose the sun to be at S, figure [17.1]; the fixed stars to be dispersed in all possible directions and distances around at *s, s, s, s,* etc. Now, setting aside the proper motion of the stars, let us first consider what will be the consequence of a proper motion in the sun and let it move in a direction from A towards B. Suppose it now arrived at C. Here, by a mere inspection of the figure, it will be evident that the stars *s, s, s,* which were seen at *a, a, a,* will now by the motion of the sun from S to C appear to have gone in a contrary direction and be seen at *b, b, b.* That is to say every star will appear more or less to have receded from the point B. . . . The sun must have a direct motion towards the point B to occasion all these appearances. . . .

. . . Astronomers have already observed what they call a proper motion in several of the fixed stars. . . . We ought, therefore, to resolve that which is common to all the stars, which are found to have what has been called a proper motion, into a single real motion of the solar system, as far as that

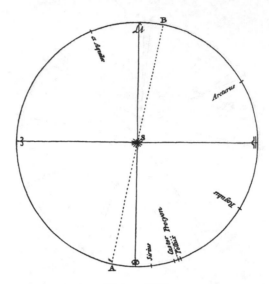

Figure 17.2

will answer the known facts and only to attribute to the proper motion of each particular star the deviations from the general law the stars seem to follow in those movements.

By Dr. Maskelyne's account of the proper motion of some principal stars, we find that Sirius, Castor, Procyon, Pollux, Regulus, Arcturus, and α Aquilae appear to have respectively the following [annual] proper motions in right ascension: −0".63; −0".28; −0".80; −0".93; −0".41; −1".40; and +0".57.* And two of them, Sirius and Arcturus, in declination, *viz.* 1".20 and 2".01, both southward. Let figure [17.2] represent an equatorial zone with the above mentioned stars referred to it, according to their respective right ascensions, having the solar system in the center. Assume the direction AB from a point somewhere not far from the 77th degree of right ascension to its opposite 257th degree, and suppose the sun to move in that direction from S towards B. Then will that one motion answer that of all the stars together: for if the supposition be true, Arcturus, Regulus, Pollux, Procyon, Castor, and Sirius should appear to decrease in right ascension, while α Aquilae, on the contrary, should appear to increase. Moreover, suppose the sun to ascend at the same time in the same direction towards some point in the northern hemisphere, for instance towards the constellation of Hercules. Then will also the observed change of declina-

* Nevil Maskelyne, then Great Britain's astronomer royal.

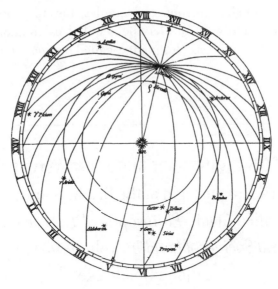

Figure 17.3

tion of Sirius and Arcturus be resolved into the single motion of the solar system. . . . But lest I should be censured for admitting so new and capital a motion upon too slight a foundation, I must observe that the concurrence of those seven principal stars cannot but give some value to an hypothesis that will simplify the celestial motions in general. We know that the sun, at the distance of a fixed star, would appear like one of them; and from analogy we conclude the stars to be suns. Now, since the apparent motions of these seven stars may be accounted for, either by supposing them to move just in the manner they appear to do or else by supposing the sun alone to have a motion in a direction, somehow not far from that which I have assigned to it, I think we are no more authorized to suppose the sun at rest than we should be to deny the diurnal motion of the earth, except in this respect, that the proofs of the latter are very numerous, whereas the former rests only on a few though capital testimonies. . . .

[Here, at the end of his paper, Herschel presents further data from a French survey that offers the proper motions of twelve stars, six of them new to his consideration. He concludes they support his contention that the Sun is moving towards the constellation Hercules, specifically toward a point near the star λ Herculis (see Figure 17.3 where the lines converge). Today, astronomers agree that the Sun is moving at a net speed of about 12 miles per second (relative to the nearby stars) in the direction of Hercules, toward a point about seven degrees northwest of Herschel's initial calculation, not far from the bright star Vega. This point is called the *solar apex*.]

"A Series of Observations on, and a Discovery of, the Period of the Variation of the Light of the Bright Star in the Head of Medusa, Called Algol." *Philosophical Transactions of the Royal Society of London*, Volume 73 (1783) by John Goodricke

A Letter from John Goodricke, Esq. to the Rev. Anthony Shepherd, Plumian Professor at Cambridge.

Sir,

I take the liberty to transmit to you the following account of a very singular variation in Algol or β Persei, which you will oblige me by presenting to the Royal Society, if you think it deserving that notice. . . . The following observations, lately made, exhibit a regular and periodical variation in that star of a nature hitherto, I believe, unnoticed.

The first time I saw it vary was on the 12th of November 1782, between eight and nine o'clock at night, when it appeared of about the fourth magnitude; but the next day it was of the second magnitude, which is its usual appearance. On the 28th of December following, I perceived it to vary again thus: at 5½ h. in the evening, it was about the fourth magnitude, as on the 12th of November last; but at 8½ h. I was much surprised to find it so quickly increased as to appear of the second magnitude. My friend Mr. Edward Pigott, whom I informed of this singular phenomenon as soon as I saw it, also observed it; and I had the pleasure to find that his observations coincided with mine. . . .

[Omitted is a sample of observations from Goodricke's journal, which provides the detailed hourly changes of Algol on eleven separate days between January 14 and May 3, 1784, and chronicles the star's variations.]

The times of the above observations are nearly apparent time and were for the most part made under favorable circumstances. My friend Mr. Edward Pigott, to whom I am under great obligations on this as well as on other occasions, also observed some of the variations, and where our times of observation were the same, always agrees with me.

From an attentive comparison of all the particulars in the above observations it appears, first, that this star changes from the second to about the fourth magnitude in nearly three hours and a half, and from thence to the second magnitude again in the same space of time; so that the whole duration of this singular variation is only about *seven hours.* And, secondly, it appears also that this variation probably recurs about every *two days and twenty-one hours.** This last conclusion will be rendered more conspicuous by the following table; the first column of which shows the days and exact time of the day when Algol was observed to be very near, or at its least brightness; the second column marks the different intervals of time elapsed between the several observations; the third exhibits the quotient arising from a division of these intervals by a certain number of revolutions, each of two days and twenty-one hours, which number of revolutions are expressed in the last column.

The results in the third column agree so nearly, that there is the greatest probability, not to say certainty, that the singular and quick variation of this star, during the space of seven hours as above mentioned, recurs regularly and periodically about every two days and nearly twenty hours and three quarters.

To ascertain this period with greater accuracy and precision will require more time and observation; but I can add that I have constantly observed Algol at different times every night when the weather permitted, ever since the 28th of December last; and upon accurately examining all these observations in my journal, I find that so far from containing any appearances the least contrary to the above conclusion, they strongly corroborate it, since I never observed that star varied in any of those days which, according to that theory, were the intervals between its variations. All Mr. Edward Pigott's observations, even at different times from mine, tend to confirm the same conclusion.

---

* Astronomers today measure Algol, 93 light-years distant, varying in magnitude from 2.1 at maximum to 3.4 at primary minimum, with a period of 2.867 days, a period that is slowly lengthening. The primary eclipse occurs when the fainter star of the binary passes in front of the brighter star and lasts for some ten hours in total. The main star is a B star, three times as large as our Sun, and the secondary star is a K star.

Table 17.1

The day and time when Algol was observed at or near its least brightness			The different intervals between the several observations		The quotients of the divisions of the 2nd column by the 4th		Number of revolutions
		d. h.	d.	h.	d.	h.	
1782	Nov.	12 8½					
	Dec.	28 5½	45	21	2	20.8	16
1783	Jan.	14 9¼	17	3¾	2	20.6	6
		31 14¼	17	5	2	20.8	6
	Feb.	6 8	5	17¾	2	21	2
		23 12+	17	4	2	20.6	6
		26 9½	2	21½	2	21.5	1
	Mar.	21 8½	22	23	2	20.9	8
	April	10 10+	20	1½	2	20.8	7
		13 8	2	22	2	22*	1
	May	3 9¼	20	1	2	20.7	7

*The difference of upwards of an hour in this quotient will easily be reduced to the others by remarking that Algol was observed on the 10th and 13th of April not when it was at but only near, its least brightness. And, indeed, all the little differences of the rest will vanish by making a reasonable allowance of the same kind.

Whether this singular phenomenon is always the same or whether it occurs only some years and ceases entirely in others . . . and whether in this case it recurs in regular periods of time or otherwise, are curious objects of investigation, which can only be determined by a long and regular course of observations for many years.

If it were not perhaps too early to hazard even a conjecture on the cause of this variation, I should imagine it could hardly be accounted for otherwise than either by the interposition of a large body revolving round Algol or some kind of motion of its own, whereby part of its body, covered with spots or such like matter, is periodically turned towards the earth. But the intention of this paper is to communicate facts, not conjectures; and I flatter myself that the former are remarkable enough to deserve the attention and farther investigation of astronomers.

18 / The First Asteroid

A stronomers since the time of Kepler sought an underlying pattern to the spacing of the planets in the solar system. They were particularly disturbed by the yawning gap between Mars and Jupiter, which somehow broke the progression of orbits outward from the Sun.

In 1766 the Prussian scientist Johann Daniel Titius of Wittenberg developed an elaborate mathematical rule (based on earlier work by Oxford professor David Gregory in 1702) that, though convoluted, did roughly account for the positions of the planets. Six years later the director of the Berlin Observatory, Johann Bode, in a new edition of a popular book on astronomy he wrote, drew attention to the pattern, and it soon became known as "Bode's law" (see Table 18.1). Although there was no physics involved in formulating the law, its influence was substantial and immediately emphasized the planet "missing" between Mars and Jupiter in the overall scheme. And when the planet Uranus was discovered in 1781, in the very position that continued the pattern, the sway of Bode's law became near-mystical. Eventually astronomers around Europe teamed up to discover the elusive body past Mars. This team divided the sky into twenty-four zones, each to be explored by one of the astronomers. They jokingly referred to themselves as the "celestial police."[5] The discovery, however, was made by someone outside this circle.

Working from the new observatory in Palermo, Sicily, that he established, Giuseppe Piazzi was assembling a star catalog, the most accurate in its day. On New Year's Day in 1801 he measured the position of an eighth-magnitude star. Following his traditional procedure, he measured the star again the following night, but it had shifted. He kept watch over its movements over subsequent nights. Working out its orbit, he saw that the path was not elongated, like a comet's, but more circular. He harbored private suspicions that the object might be something special, especially since it seemed to be located in the region of the missing planet. "I have announced this star as a comet," he wrote to a colleague, "but since it is not accompanied by any nebulosity and, further, since its movement is so slow and rather uniform, it has occurred to me several times that it might be something better than a comet. But I have been careful not to advance this supposition to the public."[6]

Seized by an illness in February that made him unable to continue his observations, he communicated his find to other astronomers in Europe. His news, though, didn't reach the observers until the object was lost in the glare of the Sun. The noted German mathematician Carl Friedrich Gauss developed a brilliant new method for calculating celestial orbits from limited data, to help astronomers refind Piazzi's discovery. It was sighted once again on December 31, near the very spot that Gauss computed. In addition, its orbital radius closely matched that predicted by Bode's law. Piazzi called the object Cerere Ferdinandea, in honor of the patron goddess of Sicily and the Sicilian king who founded his observatory.

Ceres was at first labeled a major planet, but doubts about its true nature quickly surfaced. William Herschel with his large telescope in Great Britain saw that Ceres was smaller than our Moon.* And Heinrich Olbers, a German physician and accomplished amateur astronomer, soon found a similar object in the same region, which he named Pallas. By 1807, others named Juno and Vesta were found. Being hundreds rather than thousands of miles in diameter, they appeared starlike ("asteroidic") to Herschel in his telescope, so he suggested the name *asteroid* to describe this new class of objects

---

* Ceres, a slightly flattened sphere and the largest object in the asteroid belt, has a diameter of about 590 miles.

inhabiting the solar system. It came to be believed that they were fragments of a former full-sized planet.

In the 1890s, with the development of astronomical photography, Maximilian Wolf at the Heidelberg Observatory used photographic time exposures to detect asteroids by their telltale trails of motion, increasing the known population immensely. Currently, astronomers estimate there are one million to two million rocky fragments swarming in the main asteroid belt. Rather than a former planet, though, modern astronomers believe the asteroids are a field of debris that failed to coalesce into a planet, due to the gravitational tugs of nearby Jupiter.

From *On the New Eighth Major Planet Discovered Between Mars and Jupiter* (1802)
by Johann Bode

In the second edition of my *Introduction to Knowledge of the Starry Heavens,* which I published while yet in Hamburg in the year 1772, I speak on page 462 concerning the probable existence of other planets in the solar system than had up to that time been known. Should the boundary of the solar system indeed be limited to where we see Saturn? (Since 1781 we know of Uranus at a distance double that of Saturn.) . . . And for what reason the great space which is found between Mars and Jupiter, where so far no major planet is seen? Is it not highly probable that a planet actually revolves in the orbit which the finger of the Almighty has drawn for it?

And in a note at this place: This conclusion appears to follow especially from the very remarkable relation which the six, long-known major planets observe in their distances from the Sun. If we indicate the distance of Saturn from the Sun by 100 units, Mercury is four such units from the Sun. Venus is $4 + 3 = 7$; the Earth, $4 + 6 = 10$; Mars, $4 + 12 = 16$. Now, however, there comes a gap in this regular progression. From Mars outward there follows a space of $4 + 24 = 28$ units in which, up to now, no planet has been seen. Can we believe that the Creator of the world has left this space empty? Certainly not! From here we come to the distance of Jupiter through $4 + 48 = 52$, and finally to Saturn through $4 + 96 = 100$ units (and now to that of Uranus through 4 + 192 = 196 units). . . . This progression proceeds only in small numbers and, therefore, gives only approximate results. In all my subsequent astronomical writings I have, when occasion arose, spoken of this progression, presented it in sketches, and advanced many arguments for its correctness. The discovery of Uranus was the first happy verification of it.

This law in the increasing distance of the planets from the Sun does not lend itself freely to mathematical expression; it is merely empirical and would be inferred from analysis and conclusions, but it remains an iterated indication of the harmonious order which reigns everywhere in the great works of Nature.

I found the first idea of it in Bonnet's "Observations Concerning Nature," translated by Titius, second edition, 1772, in a note by the translator on page 7. The original edition by Bonnet has nothing of it. It is noteworthy that as yet no mention has ever appeared of this progression in the astronomical work of foreigners. Only German astronomers have mentioned it after I drew attention to it in my astronomical writings.

The progression agrees very well with observations even in small numbers. If, however, we put . . . the actual mean distance of Mercury from the Sun at 387 (the distance of the Earth = 1000) and take the distance between Mercury and Venus as 293, then the relative distances of the seven known planets are still more exactly represented. The distances from the Sun are in fact as follows:

Table 18.1

		Mean distance
Mercury	387 units	387
Venus	387 + 293 = 680	723
The Earth	387 + 2 × 293 = 973	1,000
Mars	387 + 4 × 293 = 1,559	1,524
Probable planet between Mars and Jupiter	387 + 8 × 293 = 2,731	
Jupiter	387 + 16 × 293 = 5,075	5,203
Saturn	387 + 32 × 293 = 9,763	9,541
Uranus	387 + 64 × 293 = 19,139	19,082

On the 20th of March, 1801, I received from Dr. Joseph Piazzi, Royal Astronomer and Director of the Royal Observatory at Palermo, a communication dated January 24th in which he writes as follows: "On the 1st of January I discovered a comet in Taurus in right ascension 51°47', northern declination 16°8'. On the 11th it changed its heretofore (westward) retrograde motion into (eastward) direct motion; and on the 23rd was in right ascension 51°46', northern declination 17°8'. I shall continue to observe it and hope to be able to observe throughout the whole of February. It is very small, and equivalent to a star of the eighth magnitude, without any noticeable nebulosity. I beg of you to let me know whether it has already been observed by other astronomers; in this case I should save myself the trouble of computing its orbit."

In the beginning of March, I had already found a notice of the discovery in foreign journals; there was, however, as little said on the place and motion as on the appearance of this remarkable comet.

When, however, I received from the observer himself the foregoing more exact notice of the object, it struck me immediately, upon reading through his letter, as remarkable, and I was convinced that this small star without noticeable nebulosity, at one time in eastern elongation, then appearing to stand still, thereafter again moving forward toward the east, was not a comet at all; Piazzi had, indeed, here discovered a very extraordinary object. It was most probably the eighth major planet of the solar system, which already thirty years before I had announced between Mars and Jupiter, but which until now had remained undiscovered—a planet whose distance from the Sun indicated a known progression of probably 2.80, and which in four years and eight months must run its course around the Sun.

From "Results of the Observations of the New Star," Palermo Observatory, 1801 by Giuseppe Piazzi

. . . On the evening of the 1st of January of the current year, together with several other stars, I sought for the 87th of the Catalogue of the Zodiacal Stars of Mr. La Caille. I then found it was preceded by another, which, according to my custom, I observed likewise, as it did not impede the principal observation. The light was a little faint, and of the color of Jupiter, but similar to many others which generally are reckoned of the eighth magnitude. Therefore I had no doubt of its being any other than a fixed star. In the evening of the 2nd I repeated my observations, and having found that it did not correspond either in time or in distance from the zenith with the former observation, I began to entertain some doubts of its accuracy. I conceived afterwards a great suspicion that it might be a new star. The evening of the third, my suspicion was converted into certainty, being assured it was not a fixed star. Nevertheless before I made it known, I waited 'till the evening of the 4th, when I had the satisfaction to see it had moved at the same rate as on the preceding days. From the fourth to the tenth the sky was cloudy. In the evening of the 10th it appeared to me in the telescope, accompanied by four others, nearly of the same magnitude. In the uncertainty which was the new one, I observed them all, as exactly as possible, and having compared these observations with the others which I made in the evening of the 11th, by its motion I easily distinguished my star from the others. Meanwhile however I greatly wished to see it out of the meridian, to examine and to contemplate it more at leisure. But with all my labor, and that of my assistant D. Niccolò Cacciatore and [of] D. Niccolò Carioti belonging to this Royal Chapel both enjoying a sharp sight, and very expert in the knowledge of the heavens, neither with the night telescope, nor with another achromatic one of 4 inches aperture, was it possible to distinguish it from many others among which it was moving. I was therefore obliged to content myself with seeing it on the meridian, and for the short time of two minutes, that is to say the time it employed in traversing the field of the telescope; other observations, which [we] were making at the same time, not permitting the instrument to be moved from its position.

In the meantime, in order to render the observations more certain, while I was observing with the Circle, D. Niccolò Carioti observed with the transit instrument. The sky was so hazy, and often cloudy, that the observations were interrupted 'till the 11th of February; when the star having approached so near the Sun, it was not possible to see it any longer at its passage over the meridian. I intended to search for it, out of it [the meridian], by means of the Azimuth; but having fallen ill on the thirteenth of February, I was not able to make any further observations. These, however, which have been made, though they are not at the necessary distance from one another in order to assure us of the true course which the star describes in the heavens, are, notwithstanding, sufficient in my opinion, to make us know the nature of the same, as one may collect from the results, which I have deduced from them.[7]

19 / Distance to a Star

O nce Copernicus put the Earth into motion and it was established that our Sun was just the closest star, astronomers
began to think about gauging the distances to other stars. But
it was a measurement extremely difficult to carry out. In the early
decades of the nineteenth century, there were many announce-
ments that a stellar distance had been determined, but the astronomical community remained skeptical since the range of errors
was so large. Friedrich Wilhelm Bessel's great achievement in 1838
was obtaining the first measurement that dispelled all qualms.

Galileo, two centuries earlier in his *Dialogue Concerning the
Two Chief World Systems,* first discussed the technique that Bessel
used to peg his distance. Astronomers already knew they could take
advantage of parallax, the change in a star's position on the sky
when it is observed first at one end of the Earth's orbit and then six
months later at the other end. Galileo specifically suggested looking
at two stars close together on the sky and measuring how the closer
one regularly shifted in relation to the farther one, which is seem-
ingly so far away it does not move. By knowing the radius of the
Earth's orbit (1 AU or astronomical unit) and pegging the parallax
(p) of the closer star, a bit of geometric triangulation determines its
distance from the Sun. But instrumentation in Galileo's day was not
sensitive enough to discern such subtle changes. So the first estimates of stellar distance were achieved by judging how far away the

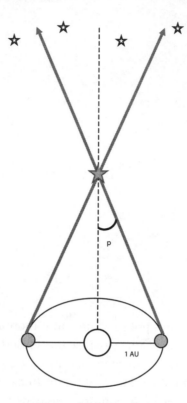

Figure 19.1. Parallax

Sun would have to be to appear as bright as a particular star. In this way, the Scottish mathematician James Gregory in 1668 arrived at a distance to the star Sirius of 83,190 AU. (Sirius is in fact more than six times farther out, at around 542,000 AU or 8.6 light-years.) The Dutch astronomer Christiaan Huygens thirty years later concluded Sirius was closer, at a distance of 27,664 AU, and Isaac Newton came in with an estimate of 1 million AU, an overshoot. Once it was realized that stars varied in brightness and so couldn't be used as "standard candles," astronomers primarily focused on using the parallax technique. Along the way, they came to understand the many problems that could affect a star's measured position, such as light aberration and nutation (see Chapter 14).

A faint double star called 61 Cygni, nearly invisible to the naked eye at fifth magnitude, came to be recognized as a good candidate for a parallax test. It was known as the "flying star" since its motion through the heavens, about five arcseconds a year, was pretty fast for a celestial object.* This suggested it was relatively close to the solar system and so would likely display a parallax. Over a fifteen-month period, Bessel painstakingly measured 61 Cygni's position, often more than a dozen times each night. A newly developed instrument for determining small angles on the sky, a heliometer, allowed him to measure the changing angular distance of 61 Cygni over the year in relation to two stars near it (labeled *a* and *b* in his report), which he assumed were much farther away and essentially motionless.

As a youth, Bessel had been a business clerk with a love for figures. Dreaming that he would someday join a maritime expedition, he studied navigation and astronomy on his own. He learned to compute the orbits of comets with such proficiency that he came to the attention of Heinrich Olbers (discoverer of the second asteroid, Pallas), who procured him a position at a private observatory. At the age of twenty-five, Bessel was appointed professor of astronomy at the University of Königsberg, where he established a major observatory that set new standards in positional astronomy.

By the end of 1838, Bessel announced that he had measured a parallax of 0.3136 arcsecond for 61 Cygni, due solely to the Earth's annual orbital motion. That corresponded to a distance of 657,700 AU from the Sun. "Light employs 10.3 years to traverse this distance," reported Bessel. It was within 10 percent of the modern measurement of 11.4 light-years. Soon the Scottish astronomer Thomas Henderson, working from an observatory at the Cape of Good Hope in South Africa, announced that one of the brightest stars in the Southern Hemisphere, Alpha Centauri, had an even larger parallax of around 1 arcsecond, which translated into a distance of 3 to 4 light-years. At the same time F. G. Wilhelm Struve, at the Dorpat Observatory in what is now Estonia, measured the parallax of the star Vega. All three were working simultaneously. Henderson, in fact, obtained his measurements of Alpha Centauri five years earlier but, concerned about the reliability of his method, held back his results until after the reports on 61 Cygni and Vega.

Bessel had acquired nineteenth-century astronomy's holy grail. "I congratulate you and myself," said John Herschel (William's son)

* 1° (degree) = 60 arcminutes; 1' (minute) = 60 arcseconds. Therefore, 1" (arcsecond) = 1/3,600 degree. For comparison, the Moon's width on the sky is roughly 1/2 degree.

upon bestowing Bessel with the Royal Astronomical Society's gold medal, "that we have lived to see the great and hitherto impassable barrier to our excursions into the sidereal universe; that barrier against which we have chafed so long and so vainly. . . . It is the greatest and most glorious triumph which practical astronomy has ever witnessed."[8] Olbers later claimed that his greatest service to astronomy was smoothing the way for Bessel's entry into the field.

"On the Parallax of 61 Cygni." *Monthly Notices of the Royal Astronomical Society,* Volume 4 (November 1838) by Friedrich Wilhelm Bessel

A letter from Professor Bessel to Sir J. Herschel, Bart., dated Konigsberg, Oct. 23, 1838.

Esteemed Sir,—Having succeeded in obtaining a long-looked-for result, and presuming that it will interest so great and zealous an explorer of the heavens as yourself, I take the liberty of making a communication to you thereupon. Should you consider this communication of sufficient importance to lay before other friends of astronomy, I not only have no objection, but request you to do so. With this view, I might have sent it to you through Mr. Baily; and I should have preferred this course, as it would have interfered less with the important affairs claiming your immediate attention on your return to England.* But, to you, I can write in my own language, and thus secure my meaning from indistinctness.

After so many unsuccessful attempts to determine the parallax of a fixed star, I thought it worth while to try what might be accomplished by means of the accuracy which my great Fraunhofer Heliometer gives to the observations. I undertook to make this investigation upon the star 61 *Cygni,* which, by reason of its great proper motion, is perhaps the best of all; which affords the advantage of being a double star, and on that account may be observed with greater accuracy. And which is so near the pole that,

---

* Francis Baily, a founder in 1820 of what became the Royal Astronomical Society. "Baily's beads," spots of brilliant light seen right before and after a total solar eclipse, were named after him. The effect occurs when beams of sunlight pass through gaps in the rugged lunar topography.

with the exception of a small part of the year, it can always be observed at night at a sufficient distance from the horizon. I began the comparisons of this star in September 1834, by measuring its distance from two small stars of the 11th magnitude, of which one precedes and the other is to the northward. But I soon perceived that the atmosphere was seldom sufficiently favorable to allow of the observation of stars so small; and, therefore, I resolved to select brighter ones, although somewhat more distant. In the year 1835, researches on the length of the pendulum at Berlin took me away for three months from the observatory; and when I returned, Halley's Comet had made its appearance and claimed all the clear nights. In 1836, I was too much occupied with the calculations of the measurement of a degree in this country, and with editing my work on the subject, to be able to prosecute the observations of α *Cygni* so uninterruptedly as was necessary, in my opinion, in order that they might afford an unequivocal result. But, in 1837 these obstacles were removed, and I thereupon resumed the hope that I should be led to the same result which Struve grounded upon his observations of α *Lyrae,* by similar observations of 61 *Cygni.*

I selected among the small stars which surround that double star, two between the 9th and 10th magnitudes; of which one (*a*) is nearly perpendicular to the line of direction of the double star; the other (*b*) nearly in this direction. I have measured with the heliometer the distances of these stars from the point which bisects the distance between the two stars of 61 *Cygni;* as I considered this kind of observation the most correct that could be obtained, I have commonly repeated the observations sixteen times every night. When the atmosphere has been unusually unsteady, I have, however, made more numerous repetitions; although, by this, I fear the result has not attained that precision which it would have possessed by fewer observations on more favorable nights. This unsteadiness of the atmosphere is the great obstacle which attaches to all the more delicate astronomical observations. In an unfavorable climate we cannot avoid its prejudicial influence, unless by observing only on the finest nights; by which, however, it would become still more difficult to collect the number of observations necessary for an investigation. The places of both stars, referred to the middle point of the double star, are for the beginning of 1838,

	Distance	Angle of Position
a	461".617	201° 29' 24"
b	706".279	109° 22' 10"

... The following tables contain all my measures of distance, freed from the effects of refraction and aberration, and reduced to the beginning of 1838. In these reductions, the annual variations employed of both distances are $= +4".3915$ and $-2".825$; which I have deduced (on the supposition that the stars a and b have no proper motions) from the mean motions of both stars of 61 *Cygni,* which M. Argelander* had lately found by comparison of my determination (from Bradley's† observations) for 1755, with his own for 1830. . . .

[Omitted here are several pages of tables, where Bessel lists the changing angular distance of his comparison stars, *a* and *b*, from 61 *Cygni* over the fifteen-month period from August 1837 to October 1838, along with some correctional data. He then discusses his method for calculating a parallax from this data and his choice to give more weight to the measurements of star *a*.]

. . . For this purpose, since both series must now be brought into connection with one another, it was necessary to deduce the *weight* of the observations contained in the second series [star *b*], the weight of those in the first series [star *a*] being taken as unit. I have found it $= 0.6889$; and hence the most probable value of the annual parallax of 61 *Cygni* $=$ $0".3136$. On this hypothesis, I find the mean distances of both stars for the beginning of 1838 to be $461".6171$ and $706".2791$; and the corrections of the assumed values of the annual variations $= -0".0293$ and $+0".2395$. The mean error of an observation of the kind of which I have assumed the weight as unit is $\pm 0".1354$, and the mean error of the annual parallax of 61 *Cygni* $= \pm 0".0202$. . . .

As the mean error of the annual parallax of 61 *Cygni* $(= 0".3136)$ is only $\pm 0".0202$, and consequently not $1/15$ of its value computed; and as these comparisons show that the progress of the influence of the parallax, which the observations indicate, follows the theory as nearly as can be expected considering its smallness, we can no longer doubt that this parallax is sensible. Assuming it $0".3136$, we find the distance of the star 61 *Cygni* from the sun 657,700 mean distances of the earth from the sun: light employs 10.3 years to traverse this distance. . . .‡

* Friedrich Wilhelm Argelander, former student of Bessel's who was then professor of astronomy at the University of Bonn.

† British astronomer James Bradley, noted for his work on stellar positions. He discovered aberration (see Chapter 14).

‡ Bessel essentially uses the formula sin p = Earth orbital radius/Distance to star. A parallax of 1 arcsecond (0.0003 degree) corresponds to a distance of 3.26 light-years, which is why astronomers call that specific distance a parsec.

I have here troubled you with many particulars; but I trust it is not necessary to offer any excuse for this, since a correct opinion as to whether the investigation of the parallax of 61 *Cygni* has already led to an approximate result, or must still be carried further before this can be affirmed of them, can only be formed from the knowledge of those particulars. Had I merely communicated to you the result, I could not have expected that you would attribute to it that certainty which, according to my own judgment, it possesses. . . .

20 / Discovery of Neptune

As soon as Uranus was discovered in 1781, its orbital path was worked out by specialists in celestial mechanics and then carefully monitored. Within a few decades, it became very obvious that Uranus was not moving as predicted. Even making adjustments for the relatively large gravitational presence of both Jupiter and Saturn nearby couldn't erase the discrepancy. This made astronomers wonder whether Uranus was accompanied by a massive satellite or had even had its trajectory disrupted by a colliding comet. More attractive was the idea that Uranus was being perturbed by another planet farther out—or, as Alexis Bouvard, the Paris Observatory's director in the early 1800s, put it, "some strange and unperceived force."[9] Two experts on Newton's laws, Urbain Jean Joseph Le Verrier of the Paris Observatory and John Couch Adams in England, decided to take up the challenge and calculate where such a hidden planet would currently reside in the sky.

Adams, newly graduated from Cambridge University as its top mathematics student, began his computations first, in 1843. By the fall of 1845, he had constructed a feasible orbit for the undiscovered planet that accounted for Uranus's errant motions. He even pinpointed a possible position for astronomers to look. But the astronomer royal at that time, George Biddell Airy, was wary of Adams's calculation and did not pursue it right away. Neither did other British astronomers—never before in astronomical history had observers been asked to use valuable telescope time to look for an object based on a theoretical prediction alone.

Working independently in France and unaware of Adams's unpublished finding, Le Verrier, an expert on gravitational perturbations in the solar system, tackled the problem with a different mathematical approach and published a series of papers on his progress. In August 1846 he announced his most precise determination of the new planet's position to date (almost identical to that of Adams) and convinced astronomers at the Berlin Observatory to look for it in the constellation Aquarius. He wrote that they should hunt for an object as bright as a star of the eighth magnitude but appearing more as a visible disk. The Berlin astronomers had an advantage not available at other observatories: a new star chart that encompassed the region being searched, which made it far easier to spot a new intruder. Indeed, Johann Galle and his assistant Heinrich d'Arrest made the discovery on the very day—September 23, 1846—they received Le Verrier's letter. Within an hour of beginning their search, the new planet was found. "Not at first glance, to tell the truth," said Galle, "but after several comparisons. Its absence from the chart was so obvious. . . ."[10] It was less than a degree from Le Verrier's predicted position. Modest about his role, Galle had the observatory director Johann Encke notify the journals.

British astronomers, who had finally initiated a search the previous July, found it six days later. Only after hearing the news from Germany did James Challis at Cambridge University compare the Berlin data with his own. He noticed that a star marked number 49 in a region he mapped on August 12 was missing in a map of the same region made on July 30. If the data had been analyzed earlier, he and Adams would have secured the discovery. As it was, a bitter dispute did arise between Great Britain and France over Adams's priority in predicting the planet's position. On October 10, using a 2-foot reflector at his aptly named home of Starfield near Liverpool, England, William Lassell discovered the first moon around the far planet.

French astronomers toyed a bit with the idea of naming the planet Le Verrier, but the community of astronomers worldwide held on to the mythological tradition. Neptune became the favored name, possibly because of the planet's faint blue-green hue. Its discovery was a significant moment for physical astronomy. The traditional order of astronomical inquiry—first observation, then development of a theory to account for it—was reversed. Here the law of gravitation (and deviations from it) led to the discovery of an entirely new celestial body, 30 AU or 2.8 billion miles from the Sun. Flush with this success, astronomers soon blamed a discrepancy in Mercury's orbit on an undiscovered planet, situated nearer the Sun. The perihelion of Mercury's orbit, the orbital point closest to the Sun, was pivoting around faster than all the gravitational tugs from the other planets could account for. Le Verrier suggested a hidden planet between Mercury and the Sun or even an array of smaller bodies, a sort of inner asteroid belt. Once Le Verrier announced this supposition, reports of sighting this new planet quickly proliferated. It came to be called Vulcan. But over time Vulcan never materialized as predicted. Supposed sightings of it often turned out to be sunspots. The mystery would not be solved until 1915, when Einstein amended the laws of gravitation. No extra inner planet was needed to explain Mercury's curious movements, only a new conception of space and time (see Chapter 36).

Letter to the Editor (dated September 8, 1846), *Astronomische Nachrichten*, Volume 25 (October 12, 1846) by Urbain Jean Joseph Le Verrier

In the last letter that I had the pleasure of writing you, I reported that I had undertaken extensive researches on the motions of Uranus, and that I was coming to the conclusion that a perturbing planet existed, for which I indicated the position. I have been very busy since then in perfecting my work, and I formed the desire to carry it through before the time of opposition of the new body in order that astronomical observers could explore with ease the region of the sky called to their attention. But I had not counted on an indisposition which has much retarded me, with the result that the opposition of the planet has already passed some days ago. Happily the disadvantage, which results from the diminution of the angular distance to the Sun, will be compensated for by the very rapid decrease in the length of the day. We will be for a long time yet in a favorable situation for the physical researches which should be attempted.

I take the liberty of addressing to you an extract of my work, with the request that it be inserted in your learned journal. I hope to be able before long to publish my researches in detail—may they inspire sufficient confidence in astronomical observers to encourage them to make a careful study of the part of the sky where it will be possible without doubt to discover a planet of which the mass is very considerable.

You will see, sir, that I have supposed that the disturbing body is situated in the ecliptic. I have not yet had the leisure to examine if it will be possible to deduce from the observations any precise data concerning the latitude. But we can be sure, even at present, that this latitude will be fairly low since the latitudes of Uranus accord very approximately with the tables in use. This is, moreover, the only point which still remains for me to consider, and I shall proceed to occupy myself with it. . . .

Researches on the Motions of Uranus—I undertake, in the publication of which I present here an abstract, to investigate the nature of the irregularities in the motion of Uranus; to determine their cause, while trying to discover, from the course which they take, the direction and the magnitude of the forces which produce them.

The theory of Uranus at the present time absorbs the attention of astronomers. It has been the subject of many hypotheses, more or less plausible, which, however, aside from geometric considerations, cannot have any real value. Several societies have even proposed the theory as a subject for competition. I believe, therefore, that because of the importance of the question I should rapidly recount its history. One can then better judge the goal of my work, the course which I have traveled, and the results at which I have arrived.

In 1820, there were available regular meridian observations extending over a period of forty years. The planet had, moreover, been observed nineteen times between 1690 and 1771 by Flamsteed, Bradley, Mayer, and Lemonnier.* These astronomers had seen it as a star of the sixth magnitude. On the other hand, the analytical expressions for the perturbations which Jupiter and Saturn produce on Uranus are to be found developed in the first volume of the Mécanique Céleste. Using all these data, one should have expected to be able to construct exact tables for the planet. This is what Bouvard, Member of the Academy of Sciences, undertook. But he encountered unforeseen difficulties.

---

* John Flamsteed, James Bradley, Johann Tobias Mayer, and Pierre-Charles Lemonnier

When the tables of a planet are based on too few observations it may happen that these tables, in the course of time, no longer give correctly the position of the planet. But at least the observations used are represented by the tables with all the rigor which they demand; it can even be said that the fewer the observations, the more easily they can be represented.

This was not the case, however, in the construction of the tables of Uranus. It was found impossible to represent at the same time the nineteen older observations and the numerous modern ones. In this embarrassing situation the learned member of the academy throws doubt upon the accuracy of the older observations; he discards them completely and takes into account only the modern observations. But one should note that though the observations of Flamsteed, Bradley, Mayer, and Lemonnier are not as exact as those of the astronomers of our epoch, one may not with any plausibility be allowed to consider them infested with such enormous errors as those of which the present tables accuse them. The author of these tables actually suggests, however, that this is his opinion, although he adds, in reviewing the difficulties which he had encountered:

"Such is then the alternative which the formation of the tables of the planet Uranus presents, that if one combines the older observations with the modern ones the former will be passably represented while the latter will not be represented with the precision they demand; and if one rejects the older observations so as to use only the modern ones, the result will be tables which will have all the desirable accuracy relative to the modern observations, but which will not be able to satisfy sufficiently the older observations. It is necessary to choose between the alternatives; I have thought best to hold to the second, as being the one which has the greater probability of truth; and the future shall have the burden of demonstrating whether the difficulty of reconciling the two systems is really connected with the inaccuracy of observations, or whether it depends on some strange and unperceived force which may be exerted on the planet."

The twenty-five years which have elapsed since that epoch have shown us that the present tables, which do not represent the older positions, are in no better agreement with the positions observed in 1845. May this disagreement be attributed to lack of precision in the theory? Or rather has not the theory been applied to the observations with sufficient exactitude in the work which has served as a basis for the present tables? Or finally, might it be that Uranus is subjected to other influences besides those which result from the action of the Sun, of Jupiter, and of Saturn? And, in this case, might one succeed, by a careful study of the disturbed motion of the planet, in determining the cause of these unforeseen irregularities? And could one come to the point of fixing the spot in the sky where the investigations of observing astronomers ought to discover the strange body, the source of all the difficulties?

In the course of the summer of the year 1845, M. Arago persuaded me that the importance of this question made it the duty of every astronomer to cooperate, as much as possible, in clearing up some point of the difficulty.* I abandoned, then, immediately the researches on comets which I had undertaken, of which several fragments have already been published, so as to occupy myself with Uranus. Such is the origin of the present research.

Letter from J. S. Encke to the Editor, *Astronomische Nachrichten*, Volume 25 (October 12, 1846)

No mail went to Hamburg yesterday and, therefore, I could not announce to you the discovery of the Le Verrier planet. Accordingly, I can today give you more information. In the *Comptes Rendus* for August 31, 1846, M. Le Verrier has given the following elements, deduced from the deviations of Uranus from its orbit, computed on the basis of the known masses:

Semi-major axis	36.154
Period of revolution	217.387 years (sidereal)
Eccentricity	0.10761
Perihelion	284°45'
Mean longitude on January 1, 1847	318°47'
Mass	$1/9{,}300$

And from this it follows:

Heliocentric True Longitude, January 1, 1847	326°32'
Distance from the Sun	33.06

In a letter which arrived on September 23, M. Le Verrier especially urged Dr. Galle to search for the planet. Probably he was guided by the supposition mentioned in his article that the planet could be identified through showing a disk. The same evening Galle compared with the sky

---

* François Arago, then director of the Paris Observatory and Le Verrier's mentor.

the excellent maps which Dr. Bremiker has plotted (Hour XXI of the Academy Star Charts), and almost immediately noticed, very near to the position which Le Verrier predicts, a star of the eighth magnitude which was missing on the chart. It was immediately measured three different times by Galle with reference to a star in Bessel's catalogue (each measure consisting of five observations), and was once measured by me. The results of these comparisons are as follows:

Sidereal Time 22^h 52^m	R. A. diff. + 1^m 25^s.84	Dec. diff. + 1'35".9	Galle
23 47	25.30	37.9	Galle
0 52	25.34	35.9	Galle
1 8	25.26	37.3	Encke

Although on the whole there is shown here a progression, nevertheless, the discrepancies in this first series were so noticeable that it cannot be depended upon. Therefore, we waited until the next evening. At that time, to be sure, the weather interfered, cloudiness interrupting the observations. Nevertheless, motion exactly in the direction of the Le Verrier elements was decisive, for we found, using the same star,

Sept. 24	20^h 7^m	+ 1^m 21^s.56	+ 1'16".4	Galle	5 Observations
	21 11	21 .30	14 .8	Galle	5
	22 20	21 .08	14 .4	Encke	4

Similarly, on the 25th of September, when Galle compared the star five times and I, ten times, the motion was confirmed. . . .

The star seemed to be only a trifle fainter than Piazzi XXI, 344, and, therefore, fully as bright as the eighth magnitude. Yesterday the atmospheric conditions were favorable. We recognized a disk, the diameter of which, using bright cross wires and a magnification of 320, I found to be 2".9; Galle found 2".7. When we subsequently used a bright field, I measured the planet greater than 3".2, and Galle considerably smaller than 2".2; but by this time the air had become much more unfavorable so that the first measurements are more to be trusted. I believe that the diameter is probably 2".5, or perhaps somewhat greater, but not as large as 3".0. In this respect also the prediction of Le Verrier, who assumed 3".3, is fully confirmed.*

* The value now adopted is around 2.3 arcseconds.

It would be superfluous to add anything more. This is the most brilliant of all planetary discoveries, because purely theoretical researches have enabled Le Verrier to predict the existence and the position of a new planet. Permit me merely to add that the prompt discovery was possible only because of the excellent Academy Star Charts by Bremiker; the disk can be recognized only when one knows that it exists.

21 / The Shape of the Milky Way

I. Thomas Wright and the Via Lactea

In 1750 Thomas Wright of Durham, England, published his cosmological credo in a handsomely designed volume augustly entitled *An Original Theory or New Hypothesis of the Universe Founded Upon the Laws of Nature, and Solving by Mathematical Principles the General Phaenomena of the Visible Creation; and Particularly the Via Lactea.* His ideas would have likely been buried in obscurity if not for the fact that others widely disseminated what they *thought* he meant, which led to his reputation as the first to postulate a model of the Via Lactea—Milky Way—as a disk of stars.

Once the solar system was found to be organized according to Newton's laws, a number of investigators began to consider whether the stars as well were arranged in a particular way. And, if so, what was the Sun's position within such a structure? Pondering cosmic models was a favorite pastime for Wright as a youth. The third son of a well-to-do carpenter, he taught himself mathematics and was so obsessive about studying astronomy that his father at one point burned his books. Though limited in formal education, he was always concocting grand intellectual schemes and wrote that his dream was to produce "an integrated picture of natural and supernatural, of creation and Creator."[11] Apprenticed to a clockmaker at the age of thirteen, Wright later taught navigation to seamen. His abilities were noted by the aristocracy, and he eventually made a comfortable living giving lectures and private instruction to noble English families. It was through such aristocratic benevolence that he was largely able to publish his most famous work, lavishly illustrated with thirty-two engravings, at the age of thirty-nine.

The central theme of his *Original Theory*, composed as a series of letters, was that the Sun was just one of many stars revolving around a common center of gravity. This was a rational assumption, as Edmond Halley had recently detected motion in the so-called fixed stars (see Chapter 17), and a rotation would prevent the stars from gravitationally collapsing toward one another. At one point in his text Wright illustrates the stars moving in a ring (or series of rings), much like the rings of Saturn. But, strongly guided by religious views, he preferred to think of the Milky Way as a spherical shell of stars, with the Sun off to one side and the Eye of Providence, the "agent of creation," residing in the center. Wright's diagram of the Milky Way as a flat layer of stars, a familiar illustration in astronomy textbooks, was actually a first step in helping his readers picture this shell—so vast in size that the small segment in which we reside would appear to be a flat plane. "I don't mean to affirm that [the plane] really is so in fact," he writes, "but only state the question thus to help your imagination to conceive more aptly what I would explain."[12]

Despite his awkward mix of theology and science, Wright does deserve credit for his insight that the Milky Way was an optical effect, the result of the solar system being immersed in an assembly of stars. He introduced the idea, now viewed as common sense, that our position in space affects how we perceive our celestial environment. The Milky Way appears as a band, he mused, because we observe a thin layer of stars edge-on, its combined light producing the milk-white appearance. Such a structure also explains why, when looking perpendicular to the band, stargazers see fewer stars.

Wright went on to speculate that the cloudy spots then being observed by astronomers in greater numbers might be external creations, "bordering upon the known one, too remote for even our telescopes to reach."[13] This idea was amplified in 1755 by the philosopher Immanuel Kant and independently suggested six years later by the German mathematician Johann Heinrich Lambert in his *Cosmological Letters*.

From *An Original Theory or New Hypothesis of the Universe* (1750) by Thomas Wright

Letter the Seventh

... When we reflect upon the various aspects and perpetual changes of the planets, both with regard to their heliocentric and geocentric motion, we may readily imagine that nothing but a like eccentric position of the stars could any way produce such an apparently promiscuous difference in such otherwise regular bodies. And that in like manner, as the planets would, if viewed from the Sun, there may be one place in the universe to which their order and primary motions must appear most regular and most beautiful. Such a point, I may presume, is not unnatural to be supposed, although hitherto we have not been able to produce any absolute proof of it.

This is the great Order of Nature, which I shall now endeavor to prove and thereby solve the phenomena of the *Via Lactea* [Milky Way]; and in order thereto, I want nothing to be granted but what may easily be allowed, namely that the Milky Way is formed of an infinite number of small stars.

Let us imagine a vast infinite gulf, or medium, every way extended like a plane and enclosed between two surfaces nearly even on both sides, but of such a depth or thickness as to occupy a space equal to the double radius or diameter of the visible creation, that is to take in one of the smallest stars each way, from the middle station, perpendicular to the plane's direction, and as near as possible according to our idea of their true distance.

But to bring this image a little lower, and as near as possible level to every capacity, I mean such as cannot conceive this kind of continued zodiac, let us suppose the whole frame of nature in the form of an artificial horizon of a globe. I don't mean to affirm that it really is so in fact, but only state the question thus to help your imagination to conceive more aptly what I would explain. [Figure 21.1] will then represent a just section of it. Now in this space let us imagine all the stars scattered promiscuously, but at such a distance from one another as to fill up the whole medium with a kind of regular irregularity of objects. And next let us consider what the consequence would be to an eye situated near the center point, or anywhere about the middle plane, as at the point A. Is it not, think you, very evident that the stars would there appear promiscuously dispersed on each side,

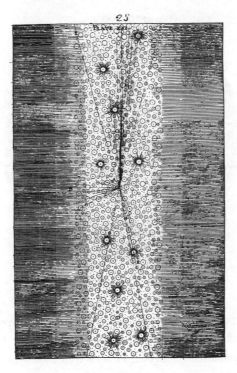

Figure 21.1

and more and more inclining to disorder as the observer would advance his station towards either surface and nearer to B or C, but in the direction of the general plane towards H or D, by the continual approximation of the visual rays crowding together as at H between the limits D and G, they must infallibly terminate in the utmost confusion. If your optics fails you before you arrive at these external regions, only imagine how infinitely greater the number of stars would be in those remote parts, arising thus from their continual crowding behind one another, as all other objects do towards the horizon point of their perspective, which ends but with infinity: Thus, all their rays at last so near uniting must meeting in the eye appear, as almost, in contact and form a perfect zone of light; this I take to be the real case and the true nature of our Milky Way, and all the irregularity we observe in it at the Earth, I judge to be entirely owing to our Sun's position in this great firmament and may easily be solved by his eccentricity and the diversity of motion that may naturally be conceived amongst the stars themselves, which may here and there, in different parts of the heavens, occasion a cloudy knot of stars, as perhaps at E.

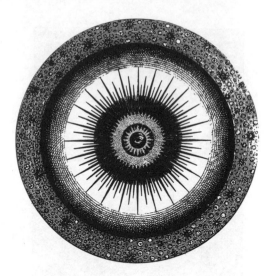

Figure 21.2

But now to apply this hypothesis to our present purpose and reconcile it to our ideas of a circular creation and the known laws of orbicular motion, so as to make the beauty and harmony of the whole consistent with the visible order of its parts, our reason must now have recourse to the analogy of things. It being once agreed that the stars are in motion, which, as I have endeavored in my last letter to show is not far from an undeniable truth, we must next consider in what manner they move. First then, to suppose them to move in right lines, you know is contrary to all the laws and principles we at present know of; . . . it must of course be the other, *i.e.* in an orbit; and consequently, were we able to view them from their middle portion, as from the Eye seated in the center of [Figure 21.2] we might expect to find them separately moving in all manner of directions round a general center, such as is there represented. It only now remains to show how a number of stars, so disposed in a circular manner round any given center, may solve the phenomena before us. . . . The first is in the manner I have above described, *i.e.* all moving the same way, and not much deviating from the same plane, as the planets in their heliocentric motion do round the solar body. . . .

The second method of solving this phenomena is by a spherical order of the stars, all moving with different direction round one common center, as the planets and comets together round the Sun, but in a kind of shell or concave orb. . . .

Hence we may imagine some creations of stars may move in the direction of perfect spheres, all variously inclined, direct and retrograde; others

Figure I.

Fig. II.

Fig. III.

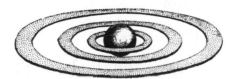

Figure 21.3

again, as the primary planets do, in a general zone or zodiac, or more properly in the manner of Saturn's rings, nay, perhaps ring within ring, to a third
or fourth order, as shown in [Figure 21.3], nothing being more evident than
that if all the stars we see moved in one vast ring, like those of Saturn,
round any central body or point, the general phenomena of our stars would
be solved by it. . . . Not only the phenomena of the Milky Way may be thus
accounted for, but also all the cloudy spots and irregular distribution of
them; and I cannot help being of [the] opinion that could we view Saturn
through a telescope capable of it, we should find his rings no other than an
infinite number of lesser planets, inferior to those we call his satellites:
What inclines me to believe it is this; this ring or collection of small bodies
appears to be sometimes very eccentric, that is more distant from Saturn's
body on one side than on the other and as visibly leaving a larger space
between the body and the ring; which would hardly be the case if the ring,
or rings, were connected or solid, since we have good reason to suppose it
would be equally attracted on all sides by the body of Saturn, and by that

means preserve everywhere an equal distance from him; but if they are really little planets, it is clearly demonstrable from our own in like cases, that there may be frequently more of them on one side than on the other, and but very rarely, if ever, an equal distribution of them all round the Saturnian globe.

How much a confirmation of this is to be wished, your own curiosity may make you judge, and here I leave it for the opticians to determine. I shall content myself with observing that Nature never leaves us without a sufficient guide to conduct us through all the necessary paths of knowledge; and it is far from absurd to suppose Providence may have everywhere throughout the whole universe, interspersed modules of every creation, as our Divines tell us, Man is the image of God himself.

Thus, Sir, you have had my full opinion, without the least reserve concerning the visible creation, considered as part of the finite universe; how far I have succeeded in my designed solution of the *Via Lactea,* upon which the theory of the whole is formed, is a thing that will hardly be known in the present century, as in all probability it may require some ages of observation to discover the truth of it. . . .

II. Immanuel Kant and the Island Universes

As a mixture of theology and cosmic speculation, Thomas Wright's book held little scientific importance at first, but it gained notoriety because of a unique sequence of events. A few months after the publication of *An Original Theory or New Hypothesis,* Wright's ideas were summarized in the Hamburg journal *Freie Urteile.* The reviewer did not stress Wright's description of the Milky Way as a spherical shell but rather its image as a flat ring of stars, "all moving the same way, and not much deviating from the same plane, as the planets in their heliocentric motion do round the solar body."[14] Immanuel Kant, then an aspiring scientist and private tutor on a nobleman's estate in Prussia, was greatly influenced by this summary and went on to imagine that Wright's ring of stars was a continuous disk, especially because of observational evidence. The French scientist Pierre de Maupertuis had been observing dim objects in the sky that appeared elliptical in shape, the very way a disk would look at an angle. By this reasoning, Kant arrived at the correct image of our galaxy's basic structure.

In 1755, at the age of thirty-one, Kant published his theory of the Milky Way in a book entitled *Allgemeine Naturgeschichte und Theorie des Himmels* (Universal Natural History and Theory of the Heavens), where he generously credited Wright for his inspiration. In this same work, he also anticipated Pierre-Simon de Laplace by outlining a nebular theory of the solar system's origin (see Chapter 15). But Kant's *Theory of the Heavens* is probably best known for its introduction of the "island universe" model, whereby our galaxy is just one of many other disks of stars inhabiting the ocean of space. (The German scientist Alexander von Humboldt actually first applied the term "island universe" to describe Kant's theory in his book *Kosmos*, published in 1845.) The pale patches of light sighted by astronomers, wrote Kant, "are just universes and, so to speak, Milky Ways, like those whose constitution we have just unfolded. . . . These higher universes are not without relation to one another, and by this mutual relationship they constitute again a still more immense system."[15]

Kant's book on cosmology was virtually ignored until, years later, he achieved considerable fame as one of the great Western philosophers. But it would take much longer for his island universe model to be widely accepted. Given the quality of astronomical observations in the 1700s, those enigmatic spots could just as easily have been celestial clouds caught at the borders of the Milky Way's disk.

From *Universal Natural History and Theory of the Heavens; or an Essay on the Constitution and Mechanical Origin of the Whole Universe* (1755) by Immanuel Kant

Preface

Mr. Wright of Durham, whose treatise I have come to know from the Hamburg publication entitled the *Frieie Urteile,* of 1751, first suggested ideas that led me to regard the fixed stars not as a mere swarm scattered without visible order, but as a system which has the greatest resemblance with that of the planets; so that just as the planets in their system are found very nearly in a common plane, the fixed stars are also related in their positions, as nearly as possible, to a certain plane which must be conceived as drawn through the whole heavens, and by their being very closely massed in it they present that streak of light which is called the Milky Way. I have become persuaded that because this zone, illuminated by innumerable suns, has very exactly the form of a great circle, our sun must be situated very near this great plane. In exploring the causes of this arrangement, I have found the view to be very probable that the so-called fixed stars may really be slowly moving, wandering stars of a higher order. . . .

I cannot exactly define the boundaries which lie between Mr. Wright's system and my own; nor can I point out in what details I have merely imitated his sketch or have carried it out further. Nevertheless, I found afterwards valid reasons for considerably expanding it on one side. I considered the species of nebulous stars, of which De Maupertuis makes mention in his treatise *On the Figure of the Fixed Stars,* which present the form of more or less open ellipses; and I easily persuaded myself that these stars can be nothing else than a mass of many fixed stars.* The regular constant roundness of these figures, taught me that an inconceivably numerous host of stars must be here arranged together and grouped around a common center, because their free positions towards each other would otherwise have presented irregular forms and not exact figures. I also perceived that they must be limited mainly to one plane in the system in which they are found united, because they do not exhibit circular but elliptical figures. And I further saw that, on account of their feeble light, they are removed to an inconceivable distance from us. What I have inferred from these analogies, is presented in the following treatise for the examination of the unprejudiced reader. . . .

Of the Systematic Constitution Among the Fixed Stars

The scientific theory of the Universal Constitution of the World has obtained no remarkable addition since the time of Huygens. At the present time nothing more is known than what was already known then, namely, that six planets with ten satellites, all performing the circle of their revolution almost in one plane, and the eternal comets which sweep out on all sides, constitute a system whose center is the sun, towards which they all fall, around which they perform their movements, and by which they are illuminated, heated, and vivified; finally, that the fixed stars are so many

* "Nebulous stars" in this case refers to the nebulae that were later found to be distant galaxies.

suns, centers of similar systems, in which everything may be arranged just as grandly and with as much order as in our system; and that the infinite space swarms with worlds, whose number and excellency have a relation to the immensity of their Creator.

The systematic arrangement which was found in the combination of the planets which move around their sun, seemed in the view of astronomers of that time to disappear in the multitude of the fixed stars; and it appeared as if the regulated relation which is found in the smaller solar system, did not rule among the members of the universe as a whole. The fixed stars exhibited no law by which their positions were bounded in relation to each other; and they were looked upon as filling all the heavens and the heaven of heavens without order and without intention. Since the curiosity of man set these limits to itself, he has done nothing further than from these facts to infer, and to admire, the greatness of Him who has revealed Himself in works so inconceivably great.

It was reserved for an Englishman, Mr. Wright of Durham, to make a happy step with a remark which does not seem to have been used by himself for any very important purpose, and the useful application of which he has not sufficiently observed. He regarded the Fixed Stars not as a mere swarm scattered without order and without design, but found a systematic constitution in the whole universe and a universal relation of these stars to the ground-plan of the regions of space which they occupy. We would attempt to improve the thought which he thus indicated, and to give to it that modification by which it may become fruitful in important consequences whose complete verification is reserved for future times.

Whoever turns his eye to the starry heavens on a clear night, will perceive that streak or band of light which on account of the multitude of stars that are accumulated there more than elsewhere, and by their getting perceptibly lost in the great distance, presents a uniform light which has been designated by the name *Milky Way*. It is astonishing that the observers of the heavens have not long since been moved by the character of this perceptibly distinctive zone in the heavens, to deduce from its special determinations regarding the position and distribution of the fixed stars. For it is seen to occupy the direction of a great circle, and to pass in uninterrupted connection round the whole heavens: two conditions which imply such a precise destination and present marks so perceptibly different from the indefiniteness of chance, that attentive astronomers ought to have been thereby led, as a matter of course, to seek carefully for the explanation of such a phenomenon.

As the stars are not placed on the apparent hollow sphere of the heavens, and as some are more distant than others from our point of view and are lost in the depths of the heavens, it follows from this, that at the distances at which they are situated away from us, one behind the other, they are not indifferently scattered on all sides, but must have a predominant relation to a certain plane which passes through our point of view and to which they are arranged so as to be found as near it as possible. This relation is such an undoubted phenomenon that even the other stars which are not included in the whitish streak, are yet seen to be more accumulated and closer the nearer their places are to the circle of the Milky Way; so that of the two thousand stars which are perceived by the naked eye, the greatest part of them are found in a not very broad zone whose center is occupied by the Milky Way.

If we now imagine a plane drawn through the starry heavens and produced indefinitely, and suppose that all the fixed stars and systems have a general relation in their places to this plane so as to be found nearer to it than to other regions, then the eye which is situated in this plane when it looks out to the field of stars, will perceive on the spherical concavity of the firmament the densest accumulation of stars in the direction of such a plane under the form of a zone illuminated by varied light. This streak of light will advance as a luminous band in the direction of the great circle, because the position of the spectator is in the plane itself. This zone will swarm with stars which, on account of the indistinguishable minuteness of their clear points that cannot be severally discerned and their apparent denseness, will present a uniformly whitish glimmer,—in a word, a Milky Way. The rest of the heavenly host whose relation to the plane described gradually diminishes, or which are situated nearer the position of the spectator, are more scattered, although they are seen to be massed relatively to this same plane. Finally, it follows from all this that our solar world, seeing that this system of the fixed stars is seen from it in the direction of a great circle, is situated in the same great plane and constitutes a system along with the other stars. . . .

I come now to that part of my theory which gives it its greatest charm, by the sublime idea which it presents of the plan of the creation. The train of thought which has led me to it is short and natural; it consists of the following ideas. If a system of fixed stars which are related in their positions to a common plane, as we have delineated the Milky Way to be, be so far removed from us that the individual stars of which it consists are no longer sensibly distinguishable even by the telescope; if its distance has the same ratio to the distance of the stars of the Milky Way as that of the latter has to the distance of the sun; in short, if such a world of fixed stars is beheld at such an immense distance from the eye of the spectator situated outside of it, then this world will appear under a small angle as a patch of space whose figure will be circular if its plane is presented directly to the eye, and elliptical if it is seen from the side or obliquely. The feebleness of its light, its figure, and the apparent size of its diameter will clearly distinguish such a phenomenon when it is presented, from all the stars that are seen single.

We do not need to look long for this phenomenon among the observations of the astronomers. It has been distinctly perceived by different observers. They have been astonished at its strangeness; and it has given occasion for conjectures, sometimes to strange hypotheses, and at other times to probably conceptions which, however, were just as groundless as the former. It is the "nebulous" stars which we refer to, or rather a species of them, which M. de Maupertuis thus describes: "They are," he says, "small luminous patches, only a little more brilliant than the dark background of the heavens; they are presented in all quarters; they present the figure of ellipses more or less open; and their light is much feebler than that of any other object we can perceive in the heavens."*

The author of the *Astro-Theology* imagined that they were openings in the firmament through which he believed he saw the Empyrean. A philosopher of more enlightened views, M. de Maupertuis, already referred to, in view of their figure and perceptible diameter, holds them to be heavenly bodies of astonishing magnitude which, on account of their great flattening, caused by the rotatory impulse, present elliptical forms when seen obliquely.

Any one will be easily convinced that this latter explanation is likewise untenable. As these nebulous stars must undoubtedly be removed at least as far from us as the other fixed stars, it is not only their magnitude which would be so astonishing—seeing that it would necessarily exceed that of the largest stars many thousand times—but it would be strangest of all that, being self-luminous bodies and suns, they should still with this extraordinary magnitude show the dullest and feeblest light.

It is far more natural and conceivable to regard them as being not such enormous single stars but systems of many stars, whose distance presents them in such a narrow space that the light which is individually imperceptible from each of them, reaches us, on account of their immense multi-

* *Discours sur la figure des astres*. Paris, 1742.

tude, in a uniform pale glimmer. Their analogy with the stellar system in which we find ourselves, their shape, which is just what it ought to be according to our theory, the feebleness of their light which demands a presupposed infinite distance: all this is in perfect harmony with the view that these elliptical figures are just universes and, so to speak, Milky Ways, like those whose constitution we have just unfolded. And if conjectures, with which analogy and observation perfectly agree in supporting each other, have the same value as formal proofs, then the certainty of these systems must be regarded as established.

The attention of the observers of the heavens, has thus motives enough for occupying itself with this subject. The fixed stars, as we know, are all related to a common plane and thereby form a co-ordinated whole, which is a World of worlds. We see that at immense distances there are more of such star-systems, and that the creation in all the infinite extent of its vastness is everywhere systematic and related in all its members.

It might further be conjectured that these higher universes are not without relation to one another, and that by this mutual relationship they constitute again a still more immense system. In fact, we see that the elliptical figures of these species of nebulous stars, as represented by M. de Maupertuis, have a very near relation to the plane of the Milky Way. Here a wide field is open for discovery, for which observation must give the key. The Nebulous Stars, properly so called, and those about which there is still dispute as to whether they should be so designated, must be examined and tested under the guidance of this theory. When the parts of nature are considered according to their design and a discovered plan, there emerge certain properties in it which are otherwise overlooked and which remain concealed when observation is scattered without guidance over all sorts of objects. . . . If the grandeur of a planetary world in which the earth, as a grain of sand, is scarcely perceived, fills the understanding with wonder; with what astonishment are we transported when we behold the infinite multitude of worlds and systems which fill the extension of the Milky Way! But how is this astonishment increased, when we become aware of the fact that all these immense orders of star-worlds again form but one of a number whose termination we do not know, and which perhaps, like the former, is a system inconceivably vast—and yet again but one member in a new combination of numbers! We see the first members of a progressive relationship of worlds and systems; and the first part of this infinite progression enables us already to recognize what must be conjectured of the whole. There is here no end but an abyss of a real immensity, in . . . which all the capability of human conception sinks exhausted, although it is supported by the aid of the science of number. The Wisdom, the Goodness, the Power which have been revealed is infinite; and in the very same proportion are they fruitful and active. The plan of their revelation must therefore, like themselves, be infinite and without bounds.

III. William Herschel's Construction of the Heavens

A more rigorous, scientific proof of the Milky Way's structure was obtained just a few decades after Wright's and Kant's speculations by William Herschel, eighteenth-century England's prince of astronomy. It might be said that he founded the field of observational cosmology by embarking on a long-term program to investigate the distribution of the stars. Herschel was ahead of his time. While other astronomers in the 1700s primarily focused their attention on the inhabitants of the solar system and the precise measurement of planetary movements, Herschel was avidly interested in the universe's overall construction.

Herschel, too, surmised that the Milky Way was an optical effect, the result of our residing within a grouping of stars, but it was his philosophy that the true shape of such a system should be "confirmed and established by a series of observations."[16] He did this by working around the celestial sphere on a great circle—in more than six hundred distinct regions—and counting the stars that were visible in each sector. He called it "gauging the heavens." He made the simplifying assumption that a high count was evidence of a greater distance to the border of the Milky Way in that direction. He noticed, of course, that stars were bountiful in or near the band of the Milky Way and diminished in number away from it. By 1785 he was able to report to the Royal Society of London that our system was a "compound nebula of the third form." In other words, it resembled a gigantic convex lens that was filled with millions of stars. Its essential feature was its flatness.

His report included an illustration, one often reproduced, that shows a disk bifurcated on one end (a cleft now known to be caused by dark obscuring clouds that extend from the constellation Cygnus to the southern latitudes). It seemed to buttress the idea that the Milky Way essentially resembled a disk or grindstone. Its diameter was five times its thickness. He estimated the total span was around 850 times the distance between the bright star Sirius and our Sun. Now knowing that Sirius is nearly 9 light-years away, Herschel was estimating the Milky Way's width to be nearly 10,000 light-years from end to end. That is less than a tenth of the current estimate of the Milky Way's extent but a monumental dimension for its day.

"On the Construction of the Heavens."
Philosophical Transactions of the Royal Society of London,
Volume 75 (1785)
by William Herschel

The subject of the construction of the heavens, on which I have so lately ventured to deliver my thoughts to this Society, is of so extensive and important a nature, that we cannot exert too much attention in our endeavors to throw all possible light upon it. . . .

By continuing to observe the heavens with my last constructed, and since that time much improved instrument, I am now enabled to bring more confirmation to several parts that were before but weakly supported, and also to offer a few still further extended hints, such as they present themselves to my present view. But first let me mention that, if we would hope to make any progress in an investigation of this delicate nature, we ought to avoid two opposite extremes, of which I can hardly say which is the most dangerous. If we indulge a fanciful imagination and build worlds of our own, we must not wonder at our going wide from the path of truth and nature; but these will vanish like the Cartesian vortices that soon gave way when better theories were offered. On the other hand, if we add observation to observation, without attempting to draw not only certain conclusions, but also conjectural views from them, we offend against the very end for which only observations ought to be made. I will endeavor to keep a proper medium; but if I should deviate from that, I could wish not to fall into the latter error.

That the Milky Way is a most extensive stratum of stars of various sizes admits no longer of the least doubt; and that our sun is actually one of the heavenly bodies belonging to it is as evident. I have now viewed and gauged this shining zone in almost every direction and find it composed of stars whose number, by the account of these gauges, constantly increases and decreases in proportion to its apparent brightness to the naked eye. But in order to develop the ideas of the universe, that have been suggested by my late observations, it will be best to take the subject from a point of view at a considerable distance both of space and of time.

Theoretical view

Let us then suppose numberless stars of various sizes, scattered over an indefinite portion of space in such a manner as to be almost equally distributed throughout the whole. The laws of attraction, which no doubt extend to the remotest regions of the fixed stars, will operate in such a manner as most probably to produce the following remarkable effects.

Formation of nebulae

Form I.   In the first place, since we have supposed the stars to be of various sizes, it will frequently happen that a star, being considerably larger than its neighboring ones, will attract them more than they will be attracted by others that are immediately around them; by which means they will be, in time as it were, condensed about a center; or, in other words, form themselves into a cluster of stars of almost a globular figure, more or less regularly so, according to the size and original distance of the surrounding stars. The perturbations of these mutual attractions must undoubtedly be very intricate, as we may easily comprehend by considering what Sir Isaac Newton says in the first book of his *Principia.* . . .

Form II.   The next case, which will also happen almost as frequently as the former, is where a few stars, though not superior in size to the rest, may chance to be rather nearer each other than the surrounding ones; for here also will be formed a prevailing attraction in the combined center of gravity of them all, which will occasion the neighboring stars to draw together; not indeed so as to form a regular or globular figure but, however, in such a manner as to be condensed towards the common center of gravity of the whole irregular cluster. . . .

Form III.   From the composition and repeated conjunction of both the foregoing forms, a third may be derived, when many large stars or combined small ones are situated in long extended, regular, or crooked rows, hooks, or branches. . . .

Form IV. We may likewise admit of still more extensive combinations; when, at the same time that a cluster of stars is forming in one part of space, there may be another collecting in a different but perhaps not far distant quarter, which may occasion a mutual approach towards their common center of gravity.

Form V. In the last place, as a natural consequence of the former cases, there will be formed great cavities or vacancies by the retreat of the stars towards the various centers which attract them; so that upon the whole there is evidently a field of the greatest variety for the mutual and combined attractions of the heavenly bodies to exert themselves in. I shall, therefore, without extending myself farther upon this subject, proceed to a few considerations that will naturally occur to everyone who may view this subject in the light I have here done.

Objections considered

At first sight then it will seem as if a system, such as it has been displayed in the foregoing paragraphs, would evidently tend to a general destruction by the shock of one star's falling upon another. It would here be a sufficient answer to say that if observation should prove this really to be the system of the universe, there is no doubt but that the great Author of it has amply provided for the preservation of the whole, though it should not appear to us in what manner this is affected. . . . And here I must observe, that though I have before, by way of rendering the case more simple, considered the stars as being originally at rest, I intended not to exclude projectile forces; and the admission of them will prove such a barrier against the seeming destructive power of attraction as to secure from it all the stars belonging to a cluster, if not forever, at least for millions of ages. Besides, we ought perhaps to look upon such clusters, and the destruction of now and then a star in some thousands of ages, as perhaps the very means by which the whole is preserved and renewed. These clusters may be the *Laboratories* of the universe, if I may so express myself, wherein the most salutary remedies for the decay of the whole are prepared.

Optical appearances

From this theoretical view of the heavens, which has been taken as we observed, from a point not less distant in time than in space, we will now retreat to our own retired station, in one of the planets attending a star in its great combination with numberless others; and in order to investigate what will be the appearances from this contracted situation, let us begin with the naked eye. The stars of the first magnitude being in all probability the nearest, will furnish us with a step to begin our scale; setting off, therefore, with the distance of Sirius or Arcturus, for instance, as unity, we will at present suppose that those of the second magnitude are at double and those of the third at treble the distance, and so forth. . . . Taking it then for granted that a star of the seventh magnitude is about seven times as far as one of the first, it follows that an observer, who is enclosed in a globular cluster of stars and not far from the center, will never be able with the naked eye to see to the end of it: for since, according to the above estimations, he can only extend his view to about seven times the distance of Sirius, it cannot be expected that his eyes should reach the borders of a cluster which has perhaps not less than fifty stars in depth everywhere around him. The whole universe, therefore, to him will be comprised in a set of constellations, richly ornamented with scattered stars of all sizes. Or if the united brightness of a neighboring cluster of stars should, in a remarkable clear night, reach his sight, it will put on the appearance of a small, faint, whitish, nebulous cloud, not to be perceived without the greatest attention. To pass by other situations, let him be placed in a much extended stratum, or branching cluster of millions of stars, such as may fall under the third form of nebulae considered in a foregoing paragraph. Here also the heavens will not only be richly scattered over with brilliant constellations, but a shining zone or milky way will be perceived to surround the whole sphere of the heavens, owing to the combined light of those stars which are too small, that is, too remote to be seen. Our observer's sight will be so confined that he will imagine this single collection of stars, of which he does not even perceive the thousandth part, to be the whole contents of the heavens. Allowing him now the use of a common telescope, he begins to suspect that all the milkiness of the bright path which surrounds the sphere may be owing to stars. He perceives a few clusters of them in various parts of the heavens and finds also that there are a kind of nebulous patches; but still his views are not extended so far as to reach to the end of the stratum in which he is situated, so that he looks upon these patches as belonging to that system which to him seems to comprehend every celestial object. He now increases his power of vision, and, applying himself to a close observation, finds that the Milky Way is indeed no other than a collection of very small stars. He perceives that those objects which had been called nebulae are evidently nothing but clusters of stars. He finds their number increase upon him, and when he resolves one nebula into stars he discovers ten new ones which he cannot resolve. He then forms the idea of immense strata of fixed stars, of clusters of stars and of nebulae; till going on with such interesting observations, he now perceives that all these appearances must naturally arise from the confined situation in which we are placed. *Confined* it may justly be called, though in no less a space than what before appeared to be the whole region of the fixed stars; but which now has assumed the shape of a crookedly branching nebula; not, indeed, one of the least, but perhaps very far from being the most considerable of those numberless clusters that enter into the construction of the heavens.

Result of observations

I shall now endeavor to show that the theoretical view of the system of the universe, which has been exposed in the foregoing part of this paper, is perfectly consistent with facts and seems to be confirmed and established by a series of observations. It will appear that many hundreds of nebulae of the first and second forms are actually to be seen in the heavens, and their places will hereafter be pointed out. Many of the third form will be described, and instances of the fourth related. A few of the cavities mentioned in the fifth will be particularized, though many more have already been observed; so that, upon the whole, I believe it will be found that the foregoing theoretical view, with all its consequential appearances, as seen by an eye enclosed in one of the nebulae, is no other than a drawing from nature, wherein the features of the original have been closely copied; and I hope the resemblance will not be called a bad one, when it shall be considered how very limited must be the pencil of an inhabitant of so small and retired a portion of an indefinite system in attempting the picture of so unbounded an extent. . . .

[Omitted here are twenty pages of tables, in which Herschel lists his star counts over the celestial sphere, and the description of his methods for reducing his data to reach his conclusion about the shape of our sidereal system.]

We inhabit the planet of a star belonging to a compound nebula of the third form

. . . It is true that it would not be consistent confidently to affirm that we were on an island unless we had actually found ourselves everywhere bounded by the ocean, and therefore I shall go no further than the gauges will authorize; but considering the little depth of the stratum in all those places which have been actually gauged, to which must be added all the intermediate parts that have been viewed and found to be much like the rest, there is but little room to expect a connection between our nebula and any of the neighboring ones. I ought also to add that a telescope with a much larger aperture than my present one, grasping together a greater quantity of light and thereby enabling us to see farther into space, will be the surest means of completing and establishing the arguments that have been used: for if our nebula is not absolutely a detached one, I am firmly persuaded that an instrument may be made large enough to discover the places where the stars continue onwards. . . .

Section of our sidereal system

. . . The section represented in figure [21.4] is one which makes an angle of 35 degrees with our equator, crossing it in 124½ and 304½ degrees. A celestial globe, adjusted to the latitude of 55° north and having σ Ceti near the meridian, will have the plane of this section pointed out by the horizon. . . . From this figure . . . which I hope is not a very inaccurate one, we may see that our nebula, as we observed before, is of the third form; that is: *A very extensive, branching, compound congeries of many millions of stars;* which most probably owes its origin to many remarkably large as well as pretty closely scattered small stars, that may have drawn together the rest. Now, to have some idea of the wonderful extent of this system, I must observe that this section of it is drawn upon a scale where

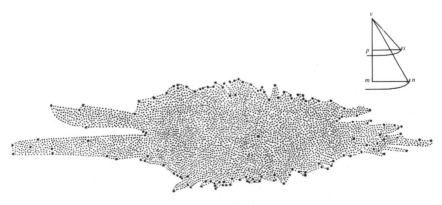

Figure 21.4

the distance of Sirius is no more than the 80th part of an inch; so that probably all the stars, which in the finest nights we are able to distinguish with
the naked eye, may be comprehended within a sphere drawn around the
large star near the middle, representing our situation in the nebula, of less
than half a quarter of an inch radius. . . .

22 / Spiraling Nebulae

A ncient observers knew there were objects in the sky that did not fit the standard description of a star. Ptolemy noted in his star catalog that five of his stellar objects were "cloudy stars," because they appeared hazier. Persian astronomers recorded a misty patch in the Andromeda constellation on their star maps in the tenth century, which the German astronomer Simon Marius described in 1612 as appearing like a "candle shining through horn." With the telescope, other amorphous shapes were sighted, such as the Orion nebula. Around 1714, Edmond Halley tallied up a list of six such nebulae. Many at the time held the ancient belief that these pale entities were breaks in the celestial sphere through which the light of the Empyrean was shining. Others suggested they were atmospheres surrounding distant stars. But Halley thought of these nebulae as distinct objects and was the first to try to explain their unique nature. They "appear to the naked eye like small fixed stars," wrote Halley, "but in reality are nothing else but the light coming from an extraordinary great space in the ether; through which a lucid *medium* is diffused, that shines with its own proper lustre."[17]

By the middle of the eighteenth century, astronomers had divided nebulae into two categories: the nebulae that eventually resolved themselves into clusters of stars and those that continued to appear as white clouds when viewed through the most powerful telescopes. In 1782 Charles Messier, the French comet hunter, published his famous list of 103 nebulae (still used today) so that he and others could avoid mistaking them for comets. M31, for example, is the Andromeda nebula. The following year, William Herschel in England began a grand sweep of the heavens and recorded a thousand new nebulae and star clusters. Later, he cataloged another fifteen hundred. His telescope, the most powerful in its day, was able to resolve some of the cloudy nebulae into stars. They appeared cloudy only because of the difficulty of discerning individual stars over such vast distances. This led Herschel to believe for a while that *all* nebulae were distant systems of stars. Like Immanuel Kant (see Chapter 21), he came to think of them as other Milky Ways. He even boasted of his discovering fifteen hundred new universes.

But Herschel changed his mind when he uncovered an example of a true nebulosity, what he called a "planetary nebula" for its resemblance to a planetary disk. It was a central star surrounded by a hazy mist.* "Cast your eye," he said, "on this cloudy star, and the result will be no less decisive . . . that the nebulosity about the star is not of a starry nature."[18] It seemed to be a shining matter, perhaps the stuff out of which stars condense.

Not until 1839 could the investigation of nebulae be advanced, when telescopes were at last built that surpassed Herschel's instrumentation. Herschel's monopoly was broken by William Parsons, the third earl of Rosse. On the grounds of Birr Castle in central Ireland, he built a reflector with a mirror 3 feet wide, twice the diameter of Herschel's most productive telescope. It provided four times the surface area for collecting the faint light from the heavens. Soon after, Rosse built an even bigger one with a 6-foot mirror. Called the "Leviathan of Parsontown" and mounted between two brick walls some 50 feet high, it began operation in 1845. Rosse was determined to solve the problem of the nebulae once and for all. Were they all just distant star clusters or were they something different? Within weeks of "first light," when Rosse first perused the sky with his massive new instrument, he announced he was able to discern more structure in M51, the fifty-first nebula in Messier's catalog

---

* Astronomers today know that a planetary nebula is an aging giant star that is shedding its outer envelope of gas. The remaining stellar core settles down as a white dwarf star.

(and now known as the Whirlpool galaxy). John Herschel (William's son) had earlier noticed that M51 looked like a central cluster of stars surrounded by a ring. "Perhaps this is our brother system," he said.[19] With the added magnification of his gargantuan telescope, Rosse was the first to see that M51 was actually spiral in shape. He went on to discern more than a dozen spiral nebulae. With photography not yet introduced to astronomy, Rosse captured the swirling patterns in drawings of splendid detail. For the most part, astronomers came to assume that these spirals were simply embryonic globular clusters or planetary systems in the making. But for some, it renewed their speculation that other systems of stars resided outside the borders of the Milky Way.

"Observations on the Nebulae." *Philosophical Transactions of the Royal Society of London*, Volume 140 (1850) by the Earl of Rosse

In laying before the Royal Society an account of the progress which has been made up to the present date in the re-examination of Sir John Herschel's Catalogue of Nebulae published in the *Philosophical Transactions* for 1833, it will be necessary to say something of the qualities of the instrument employed.

The telescope has a clear aperture of 6 feet, and a focal length of 53 feet. It has hitherto been used as a Newtonian, but in constructing the galleries provision was made for the easy application of a little additional apparatus to change the height of the observer, so that the focal length of the speculum remaining the same, the instrument could be conveniently worked as a Herschelian.

Although with an aperture so great in proportion to the focal length, the performance of a parabolic speculum placed obliquely would no doubt be very unsatisfactory, still additional light is so important in bringing out faint details, that it is not improbable in the further examination of the objects of most promise with the full light of the speculum, *undiminished by a second reflection,* some additional features of interest will come out. . . .

When the air is unsteady, minute stars are no longer points, the diffused image is much fainter, and single stars, easily seen when the air is steady, are no longer visible. When many minute stars are crowded together the whole become blended, and instead of a resolved nebula we have merely a diffused, perhaps bright nebulosity. The transparency of the air varies also quite as much; and the aspect of the nebulae changes from night to night, just as the appearance of a distant building alters as the details of the architecture are more or less obscured by the intervening mist. With these facts, the Society will not be surprised should it be in our power at a future time to communicate some additional particulars, even as to the nebulae which have been the most frequently observed.

The sketches which accompany this paper are on a very small scale, but they are sufficient to convey a pretty accurate idea of the peculiarities of structure which have gradually become known to us: in many of the nebulae they are very remarkable, and seem even to indicate the presence of dynamical laws we may perhaps fancy to be almost within our grasp. To have made full-sized copies of the original sketches would have been useless, as many micrometrical measures are still wanting, and there are many matters of detail to be worked in before they will be entitled to rank as astronomical records, to be referred to as evidence of change, should there hereafter be any reason to suspect it.

Much however as the discovery of these strange forms may be calculated to excite our curiosity, and to awaken an intense desire to learn something of the laws which give order to these wonderful systems, as yet, I think, we have no fair ground even for plausible conjecture; and as observations have accumulated the subject has become, to my mind at least, more mysterious and more inapproachable. There has therefore been little temptation to indulge in speculation, and consequently there can have been but little danger of bias in seeking for the facts. When certain phenomena can only be seen with great difficulty, the eye may imperceptibly be in some degree influenced by the mind; therefore a preconceived theory may mislead, and speculations are not without danger. On the other hand, speculations may render important service by directing attention to phenomena which otherwise would escape observation, just as we are sometimes enabled to recognize a faint object with a small instrument, having had our attention previously directed to it by an instrument of greater power. The conjectures therefore of men of science are always to be invited as aids during the active prosecution of research.

It will be at once remarked, that the spiral arrangement so strongly developed in [Figure 22.1], 51 *Messier,* is traceable, more or less distinctly, in several of the sketches. More frequently indeed there is a nearer approach to a kind of irregular interrupted annular disposition of the luminous material than to the regularity so striking in 51 *Messier;* but it can scarcely be doubted that these nebulae are systems of a very similar nature, seen more or less perfectly, and variously placed to the line of sight. In general the details which characterize objects of this class are extremely faint, scarcely perhaps to be seen with certainty on a moderately good night with less than the full aperture of 6 feet: in 51 *Messier,* however, and perhaps a few more, it is not so. A 6-feet aperture so strikingly brings out the characteristic features of 51 *Messier,* that I think considerably less power would suffice, on a very fine night, to bring out the principal convolutions. This nebula has been seen by a great many visitors, and its general resemblance to the sketch at once recognized, even by unpracticed eyes. Messier describes this object as a double nebula without stars; Sir William Herschel as a bright round nebula, surrounded by a halo or glory at a distance from it, and accompanied by a companion; and Sir John Herschel observed the partial subdivision of the *s. f.* limb of the ring into two branches. Taking Sir J. Herschel's figure, and placing it as it would be if seen with a Newtonian telescope, we shall at once recognize the bright convolutions of the spiral, which were seen by him as a divided ring. We thus observe, that with each successive increase of optical power, the structure has become more complicated and more unlike anything which we could picture to ourselves as the result of any form of dynamical law, of which we find a counterpart in our system. The connection of the companion with the greater nebula, of which there is not the least doubt, and in the way represented in the sketch, adds, as it appears to me, if possible, to the difficulty of forming any conceivable hypothesis. That such a system should exist, without internal movement, seems to be in the highest degree improbable: we may possibly aid our conceptions by coupling with the idea of motion that of a resisting medium; but we cannot regard such a system in any way as a case of mere statical equilibrium. Measurements therefore are of the highest interest, but unfortunately they are attended with great difficulties. Measurements of the points of maximum brightness in the mottling of the different convolutions must necessarily be very loose; for although on the finest nights we see them breaking up into stars, the exceedingly minute stars cannot be seen steadily, and to identify one in each case would be impossible with our present means. The nebula itself, however, is pretty well studded with stars, which can be distinctly seen of various sizes, and of a few of these, with reference to the principal nucleus, measurements were taken by my assistant, Mr. Johnstone Stoney, in the spring of 1849, during

Figures 22.1 (top) and 22.2 (bottom)

my absence in London; for some time before the weather had been continually cloudy. These measurements have been again repeated by him this
year, 1850, during the months of April and May. Just as was the case last
year, in February and March the sky was almost constantly overcast. He
has also taken some measures from the center of the principal nucleus to
the apparent boundary of the coils, in different angles of position. The
micrometer employed was furnished with broad lines formed of a coil of
silver wire in the way I have described, seen without illumination. Some of
the stars in the nebula are so bright, I have little doubt they would bear illumination; if so, their positions with respect to some one star might be
obtained with great accuracy of course by employing spiders' lines; this
season however it is too late to make the attempt. Several of these stars are
no doubt within the reach of the great instruments at Pulkova and at Cam-

bridge, U.S., and I hope the distinguished astronomers who have charge of them will consider the subject worthy of their attention. Their better climate gives them many advantages, of which not the least is the opportunity of devoting time to measurements without any serious interruption to other work. I need perhaps hardly add, that measurements taken from the estimated center of a nucleus, and still more from the estimated termination of nebulosity, are but the roughest approximations; they are however the only measurements nebulosity admits of, and if sufficiently numerous, I think they will bring to light any considerable change of place, or form, which may occur.

The spiral arrangement of 51 *Messier* was detected in the spring of 1845. In the following spring an arrangement, also spiral but of a different character, was detected in 99 *Messier* [Figure 22.2]. This object is also easily seen, and probably a smaller instrument, under favorable circumstances, would show everything in the sketch. Numbers 3239 and 2370 of Herschel's Southern Catalogue are very probably objects of a similar character, and as the same instrument does not seem to have revealed any trace of the form of 99 *Messier,* they are no doubt much more conspicuous. It is not therefore unreasonable to hope, that whenever the southern hemisphere shall be re-examined with instruments of great power, these two remarkable nebulae will yield some interesting result.

The other spiral nebulae discovered up to the present time are comparatively difficult to be seen, and the full power of the instrument is required, at least in our climate, to bring out the details. It should be observed that we are in the habit of calling all objects spirals in which we have detected a curvilinear arrangement not consisting of regular re-entering curves; it is convenient to class them under a common name, though we have not the means of proving that they are similar systems. They at present amount to fourteen, four of which have been discovered this spring; there are besides other nebulae in which indications of the same character have been observed, but they are still marked doubtful in our working list, having been seen when the air was not very transparent; 51 *Messier* is the most conspicuous object of that class. . . .

IV

Touching the Heavens

In the second half of the nineteenth century, a dramatic shift took place in astronomy. Up until that time, astronomers focused primarily on the solar system, emphasizing the cyclic beauty and harmony in the movement of the Sun, Moon, and planets against the background of the fixed stars. With a few exceptions (such as William Herschel in Great Britain), astronomers were essentially geometers. They concentrated on showing how planetary motions followed Newton's distinct mathematical rules, from which they could predict future celestial behavior for help in navigating the seas, forecasting the tides, and regulating time. And all the while, observers continued their surveys of the heavens to produce more accurate catalogs of stellar positions and magnitudes. Stars were simply dynamical objects, convenient entities for testing Newton's laws of gravity. Astronomy's knowledge of the stars at this time could be summed up in this way, said British astronomer William Huggins: "That they shine; that they are immensely distant; that the motions of some of them show them to be composed of matter endowed with a power of mutual attraction."[1]

Stars seemed so inaccessible to direct study that Auguste Comte, the French philosopher of positivism, which stressed that true science must be based on experience, concluded around 1835 that humanity would be forever barred from discerning the chemical and physical makeup of celestial bodies. "Never, by any means, will we be able to study their chemical composition, their mineralogic structure," he wrote with firm assurance in his *Cours de philosophie positive*. "I persist in the opinion that every notion of the true mean temperature of the stars will necessarily always be concealed from us."[2]

What Comte did not anticipate at the time of his bold assertion was the development within two decades of new techniques and equipment that would greatly alter the course of astronomy. With the introduction of the spectroscope, an instrument that separates light into its component wavelengths, an entirely new arena opened up to astronomers. Studies of the positions, magnitudes, and motions of astronomical bodies, which had been the major concern of astronomy for millennia, were redirected to analyses of stellar compositions, laying the groundwork for a series of astronomical breakthroughs: recognition of the wide variety of stellar masses, sizes, and luminosities, the true nature of nebulae, and the evolution of stars.

Before spectroscopy, astronomers considered it possible that conditions were so vastly different in other regions of space that the alien surroundings gave rise to substances bearing little resemblance to earthly chemicals. With a spectroscope, though, astronomers could at last "touch" the heavens and directly assess its physical characteristics by interpreting the light waves emitted uniquely by each element. Such familiar terrestrial substances as sodium, iron, and calcium were found in the Sun; comets were seen to emit glowing gases composed of carbon dioxide and various hydrocarbons; ammonia and methane were detected in the atmospheres of the giant gas planets, and nitrogen in far-off nebulae. Aristotle's preference for an ethereal cosmos, separate and apart from earthly materials, was repudiated once and for all. The elements of Earth were clearly identical to the substances shining in the heavens.

Spectroscopy was at first called the "new astronomy" and later "astrophysics." It was a technique initially adopted by wealthy amateurs or supported by private means, because traditional astronomers were resistant to the newcomer. Simon Newcomb, the dean of classical astronomy in the United States in the late nineteenth century, noted that "the cultivators of the older astronomy have sometimes looked askance upon this youthful competitor as upon one that has not yet attained the dignity of the older science and have therefore been quite satisfied to make a distinction between the two classes, that of astronomers and that of astrophysicists."[3] Not until 1916 in the United States did the two camps fuse, when the original Astronomical and Astrophysical Society of America became simply the American Astronomical Society.

Other innovations, such as photography, accelerated astronomy's transformation. The Sun was the first popular subject for imaging, since its bright luminosity was necessary when early photographic processes were slow and inefficient. But photographic plates were later made more sensitive to record the faint light from stars and nebulae. The first photograph of a star (Vega), a 100-second exposure, was taken by J. A. Whipple and William C. Bond at the Harvard College Observatory on July 17, 1850. Within a few decades, the first photographic survey of the heavens was organized. Plate holders and spectrographs became standard equipment at observatories, changing an astronomer's lifestyle. No longer restricted to nighttime hours, astronomers could carefully examine images and spectra at their leisure, with timed exposures extending the limitations of their eyes alone.

Along with the introduction of photography, telescopes themselves underwent tremendous improvements. Refractors grew from 9½-inch apertures to 40 inches, the diameter of the lens built for the Yerkes Observatory of the University of Chicago in 1897. At the same time, the technology of reflecting telescopes advanced—so greatly that it soon became the telescope of choice for serious astronomical work. Mirrors were originally constructed of metal, which was hard to shape. But once a method was developed to deposit silver on glass, the sizes of reflectors grew rapidly, reaching 100 inches at the Mount Wilson Observatory in California by 1917.

All these technological advancements in the nineteenth century—spectroscopy, bigger telescopes, photography—helped move astronomy from its traditional practice of tracking celestial objects toward discerning and understanding their physical nature.

23 / Spectral Lines

In 1814 Joseph Fraunhofer constructed the first astronomical spectroscope, an invention that would come to revolutionize the field of astronomy as much as Galileo's telescope. It allowed astronomers to directly assess the physical characteristics of celestial bodies—in effect, to discern the chemistry of the heavens.

The origins of spectroscopy can be traced to 1666, when Isaac Newton, at the age of twenty-three, sat in a darkened room and let a small stream of sunlight enter through a hole in his window shutter and pass through a triangular prism of glass. On the wall behind him, Newton beheld a rainbow of colors—red, orange, yellow, green, blue, and violet—a captivating effect that had been observed with pieces of glass since antiquity. Newton demonstrated that white light was a mixture of these hues, with each color getting bent, or refracted, by the glass to a different degree. This allowed the white light to be separated into its varied components. In his report to the *Philosophical Transactions* (the first major scientific discovery announced in a journal rather than a book), Newton dubbed the multicolored display a *spectrum*, Latin for "apparition" or "spector."[4]

Newton's experiments were not greatly extended until 1802, when the English experimentalist William Hyde Wollaston sent sunlight through a narrow slit instead of a pinhole. Examining the light through a prism, he found several black lines in the solar spectrum.[5] He just assumed these dark gaps marked off the natural boundaries of light's various colors. Fraunhofer, a Bavarian instrument maker, would prove that was not the case.

Orphaned as a young boy, Fraunhofer was apprenticed to a tyrannical looking-glass manufacturer, who overworked and underfed the frail lad. Fraunhofer was liberated in 1801 at the age of fourteen when his slum tenement house collapsed and he was rescued from the wreckage as the sole survivor. Given a large sum of money from a sympathetic government official, the gifted teenager was able to pursue optics on his own. Becoming a master optician within a decade, Fraunhofer began studying prisms in the hope of getting rid of the rainbow effect in lenses, their tendency to produce fringes of color in an image. In the course of this investigation, he constructed the first spectroscope—a small telescope focused on a narrow slit, through which sunlight entered and passed through a prism. To his surprise, when he looked through the telescopic eyepiece, he saw *hundreds* of lines, both strong and weak, in the solar spectrum; some of the lines were almost perfectly black, as if ebony threads had been sewn across a rainbow. Following Wollaston's lead, he labeled the most prominent solar lines with letters, starting with A at the red end of the spectrum to H and I in the violet. Today we know them as Fraunhofer lines. The position of the D lines, he noticed, seemed to coincide with a bright double line in the orange that appeared in the spectra of laboratory flames.

Fraunhofer eventually convinced himself (but not others) that the dark streaks were part of the sunlight itself and not just an optical or atmospheric effect. He went on to use his spectroscope on the Moon and bright planets, which displayed the same fixed lines in their spectra as the Sun. This proved, once and for all, that planetary bodies shine by reflected sunlight. But when aiming his spectroscope at the most luminous stars, such as Castor, Betelgeuse, and Sirius, he noticed they exhibited their own unique spectral "fingerprints." Fraunhofer reported, for example, that the star Sirius had "three broad bands which appear to have no connection with those of sunlight; one of these bands is in the green, two are in the blue."[6]

What was the reason for these mysterious dark lines in the solar and stellar spectra? Fraunhofer did not have time to find out. Sickly from his early years of deprivation, he died of tuberculosis at the age of thirty-nine. On his tombstone in Munich are carved the words *Approximavit sidera*, "He approached the stars." Three decades would pass before the true nature of the dark lines would be revealed, leading to the birth of astrophysics (see Chapter 24).

Denkschriften der königlichen Akademie der Wissenschaften zu München [Memoranda of the Royal Academy of Sciences in Munich], Volume 5 (1817) and the *Edinburgh Journal of Science,* Volume 8 (1828) by Joseph Fraunhofer

. . . In the window-shutter of a darkened room I made a narrow opening— about 15 seconds broad and 36 minutes high—and through this I allowed sunlight to fall on a prism of flint-glass which stood upon the theodolite described before. The theodolite was 24 feet from the window, and the angle of the prism was about 60°. The prism was so placed in front of the objective of the theodolite-telescope that the angle of incidence of the light was equal to the angle at which the beam emerged. I wished to see if in the color-image from sunlight there was a bright band similar to that observed in the color-image of lamplight. But instead of this I saw with the telescope an almost countless number of strong and weak vertical lines, which are, however, darker than the rest of the color-image; some appeared to be almost perfectly black. If the prism was turned so as to increase the angle of incidence, these lines vanished; they disappear also if the angle of incidence is made smaller. For increased angle of incidence, however, these lines become visible again if the telescope is made shorter; while, for a smaller angle of incidence, the eye-piece must be drawn out considerably in order to make the lines reappear. If the eye-piece was so placed that the lines in the red portion of the color-image could be plainly seen, then, in order to see the lines in the violet portion, it must be pushed in slightly. If the opening through which the light entered was made broader, the fine lines ceased to be clearly seen, and vanished entirely if the opening was made 40 seconds wide. If the opening was made 1 minute wide, even the broad lines could not be seen plainly. The distances apart of the lines, and all their relations to each other, remained unchanged, both when the width of the opening in the window-shutter was altered and when the distance of the theodolite from the opening was changed. The prism could be of any

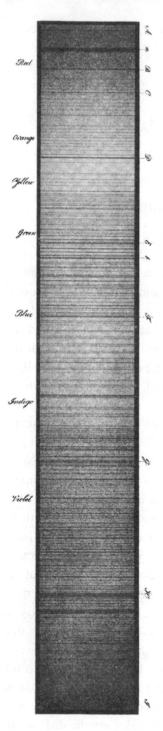

Figure 23.1

kind of refractive material, and its angle might be large or small; yet the lines remained always visible, and only in proportion to the size of the color-image did they become stronger or weaker, and therefore were observed more easily or with more difficulty.

The relations of these lines and streaks among themselves appeared to be the same with every refracting substance; so that, for instance, one particular band is found in every case only in the blue; another is found only in the red; and one can, therefore, at once recognize which line he is observing. These lines can be recognized also in the spectra formed by both the ordinary and the extraordinary rays of Iceland spar. The strongest lines do not in any way mark the limits of the various colors; there is almost always the same color on both sides of a line, and the passage from one color into another cannot be noted.

With reference to these lines the color-image is as shown in [Figure 23.1]. It is, however, impossible to show on this scale all the lines and their intensities. (The red end of the color-image is in the neighborhood of A; the violet end is near I.) It is, however, impossible to set a definite limit at either end, although it is easier at the red than at the violet. Direct sunlight, or sunlight reflected by a mirror, seems to have its limits on the one hand, somewhere between G and H; on the other, at B; yet with sunlight of great intensity the color-image becomes half again as long. In order, however, to see this great spreading-out of the spectrum, the light from the space between C and G must be prevented from entering the eye, because the impression which the light from the extremities of the color-image makes upon the eye is very weak, and is destroyed by the rest of the light. At *A* there is easily recognized a sharply defined line; yet this is not the limit of the red color, for it proceeds much beyond. At *a* there are heaped together many lines which form a band; *B* is sharply defined and is of noticeable thickness. In the space between *B* and *C* there can be counted 9 very fine, sharply defined lines. The line *C* is of considerable strength, and, like *B,* is very black. In the space between *C* and *D* there can be counted 30 very fine lines; but these (with two exceptions), like those between *B* and *C,* can be plainly seen only with strong magnification or with prisms which have great dispersion; they are, moreover, very sharply defined. *D* consists of two strong lines which are separated by a bright line. Between *D* and *E* there can be counted some 84 lines of varying intensities. *E* itself consists of several lines, of which the one in the middle is somewhat stronger than the rest. Between *E* and *b* are about 24 lines. At *b* there are 3 very strong lines, two of which are separated by only a narrow bright line; they are among the strongest lines in the spectrum. In the space between *b* and *F* there can be counted about 52 lines; *F* is fairly strong. Between *F* and *G* there are about 185 lines of different strengths. At *G* there are massed together many lines, among which several are distinguished by their intensity. In the space between *G* and *H* there are about 190 lines, whose intensities differ greatly. The two bands at *H* are most remarkable; they are almost exactly equal, and each consists of many lines; in the middle of each there is a strong line which is very black. From *H* to *I* the lines are equally numerous. . . .

I have convinced myself by many experiments and by varying the methods that these lines and bands are due to the nature of sunlight, and do not arise from diffraction, illusion, etc. If light from a lamp is allowed to pass through the same narrow opening in the window-shutter, none of these lines are observed, only the bright line *R,* which, however, comes exactly in the same place as the line *D,* so that the indices of refraction of the rays *D* and *R* are the same. The reason why the lines fade away, or even entirely vanish, when the opening at the window is made too wide is not difficult to give. The stronger lines have a width of from five to ten seconds; so, if the opening of the window is not so narrow that the light which passes through can be regarded as belonging to one ray, or if the angular width of the opening is much more than that of the line, the image of one and the same line is repeated several times side by side, and consequently becomes indistinct, or vanishes entirely if the opening is made too wide.

[*Spectra of Venus and Sirius*] I applied this form of apparatus at night-time to observe Venus directly, *without making the light pass through a small opening;* and I discovered in the spectrum of this light the same lines as those which appear in sunlight. Since, however, the light from Venus is feeble in comparison with sunlight reflected from a mirror, the intensity of the violet and the extreme red rays are very weak; and on this account even the stronger lines in both these colors are recognized only with difficulty, but in the other colors they are very easily distinguished. I have seen the lines *D, E, b, F* perfectly defined, and have even recognized that *b* consists of two lines, one weak and one strong; but the fact that the stronger one itself consists of two I could not verify owing to lack of light. For the same reason the other finer lines could not be distinguished satisfactorily. I have convinced myself by an approximate measurement of the arcs *DE* and *EF* that the light from Venus is in this respect of the same nature as sunlight.

With this same apparatus I made observations also on the light of some fixed stars of the first magnitude. Since, however, the light of these stars is much weaker than that of Venus, it is natural that the brightness of the spectrum should be much less. In spite of this I have seen with certainty in the spectrum of Sirius three broad bands which appear to have no connection with those of sunlight; one of these bands is in the green, two are in the blue. In the spectra of other fixed stars of the first magnitude one can recognize bands; yet these stars, with respect to these bands, seem to differ among themselves. Since the objective of the telescope has an aperture of only 13 lines [1 centimeter equals 4.43296 lines], it is clear that these observations can be repeated with much greater accuracy. I intend to repeat them with suitable alterations, and with a larger objective, in order to induce, perhaps, some skilled investigator to continue the experiments. Such a continuation is all the more to be desired, because the experiments would serve at the same time for the accurate comparison of the refraction of the light of the fixed stars with that of sunlight.

[*Spectra of the Moon and Starlight*] As is well known, the prismatic *color-spectrum* of the light coming from a *flame* (*lamplight*) does not show the dark fixed lines which are present in the spectrum of sunlight; instead of them there is in the orange a bright line which is prominent above the rest of the spectrum, is double, and is at the same place where in sunlight the double line *D* is found. The spectrum obtained from the light of a flame which is blown with a *blast-tube* contains several prominent bright lines.

Of still greater interest for optical experiments is the fact that, by skillful blowing of the flame, the light of the *front half* of the flame can be dispersed no further by the prism, and, consequently, is simple homogeneous light. This light has, so far as I have investigated it, the same refrangibility as the *D* ray of sunlight. Simple homogeneous light which proceeds in all directions is, for known reasons, very difficult to produce, and can never be obtained with prisms directly; therefore, this flame is of great use in many experiments. . . .

The light of the *moon* gave me a spectrum which showed in the brightest colors the same fixed lines as did sunlight, and in exactly the same places.

To observe the *spectra of the light of the fixed stars,* and at the same time to *determine the refrangibility of this light,* I prepared a short time ago a suitable apparatus specially adapted to this end, the telescope belonging to it having an objective of 4 inches' aperture. . . .

Up to the present we have found no *fixed star* whose light, so far as its *refrangibility* is concerned, is sensibly different from that of the *planets.* When the fixed lines of the spectra are seen plainly, one can be certain with this instrument to 10 seconds; and when the fixed lines cannot be seen, one can still be certain for the *orange light* to $\frac{1}{2}$ minute. Since the total refraction through the prism is 26°, a difference amounting to $\frac{1}{9360}$ of the whole refraction could still be noticed with this instrument, a difference which even with the horizontal refraction in the atmosphere did not amount to $\frac{1}{4}$ second. Up to this time, as is well known, some astronomers have doubted whether the refraction tables for different stars should not be somewhat different; therefore, this doubt seems to be removed by the experiment noted. The continuation of this investigation will lead us, I hope, to more complete knowledge.

In order to see the *fixed lines of* the different stars (with this large instrument) the air must be most favorable—a condition which happens rarely to a sufficient extent. The spectra of the light from *Mars* and *Venus* contain the same fixed lines as does sunlight, and in exactly the same places, at least so far as the lines *D, E, b,* and *F* are concerned, whose relative positions can be exactly determined. In the spectrum of the light from *Sirius* I could not distinguish fixed lines in the orange and yellow; in the green, however, there is seen a very strong streak; and in the blue there are two other unusually strong streaks, which seem to be unlike any of the lines of planetary light. We have determined their positions with the micrometer. *Castor* gives a spectrum which is like that of Sirius; the streak in the green, in spite of the weak light, was intense enough for me to be able to measure it; and I found it in exactly the same place as it was with Sirius. I could also distinguish the streaks in the blue; but the light was too feeble to allow of measurement. In the spectrum of *Pollux,* I recognized many fixed lines which resembled those of Venus; but all were weak. I saw the *D* line quite plainly, in exactly the same position as with planetary light. *Capella* gives a spectrum in which, at the places *D* and *b,* the same fixed lines are seen as in sunlight. The spectrum of *Betelgeux* (α Orionis) contains countless fixed lines which, with a good atmosphere, are sharply defined; and, although at first sight it seems to have no resemblance to the spectrum of Venus, yet similar lines are found in the spectrum of this fixed star in exactly the places where with sunlight *D* and *b* come. Some lines can be distinguished in the spectrum of *Procyon;* but they are seen with difficulty, and so indistinctly that their positions cannot be determined with certainty. I think I saw a line at the position *D* in the orange.

24 / Deciphering the Solar Spectrum

Starting in the mid-eighteenth century, chemists began to notice that hot flames contaminated with metals or salts produced special kinds of spectra—discrete lines of color, resembling a picket fence with colorful posts. Whereas Joseph Fraunhofer discovered that the solar spectrum was a continuous rainbow riddled with dark lines (see Chapter 23), these laboratory spectra were the exact opposite: bright lines set against a dark background. Was there a connection?

Work on this problem proceeded in a number of countries and was finally deciphered in Germany around 1859 by Gustav Kirchhoff, a professor of physics at the University of Heidelberg, and chemist Robert Bunsen, creator of the famous laboratory burner. Bunsen had become interested in identifying substances by the specific light they emitted during chemical reactions or when burning. Kirchhoff, Bunsen's friend and colleague at Heidelberg, suggested that he use a prism and slit—a spectroscope—to distinguish the colorful emissions with more assurance. It was the start of a fruitful collaboration.

With the clear hot flame of Bunsen's improved burner, free of the contaminations that misled earlier researchers, it soon became apparent that each chemical element did indeed produce a characteristic pattern of colored lines when heated and viewed through a spectroscope. The two investigators even came across spectra never before recorded, which led to their discovery of two new elements, the metals cesium (from the Latin for "bluish gray," the color of its most prominent lines) and rubidium (Latin for "red," its distinctive spectral feature).

A dramatic event turned their attention to the stars. Using their spectroscope one evening to peer at a distant fire, visible across the Rhine plain from their laboratory window, Kirchhoff and Bunsen detected the spectral signatures of barium and strontium in the roaring blaze. Afterward, Bunsen wondered if they could analyze the Sun's light in a comparable manner. Other scientists, such as George Stokes in Great Britain and Jean Foucault in France, had expressed similar suspicions. Light knows no distance in space; electromagnetic waves can be effectively studied whether the light originates from a distance of 1 foot or 1 million light-years.

In the course of these investigations, Kirchhoff carried out a series of elegant experiments, arranging the equipment in such a way that he could simultaneously compare the solar and laboratory spectra. From this and other research, he concluded that the dark lines discovered by Fraunhofer were generated as each element in the Sun's cooler outer atmosphere *absorbed* specific wavelengths from the Sun's hot inner glow. The dark lines, consequently, came to be called "absorption lines." Elements in the Sun's cool layers, in a sense, were robbing the sunshine of selected wavelengths before the light continued on its journey outward. The bright lines observed in laboratory flames are simply the reverse of this process: the elements emitting those select wavelengths of light as they fiercely burn. That explained why the dark Fraunhofer D lines, a distinctive double in the solar spectrum, precisely matched two bright lines emitted by the element sodium. A substance that emits specific wavelengths of light, said Kirchhoff, can also absorb them under the right conditions. (The exact mechanism behind this effect would not be understood until the twentieth century with the development of atomic theory.)

By matching the pattern of bright lines emitted by a substance heated in a laboratory with the dark lines observed in the solar spectrum, Kirchhoff was able to identify a number of elements in the sun's atmosphere; besides sodium, there was also iron, calcium, magnesium, chromium, barium, copper, zinc, and nickel. Here was definitive proof that the chemistry of the Earth was identical to the chemistry of the heavens. The long-standing Aristotelian belief that cosmic matter differed from the terrestrial elements was finally abolished. Others had suspected that was the case, but it had seemed impossible to test. The work of Bunsen and Kirchhoff at last provided the means, an achievement that marked the birth of astrophysics. "Spectrum analysis, which . . . offers a wonderfully simple means for discovering the smallest traces of certain elements in terrestrial substances," reported Kirchhoff and Bunsen in 1860, "also opens to chemical research a hitherto completely closed region extending far beyond the limits of the earth and even of the solar system."[7]

Within a few years the most adventurous astronomers, particularly William Huggins in Great Britain (see Chapter 25), began to utilize the spectroscope regularly and extend Kirchhoff's spectral interpretations to the stars. In 1863, Huggins and his colleague William Miller reported the spectral lines they observed in the bright stars Sirius, Betelgeuse, and Aldebaran.[8] At the same time, similar observations were carried out by Lewis Rutherfurd in the United States,[9] Hermann Vogel in Germany, and Angelo Secchi in Italy.[10] Practitioners of spectroscopy would lead astronomy into rich and fertile new territories. In the twentieth century, spectral analysis allowed astronomers to discern the evolution of stars, to reveal the source of stellar power, and to unmask a universe where billions of galaxies are speeding away from one another in a grand cosmic expansion.

From *Researches on the Solar Spectrum and the Spectra of the Chemical Elements* by Gustav Kirchhoff

Translated by Henry E. Roscoe

The dark lines of the solar spectrum afford invaluable assistance in determining the position of the bright lines of the various elementary bodies. In order to make use of these dark lines I have fixed on to the upper half of the slit in the apparatus above described two small rectangular glass prisms, so arranged that whilst direct sunlight can enter the lower half of the slit, the rays from an artificial source of light placed at one side can reach the large prisms after twice suffering total reflection. The small prisms were placed upon each other so that their hypotenuse faces were parallel, and after the one had been turned round the axis perpendicular to the surfaces in contact, through an angle of about 15°, they were cemented together with rosin, and in this position fastened before the slit. . . . In this way, whilst in the upper half of the field of the (astronomical) telescope the solar spectrum is seen, in the lower half, but in immediate contact with the other, the spectrum of the artificial source of light becomes apparent, and the positions of the bright lines in the later spectrum can be accurately compared with those of the dark lines in the solar spectrum. In order to obtain the spectra of the metals, I have almost invariably employed the electric spark, chiefly owing to its great luminous intensity. . . .

In the course of the experiments already alluded to, which Foucault instituted on the spectrum of the electric arc formed between the carbon points, this physicist observed that the bright sodium lines present were changed into dark bands in the spectrum produced by the light from one of the carbon poles, which had been allowed to pass through the luminous arc; and when he passed direct sunlight through the arc he noticed that the double D line was seen with an unusual degree of distinctness. No attempt was made to explain or to increase these observations either by Foucault or by any other physicist, and they remained unnoticed by the greatest number of experimentalists. They were unknown to me when Bunsen and I, in the year 1859, commenced our investigations on the spectra of colored flames.

In order to test in the most direct manner possible the truth of the frequently asserted fact of the coincidence of the sodium lines with the lines $D,$ I obtained a tolerably bright solar spectrum, and brought a flame colored by sodium vapor in front of the slit. I then saw the dark lines D change into bright ones. The flame of a Bunsen's lamp threw the bright sodium lines upon the solar spectrum with unexpected brilliancy. In order to find out the extent to which the intensity of the solar spectrum could be increased, without impairing the distinctness of the sodium lines, I allowed the full sunlight to shine through the sodium flame upon the slit, and, to my astonishment, I saw that the dark lines D appeared with an extraordinary degree of clearness. I then exchanged the sunlight for the Drummond's or oxyhydrogen limelight, which, like that of all incandescent solid or liquid bodies, gives a spectrum containing no dark lines. When this light was allowed to fall through a suitable flame colored by common salt, dark lines were seen in the spectrum in the position of the sodium lines. The same phenomenon was observed if instead of the incandescent lime a platinum wire was used, which being heated in a flame was brought to a temperature near to its melting point by passing an electric current through it.

The phenomenon in question is easily explained upon the supposition that the sodium flame absorbs rays of the same degree of refrangibility as those it emits, whilst it is perfectly transparent for all other rays. This supposition is rendered probable by the fact, which has long been known, that certain gases, as for instance, nitrous acid and iodine vapor, possess at low temperatures the property of such a selective absorption. The following considerations show that this is the true explanation of the phenomenon. If a sodium flame be held before an incandescent platinum wire whose spectrum is being examined, the brightness of the light in the neighborhood of the sodium lines would, according to the above supposition, *not* be altered; in the position of the sodium lines themselves, however, the brightness *is* altered, for two reasons; in the first place, the intensity of light emitted by the platinum wire is reduced to a certain fraction of its original amount by absorption in the flame, and secondly, the light of the flame itself is added to that from the wire. It is plain that if the platinum wire emits a sufficient amount of light, the loss of light occasioned by absorption in the flame must be greater than the gain of light from the luminosity of the flame; the sodium lines must then appear darker than the surrounding parts, and by contrast with the neighboring parts they may seem to be quite black, although their degree of luminosity is necessarily greater than that which the sodium flame alone would have produced.

The absorptive power of sodium vapor becomes most apparent when its luminosity is smallest, or when its temperature is lowest. In fact we were unable to produce the dark sodium lines in the spectrum of a Drummond's light, or in that of an incandescent wire, by means of a Bunsen's gas-flame in which common salt was placed; but the experiment succeeded with a flame of aqueous alcohol containing common salt. The following experiment proposed by Crookes likewise very clearly shows this influence of temperature. If a piece of sodium is burnt in a room, and the air thus filled with the vapor of sodium compounds, every flame is seen to burn with the characteristic yellow light. If a small flame in which a bead of soda salt is placed be now fixed in front of a large one, so that the former is seen projected on the latter as a background, the small flame appears to be surrounded with a black smoky mantle. This dark mantle is produced by

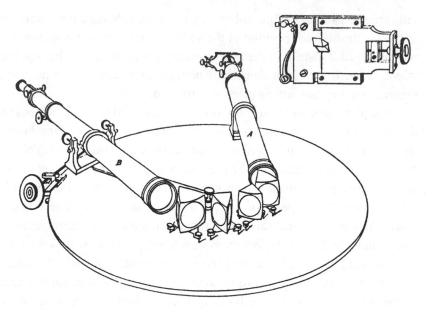

Figure 24.1: The apparatus employed by Kirchhoff to observe the solar spectrum.

the absorptive action of the sodium vapors in the outer part of the flame, which are cooler than those in the flame itself. Bunsen and I have produced the dark lines in the spectrum of a common candle-flame, by allowing the rays to pass through a test tube containing a small quantity of sodium-amalgam, which we heated to boiling; so that the sodium vapor effecting the absorption had in this case possessed a temperature far below the red-heat. The same phenomenon is observed in a much more striking manner if a glass tube is used containing some small pieces of sodium first filled with hydrogen, and then rendered vacuous and sealed. The lower end of the tube can be heated so as to vaporize the sodium. By means of this arrangement, which was proposed by Roscoe, the heated vapor of the sodium, when viewed by the sodium-light, is seen as a dark black smoke which throws a deep shadow, but is perfectly invisible when observed by the ordinary gaslight. . . .

The sodium flame is characterized beyond that of any other colored flame by the intensity of the lines in its spectrum. Next to it in this respect comes the lithium flame. It is just as easy to reverse the red lithium line, that is, to turn the bright line into a dark one, as it is to reverse the sodium line. If direct sunlight be allowed to pass through a lithium flame, the spectrum exhibits in the place of the red lithium band a black line which in distinctness bears comparison with the most remarkable of Fraunhofer's lines, and disappears when the flame is withdrawn. It is not so easy to obtain the reversal of the spectra of the other metals; nevertheless Bunsen and I have succeeded in reversing the brightest lines of potassium, strontium, calcium, and barium, by exploding mixtures of the chlorates of these metals and milk-sugar in front of the slit of our apparatus whilst the direct solar rays fell on the instrument.

These facts would appear to justify the supposition that each incandescent gas diminishes by absorption the intensity of those rays only which possess degrees of refrangibility equal to those of the rays which it emits; or, in other words, that the spectrum of every incandescent gas must be reversed, when it is penetrated by the rays of a source of light of sufficient intensity giving a continuous spectrum. . . .

. . . It is especially remarkable that, coincident with the positions of all the bright iron lines which I have observed, well-defined dark lines occur in the solar spectrum. By the help of the very delicate method of observation which I have employed, I believe that each coincidence observed by me between an iron line and a line in the solar spectrum, may be considered to be at least as well established as the coincidence of the sodium lines and the lines *D* was up to the present time. . . .

As soon as the presence of *one* terrestrial element in the solar atmosphere was thus determined, and thereby the existence of a large number of Fraunhofer's lines explained, it seemed reasonable to suppose that other terrestrial bodies occur there, and that by exerting their absorptive power, they may cause the production of other Fraunhofer's lines. For it is very probable that elementary bodies which occur in large quantities on the earth, and are likewise distinguished by special bright lines in their spectra, will, like iron, be visible in the solar atmosphere. This is found to be the case with calcium, magnesium, and sodium. . . .

25 / Gaseous Nebulae

The mystery of the nebulae, pale patches of light scattered over the nighttime sky, occupied astronomers for some two hundred years. In the 1700s Immanuel Kant and others wondered if they were separate islands of stars, distant congregations similar to our own Milky Way galaxy. When the British astronomer William Herschel surveyed the heavens with his telescope, unmatched at the time for its superior magnification, he discovered hundreds more than had ever been known. Like Kant, he at first believed they were distant collections of stars, but reversed his opinion when he came across certain clouds—what he dubbed planetary nebulae— that he concluded were composed not of stars but of luminous matter (see Chapter 22). This was not direct proof, though, only a visual assessment.

In the mid-1800s, the problem remained as vexing. Astronomers generally recognized that there were irregular clouds and planetary nebulae that tended to be situated in the plane of the Milky Way (and so labeled galactic nebulae) and then there were others that primarily crowded around the poles of the Milky Way, away from its plane. These came to be called extragalactic nebulae, a name that would take on a far deeper significance in the twentieth century (see Chapter 51). Were these nebulae in the end swarms of suns, too distant to distinguish, as first suspected? Or were they, as Herschel later suggested, regions of diffuse glowing matter within the Milky Way?

Using the new technique of spectroscopy, William Huggins in 1864 was able to conclusively prove that a large fraction of the nebulae were composed of gas after all.

In its early days, spectroscopy was a technique particularly favored by amateur astronomers, such as Huggins, who lacked formal training in classical astronomy. Passionate about science since childhood, Huggins sold his mercer's business at the age of thirty in 1854 and turned to astronomy full time at his private observatory at Tulse Hill, then a rural area south of London. Soon tiring of routine tasks, such as transits and planet drawings, he was inspired by news of Kirchhoff and Bunsen's spectroscopic discoveries (see Chapter 24), which he compared to "coming upon a spring of water in a dry and thirsty land."[11] His first success in 1862, determining that the elements found in the Sun also dwelled in the distant stars, was done in collaboration with W. Allen Miller, professor of chemistry at King's College, who assisted Huggins with his instrumentation.[12] It was an impressive accomplishment, since the light arriving from a bright star such as Vega is less than a billionth of the radiation received from the Sun. "The chemistry of the solar system was shown to prevail . . . ," wrote Huggins, "wherever a star twinkles."[13]

In 1864, Huggins shifted his focus from stars to nebulae. On the evening of August 29, he aimed his telescope at a roundish nebula in the Draco constellation. In a memoir, he remembered feeling "excited suspense, mingled with a degree of awe" as he put his eye to the spectroscope.[14] The spectrum he beheld was a surprise: "A single bright line only!" he recalled. "At first I suspected some displacement of the prism, and that I was looking at a reflection of the illuminated slit. . . . This thought was scarcely more than momentary; then the true interpretation flashed upon me. . . . The riddle of the nebulae was solved. The answer, which had come to us in the light itself, read: Not an aggregation of stars, but a luminous gas."[15] His report to the Royal Society of London included spectral descriptions of Draco, five other planetary nebulae, and the Dumbbell nebula. He also listed six other objects whose spectral signatures differed from them. Within four years, Huggins examined around seventy nebulae. One-third were clearly composed of gas, while the remaining nebulae displayed starlike spectra. Up until that time, astronomers generally believed that nebulae were either one thing or another; Huggins discovered there were two classes, and the spectroscope offered the means to distinguish between the two, when a telescope alone was not sufficient to denote the difference.

Along the way (in collaboration with his wife and observational partner, Margaret), Huggins detected a number of spectral lines that he could not identify, which he boldly suggested might indicate the presence of matter not yet recognized by earthbound chemists. There were mysterious lines, for example, in the green band of the spectrum that gave such objects as the Orion nebula their ghostly greenish glow. Huggins postulated a new element, which he dubbed "nebulium," as the source of this pale green light. Huggins died in 1910 at the age of eighty-six never knowing that nebulium was not a new element after all. In 1928 Ira Bowen, an astrophysicist with the California Institute of Technology, reported that Huggins's green-tinged radiation was being emitted by oxygen and nitrogen atoms that had lost some of their electrons and become highly ionized.[16]

"On the Spectra of Some of the Nebulae."
Philosophical Transactions of the Royal Society of London, Volume 154 (1864)
by William Huggins

. . . Prismatic analysis, if it could be successfully applied to objects so faint, seemed to be a method of observation specially suitable for determining whether any essential physical distinction separates the nebulae from the stars, either in the nature of the matter of which they are composed, or in the conditions under which they exist as sources of light. The importance of bringing analysis by the prism to bear upon the nebulae is seen to be greater by the consideration that increase of optical power alone would probably fail to give the desired information; for, as the important researches of Lord Rosse have shown, at the same time that the number of the clusters may be increased by the resolution of supposed nebulae, other nebulous objects are revealed, and fantastic wisps and diffuse patches of light are seen, which it would be assumption to regard as due in all cases to the united glare of suns still more remote.

Some of the most enigmatical of these wondrous objects are those which present in the telescope small round or slightly oval disks. For this reason they were placed by Sir William Herschel in a class by themselves under the name of Planetary Nebulae. They present but little indication of resolvability. The color of their light, which in the case of several is blue tinted with green, is remarkable, since this is a color extremely rare amongst single stars. These nebulae, too, agree in showing no indication of central condensation. By these appearances the planetary nebulae are specially marked as objects which probably present phenomena of an order altogether different from those which characterize the sun and the fixed stars. On this account, as well as because of their brightness, I selected these nebulae as the most suitable for examination with the prism. . . .

. . . The numbers and descriptions of the nebulae, and their places for the epoch 1860, January 0, included within brackets, are taken from the last Catalogue of Sir John Herschel.

[No. 4373. 37 H. IV. R.A. $17^h 58^m 20^s$. N.P.D. $23° 22' 9''.5$. A planetary nebula; very bright; pretty small; suddenly brighter in the middle, very small nucleus.] In Draco.

On August 29, 1864, I directed the telescope armed with the spectrum apparatus to this nebula. At first I suspected some derangement of the instrument had taken place; for no spectrum was seen, but only a short line of light perpendicular to the direction of dispersion. I then found that the light of this nebula, unlike any other ex-terrestrial light which had yet been subjected by me to prismatic analysis, was not composed of light of different refrangibilities, and therefore could not form a spectrum. A great part of the light from this nebula is monochromatic, and after passing through the prisms remains concentrated in a bright line occupying in the instrument the position of that part of the spectrum to which its light corresponds in refrangibility. A more careful examination with a narrower slit, however, showed that, a little more refrangible than the bright line, and separated from it by a dark interval, a narrower and much fainter line occurs. Beyond this, again, at about three times the distance of the second line, a third, exceedingly faint line was seen. The positions of these lines in the spectrum were determined by a simultaneous comparison of them in the instrument with the spectrum of the induction spark taken between electrodes of magnesium. The strongest line coincides in position with the brightest of the air lines. This line is due to nitrogen, and occurs in the spectrum about midway between *b* and *F* of the solar spectrum. . . .

The color of this nebula is greenish blue.

[No. 4390. 2000 h. Σ 6. R.A. 18^h 5^m $17^s.8$. N.P.D. 83° 10' 53".5. A planetary nebula; very bright; very small; round; little hazy.] In Taurus Poniatowskii.

The spectrum is essentially the same as that of No. 4373.

The three bright lines occupy the same positions in the spectrum, which was determined by direct comparison with the spectrum of the induction spark. These lines have also the same relative intensity. They are exceedingly sharp and well defined. The presence of an extremely faint spectrum was suspected. In connection with this it is important to remark that this nebula does not possess a distinct nucleus.

The color of this nebula is greenish blue.

[Omitted here are similar spectral descriptions of five other planetary nebulae, designated 73 H. IV., 51 H. IV., 1 H. IV., 57 M., and 18 H. IV., and the Dumbbell nebula, 27 M., in the constellation Vulpecula.]

In addition to these objects the following were also observed:—

[No. 4294. 92 M. R.A. 17^h 12^m $56^s.9$. N.P.D. 46° 43' 31".2] In Hercules. Very bright globular cluster of stars. The bright central portion was brought upon the slit. A faint spectrum similar to that of a star. The light could be traced from between C and D to about G.

Too faint for the observation of lines of absorption.

[No. 4244. 50 H. IV. R.A. 16^h 43^m $6^s.4$ N.P.D. 42° 8' 38".8. Very bright; large; round.] In Hercules. The spectrum similar to that of a faint star. No indication of bright lines.

[No. 116. 50 h. 31 M. R.A. 0^h 35^m $3^s.9$. N.P.D. 49° 29' 45".7.] The brightest part of the great nebula in Andromeda was brought upon the slit.

The spectrum could be traced from about D to F. The light appeared to cease very abruptly in the orange; this may be due to the smaller luminosity of this part of the spectrum. No indication of the bright lines.

[No. 117. 51 h. 32 M. R.A. 0^h 35^m $5^s.3$. N.P.D. 49° 54' 12".7. Very very bright; large; round; pretty suddenly much brighter in the middle.]

This small but very bright companion of the great nebula in Andromeda presents a spectrum apparently exactly similar to that of 31 M.

The spectrum appears to end abruptly in the orange; and throughout its length is not uniform, but is evidently crossed either by lines of absorption or by bright lines.

[No. 428. 55 Androm. R.A. 1^h 44^m 55^s.9. N.P.D. 49° 57′ 41″.5. Fine nebulous star with strong atmosphere.] The spectrum apparently similar to that of an ordinary star.

[No. 826. 2618 h. 26 IV. R.A. 4^h 7^m 50^s.8. N.P.D. 103° 5′ 32″.2. Very bright cluster.] In Eridanus. The spectrum could be traced from the orange to about the blue. No indication of the bright lines.

Several other nebulae were observed, but of these the light was found to be too faint to admit of satisfactory examination with the spectrum apparatus.

It is obvious that the nebulae 37 H. IV., 6 Σ., 73 H. IV., 51 H. IV., 1 H. IV., 57 M., 18 H. IV. and 27 M. can no longer be regarded as aggregations of suns after the order to which our own sun and the fixed stars belong. . . . [We] find ourselves in the presence of objects possessing a distinct and peculiar plan of structure.

In place of an incandescent solid or liquid body transmitting light of all refrangibilities through an atmosphere which intercepts by absorption a certain number of them, such as our sun appears to be, we must probably regard these objects, or at least their photo-surfaces, as enormous masses of luminous gas or vapor. For it is alone from matter in the gaseous state that light consisting of certain definite refrangibilities only, as is the case with the light of these nebulae, is known to be emitted.

It is indeed *possible* that suns endowed with these peculiar conditions of luminosity may exist, and that these bodies are clusters of such suns. There are, however, some considerations, especially in the case of the planetary nebulae, which are scarcely in accordance with the opinion that they are clusters of stars.

Sir John Herschel remarks of one of this class, in reference to the absence of central condensation, "Such an appearance would not be presented by a globular space uniformly filled with stars or luminous matter, which structure would necessarily give rise to an apparent increase of brightness towards the center in proportion to the thickness traversed by the visual ray. We might therefore be inclined to conclude its real constitution to be either that of a hollow spherical shell or of a flat disk presented to us (by a highly improbable coincidence) in a plane precisely perpendicular to the visual ray." This absence of condensation admits of explanation, without recourse to the supposition of a shell or of a flat disk, if we consider them to be masses of glowing gas. For supposing, as we probably must do, that the whole mass of the gas is luminous, yet it would follow, by the law which results from the investigations of Kirchhoff, that the light emitted by the portions of gas beyond the surface visible to us, would be in great measure, if not wholly, absorbed by the portion of gas through which it would have to pass, and for this reason there would be presented to us a *luminous surface* only.

Sir John Herschel further remarks, "Whatever idea we may form of the real nature of the planetary nebulae, which all agree in the absence of central condensation, it is evident that the intrinsic splendor of their surfaces, *if continuous,* must be almost infinitely less than that of the sun. A circular portion of the sun's disk, subtending an angle of 1', would give a light equal to that of 780 full moons, while among all the objects in question there is not one which can be seen with the naked eye." The small brilliancy of these nebulae is in accordance with the conclusions suggested by the observations of this paper; for, reasoning by analogy from terrestrial physics, glowing or luminous gas would be very inferior in splendor to incandescent solid or liquid matter.

Such gaseous masses would be doubtless, from many causes, unequally dense in different portions; and if matter condensed into the liquid or solid state were also present, it would, from its superior splendor, be visible as a bright point or points within the disk of the nebula. These suggestions are in close accordance with the observations of Lord Rosse.

Another consideration which opposes the notion that these nebulae are clusters of stars is found in the extreme simplicity of constitution which the three bright lines suggest, whether or not we regard these lines as indicating the presence of nitrogen, hydrogen, and a substance unknown.

It is perhaps of importance to state that, except nitrogen, no one of thirty of the chemical elements the spectra of which I have measured has a strong line very near the bright line of the nebulae. If, however, this line were due to nitrogen, we ought to see other lines as well; for there are specially two strong double lines in the spectrum of nitrogen, one at least of which, if they existed in the light of the nebulae, would be easily visible. In my experiments on the spectrum of nitrogen, I found that the character of the brightest of the lines of nitrogen, that with which the line in the nebulae coincides, differs from that of the two double lines next in brilliancy. This line is more nebulous at the edges, even when the slit is narrow and the other lines are thin and sharp. The same phenomenon was observed with some of the other elements. We do not yet know the origin of this difference of character observable among lines of the same element. May it not indicate a physical difference in the atoms, in connection with the vibrations of which the lines are probably produced? The speculation presents itself, whether the occurrence of this one line only in the nebulae may not indicate a form of matter more elementary than nitrogen, and which our analysis has not yet enabled us to detect.

Observations on other nebulae which I hope to make, may throw light upon these and other considerations connected with these wonderful objects.

26 / Doppler Shifts and Spectroscopic Binaries

I n the latter half of the nineteenth century, conventional astronomers were more concerned with the positions of the stars than with determining their chemistry, and so were wary of the art of spectroscopy, just recently introduced. But they conceded there was a practical use for a spectroscope upon learning it could measure a star's motion in the "line of sight" (that is, either toward or away from our solar system).

In the 1840s the Austrian physicist Christian Doppler had surmised that the length of a wave, such as the tone of a sound or the color of a light wave, would be altered whenever the source of the wave moved. Doppler suggested that a star receding from us would have its light waves stretched out, making the star appear redder. Conversely, a star coming toward us would look more blue, as its light waves crowded together. Doppler wondered whether this explained the different colors of the stars. He was wrong (stellar wavelengths formerly invisible would just shift into the visible, maintaining the overall color), but the French physicist Hippolyte Fizeau realized that the dark absorption lines in a star's spectrum, first observed by Joseph Fraunhofer (see Chapter 23), would change position due to the Doppler effect.[17] The amount a line shifted would reveal exactly how fast the star was moving and in what direction. A shift toward the red meant the star was receding, while a shift toward the blue indicated it was approaching us. This straightforward transformation became one of the most valuable implements in astronomy's toolbox for studying the dynamics of the universe.

Pioneering spectroscopists, such as William Huggins in Great Britain and Angelo Secchi in Italy, made some valiant attempts to peg a star's velocity using a spectroscope and judging the spectral line shifts by eye alone, but the shimmering sky made these initial reckonings notoriously undependable. Determining doppler shifts became far more reliable with the introduction of astrophotography to record the spectrum.

The first successful measurements were carried out around 1890 by Hermann Vogel and Julius Scheiner at the relatively new Potsdam Astrophysical Observatory in Germany (one of the first built specifically for spectroscopic work) and by James Keeler at the Lick Observatory in California. Keeler chose planetary nebulae as his targets. The swiftest was a greenish cloud in the constellation Aquila, racing away from the Earth at nearly 40 miles a second.[18]

Vogel and his assistant focused on stars, publishing the velocities of fifty-one stars in 1892 after four years of observations. While carrying out this program, Vogel made his most spectacular discovery—spectroscopic binaries, double-star systems revealed only through a spectroscope. While monitoring the star Algol, he could see from the changing spectral shifts in a strong line produced by hydrogen that the star was periodically moving back and forth as a close invisible companion revolved about it. This was definitive proof that Algol was an eclipsing binary, which the British astronomers John Goodricke and Edward Pigott first suspected a century earlier (see Chapter 17). Soon after, Vogel found the star Spica to be a spectroscopic binary as well. Although the hidden companions in these binaries could not be seen directly, the spectral information allowed Vogel to determine a variety of details about each system: the diameters of both the visible star and its satellite, the distance between them, their masses, and their orbital velocities. Edward Pickering and Antonia Maury at the Harvard College Observatory independently discovered this new class of binaries around the same time by noticing periodic changes in the spectra of the stars Mizar and β Aurigae.[19] The spectral absorption lines would at times split into two, which they figured were generated as one star in the close binary moved toward the Earth, while the other receded, shifting each star's lines apart. Careful monitoring of such doppler shifts provided the means for astronomers in the late twenti-eth century to discover the first extrasolar planets (see Chapter 74) and also allowed observers in the 1920s to behold an entire universe expanding (see Chapter 52).

"On the Spectrographic Method of Determining the Velocity of Stars in the Line of Sight." *Monthly Notices of the Royal Astronomical Society,* Volume 52 (1892) by H. C. Vogel

The experiments made at Potsdam in 1887 showed that, as a result of the extremely sensitive photographic methods employed, a sufficiently great dispersion could be made use of to readily detect and measure the dis-placement of the spectral lines produced by the motion of the stars in the line of sight. It very soon became clear that the measurement of the stellar spectra admitted of a far greater exactness than the direct observations, and that the disturbances of the atmosphere—the chief cause of the difficulties of the direct method—exert their influence in a lesser degree on the photo-graph. The very numerous measurements on more than two hundred nega-tives of forty-seven stars, which are now available, have confirmed this result, and show further that the exactness of the measurements far sur-passes the expectations based on the first plates taken with a provisional apparatus, and that the definitive observations have reached a degree of accuracy which in some cases is surprising.

This great accuracy has been secured by an advantageous construction of the apparatus, by its very exact adjustment, and especially by the pecu-liar methods adopted in measuring the photographs. . . .

The first result of any importance which the spectrographic method furnished was the proof of the influence of the Earth's motion on the dis-placement, which the earlier direct observations had failed to show with certainty. I append here [in Table 26.1] a few examples [velocities are in miles].

A further result of the new method was the discovery of the changes in the motion of *Algol* [see paper below], and thereby the proof of the exis-tence of a dark satellite, for the determination of which the most delicate

Table 26.1

Date	Obs. Vel.	Earth's Vel.	Vel. of Star Relative to Sun
*α Aurigae**			
1888 Oct 6	−3.5	−15.4	+11.9
22	+2.9	−13.0	+15.9
24	+3.8	−12.6	+16.4
25	+3.5	−12.4	+15.9
28	+3.8	−11.8	+15.6
Nov 9	+9.2	−8.9	+18.1
Dec 1	+10.8	−2.9	+13.7
13	+15.6	+0.7	+14.9
1889 Jan 2	+20.2	+6.6	+13.6
Feb 5	+30.8	+14.3	+16.5
May 6	+33.2	+17.0	+16.2
Sep 15	−3.6	−16.8	+13.2
α Tauri			
1888 Oct 28	+18.1	−9.5	+27.6
Nov 10	+24.9	−5.9	+30.8
Dec 4	+30.6	+1.8	+28.8
1890 Jan 9	+43.7	+12.3	+31.4
α Ophiuchi			
1888 Sep 30	+27.6	+14.1	+13.5
1889 Jun 7	+9.7	−1.0	+10.7
α Ursae Majoris			
1888 Nov 7	−17.0	−11.9	−5.1
9	−18.1	−11.9	−6.2
1889 May 4	+5.2	+11.8	−6.6
22	+3.4	+11.2	−7.8

*A positive sign signifies the star is moving away from our Sun; with a negative sign it is approaching.

measurements were necessary The discovery of the periodic motion of α *Virginis* [Spica] then followed. . . .

I remark, further, that the observations of *Sirius* by the method of stars for the second class give 7.3 miles, and with the aid of the iron comparison spectrum 9.0 miles as the rate of approach towards the Sun. . . .

Each star has on the average been observed 3.3 times, and the measurements have been made independently by myself and by Dr. Scheiner. It may, therefore, be concluded that the probable error of the definitive values for both spectral classes will amount to less than one mile.

I intend after the definite completion of the measurements to communicate to the Society a list of the observed velocities, and will remark in conclusion that the velocities of the stars have proved to be much smaller than was to be expected from the direct observations. The mean result for forty-seven stars is 10.6 English miles.

Among them six have a velocity less than 2, and five greater than 20 miles; the greatest is that of α *Tauri,* about 30 miles. Fifteen of the stars have a positive and thirty-two a negative motion.

"List of the Proper Motions in the Line of Sight of Fifty-One Stars." *Monthly Notices of the Royal Astronomical Society,* Volume 52 (1892) by H. C. Vogel

In continuation of my communication of 1891 December, on the spectrographic method . . . I hereby transmit [see Table 26.2] the definitive results of that investigation, the observations having been meanwhile brought to a close. . . .

Table 26.2

No.	Star.	Epoch.	No. of Plates.	Velocity relative to Sun. (English Miles)		Mean.
				Vogel.	Scheiner.	
1	α Andromedæ	1889.93	2	+1.2	+4.4	+2.8
2	β Cassiopeiæ	1889.04	2	+0.8	+5.6	+3.2
3	α Cassiopeiæ	1890.14	2	−9.3	−9.7	−9.5
4	γ Cassiopeiæ	1888.89	2	+2.5	−6.9	−2.2
5	β Andromedæ	1889.26	2	+5.6	+8.3	+7.0
6	α Ursæ minoris	1888.90	2	−15.8	−16.3	−16.1
7	γ Andromedæ	1889.34	2	−4.9	−11.1	−8.0
8	α Arietis	1889.69	3	−9.0	−9.3	−9.2
9	β Persei[†]	1889.94	12	. . .	. . .	−1.0
10	α Persei	1888.93	2	−6.7	−6.1	−6.4
11	α Tauri	1889.16	4	+29.6	+30.7	+30.2
12	α Aurigæ	1888.98	11	+15.4	+15.0	+15.2
13	β Orionis	1889.24	14	+10.9	+9.5	+10.2

No.	Star.	Epoch.	No. of Plates.	Velocity relative to Sun. (English Miles)		Mean.
				Vogel.	Scheiner.	
14	γ Orionis	1890.37	3	+8.0	+3.4	+5.7
15	β Tauri	1889.65	3	+5.6	+4.4	+5.0
16	δ Orionis	1890.07	4	−0.1	+1.3	+0.6
17	ε Orionis	1889.00	3	+17.3	+15.6	+16.5
18	ζ Orionis	1889.00	2	+10.7	+7.8	+9.3
19	α Orionis	1889.32	2	+9.7	+11.7	+10.7
20	β Aurigæ[†]	1890.50	6	−16.0	−18.9	−17.5
21	γ Geminorum	1889.83	4	−9.7	−10.8	−10.3
22	α Canis majoris	1890.09	10	−8.4[‡]	−12.5	−9.8
23	α Geminorum*	1889.16	3	−18.4:	−18.4:	−18.4:
24	α Canis minoris	1889.68	3	−4.9	−6.5	−5.7
25	β Geminorum	1889.06	2	+1.2	+0.2	+0.7
26	α Leonis	1889.22	2	−5.3	−6.1	−5.7
27	γ Leonis	1889.76	2	−22.7	−25.2	−24.0
28	β Ursæ majoris	1889.39	2	−18.8	−17.6	−18.2
29	α Ursæ majoris	1889.11	4	−6.4	−7.9	−7.2
30	δ Leonis	1889.94	3	−9.3	−8.6	−8.9
31	β Leonis	1899.59	3	−8.6	−6.5	−7.6
32	γ Ursæ majoris	1889.40	2	−18.6	−14.4	−16.5
33	ε Ursæ majoris	1889.39	2	−21.3	−16.2	−18.8
34	α Virginis[†]	1890.34	27	. . .	. . .	−9.2
35	ζ Ursæ majoris*[†]	1890.33	8	−20.2	−18.5	−19.4
36	η Ursæ majoris	1889.83	2	−17.8	−14.8	−16.3
37	α Bootis	1889.57	6	−4.4	−5.2	−4.8
38	ε Bootis	1889.36	2	−10.4	−9.7	−10.1
39	β Ursæ minoris	1889.24	4	+8.9	+8.8	+8.9
40	β Libræ	1889.34	1	−6.0:		−6.0:
41	α Coronæ Borealis	1890.91	5	+19.7	+20.0	+19.9
42	α Serpentis	1889.36	1	+14::		+14::
43	β Herculis	1889.46	2	−21.3	−22.6	−22.0
44	α Ophiuchi	1889.09	2	+12.9	+10.9	+11.9
45	α Lyræ	1889.64	8	−8.7	−10.2	−9.5
46	α Aquilæ	1888.81	3	−24.7	−21.1	−22.9
47	γ Cygni	1888.93	2	−3.6	−4.3	−4.0
48	α Cygni	1888.99	4	−3.7	−6.2	5.0
49	ε Pegasi	1888.81	2	+4.6	+5.4	+5.0
50	β Pegasi	1889.90	1	+4.1:		+4.1:
51	α Pegasi	1888.81	2	+1.1	+0.4	+0.8

*Brightest component. [†]Motion of the system. [‡]Weight 2. : denotes less certain, and :: uncertain.

Greatest observed velocity . . . +30.2 miles (α Tauri); −24.0 miles (γ Leonis)

Average velocity	10.4 miles
No. of stars with positive velocity greater than 10.4 miles	7
No. of stars with negative velocity greater than 10.4 miles	11
Average probable error of the measurements for a single plate and one observer	±1.6 mile

"Orbit and Mass of the Variable Star *Algol* (*β Persei*)."
Publications of the Astronomical Society of the Pacific, Volume 2 (January, 1890) by H. C. Vogel and J. Scheiner

On the 28th of November a very important discovery was communicated to the Academy of Sciences of Berlin by Professor H. C. Vogel, Director, and Dr. Scheiner, Astronomer of the Astrophysikalisches Observatorium of Potsdam. I condense from the *Sitzungsberichte* of the Academy, 1889, (page 1045), the following:—

> Three photographic negatives of the spectrum of *Algol* taken during the winter of 1888–9 showed that before a minimum *Algol* was moving away from the Sun, and after a minimum it was moving towards it. Three new exposures of November, 1889, confirm this result. The observations taken together afford a very strong support to the theory that the cause of the variations in the light of *Algol* is to be found in the eclipses of this star by a dark (invisible) satellite revolving about it. The phenomena can be explained by assuming the following particulars of the dimensions of the two bodies:—

Table 26.3

Diameter of *Algol*	= 230,000 geographical miles.
Diameter of the invisible satellite	= 180,000 " "
Distance between their centers	= 700,000 " "
Satellite's velocity in orbit	= 12.0 " "
Mass of *Algol*	= 4/9 of the Sun's mass.
Mass of the satellite	= 2/9 " "

Motion of both bodies in the line of sight (toward the Sun) 0.5 geographical miles.

27 / Classification of the Stars

The ancients could see with their eyes that stars came in diverse colors—blue-white Sirius was starkly different from blood-red Arcturus. But it was not until the nineteenth century, with the development of thermodynamics—the physics of heat and energy— that astronomers began to slowly recognize the reasons for these variations. The color and spectrum of a star provide information on its temperature and physical condition.

To learn why stars differed, astronomers began to classify them based on their spectra, much the way early biologists first arranged flora and fauna into separate categories. Between 1863 and 1867 Angelo Secchi, a priest-astronomer at the Collegio Romano in Italy, painstakingly examined the spectral lines and colors of some four hundred stars by eye and proposed four classes of stars: (I) the white or bluish stars, such as Sirius and Vega, which show four strong dark lines due to hydrogen, as well as faint metal lines*; (II) stars such as our Sun, which predominantly shine in the middle part of the spectrum—yellow—and which display many fine dark lines; (III) orange-red stars, such as Betelgeuse and Antares, with broad bands of light, stronger at the red end of the spectrum and fading out in the blue; and (IV) dimmer stars of a deep red hue, described as gleaming "like rubies among the other stars."[20] Similar groupings were set

* In astronomy, all elements heavier than helium are considered metals.

up independently by Hermann Vogel in Germany and Lewis Rutherfurd in the United States.[21]

Photography helped astronomers refine and expand these early classifications. Henry Draper, a wealthy New York physician working from his private observatory, took the first successful photograph of a star's spectrum in 1872. Upon his death a decade later, his widow endowed a fund at the Harvard College Observatory to support a program in stellar spectroscopy. With this money Edward C. Pickering, observatory director from 1877 to 1919 and a pioneer in moving astronomical research into astrophysics, initiated an ambitious, decades-long program that ultimately photographed and classified the spectra of a quarter million stars in both the Northern and Southern Hemispheres. He strip-mined the sky spectrally. With a prism in front of the telescope's lens, every star in the telescope's sight was photographed directly as a spectrum. A photographic plate contained the spectra of hundreds of stars at once. Results were published periodically over the years; by 1924 the Henry Draper Catalogue filled nine volumes of the Harvard Observatory annals.

To accomplish this immense task, Pickering hired a corps of women, called "computers," who were trained to swiftly and accurately categorize the spectra. Annie Jump Cannon, who studied physics and astronomy in college, signed on in 1896 and ultimately established the stellar classification scheme still used today. Pickering had first used a simple lettering system from A to Q, extending Secchi's original groups into more detailed classes. Over time, groups were combined, some letters deleted, and others shuffled around. After much jockeying (with some of the reasons described in the excerpt below), Cannon settled on the stellar sequence O, B, A, F, G, K, M, which organized the stars by descending temperature. Generations of astronomy students have been taught to remember the lineup with the refrain, "Oh, Be A Fine Girl, Kiss Me." Even finer divisions are labeled by numbers 0 to 9; our Sun, for example, is a G2 star. The groups R, N, and S were later added to the end as special types ("Right Now, Smack").

Early spectroscopists had wondered whether the various spectral types might be the result of differing compositions among the stars. But the work at Harvard supported the growing realization that a spectrum reflects a star's physical condition, particularly its surface temperature. Cannon's sequence commences with the blue-white

O stars, immensely hot at 10,000 degrees or more on the Kelvin scale, and works its way down the temperature range, through yellow and orange, to the relatively cool M stars at 3,000 K. "It was almost as if the distant stars had really acquired speech," said Cannon, "and were able to tell of their constitution and physical condition."[22]

Cannon was the consummate cataloger and never attempted (or desired) to postulate a theory about her classifications, but others did begin to wonder whether the sequence somehow represented the evolution of a star. Classification was a first step. More work, however, had to be done before astronomers could reach a complete understanding of a star's life cycle. Many of the clues resided within the Draper catalog (see Chapter 28).

"Spectra of Bright Southern Stars." *Annals of the Astronomical Observatory of Harvard College,* Volume 28, Part II (1901) by Annie J. Cannon

Introductory Note

. . . It was deemed best that the observer should place together all stars having similar spectra and thus form an arbitrary classification rather than be hampered by any preconceived theoretical ideas, or by the previous study of visual spectra by other astronomers. If spectra which are absolutely identical can be placed together it makes but little difference what name is assigned to them, since in any future classification it is only necessary to take one star of each class and arrange them in any order, or give to them any nomenclature that may be desired. . . . It is believed that the present volume will furnish the principal facts regarding the spectra of all the brighter stars from the North to the South Pole, so that the reader can classify them according to any system he may choose, without the necessity of referring to the spectra themselves in each particular case. . . .

Edward C. Pickering
Director of the Observatory of Harvard College

The following pages contain a classification of 1,122 stars by means of their photographic spectra. These spectra have been examined on 5,961 plates photographed at Arequipa, Peru, with the Boyden telescope of this Observatory. The first plate was taken on November 29, 1891, the last on December 6, 1899. The telescope has an aperture of 13 inches (33 cm.), and a focal length of 16 feet ([488] cm.). The photographs of the spectra were made by placing one, two, or three prisms in front of the object glass. The dispersion of the prisms is such that the spectra measure from Hε to Hβ, 2.24, 4.86, and 7.43 cm., for one, two, and three prisms, respectively. An appreciable width, generally not less than 0.5 cm. but varying according to the magnitude of the star, was given to the spectra by attaching different weights to the pendulum of the clock controlling the motion of the telescope. The time of exposure was generally about one hour. The stars classified include, first, all those south of −30° in declination whose photometric magnitude is 5.00 or brighter; second, numerous fainter stars south of −30° in declination; third, numerous stars whose declinations are included between 0° and −30°; fourth, a few northern stars.

. . . All photographs of stellar spectra were taken from this series, and arranged in boxes according to the type of each spectrum following the classification of the Draper Catalogue. In cases where a plate showed more than one spectrum, the brightest only was considered in this preliminary examination. Thus, all plates on which the spectrum, or the brightest spectrum, was of the first type with the Orion lines present, were placed together and marked "B."* In like manner those whose spectra were of the first type without Orion lines were marked "A"; those whose spectra appeared intermediate between the first and the second types were marked "F"; those whose spectra were of the second type were marked "G"; those whose spectra were of the third type showing sudden changes in intensity at the end of greater wavelength were marked "M"; and, lastly, all those having bright lines were marked "Bright Line Stars."

. . . Each plate was placed on a stand and inclined at an angle of 45°. The light of the sky was reflected through it by means of a horizontal mirror. The spectrum was compared with the three typical spectra, and the plate was then superposed film to film on that which it most nearly resembled, so that the ends of the lines coincided. A positive eye-piece having a focal length of two inches was used in this examination. A record was

---

* Spectral absorption lines discovered to be produced by helium and often seen in stars within the Orion constellation.

made of the number and quality of the plate, the number of prisms used, the name or catalogue number of the star, the name of the typical star it resembled, and remarks were recorded concerning peculiarities, in cases where the spectrum differed from that of the typical star of the class. The same method of examination was later extended to the other spectra in the following order: "M," "F," "A," "B," "Bright Line Stars." After the general classification was outlined, a detailed study of the lines of each typical star and of each peculiar star was made. . . .

. . . Each spectrum was identified twice independently, once by Miss L. D. Wells and once by the writer.

Of the 1,122 stars classified from these plates, 41 have been photographed with three prisms, 268 with two prisms, and 813 with one only. The plates taken with one prism were found most useful in making the general classification, those with two or three prisms were found necessary for detailed study of peculiarities and intensities of lines. . . .

The letters of the Draper Catalogue are used in the following discussion to denote the various classes of spectra. The relation between the letters of the Draper Catalogue and the five types in ordinary use* for visual spectra is as follows:—

Type.	Letter.	Type.	Letter.
I	A, B	III	M
I–II	F	IV	N
II	G	V	O
II–III	K		

When the letters of the Draper Catalogue were adopted as the symbols of classification in this discussion, it was soon found that many subdivisions must be made to suit the varieties of spectra seen on plates of greater dispersion. Thus, the letter B is used in the Draper Catalogue to represent all the spectra showing the dark hydrogen lines together with the Orion lines, those at wavelengths 4026.4 and 4471.8 being, it is stated, the only lines commonly seen with the small dispersion employed.† It is obvious, however, that the letter B cannot represent all the varieties of spectra photographed with a dispersion which shows eighty or more Orion lines with many combinations of intensities, as on plates taken with three prisms. It

---

* Similar to the types Angelo Secchi first established in the 1860s.

† Wavelength in angstroms; 1 angstrom = 10^{-10} meter, roughly the width of an atom.

was therefore decided that the letter B should be used to indicate stars of the Orion type in which some of the Orion lines are as intense as the hydrogen lines. The gradual decrease in the intensities of the Orion lines and the increase in the hydrogen lines were next made the basis of a series of subdivisions of Orion stars in which the interval between Class B and Class A is estimated in tenths. Thus the spectra of Class B 1 A are very nearly like those of Class B, while those called Class B 9 A show only slight differences from those of Class A. Those of Class B 5 A appear to be about midway between the two classes. The evidence that the Orion spectra precede the Sirian is as good as that the Sirian precede the solar. The gradual decrease in the intensities of the Orion lines is accompanied by gradual increase in the hydrogen lines, and by the incoming of faint solar lines, so that in spectra of Classes B 8 A and B 9 A, solar and Orion lines are commingled. Hence, it was necessary either to interchange the letters B and A of the Draper Catalogue or to place the letter B before the letter A. The first alternative would prove confusing. The second presents no real difficulties since the letters are merely symbols to express an observed condition.

The letter A in this classification represents spectra of the Sirian type, of which α Canis Majoris [Sirius] and α Lyrae [Vega] are examples. These spectra may be defined as those in which the Orion lines in general are absent, the line K and the solar lines are faint, and the hydrogen lines are of great intensity.

The letter F represents spectra in which the wide bands of calcium, K and H, are the most conspicuous features, while the hydrogen lines are still more intense than any solar lines. The gradations of spectra found between Classes A and F are indicated by the combinations A 2 F, A 3 F, and A 5 F.

The letter G represents spectra of the characteristic solar type, of which α Aurigae [Capella] has been used as the best example. The spectra of Class G may be defined as those in which the lines K and H of calcium and the band G are the most conspicuous features, while the hydrogen lines are still as intense as any of the solar lines. Spectra intermediate between Classes F and G are indicated as F 2 G, F 5 G, and F 8 G. The letter K represents spectra of the later second type, or intermediate between the second and third types. The letter K may be briefly described as representing those spectra in which the bands K and H, the band G, and the line 4227.0 are the most conspicuous features, and in which the end of shorter wavelength is faint, and the distribution of light is not uniform in different parts of the spectrum. The hydrogen lines in this class are fainter than numerous solar lines. Spectra intermediate between Classes G and K are designated by the letters G 5 K. The letter M represents, in general, the spectra which differ from those of Class K mainly in showing sudden diminutions in intensity as the wavelength increases, at 4762, 4954, 5168, and 5445. Spectra coming between Classes K and M are indicated as K 2 M and K 5 M. Since no spectra were found which followed those of Class M, and into which this type of spectrum appeared to merge, the variations of these spectra could not be expressed in intervals between M and any other letter. The two divisions of stars of Class M are therefore indicated by the letters Ma and Mb. With the spectra of Class Mb, the series in which the various spectra merge almost insensibly from one type to another is concluded as far as it has been observed on these plates. The letters Md represent spectra of the third type showing one or more bright hydrogen lines. Spectra of the fourth type, for which the letter N is used in the Draper Catalogue, do not appear on any plates in the series examined for this classification. Thus, with the symbols B, A, F, G, K, Ma, and Mb, or with combinations of these symbols, and by the aid of remarks to explain the peculiarities of those spectra that vary slightly from the typical stars, all spectra with wholly dark lines can be provided for with one exception.

A few spectra of the Orion type were found which clearly precede those of the class called B. These spectra might have been called by the letter B, and those now called B passed on to B I A, and so on. But when it was found that these spectra were intermediate between a class of spectra showing bright lines and those of Class B, it was deemed advisable to express that fact in the symbol used to designate them. Stars of the fifth type are those whose spectra consist mainly of bright lines. The spectra of these stars are characterized by the bright bands at wavelengths 4633 and 4688, and the line at 5007 characteristic of gaseous nebulae, is sometimes present. Stars whose spectra are of the fifth type are called O in the Draper Catalogue. . . . These spectra, in combination with those of Class Oe, appear to establish the position of spectra of the fifth type as preceding those of the Orion type. The letter O has been placed, therefore, in this classification, before the letter B instead of after the letter M. . . .

As a result of this classification it is found that most of these 1,122 spectra can be arranged in a sequence. The spectra of Class Oe, or possibly those of Class Ob, are at one end of the sequence, while the spectra of Class Mb are at the other end. The order of the development is not indicated, and the series might proceed from Class Mb to Class Oe, instead of from Class Oe to Class Mb. The latter seems more probable, perhaps owing to its agreement with Laplace's theory of stellar development. The progressive changes in the spectra are at times so slight as to be almost imperceptible, and so gradual as to make exact differentiation difficult, while again the changes are somewhat abrupt, and it appears as if intermediate forms as yet undiscovered might exist. . . .

28 / Giant Stars and Dwarf Stars

The great surge at the end of the nineteenth century in generating catalogs of stellar spectra, proper motions, and parallaxes finally allowed astronomers to pick out groups of stars for comparison. Stars were beginning to be seen as separate individuals, displaying a variety of temperatures, brightnesses, and colors. Was there a pattern in the way stars varied?

A pioneer in this endeavor was the Danish astronomer Ejnar Hertzsprung. While at the observatory of the University of Copenhagen, he was inspired by a discovery made by Antonia Maury, one of the women at the Harvard College Observatory who were involved in the most extensive survey of stellar spectra in its day (see Chapter 27). Maury had noticed that the spectra of two stars could be perfectly identical, except that the widths of their spectral lines differed. One would have broad lines; the other, very sharp and narrow features. Carrying out a detailed analysis of stellar spectra in the early 1900s, Hertzsprung noticed that the red stars displaying Maury's narrow lines (a subdivision denoted by the letter *c*) tended to have smaller movements across the sky (proper motions) than the red stars with broad lines (which Maury labeled as *a* and *b*). That meant the *c* stars had to be more distant than their broad-lined cousins, which suggested they were far bigger and more luminous, otherwise they couldn't be seen from so far away. He soon wrote the Harvard College Observatory, then the prime center for stellar spectra research, urging them to use Maury's classifications (by then ignored) for finding stars of great luminosity. "To neglect the c-properties in classifying stellar spectra . . . ," wrote Hertzsprung, "is nearly the same thing as if the zoologist, who has detected the deciding differences between a whale and a fish, would continue in classifying them together."[23] But Harvard Observatory director Edward Pickering remained skeptical.

Hertzsprung also recognized that there was a direct connection between a star's absolute brightness and its color: stars got progressively dimmer as you proceeded from the brilliant blue-white O and B stars, down through the A, F, G, K stars, to the cool red M stars at the end of the sequence. Hertzsprung's analyses languished in obscurity, though, because he published his findings in a minor German journal for scientific photography in 1905 and 1907.

Hertzsprung's findings became more widely known once Henry Norris Russell at Princeton University, who later became the dean of American astronomy, independently arrived at the same conclusions. In 1913 Russell created a diagram in which he tracked a star's spectral type and luminosity. He first published this graph in an article in *Popular Astronomy*, although he had been gradually disseminating his ideas on stellar evolution earlier. Like Hertzsprung, he saw that the stars tended to line up along a diagonal band, with the luminosity of a star diminishing as the star's temperature decreased. There was an exception: certain stars of great luminosity, a rarer species, stood alone and isolated in a corner of the chart. From studies of binary stars, Russell realized these stars were not large in mass but large in size and low in density. It was an immense surface area that made these stars so bright. They were huge compared to the stars along the main sequence of his diagram and came to be called "giants." The red-orange giant Aldebaran, the eye of Taurus the Bull, for example, is 46 times wider than our Sun (half the size of Mercury's orbit) and more than a hundred times brighter. Other giant stars, such as Betelgeuse in Orion, would extend to the orbit of Jupiter.* The smaller stars, by contrast, were named "dwarfs."

* This was verified in 1920 when Albert A. Michelson and Francis G. Pease used an interferometer, mounted on Mount Wilson Observatory's 100-inch telescope, to make the first successful measurement of a star's diameter. Betelgeuse was their target.[24]

The graph displaying these stellar relationships is now known as the Hertzsprung-Russell diagram, which became the cornerstone of astronomical research regarding the evolution of stars (see Figure 28.1). With the H-R diagram in front of them, astronomers could imagine the stars as ever-changing over cosmic time scales. Extending an idea first suggested by the British astronomer Norman Lockyer in the 1880s, Russell initially wondered whether the diagram represented different stages of a star's life: a giant reddish star first condenses out of a diffuse nebula, grows luminous as it contracts and heats up to a blue-white brilliance, and then eventually extinguishes itself—from yellow, to orange, and back to red once again—as it shrinks and cools at the end of its life. This scenario made sense in an era when astronomers believed that a star derived its energy from gravitational contraction alone, but this scheme is now vastly outdated. Stars do not simply evolve from hot O to cool M. A full understanding of a star's development was achieved only after astronomers realized how a star is powered by nuclear fusion (see Chapters 43 and 47).

"On the Radiation of Stars."
Zeitschrift für wissenschaftliche Photographie [Magazine for Scientific Photography]. Volume 3 (1905) by Ejnar Hertzsprung

In volume 28 of the "Annals of the Astronomical Observatory of Harvard College" a detailed survey of the spectra is given for northern and southern bright stars by Antonia C. Maury and Annie J. Cannon, respectively.

. . . Here we can find room for only a few words concerning the three sub-classifications *b, a,* and *c.* The *b* stars have broader lines than those of "division" *a.* The relative intensities of the lines seem, however, to be equal for *a-* and *b*-stars "so that there appears to be no decided difference in the constitution of the stars belonging, respectively, to these two divisions." As the most important characteristics of subclass *c* we can mention, first, that the lines are unusually narrow and sharp; second, that among the "metallic" lines others occur which are not identifiable with any solar lines, and the relative intensities of the remainder do not correspond with the intensities observed in the solar spectrum. "In general, division *c* is distinguished by the strongly defined character of its lines, and it seems that stars of this division must differ more decidedly in constitution from those of division *a* than is the case with those of division *b*." Antonia C. Maury suspects that the *a*- and *b*-stars on the one hand and the *c*-stars on the other, belong to collateral series of development. That is to say not all stars have the same spectral development. What determines such a differentiation (differences in mass and constitution, etc.) is a question that remains unanswered.

The question arises how great the systematic differences of the brightness, reduced to a common distance, of stars of the different groups will be. For this purpose I have used the proper motions of the stars. . . .

[Omitted here is a table of data and a detailed discussion in which Hertzsprung shows that the absolute luminosities of the stars displaying Maury's subclasses *a* and *b* systematically decrease as one progresses through the classes of stars: O, B, A, F, G, K, M. He goes on to note that stars with narrow lines, Maury's subclass *c*, are different by being more luminous.]

In any case we may say that the annual proper motion of an average *c*- star, reduced to magnitude 0, amounts to only a few hundredths of a second. With the relatively large errors of these small values, a dependence on spectral class cannot be recognized. In other words, the *c*-stars are at least as bright as the Orion stars. In both of the spectroscopic binaries o Andromedae and β Lyrae the brightness of the *c*-star and of the companion star of the Orion type appear to be of the same order of brightness. The proper motions (not here given) are all small, according to the Auwers-Bradley Catalogue. . . . For the stars in Annie J. Cannon's listing that have narrow sharp lines, I can also find only small proper motions. This result confirms the assumption of Antonia C. Maury that the *c*-stars are something unique. . . .

"Relations Between the Spectra and Other Characteristics of the Stars." *Popular Astronomy,* Volume 22 (1914) by Henry Norris Russell

Investigations into the nature of the stars must necessarily be very largely based upon the average characteristics of groups of stars selected in various ways—as by brightness, proper motion, and the like. The publication within the last few years of a great wealth of accumulated observational material makes the compilation of such data an easy process; but some methods of grouping appear to bring out much more definite and interesting relations than others, and, of all the principles of division, that which separates the stars according to their spectral types has revealed the most remarkable differences, and those which most stimulate attempts at a theoretical explanation. . . .

Thanks to the possibility of obtaining with the objective prism photographs of the spectra of hundreds of stars on a single plate, the number of stars whose spectra have been observed and classified now exceeds one hundred thousand, and probably as many more are within the reach of existing instruments. The vast majority of these spectra show only dark lines, indicating that absorption in the outer and least dense layers of the stellar atmospheres is the main cause of their production. Even if we could not identify a single line as arising from some known constituent of these atmospheres, we could nevertheless draw from a study of the spectra, considered merely as line-patterns, a conclusion of fundamental importance.

The spectra of the stars show remarkably few radical differences in type. More than ninety-nine percent of them fall into one or other of the six great groups which, during the classic work of the Harvard College Observatory, were recognized as of fundamental importance, and received as designations, by the process of "survival of the fittest," the rather arbitrary series of letters B, A, F, G, K, and M. That there should be so very few types is noteworthy; but much more remarkable is the fact that they form a continuous series. . . . This series is not merely continuous; it is *linear.* There exist indeed slight differences between the spectra of different stars of the same spectral class, such as A0; but these relate to minor details, which usually require a trained eye for their detection, while the difference between successive classes, such as A and F, are conspicuous to the novice.

Almost all the stars of the small outstanding minority fall into three other classes, denoted by the letters O, N, and R. Of these O undoubtedly precedes B at the head of the series, while R and N, which grade into one another, come probably at its other end, though in this case the transition stages, if they exist, are not yet clearly worked out. . . .

The first great problem of stellar spectroscopy is the identification of this predominant cause of the spectral differences. The hypothesis which suggested itself immediately upon the first studies of stellar spectra was that the differences arose from variations in the chemical composition of the stars. . . . [T]o the [writer] . . . it is almost unbelievable that differences of chemical composition should reduce to a function of a single variable, and give rise to the observed linear series of spectral types.

I need not detain you with the recital of the steps by which astrophysicists have become generally convinced that the main cause of the differences of the spectral classes is difference of temperature of the stellar atmospheres. . . .

. . . I will now ask your attention in greater detail to certain relations which have been the more special objects of my study.

Let us begin with the relations between the spectra and the real brightness of the stars. These have been discussed by many investigators—notably by Kapteyn* and Hertzsprung—and many of the facts which will be brought before you are not new; but the observational material here presented is, I believe, much more extensive than has hitherto been assembled. We can only determine the real brightness of a star when we know its distance; but the recent accumulation of direct measures of parallax, and the discovery of several moving clusters of stars whose distances can be determined, put at our disposal far more extensive data than were available a few years ago.

Figure [28.1] shows graphically the results derived from all the direct measures of parallax available in the spring of 1913 (when the diagram was constructed). The spectral class appears as the horizontal coordinate, while the vertical one is the absolute magnitude, according to Kapteyn's definition—that is, the visual magnitude which each star would appear to have if it should be brought up to a standard distance, corresponding to a parallax of 0″.1 (no account being taken of any possible absorption of light in space). The absolute magnitude −5, at the top of the diagram, corresponds to a luminosity 7500 times that of the Sun, whose absolute magnitude is

* Dutch astronomer Jacobus Kapteyn, known for his study in the early 1900s of the structure of our galaxy.

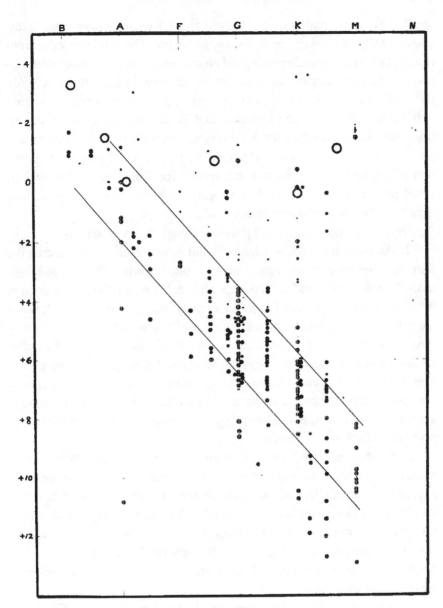

Figure 28.1: The spectrum-luminosity diagram for bright stars. Ordinates are absolute magnitudes; abscissae, spectral classes.

4.7. The absolute magnitude 14, at the bottom, corresponds to $\frac{1}{5000}$ of the Sun's luminosity. The larger dots denote the stars for which the computed probable error of the parallax is less than 42 percent of the parallax itself, so that the probable error of the resulting absolute magnitude is less than $\pm 1^{\mathrm{m}}.0$. This is a fairly tolerant criterion for a "good parallax," and the small dots, representing the results derived from the poor parallaxes, should hardly be used as a basis for any argument. The solid black dots represent stars whose parallaxes depend on the mean of two or more determinations; the open circles, those observed but once. In the latter case, only the results of those observers whose work appears to be nearly free from systematic error have been included, and in all cases the observed parallaxes have been corrected for the probable mean parallax of the comparison stars to which they were referred. The large open circles in the upper part of the diagram represent mean results for numerous bright stars of small proper motion (about 120 altogether) whose observed parallaxes hardly exceed their probable errors. . . .

Upon studying Figure [28.1], several things can be observed.

1. All the white stars, of Classes B and A, are bright, far exceeding the Sun; and all the very faint stars,—for example, those less than $1/50$ as bright as the Sun,—are red, and of Classes K and M. We may make this statement more specific by saying, as Hertzsprung does, that there is a certain limit of brightness for each spectral class, below which stars of this class are very rare, if they occur at all. Our diagram shows that this limit varies by rather more than two magnitudes from class to class. The single apparent exception is the faint double companion to o^2 Eridani, concerning whose parallax and brightness there can be no doubt, but whose spectrum, though apparently of Class A, is rendered very difficult of observation by the proximity of its far brighter primary.

2. On the other hand, there are many red stars of great brightness, such as Arcturus, Aldebaran and Antares, and these are as bright, on the average, as the stars of Class A, though probably fainter than those of Class B. Direct measures of parallax are unsuited to furnish even an estimate of the upper limit of brightness to which these stars attain, but it is clear that some stars of all the principal classes must be very bright. The range of actual brightness among the stars of each spectral class therefore increases steadily with increasing redness.

3. But it is further noteworthy that all the stars of Classes K5 and M which appear on our diagram are either very bright or very faint. There are none comparable with the Sun in brightness. We must be very careful here not to be misled by the results of the methods of selection employed by observers of stellar parallax. They have for the most part observed either the stars which appear brightest to the naked eye or stars of large proper motion. In the first case, the method of selection gives an enormous preference to stars of great luminosity, and, in the second, to the nearest and most rapidly moving stars, without much regard to their actual brightness. It is not sur-

prising, therefore, that the stars picked out in the first way (and represented by the large circles in Figure [28.1]) should be much brighter than those picked out by the second method (and represented by the smaller dots). But if we consider the lower half of the diagram alone, in which all the stars have been picked out for proper-motion, we find that there are no very faint stars of Class G, and no relatively bright ones of Class M. As these stars were selected for observation entirely without consideration of their spectra (most of which were then unknown), it seems clear that this difference, at least, is real, and that there is a real lack of red stars comparable in brightness to the Sun, relatively to the number of those 100 times fainter.

The appearance of Figure [28.1] therefore suggests the hypothesis that, if we could put on it some thousands of stars, instead of the 300 now available, and plot their absolute magnitudes without uncertainty arising from observational error, we would find the points representing them clustered principally close to two lines, one descending sharply along the diagonal, from B to M, the other starting also at B, but running almost horizontally. The individual points, though thickest near the diagonal line, would scatter above and below it to a vertical distance corresponding to at least two magnitudes, and similarly would be thickest near the horizontal line, but scatter above and below it to a distance which cannot so far be definitely specified, so that there would be two fairly broad bands in which most of the points lay. For Classes A and F, these two zones would overlap, while their outliers would still intermingle in Class G, and probably even in Class K. There would however be left a triangular space between the two zones, at the right-hand edge of the diagram, where very few, if any, points appeared; and the lower left-hand corner would be still more nearly vacant.

We may express this hypothesis in another form by saying that there are two great classes of stars—the one of great brightness (averaging perhaps a hundred times as bright as the Sun) and varying very little in brightness from one class of spectrum to another; the other of smaller brightness, which falls off very rapidly with increasing redness. These two classes of stars were first noticed by Hertzsprung, who has applied to them the excellent names of *giant* and *dwarf* stars. . . .*

---

* Actually, Hertzsprung never coined the terms *giants* and *dwarfs*. In the 1890s evidence was emerging that there were stars bigger than our Sun; some called them "giant stars." Later the German theorist Karl Schwarzschild, trying to make astronomers aware of the findings of Hertzsprung, lectured on his friend's discoveries and also called Hertzsprung's highly luminous stars giants.[25]

29 / Hydrogen: The Prime Element

Hydrogen is now recognized as the predominant element in the universe. Until 1925, no one knew this.

Starting in the 1860s astronomers were identifying the various elements within the Sun and the stars by translating the distinctive patterns within their spectra (see Chapter 24). But the actual origin of these spectral lines was a mystery. The mechanism was at last revealed with the introduction of quantum mechanics at the start of the twentieth century. The foundation was set in 1900 when the German physicist Max Planck originated the idea that electromagnetic radiation is only absorbed or emitted by matter in discrete packets—the quantum. By 1913, inspired by the nuclear model of the atom established a few years earlier by Ernest Rutherford in Great Britain, the Danish physicist Niels Bohr figured that an atomic spectrum is generated as the electrons in an atom—then thought to be circling a cluster of protons like planets in a solar system—were jumping from one orbit to another, emitting or absorbing specific packets of light (photons) along the way. Each element has its own distinctive pattern of spectral absorptions and emissions because each material has unique tiers of orbits between which the electrons jump to and fro.

In 1920 the Indian physicist Meghnad Saha combined his expertise in thermodynamics and quantum mechanics to show how stellar spectra could be used to diagnose a star's physical condition.

Early spectroscopists had believed that the intensity of a star's spectral lines somehow indicated its composition. But in a famous paper that inspired a generation of astrophysicists, Saha argued that an element will either stand out in a spectrum or remain hidden depending on a star's temperature, pressure, and density.[26] He concluded that different types of stars, from O to M, generally share the same composition; their spectra are dissimilar because of their differing physical states. The cool M stars, whose surface temperatures are a mere 3,000 K, display a preponderance of neutral atoms in their spectra. The intensely hot O stars, on the other hand, with surface temperatures of tens of thousands of degrees, are dominated by ionized atoms—that is, electrically charged atoms that have had electrons stripped off by the high energies.

Cecilia Payne (later Payne-Gaposchkin) used Saha's theory, as well as the statistical theories of Ralph H. Fowler and Edward Milne, to determine the relative abundance of elements within stars as part of her graduate work at the Harvard College Observatory in the early 1920s. Born in England in 1900, she had gone to the United States for her doctorate, because women there at the time had better opportunities in astronomy. A pioneer in drawing on the new physics to solve astrophysical problems, she uncovered the very first hint that the simplest element—hydrogen—was the most abundant substance in the universe. This essential fact echoed long and hard down the corridors of astronomy in later years: here was the abundant fuel for a star's persistent burning (see Chapter 43) and the remnant debris from the first few minutes of the universe's creation.

Yet her famous 1925 dissertation (described as "the most brilliant Ph.D. thesis ever written in astronomy") does not make this important announcement about hydrogen.[27] In the course of her calculations, Payne had noticed that the common elements in the Earth's crust were also present in the stars. For elements such as silicon and carbon, for example, she found roughly the same relative proportions as seen in the Earth. In the case of our Sun, it's the signature of our common origin out of a swirling cloud of interstellar matter some five billion years ago. The similarities, though, were marred by two glaring exceptions: hydrogen and helium. In her table of abundances (see paper excerpted below), Payne showed on a logarithmic scale that hydrogen was as much as a million times more plentiful in the stars than on the Earth or in meteorites. The neutral helium in the stars she investigated was about a thousand times more abundant than the other heavier elements. Before publication, though, her preliminary results were sent to Henry Norris Russell at Princeton, a man then at the vanguard of incorporating modern physics into astronomy. Cautious of upsetting respected solar models, he told Payne that "it is clearly impossible that hydrogen should be a million times more abundant than the metals [elements heavier than helium]."[28] With atomic theory so new, there were worries that hydrogen, the simplest element, was exhibiting abnormal spectral behavior that skewed her results. As a result, Payne amended her report to say that the abundances for hydrogen and helium "are regarded as spurious . . . almost certainly not real."[29]

Within a few years Albrecht Unsöld[30] in Germany and William McCrea[31] in Great Britain used other methods to show that hydrogen was indeed an abundant element in the Sun. Ironically, just four years after Payne's inaugural foray into stellar abundances, it was Russell himself who principally convinced astronomers of hydrogen's overwhelming presence in the Sun and stars. After collecting more detailed observations of the Sun, with the help of his steadfast assistant Charlotte Moore, Russell concluded that many difficulties in interpreting the solar and stellar spectra could be overcome if their atmospheres really do consist "mainly of hydrogen . . . If this is true, the outer portions of these stars must be almost pure hydrogen, with hardly more than a smell of metallic vapors in it."[32] He compared his findings on the Sun's hydrogen with the abundances calculated by Payne for hotter stars and mentioned in his landmark paper that there was "a very gratifying agreement" with Payne's numbers (without noting that it had been labeled "spurious data" in her original paper).[33] By 1932 the Danish theorist Bengt Strömgren proved that the high hydrogen content existed throughout a star, not just in the stellar atmosphere.[34] In the majority of stars hydrogen, by number of atoms, dominates about 90 percent of a star's composition; helium follows at nearly 10 percent, with all the other elements together making up the remaining 0.1 percent, bits of trace "dirt" in the celestial mix.

From *Stellar Atmospheres*
by Cecilia H. Payne

The Relative Abundance of the Elements

The relative frequency of atomic species has for some time been of recognized significance. Numerous deductions have been based upon the observed terrestrial distribution of the elements; for example, attention has been drawn to the preponderance of the lighter elements (comprising those of atomic number less than thirty), to the "law of even numbers," which states that elements of even atomic number are far more frequent than elements of odd atomic number, and to the high frequency of atoms with an atomic weight that is a multiple of four.

The existence of these general relations for the atoms that occur in the crust of the earth is in itself a fact of the highest interest, but the considerations contained in the present chapter indicate that such relations also hold for the atoms that constitute the stellar atmospheres and therefore have an even deeper significance than was at first supposed. Data on the subject of the relative frequency of the different species of atoms contain a possible key to the problem of the evolution and stability of the elements. Though the time does not as yet seem ripe for an interpretation of the facts, the collection of data on a comprehensive scale will prepare the way for theory, and will help to place it, when it comes, on a sound observational basis.

The intensity of the absorption lines associated with an element immediately suggests itself as a possible source of information on relative abundance. But the same species of atom gives rise simultaneously to lines of different intensities belonging to the same series, and also to different series, which change in intensity relative to one another according to the temperature of the star. The intensity of the absorption line is, of course, a very complex function of the temperature, the pressure, and the atomic constants. . . .

The observed intensity can therefore be used *directly* for only a crude estimate of abundance. Roughly speaking, the lines of the lighter elements predominate in the spectra of stellar atmospheres, and probably the corresponding atoms constitute the greater part of the atmosphere of the star, as they do of the earth's crust. Beyond a general inference such as this, few direct conclusions can be drawn from line-intensities. Russell made the solar spectrum the basis of a discussion in which he pointed out the apparent similarity in composition between the crust of the earth, the atmosphere of the star, and the meteorites of the stony variety.* The method used by him should be expected, in the light of subsequent work, to yield only qualitative results, since it took no account of the relative probabilities of the atomic states corresponding to different lines in the spectrum.

Uniformity of Composition of the Stellar Atmosphere

The possibility of arranging the majority of stellar spectra in homogeneous classes that constitute a continuous series, is an indication that the composition of the stars is remarkably uniform—at least in regard to the portion that can be examined spectroscopically. The fact that so many stars have *identical* spectra is in itself a fact suggesting uniformity of composition; and the success of the theory of thermal ionization in predicting the spectral changes that occur from class to class is a further indication in the same direction.

If departures from uniform distribution did occur from one class to another, they might conceivably be masked by the thermal changes of intensity. But it is exceedingly improbable that a lack of uniformity in distribution would *in every case* be thus concealed. It is also unlikely, though possible, that a departure from uniformity would affect equally and solely the stars of one spectral class. Any such departure, if found, would indicate that the presence of abnormal quantities of certain elements was an effect of temperature. This explanation appears, however, to be neither justified nor necessary; there is no reason to assume a sensible departure from uniform composition for members of the normal stellar sequence.

[Omitted here are Payne's comments on the "marginal appearance" of spectrum lines. Based on the work of British theorists R. H. Fowler and Edward Milne, Payne assumed that the number of atoms required to make an absorption line just barely visible in a stellar spectrum is the same for the lines of all elements, and she used this information to calculate the relative abundance of the elements in stars.]

* "Henry Norris Russell, 'The Solar Spectrum and the Earth's Crust,' *Science* 39 (1914): 791–94."

Table 29.1

Atomic Number	Atom	Log a$_r$	Atomic Number	Atom	Log a$_r$	Atomic Number	Atom	Log a$_r$
1	H	11	13	Al	5.0	23	V	3.0
2	He	8.3	14	Si	4.8	24	Cr	3.9
	He+	12		Si+	4.9	25	Mn	4.6
3	Li	0.0		Si+++	6.0	26	Fe	4.8
6	C+	4.5	19	K	3.5	30	Zn	4.2
11	Na	5.2	20	Ca	4.8	38	Sr	1.8
12	Mg	5.6		Ca+	5.0		Sr+	1.5
	Mg+	5.5	22	Ti	4.1	54	Ba+	1.1

Method of Estimating Relative Abundances

. . . [T]he relative abundances of the atoms are given directly by the reciprocals of the respective fractional concentrations at marginal appearance. The values of the relative abundance thus deduced are contained in Table [29.1]. Successive columns give the atomic number, the atom, and the logarithm of the relative abundance, a$_r$.

Comparison of Stellar Atmosphere and Earth's Crust

The preponderance of the lighter elements in stellar atmospheres is a striking aspect of the results, and recalls the similar feature that is conspicuous in analyses of the crust of the earth. A distinct parallelism in the relative frequencies of the atoms of the more abundant elements in both sources has already been suggested by Russell, and discussed by H. H. Plaskett, and the data contained in Table [29.1] confirm and amplify the similarity. . . .

The most obvious conclusion that can be drawn from Table [29.1] is that all the commoner elements found terrestrially, which could also, for spectroscopic reasons, be looked for in the stellar atmosphere, are actually observed in the stars. The twenty-four elements that are commonest in the crust of the earth, in order of atomic abundance, are oxygen, silicon, hydrogen, aluminum, sodium, calcium, iron, magnesium, potassium, titanium, carbon, chlorine, phosphorus, sulphur, nitrogen, manganese, fluorine, chromium, vanadium, lithium, barium, zirconium, nickel, and strontium.

The most abundant elements found in stellar atmospheres, also in order of abundance, are silicon, sodium, magnesium, aluminum, carbon, calcium, iron, zinc, titanium, manganese, chromium, potassium, vanadium, strontium, barium, (hydrogen, and helium). All the atoms for which quantitative estimates have been made are included in this list. Although hydrogen and helium are manifestly very abundant in stellar atmospheres, the actual values derived from the estimates of marginal appearances are regarded as spurious.

The absence from the stellar list of eight terrestrially abundant elements can be fully accounted for. The substances in question are oxygen, chlorine, phosphorus, sulphur, nitrogen, fluorine, zirconium, and nickel, and none of these elements gives lines of known series relations in the region ordinarily photographed. . . .

The outstanding discrepancies between the astrophysical and terrestrial abundances are displayed for hydrogen and helium. The enormous abundance derived for these elements in the stellar atmosphere is almost certainly not real. Probably the result may be considered, for hydrogen, as another aspect of its abnormal behavior, already alluded to; and helium, which has some features of astrophysical behavior in common with hydrogen, possibly deviates for similar reasons. The lines of both atoms appear to be far more persistent, at high and at low temperatures, than those of any other element.

The uniformity of composition of stellar atmospheres appears to be an established fact. The quantitative composition of the atmosphere of a star is derived, in the present chapter, from estimates of the "marginal appearance" of certain spectral lines, and the inferred composition displays a striking parallel with the composition of the earth.

The observations on abundance refer merely to the stellar atmosphere, and it is not possible to arrive in this way at conclusions as to internal composition. But marked differences of internal composition from star to star might be expected to affect the atmospheres to a noticeable extent, and it is therefore somewhat unlikely that such differences do occur.

30 / Stellar Mass, Luminosity, and Stability

With its tremendous mass, why doesn't a star simply collapse over time under the powerful pull of gravity? Does the mass of a star determine how bright it will be? One of the greatest achievements of Arthur Eddington was finding answers to these basic questions in astronomy.

Born in 1882 and educated at Cambridge University, Eddington was first involved in matters of practical astronomy but soon became one of the greatest theoreticians of his generation by applying his sharp mathematical skills to a wide range of astronomical problems. This included crafting the basic laws that link a star's mass to its luminosity and that show why a star remains stable.

Around 1916 Eddington began investigating the structure and physics of stars and came to realize that the tremendous force of gravity pulling inward, trying to squeeze a star's great gaseous envelope tighter and tighter, is countered by the monstrous pressure of the radiation, hot gases, and electrons that are bouncing around inside the star pushing outward. As a result, a star neither wafts away nor shrinks into oblivion. The two opposing forces—gravity and pressure—keep the star in an exquisite balance.

His investigations into the structure of stars led Eddington by 1924 to gather all the known observational data on stellar masses, which astronomers produced by closely examining binary stars. When two stars are in orbit about one another, their masses can be calculated from Newton's laws. He saw that most stars range from one-fifth the Sun's mass to about twenty-five times its mass. Others had earlier noticed from such data that a star's luminosity was closely associated with its mass, but Eddington, in what has been described as "a tour de force," was able to determine the theoretical law that governed that relationship.[35] The graph that he included in his paper on this finding displays the distinct relationship (see Figure 30.1). Each time the mass of a star doubles, the luminosity increases roughly tenfold. To find this rule, Eddington assumed that stars behave like a gas rather than a liquid, as Eddington's chief rival James Jeans was arguing at the time. As a result, he was surprised to see that his law worked not just for giant stars, as he was first expecting, but for smaller dwarf stars as well.

"On the Radiative Equilibrium of the Stars."
Monthly Notices of the Royal Astronomical Society,
Volume 77 (1916)
by Arthur S. Eddington

The theory of radiative equilibrium of a star's atmosphere was given by [Karl] Schwarzschild in 1906. He did not apply the theory to the interior of a star; but the necessary extension of the formulae (taking account of the curvature of the layers of equal temperature) is not difficult. It is found that the resulting distribution of temperature and density in the interior follows a rather simple law.

Taking a star—a "giant" star of low density, so that the laws of a perfect gas are strictly applicable—and calculating from its mass and mean density the numerical values of the temperature, we find that the temperature gradient is so great that there ought to be an outward flow of heat many million times greater than observation indicates. This contradiction is not peculiar to the radiative hypothesis; a high temperature in the interior is necessary in order that the density may have a low mean value notwithstanding the enormous pressure due to the weight of the column of material above.

There is a way out of the difficulty, however, if we are ready to admit that the radiation-pressure due to the outward flow of heat may under calculable conditions of temperature, density, and absorption nearly neutralize the weight of the column, and so reduce the pressure which would otherwise exist in the interior. . . .

We thus arrive at the theory that a rarefied gaseous star adjusts itself into a state of equilibrium such that the radiation-pressure very approximately balances gravity at interior points. This condition leads to a relation between mass and density on the one side and effective temperature on the other side, which seems to correspond roughly with observation. . . .

[Omitted here are the detailed calculations that led Eddington to his conclusions.]

"On the Relation between the Masses and Luminosities of the Stars." *Monthly Notices of the Royal Astronomical Society,* Volume 84 (1924) by Arthur S. Eddington

A theory of the stellar absorption-coefficient should, if successful, lead to formulae determining the absolute magnitude of any giant star of which the mass and effective temperature are known. I have hitherto laid most stress on whether the theory will predict the absolute magnitude of Capella. The present position of that problem was summarized in my last paper; although there appears to have been some measure of success, the final conclusion is not yet certain.*

In this paper we shall consider the differential instead of the absolute results of the theory. We are not yet certain what should be the form of the absolute factor occurring in the formula connecting total radiation and mass; but apart from this factor, the form of the law seems to be fixed within narrow limits. Instead of constructing the absolute factor from physical constants we shall be content to determine its value from the observa-

---

* *"Monthly Notices of the Royal Astronomical Society* 84 (1924): 104."

tional data for Capella; and then it ought to be possible to calculate the luminosity of any other giant star, the result depending differentially on Capella.

Using the constant determined from Capella, we shall find that the formulae of the theory appear to predict correctly the absolute magnitudes of all other ordinary stars available for the test, *regardless of whether they are giants or dwarfs.*

The evidence for this statement is shown graphically in figure [30.1]. . . .

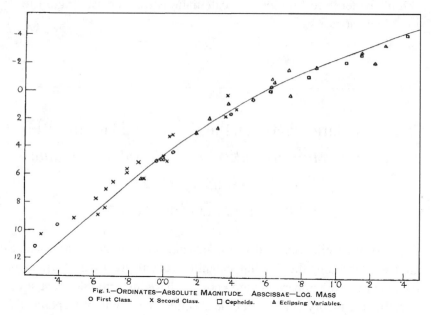

-4
-2
0
2
4
6
8
10
12
·4
·6
·8
0·0
·2
·4
·6
·8
1·0
·2
·4
Fig. 1.—ORDINATES—ABSOLUTE MAGNITUDE. ABSCISSAE—LOG. MASS
O First Class. X Second Class. □ Cepheids. Δ Eclipsing Variables.

Figure 30.1: On the vertical scale is the star's absolute magnitude; on the horizontal scale is the mass of the star in solar masses. The Sun is at log 0 (1 solar mass). A star of 25 solar masses is about 4,000 times brighter than our Sun.

31 / Sunspot Cycle, Sun/Earth Connection, and Helium

I. Solar Sunspot Cycle

With the invention of the telescope, sunspots immediately came under examination by a number of observers, including Galileo in Italy, Thomas Harriot in England, Johannes Fabricius in Holland, and Christoph Scheiner in Germany. But no great advances were made in understanding the phenomenon until 1826, when Heinrich Schwabe of Dessau, a town southwest of Berlin, acquired a small telescope and began to regularly monitor the Sun. Trained as an apothecary in the family business, he eventually devoted himself to observing full time. Schwabe originally intended to look for the notorious missing planet Vulcan, thought to be orbiting between Mercury and the Sun (see Chapter 20), but along the way became fascinated with sunspots. After eighteen years of observing, Schwabe reported that he was seeing the number of spots regularly wax and wane about every ten years. His finding, tucked away as a small note in the German journal *Astronomische Nachrichten* in 1844, was largely unappreciated until the German natural historian Alexander von Humboldt included Schwabe's statistics in the third edition of his influential book *Kosmos*, published in 1851. Six years later, Schwabe received the Royal Astronomical Society's gold medal for his groundbreaking work. "Twelve years . . . he spent to satisfy himself, six more years to satisfy, and still thirteen more to convince mankind," noted the society's president, M. J. Johnson. "For thirty years never has the Sun exhibited his disk above the horizon of Dessau without being confronted by Schwabe's imperturbable telescope. . . . The energy of one man has revealed a phenomenon that had eluded even the suspicion of astronomers for 200 years!"[36] Shortly afterward, with more averaging, the sunspot cycle was recalculated to be about eleven years.

Sunspots are relatively cooler areas (3,700 K instead of 5,700 K) on the Sun's surface where magnetic field lines, just below the visible surface (photosphere), have become twisted and poke through the surface. They appear more frequently during periods of high solar activity and are thought to be related to gradual and periodic changes in both the polarity and strength of the Sun's overall magnetic field as it rotates.

"Solar Observations in the Year 1843."
Astronomische Nachrichten, Volume 21 (1844)
by Heinrich Schwabe

The weather throughout this year was so extremely favorable that I have been able to observe the Sun clearly on 312 days; however I counted only 34 groups of sun spots. . . .

. . . From my earlier observations, which I have communicated annually to this journal, there has already appeared a certain periodicity of sun spots, and the probable periodicity increases in certainty with this year's contribution. . . . I include here a complete list of all sun spots observed by me, noting in addition to the number of sun spots also the number of days of observation and the number of days on which there were no spots.

For the number of groups does not alone give sufficient accuracy for the determination of a period, because I am convinced that at times of great frequency of sun spots the number of groups is reckoned somewhat too small, while at times of their infrequent appearance the number is judged too large. In the first case several groups often merge into a single one, and in the second case it easily happens that one group, due to the disintegration of some spots, divides into two distinct groups. For this reason I shall probably be excused for repeating the former list.

Table 31.1

Year	Groups	Number of days of no spots	Number of days of observation
1826	118	22	277
1827	161	2	273
1828	225	0	282
1829	199	0	244
1830	190	1	217
1831	149	3	239
1832	84	49	270
1833	33	139	267
1834	51	120	273
1835	173	18	244
1836	272	0	200
1837	333	0	168
1838	282	0	202
1839	162	0	205
1840	152	3	263
1841	102	15	283
1842	68	64	307
1843	34	149	324

If we compare the number of groups with the number of days free from spots, we find that the sun spots had a period of about ten years, and that throughout five years they appeared so frequently that during this time there occurred few, if any, days free from spots.

The future must decide whether this period shows constancy; whether the time of least activity of the sun in producing sun spots lasts one or two years, and whether this activity increases more rapidly than it decreases.

II. The Sun/Earth Connection

In the eighteenth century, with the invention of a sensitive compass, scientists began to notice that the Earth's magnetic field could undergo sudden and intense variations—magnetic storms. Spurred by a proposal at a scientific congress held in Berlin in 1828, a network of observatories was eventually established worldwide to gather terrestrial magnetism data. By 1851 Johann Lamont, director of the Munich observatory, had discovered that magnetic readings in Munich and Göttingen tended to vary in a distinct way over a period of ten years. Edward Sabine, a British army officer and explorer, independently recognized the same cycle by reviewing observations made at two widely separated British stations in Toronto, Canada, and Hobarton, Tasmania, which meant the changes were worldwide, not just local. More importantly, Sabine made the daring suggestion that the changes were somehow linked to the Sun's activities, pointing out that the periodic ups and downs in the number of magnetic storms were in step with the solar cycle observed earlier by Heinrich Schwabe in Germany. What Sabine had uncovered was the first evidence of a Sun/Earth connection.

A dramatic event in September 1859 only emphasized the bond. Two astronomers in England noticed bright patches of light near a large group of sunspots, a flare that lasted some five minutes. Within eighteen hours, telegraphic communications on the Earth were disrupted, and sky watchers observed a magnificent auroral display.

In the early 1900s, George Ellery Hale established the first modern solar observatory atop Mount Wilson in southern California to carry out long-term investigations of the Sun. He soon revealed the existence of strong magnetic fields in sunspots.[37] Joseph Larmor in Great Britain suggested in 1919 that the Sun produced the magnetic fields because the spinning, glowing orb was acting as a giant dynamo.[38] Others, such as Kristian Birkeland and Carl Störmer in Norway, postulated that streams of particles were being unleashed by solar flares and impacting the Earth at its poles, leading to the brilliant auroral displays in the higher latitudes and the worldwide magnetic storms.[39] From observations of the motions of comet tails (which always point away from the Sun), the German theorist Ludwig Biermann in 1957 suggested that a diffuse ionized gas—a solar wind—is continually streaming outward from the Sun, filling interplanetary space.[40] Spacecraft missions to other planets later confirmed this. Some of the solar particles were found to be trapped in a ring around the Earth, now known as the Van Allen belt (see Chapter 59). By the early 1960s astronomers were measuring oscillations on the surface of the Sun, founding the field of helioseismology.[41] These quivers and shakes allow solar observers to probe the inner depths of the Sun, source of the Sun/Earth connection, much the way seismic tremors on the Earth permit geophysicists to scan the Earth's interior.

From "On Periodical Laws Discoverable in the Mean Effects of the Larger Magnetic Disturbances—No. II." *Philosophical Transactions of the Royal Society of London,* Volume 142 (1852) by Edward Sabine

III. Variation in the numbers and aggregate values of the disturbed observations [magnetic storms] *in different years.*

Table [31.2] exhibits the ratios of the numbers and aggregate values of the disturbed observations at Toronto and Hobarton in the different years, to the average annual number and aggregate value respectively.

On the first aspect of this Table, two features of principal interest present themselves; first, there is a considerable variation in the numbers and values of the disturbed observations in different years; and second, there is a remarkable correspondence in the variation in different years at the two stations. . . .

. . . We shall be inclined perhaps to regard the accordance in the ratios at the two stations in different years as being quite as near as could be expected, even on the extreme supposition which the case will admit, namely, that of *all* disturbances being *general.* That they are so *for the most*

Table 31.2

	Numbers		Values		
Years	Toronto	Hobarton	Toronto	Hobarton	Years
1843	0.68	0.52	0.55	0.48	1843
1844	0.76	0.81	0.73	0.82	1844
1845	0.72	0.72	0.62	0.67	1845
1846	1.31	1.09	1.26	1.03	1846
1847	1.19	1.36	1.40	1.44	1847
1848	1.37	1.50	1.43	1.60	1848

part at Toronto and Hobarton, may be concluded from the circumstance, that by far the greater part of the disturbances which form the subject of discussion in this paper, occurred *on the same days at the two stations.* . . .

Recurring now to the ratios in Table [31.2], and directing our attention to the *character* of the inequality which they show to have existed in the amount of disturbance in different years, the facts which present themselves most obviously and unquestionably to our notice are, that in the years 1843, 1844 and 1845, the ratios were uniformly *considerably less than unity,* and that in the years 1846, 1847 and 1848, they were as uniformly *considerably greater than unity.* . . . Facts of such remarkable character, evidenced by the independent and concurrent testimony of so large a body of observations at stations so widely distant from each other, seem to be well deserving the consideration of magnetical physicists; more particularly of those who are disposed to regard thermometrical differences as the cause of the periodical and other magnetic variations. The ratios of disturbance in the years 1846, 1847 and 1848, were nearly *twice as great* as in the years 1843, 1844 and 1845. Did there occur any notable differences of either local or general temperature, or thermometrical peculiarities of any description, in the years in question, to which variations of such magnitude in the ratios of magnetic disturbance can be ascribed, or with which they can be connected?

We should not however derive all the advantage which an examination of the ratios in Table [31.2] seems suited to afford to those who desire to obtain an insight into the character of the variations they represent, were we to overlook the still more remarkable fact which they manifest, of a general, and with a single exception, uninterrupted *progressive increase* in the amount of disturbance from a minimum in 1843 to a maximum in 1848. . . .

The variation in the amount of disturbance in the different years . . . has certainly far more the aspect of a *periodical inequality,* than of what may be called for distinction's sake, *accidental* variation. . . .

. . . In our present ignorance of the physical agency by which the periodical magnetic variations are produced, the possibility of the discovery of some cosmical connection which may throw light on a subject as yet so obscure, should not be altogether overlooked. As the sun must be recognized as at least the *primary* source of all magnetic variations which conform to a law of local hours, it seems not unreasonable that in the case of other variations also, whether of irregular occurrence or of longer period, we should look in the first instance to any periodical variation by which we may learn that the sun is affected, to see whether any coincidence of period or epoch is traceable. Now the facts of the *solar spots,* as they have been recently made known to us by the assiduous and systematic labors of Schwabe, present us with phenomena which appear to indicate the existence of some periodical affection of an outer envelope (the photosphere) of the sun; and it is certainly a most striking coincidence, that the period, and the epochs of minima and maxima, which [Mr.] Schwabe has assigned to the variation of the solar spots, are absolutely identical with those which have been here assigned to the magnetic variations. In the third volume of *Kosmos . . .* Baron von Humboldt has published a tabular abstract supplied by [Mr.] Schwabe, of the results of that gentleman's observations of the solar spots from 1826 to 1850; from which [Mr.] Schwabe has derived the conclusion, that "the numbers in the Table leave no room to doubt that, at least from the years 1826 to 1850, the solar spots have shown a period of about ten years, with maxima in 1828, 1837 and 1848, and minima in 1833 and 1843." . . .

[Mr.] Schwabe has not been able to derive from the indications of the thermometer or barometer any sensible connection between climatic conditions and the number of spots. The same remark would of course hold good in respect to the connection of climatic conditions with the magnetic inequalities, as their periodical variation in different years corresponds with that of the solar spots. But it is quite conceivable that affections of the gaseous envelope of the sun, or causes occasioning those affections, may give rise to sensible *magnetical* effects at the surface of our planet, without producing sensible *thermic* effects.

It may be confidently anticipated that so remarkable a coincidence in the degree of energy with which the causes producing obscurations in the luminous disc of the sun, and those producing the magnetic variations at the surface of our planet, appear to have acted in the different years between 1843 and 1848, will receive due attention at those observatories which, by their more permanent character, are more particularly adapted for the investigation of problems requiring several years for their solution.

As the physical agency by which the phenomena are produced is in both cases unknown to us, our only resource for distinguishing between accidental coincidence and causal connection seems to be *perseverance in observation,* until either the inferences from a possibly too limited induction are disproved, or until a more extensive induction has sufficed to establish the existence of a connection, although its precise nature may still be imperfectly understood. . . .

III. Discovery of Helium

There was one time—and only one—when an element was detected in the heavens before it was found on the Earth. Helium was discovered in the Sun in 1868.

To examine the Sun's outer atmosphere, early solar astronomers had to wait for rare eclipses, at which time they would see great prominences—flames of red tinged with lilac arising from bright emissions of hydrogen. J. Norman Lockyer, a former clerk in the English civil service who became noted for his skills in solar astronomy, advanced the field by commissioning a special spectroscope that could examine the solar atmosphere continually, without the aid of an eclipse. On October 20, 1868, he made his first observations, quickly noticing a brilliant yellow-orange line that he had never seen before. It was situated near the line that Fraunhofer had labeled D in his spectral studies (see Chapter 23).

Lockyer was not aware that Pierre Janssen of France had carried out a similar observation from India the previous August. Janssen's findings arrived by post at the French Academy just minutes before Lockyer's results were reported to England's Royal Society.[42] Rather than turning into rivals, the two became professional friends.

Lockyer later suggested that the newly discovered spectral line, D_3, in the solar prominence was evidence of an undiscovered element. He dubbed it "helium," after *helios*, the Greek word for Sun. He was emboldened by the fact that no laboratory on Earth was able to produce the line from known elements. Lockyer became doubtful, though, when a quarter century went by without an identification, making him wonder whether the line was due to hydrogen under special conditions, as others believed.

The mystery was solved in 1895 when the Scottish chemist William Ramsay, upon examining gases released from the rare mineral cleveite, noticed a bright yellow line in the spectrum and in the very position as Lockyer's enigmatic solar line.[43] The missing element, helium, was at last found on the Earth. Helium is the second most abundant element in the universe (after hydrogen), accounting for around a quarter of the universe's ordinary matter by mass.

"Spectroscopic Observations of the Sun—No. II."
Philosophical Transactions of the Royal Society of London, Volume 159 (1869)
by J. Norman Lockyer

I began my work with the new instrument by continuing my search after the prominences. I found that the circumsolar light was now so greatly reduced that, although the lines were faintly seen on a dimly colored background, the background itself was apparently dark enough to render a bright line distinctly visible. My first attempts, however, with the new instrument, not yet in adjustment, were as unsuccessful as those made with the smaller one; and it was not till the 20th of October that, after sweeping for about an hour round the limb and arriving at the vertex of the image, near the south pole of the sun, I saw a bright line flash into the field.

My eye was so fatigued at the time that I at first doubted its evidence, although, unconsciously, I exclaimed "at last!" The line, however, remained—an exquisitely colored line absolutely coincident with the line C of the solar spectrum, and, as I saw it, a prolongation of that line. Leaving the telescope to be driven by the clock, I quitted the observatory to fetch my wife to endorse my observation.

Detail of the Observations

October 20—Having settled that the new line was absolutely coincident with C, I commenced to search for more lines. This I found very difficult, as the instrument requires several movements and adjustments for the various parts of the spectrum, and the rate of the driving-clock was not properly adjusted for the sun's motion; the prominence was therefore lost at times; moreover the observations were impeded by clouds.

I commenced the search for lines from C to A. B was first brought into the field with the newly discovered line at C. There were no new lines visible. I then made an excursion to A with no result, and returned to C to assure myself that the prominence was still on the slit.

I then worked from the line at C towards D. A little beyond D, the lines of which are widely separated in my instrument, I detected another single and less vivid line, by estimation 8° or 9° of Kirchhoff's scale more refrangible than the more refrangible of the strongest D lines. I could detect no line corresponding to it in the solar spectrum, but the definition was not good.

b was next tried, the excursion now being made from the new line near D. There was no line at *b,* though the new D line was still visible when I returned to it. . . .

[Omitted here are Lockyer's description of his observations on October 22 and 27.]

November 5—The next observations were made on this date under superb atmospheric conditions, and after an important alteration had been made in the instrument, enabling me to make the several adjustments with the utmost nicety.

After the adjustments to the sun's limb had been made, I at once saw what I imagined to be the indication of a small prominence, and swept for a development of it, thinking that the portion observed might be one of the loops or lower levels which generally separate the higher peaks. Having swept for some distance on both sides the region on which the telescope was clamped in the first instance, and finding everywhere the same uniformity of height, it at once struck me that I was in the presence of something new, and that possibly what I was seeing might indicate a solar envelope. I rapidly, therefore, tried several other parts of the limb to test the idea. It was soon established. *In every solar latitude both the C and F bright lines were seen extending above the solar spectrum.* The spectrum near D was so bright that I was compelled to refrain from examining it, but I caught the line near D once. . . .

On the Spectrum of the Prominences

The existence of three lines in the spectrum of the prominences and their approximate positions were determined and communicated to the Royal Society on the 20th of October. See [Figure 31.1, numbers 1, 2, 3].

The coincidence of one of the lines with the solar line C was at once determined.

The coincidence of another line with F at a certain distance from the sun's surface was finally determined on the 5th of November, when the fact of the widening out of the lines towards the sun was discovered.

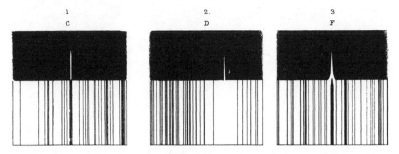

Figure 31.1: "1, 2, and 3 show the position of the lines observed on October 20, 1868, and their usual form, *i.e.* the line F is broad at the base and gradually tapers upwards, while C, and the line near D (with no corresponding absorption-line ordinarily visible) do not, as a rule, present this peculiarity with the instrument employed."

The exact position of the line near D is shown in [Figure 31.1, number 2], in which it is laid down from the mean of three careful micrometrical measurements made under far from good atmospheric conditions on the 15th of November. In Kirchhoff's map the new line falls in a region where no line was measured by him. I may also add that, by the kindness of Mr. Gassiot, I have been enabled to inspect the very elaborate maps of the spectrum constructed at Kew Observatory [outside London]. The measures above given make the new line fall between two lines of almost inconceivable faintness; in Mr. Gassiot's map, indeed, there are none but such lines for some distance on either side of the region in which the new one falls. . . .

32 / Origin of Meteors and Shooting Stars

M eteors were long regarded by observers as an atmospheric phenomenon, possibly "thunderstones" formed in violent storms or ejected from volcanoes. But brilliant fireball events, well tracked in France in 1790 and 1803 and then near Weston, Connecticut, in 1807, led to suspicions that the luminous fireballs had a cosmic origin. In 1794 the German physicist Ernst Chladni published a theory speculating that meteors were debris left over from the formation of the solar system, occasionally falling to the Earth as meteorites from interplanetary space.[44] His theory gained credence when mineralogists proved the fallen stones were very different from the Earth's crustal rock.[45]

Denison Olmsted, a professor of natural philosophy at Yale College, later gathered compelling proof that meteor showers—torrents of shooting stars seen periodically—were somehow linked to comets orbiting the Sun. Olmsted's evidence arrived on the night of November 12, 1833, when observers witnessed one of the most spectacular meteor showers ever recorded. After watching through the early morning hours, he came to suspect that the shower was emanating from a specific location in the heavens that had never changed over the night—the constellation Leo—which indicated the origin of the meteors was celestial rather than terrestrial. To resolve his suspicion, Olmsted put out a call to hear the accounts of other observers from throughout North America and around the world. He was particularly intrigued that a similar shower had occurred in the month of November in 1799. The data he collected led him to the conclusion that the shower occurred whenever the Earth periodically passed through nebulous cometary material orbiting the Sun.

This idea was not immediately accepted. Meteor observer W. A. Clarke in England continued to argue that the showers were electrical agitations in the atmosphere, linked to volcanic eruptions. "The subsequent agitations of the atmosphere in 1834, as well as in 1833, the gales that occurred, and the volcanic phenomena that preceded, all lead to the same conclusion," he wrote.[46]

Olmsted thought the comet traveled in an orbit interior to the Earth's. But Hubert Newton, also at Yale, later tracked the 1833 storm back through the centuries, as far as A.D. 902, and confirmed that it was linked to a swarm of meteoroids (mostly dust particles millimeters in size) orbiting the Sun, out to Uranus, every thirty-three years. "Until they come into the Earth's atmosphere, where they burn for an instant and are dissipated into smoke or dust," he wrote in 1864.[47] He predicted the storm would return with splendor in 1866, which occurred right on schedule.

All doubts about a cosmic origin dissipated when Giovanni Schiaparelli in Italy that same year demonstrated that the Perseid meteor showers, which regularly appear every August, shared the same orbit as a comet sighted in 1862. A year later, he specifically linked Olmsted's Leonid shower to a rather unspectacular comet known as Tempel-Tuttle, then just recently discovered.[48] The shower occurs when the Earth in its orbit crosses the comet's dust-laden path.

"Observations on the Meteors of November 13th, 1833."
The American Journal of Science and Arts,
Volumes 25 (January 1834) and 26 (July 1834)
by Denison Olmsted

The morning of November 13th, 1833, was rendered memorable by an exhibition of the phenomenon called shooting stars, which was probably more extensive and magnificent than any similar one hitherto recorded. The morning itself was, in most places where the spectacle was witnessed, remarkably beautiful. The firmament was unclouded; the air was still and mild; the stars seemed to shine with more than their wonted brilliancy, a circumstance arising not merely from the unusually transparent state of the atmosphere, but in part no doubt from the dilated state of the pupil of the eye of the spectator, emerging suddenly from a dark room; the large constellation Orion in the southwest, followed by Sirius and Procyon, formed a striking counterpart to the planets Saturn and Venus which were shining in the southeast; and, in short, the observer of the starry heavens would rarely find so much to reward his gaze, as the sky of this morning presented, independently of the magnificent spectacle which constituted its peculiar distinction.

Probably no celestial phenomenon has ever occurred in this country, since its first settlement, which was viewed with so much admiration and delight by one class of spectators, or with so much astonishment and fear by another class. For some time after the occurrence, the "Meteoric Phenomenon" was the principal topic of conversation in every circle, and the descriptions that were published by different observers, were rapidly circulated by the newspapers, through all parts of the United States. . . .

Descriptions

1. Phenomena as observed at New Haven (Lat. 41° 18′ N., Lon. 72° 58′ W.) and published in the New Haven Daily Herald.

About day break this morning, our sky presented a remarkable exhibition of fire balls, commonly called *shooting stars.* The attention of the writer was first called to the phenomenon about half past five o'clock; from which time until near sunrise, the appearance of these meteors was striking and splendid, beyond anything of the kind he has ever witnessed.

To form some idea of the phenomenon, the reader may imagine a constant succession of fire balls, resembling sky rockets, radiating in all directions from a point in the heavens, a few degrees southeast of the zenith, and following the arch of the sky towards the horizon. They commenced their progress at different distances from the radiating point, but their directions were uniformly such, that the lines they described, if produced upwards, would all have met in the same part of the heavens. Around this point, or imaginary radiant, was a circular space of several degrees, within which no meteors were observed. The balls, as they traveled down the vault, usually left after them a vivid streak of light, and just before they disappeared, exploded or suddenly resolved themselves into smoke. No report or noise of any kind was observed, although we listened attentively. . . .

A quarter before six o'clock, it appeared to the company that the point of apparent radiation was moving eastward from the zenith, when it occurred to the writer to mark its place, accurately, among the fixed stars. The point was then seen to be in the constellation Leo, within the bend of the sickle, a little to the westward of Gamma Leonis. During the hour following, the radiating point remained stationary in the same part of Leo, although the constellation in the mean time, by the diurnal revolution, moved westward to the meridian nearly 15 degrees. . . .*

The writings of Humboldt contain a description of a similar appearance observed by Bonpland, at Cumana, in 1799. It is worthy of remark, that this phenomenon was seen nearly at the same hours of the morning, and so on the 12th of November. . . .

Explanation

The principal questions involved in the present inquiry, are the following. Was the *origin* of the meteors within the atmosphere or beyond it? What was the *height* of this place above the surface of the earth? By what *force* were they drawn or impelled towards the earth? In what *directions* did they move? With what *velocity?* What was the cause of their *light* and *heat?* Of what *size* were the larger varieties? At what height above the earth did they disappear? What was the nature of the *luminous trains,* which sometimes remained behind? What *sort of bodies* were the meteors themselves—of *what kind of matter* constituted—and in what manner did they exist *before they fell to the earth?* Finally, what *relations* did the source from which they emanated sustain to our earth?

The meteors of November 13th had their origin beyond the limits of our atmosphere.

All bodies near the earth, including the atmosphere itself, have a com-

---

* "Aware of the importance of this fact to the question whether the origin of the meteors was terrestrial or not, the writer remarked it with much interest; but the advancing light of day rendered his means of observation imperfect, and he therefore felt it necessary to rely on those who saw the phenomena earlier and longer, for a confirmation of it, if the fact was so. Accordingly, in the [newspaper] of the succeeding day, he inserted a special request for information respecting this point. The same request has been addressed to observers in several places remote from each other; the result will appear in the sequel."

mon motion with the earth around its axis from west to east; but the *radiant point,* which indicated the position of the source from which the meteors emanated, followed the course of the stars from east to west: therefore, it was independent of the earth's rotation, and consequently at a great distance from it, and beyond the limits of the atmosphere.

It has been supposed that this westerly progress of the radiant point might be owing to the effects of *a strong current of wind,* in the upper regions of the atmosphere; for, although the wind at the surface was at that time in the opposite direction, namely, from west to east, yet counter currents of wind are known sometimes to exist at different elevations. But it would be very remarkable that the progress of the wind westward should *exactly keep pace* with the revolution of the earth in the opposite direction; and it is, moreover, inconceivable that the wind should blow with such a velocity—a velocity which, in our latitude, is nearly seven hundred and fifty miles an hour, while the most violent hurricanes rarely exceed one hundred miles an hour. . . .

That the source of the meteors did not partake of the earth's rotation, but that it existed in space in such a manner that places lying westward of each other came successively under it by the diurnal revolution, may be inferred from the fact, that the phenomenon, at any given stage, as at the *maximum,* for example, *occurred nearly at the same hour of the night,* at places differing greatly in longitude. . . .

The meteors consisted of combustible matter, and took fire and were consumed in traversing the atmosphere.

That these bodies underwent combustion, we had the direct evidence of the senses. We saw them glowing with intense light and heat, increasing in size and splendor as they approached the earth; we saw them extinguished in a manner in all respects resembling a combustible body like a sky rocket, burnt in the air; and in the case of the larger, we saw, for the product of combustion, a cloud of luminous vapor, which frequently spread over a great extent, and remained in sight, in some cases, for half an hour. . . .

That these bodies took fire *in the atmosphere,* we infer from the fact that they were not luminous in their original situation in space, otherwise we should have seen the cloud, or body, or whatever it was, from which they emanated; but they were not luminous except for the few seconds while they were within the atmosphere, for had they been so before, we should have seen them during the whole of their progress towards the earth. . . .

The meteors were combustible bodies and were constituted of light and transparent materials.

(1) The fact that they burned is sufficient proof that they belonged to the class of *combustible* bodies . . .

(2) They must have been composed of comparatively *light materials,* otherwise their momentum would have been sufficient to enable them to make their way through the atmosphere to the surface of the earth. . . .

(3) . . . If we were permitted to class unknown things with unknown, we should say that the cloud which produced the fiery shower consisted of nebulous matter, analogous to that which composes the tails of comets. We do not know, indeed, precisely what is the constitution of the material of which the latter are composed; but we know that it is very *light,* since it exerts no appreciable force of attraction on the planets, moving even among the satellites of Jupiter without disturbing their motions, although its own motions, in such cases, are greatly disturbed, thus proving its materiality; and we know that it is exceedingly *transparent,* since the smallest stars are visible through it. Indeed, Sir John Herschel was able to see stars through the densest part of the small comet (Biela's) which visited our planet last year. Hence, so far as we can gather any knowledge of the material of the nebulous matter of comets, and of that composing the meteors of Nov. 13th, they appear to be analogous to each other. . . .

We have finally to enter on the inquiry, *What relations did the body which afforded the meteoric shower sustain to the earth?* Was it of the nature of a *satellite,* or *terrestrial comet,* that revolves around the earth as its center of motion—was it a collection of *nebulous matter,* which the earth encountered in its annual progress—or was it a *comet,* which chanced at this time to be pursuing its path along with the earth, around their common center of motion?

We conclude that it could not have been of the nature of a satellite to the earth, because it remained so long stationary with respect to the earth. . . . Nor can we suppose that the earth, in its annual progress, came into the vicinity of a *nebula,* which was either stationary, or wandering lawless through space. Such a collection of matter could not remain stationary within the solar system, in an insulated state. . . .

We have seen that the meteors appeared to be analogous, in their constitution, to the material of which the nebulous matter of comets is composed, in all the particulars in which we can compare the two. We may be permitted, therefore, in order to avoid circumlocution, to call the body which afforded the meteoric shower, a comet. . . .

Now the comet remained apparently at rest. . . . This it could not have done, unless it had been moving in nearly the same direction as the earth, and with nearly the same angular velocity around the sun. For had it been at rest, the earth, moving at the rate of 19 miles per second, would have overtaken it in less than two minutes; or, had it been moving in the opposite direction, the meeting would have occurred in still less time; or had not the angular velocities of the two bodies been nearly equal, they could not have remained so long stationary with respect to each other. Hence we conclude, (1) *that the body was pursuing its way along with the earth around the sun.*

33 / Cosmic Rays

In a series of balloon flights in the 1910s, Austrian physicist Victor Hess gathered the first convincing evidence that a penetrating radiation measured in the Earth's atmosphere, earlier detected by Theodor Wulf from atop the Eiffel Tower in Paris, did not originate in the Earth or in the air but arrived from outer space.

Hess, an ardent amateur balloonist, was then a physicist at the Institute for Radium Research in Vienna. His moment of discovery came on August 7, 1912, during the seventh in a series of balloon flights he was conducting that year. Using a newly improved gold-leaf electrometer for measuring ionization, he detected a noticeable increase in his readings as his balloon rose to an altitude of 5,350 meters (3.3 miles), too far up to be radiation released by radioactive substances in the Earth. Hess came to call it his *Höhenstrahlung* or "radiation coming from above."

Many were not convinced, however, until Caltech physicist Robert Millikan entered the field in the 1920s and used unmanned balloons to take his instruments to far higher heights, up to 15 kilometers. He and his colleagues also took a series of measurements atop high mountains and deep underwater, confirming that the radiation did not originate in the Earth's atmosphere and was highly penetrating. Thinking at first that the radiation was electromagnetic in nature—wavelengths shorter than gamma rays—Millikan dubbed them "cosmic rays" at a 1925 meeting of the National Academy of Sciences in Wisconsin. He mistakenly theorized that these energetic photons were released when hydrogen atoms in interstellar space somehow condensed into higher elements. To Millikan, cosmic rays were the "birth cries of infant atoms."[49]

With the development of Geiger counters and cloud chambers, physicists came to realize that cosmic rays were not photons but rather charged particles. A battle between Millikan and University of Chicago physicist Arthur Compton raged for years over this point. Compton confirmed the rays' charged-particle nature in 1932 by sending teams of researchers around the globe, from Alaska to New Zealand, and demonstrating that the rays vary in intensity with latitude; going from the equator to the poles, cosmic rays increase in number as they get deflected by the Earth's magnetic field.[50] Cosmic rays can be either atomic nuclei, electrons, or protons in a range of energies; they enter the Earth's atmosphere from all directions. Upon colliding with air molecules, the primary rays generate a cascade of secondary particles that plummet to the ground. Before the era of large particle accelerators, physicists used cosmic rays as a means of studying high-energy nuclear interactions and discovering new elementary particles.

In 1934 astronomers Walter Baade of the Mount Wilson Observatory and Fritz Zwicky of Caltech suggested the rays originated in spectacular stellar blasts, explosions they dubbed "supernovae" (see Chapter 41).[51] And in 1949 Enrico Fermi proposed that the rays were whisked along through interstellar space by interacting with our galaxy's magnetic field.[52] Additional sources of cosmic rays were recognized with the discovery of pulsars and black holes.

"Concerning Observations of Penetrating Radiation on Seven Free Balloon Flights."
Physikalishe Zeitschrift, Volume 13 (1912)
by Victor Hess

By last year I had already had the opportunity to undertake two balloon flights to investigate the penetrating radiation; I have reported on the first flight at the scientific congress in Karlsruhe. In both flights no significant change in radiation from ground level to 1,100-m altitude was found. In two balloon flights Gockel also had not been able to find the expected decrease in radiation with altitude.* The inference was drawn that, in addition to the γ radiation of radioactive substances in the earth's crust, there must exist still another source of penetrating radiation.

A subvention from the Kaiserliche Akademie der Wissenschaften in Vienna made it possible this year for me to carry out a sequence of seven further balloon flights, whereby more extensive and, in several respects more extended observational material was obtained.

To observe the penetrating radiation two Wulf radiation apparatuses of 3-mm wall thickness served in the first line. These were closed in a completely airtight way and were also made to withstand all the pressure changes occurring on balloon flights. . . .

For the purpose of studying the behavior of β radiation at the same time, I utilized yet a third apparatus. This, not built airtight, was a common Wulf two-filament electrometer; inverted on it was a cylindrical ionization receptacle of 16.7-I volume, made out of the thinnest sheet zinc commercially available (wall thickness 0.188 mm). Thus, soft radiation, such as β radiation, could still be to some extent effective. . . .

According to all observers of the penetrating radiation on towers, a steady decrease in radiation had been confirmed, while Gockel and I had not been able to find, with certainty, such a decrease in free balloons. Thus, there was a need for longer flights at low altitudes to carry out measurements and thereby obtain reliable mean values. Parallel observations with the thin-walled apparatus 3 should show whether the softer radiation behaves in the same way as the γ rays. . . .

The last and most important point of my investigation was the measurement of the radiation at the greatest possible altitudes. On the six flights undertaken with Vienna as starting point, the small carrying capacity of the gas used, as well as meteorological chance, did not permit measurement at very high altitudes; but I did succeed in taking measurements up to 5,350-m altitude on an ascent begun at Aussig on the Elbe.

Before each flight, control observations were made for several hours with all three apparatuses. Here the apparatuses were fastened, exactly as on a flight itself, by means of brackets to the balloon basket. The observations before the ascents were carried out at the clubhouse of the Austrian

* Swiss physicist Albert Gockel.

Aeroclub, on a flat lawn in the Prater in Vienna. L. V. King has expressed the conjecture that the balloon observations could be disturbed by the proximity of the possibly radioactive ballast sand, but I have never found a heightening of the radiation in the immediate vicinity of greater supplies of ballast sand. . . .

[Omitted here are Hess's observations and data gathered from his first six flights.]

Flight 7

We ascended at 6:12 A.M. on August 7, 1912, from Aussig on the Elbe. We flew over the Saxony border at Peterswalde, Struppen bei Pirna, Bischofswerda, and Kottbus. In the vicinity of the Schwielochsee we reached 5,350-m altitude. At 12:15 P.M. we landed at Pieskow, 50 km east of Berlin. . . .

On this flight the weather was not completely clear. A barometric depression approaching from the west made itself noticeable through the onset of cloudiness. Yet let it be expressly noted that we never found ourselves in a cloud, indeed not once in the vicinity of a cloud; because, at the time when the cumulus clouds appeared scattered in isolated balls over the whole horizon, we were already at altitudes above 4,000 m. When we traveled at the maximum altitude, there was a thin cloud layer, still much higher, above us. Its underside must have been at least 6,000-m altitude. The sun shimmered through only very weakly.

At 1,400–2,500-m mean altitude the radiation was approximately as strong as it is usually found to be on the ground. Then, however, *a clearly noticeable rise in radiation began in both thick-walled apparatuses, 1 and 2, with increasing altitude;* at 3,600 m above the ground the values already were 4–5 ions higher than on the ground.

In the thin-walled apparatus, the rise in radiation is apparent at even lower altitudes. Because of the uncertainty previously discussed, arising from the correction of the values of this apparatus to normal pressure, one may view this conclusion as not completely certain. Moreover, the qualitative rise in the uncorrected values of q_3 should also be recognized. The readings for apparatus 3 found an unintended end at 10:45 A.M.; an unexpected shock just before the reading at maximum height caused the ionization cylinder to loosen itself, and when it touched the center post the apparatus discharged itself.

Table 33.1: Mean values (q values in ions cm^{-3} sec^{-1})

Mean altitude over the earth (m)	Observed Radiation			
	Apparatus 1	Apparatus 2	Apparatus 3	
	q_1	q_2	$q_{3\,corr}$	q_3
0	16.3(18[a])	11.8(20)	19.6(9)	19.7(9)
to 200	15.4(13)	11.1(12)	19.1(8)	18.5(8)
200–500	15.5(6)	10.4(6)	18.8(5)	17.7(5)
500–1,000	15.6(3)	10.3(4)	20.8(2)	18.5(2)
1,000–2,000	15.9(7)	12.1(8)	22.2(4)	18.7(4)
2,000–3,000	17.3(1)	13.3(1)	31.2(1)	22.5(1)
3,000–4,000	19.8(1)	16.5(1)	35.2(1)	21.8(1)
4,000–5,200	34.4(2)	27.2(1)	—	—

a. The parenthetical numbers denote the number of observations from which the corresponding mean values were constructed.

For the two γ-ray apparatuses the values at the maximum altitude are from 20 to 24 ions higher than on the ground. . . . In order to obtain an overview of the change in the penetrating radiation with height, as represented in the mean values, I have assembled in table [33.1] all eighty-eight of the radiation values I observed on the balloon, arranged according to the corresponding altitude range. . . . From table [33.1] we notice that directly above the earth the total radiation decreases a little. In mean values these decreases amount to 0.8 to 1.4 ions. Because, however, for the single flights, a decrease up to 3 ions has several times been found, for many measurements over 2 ions, we will claim something like 3 ions as the maximum value of the decrease. This decrease reaches to approximately 1,000 m above the ground. As mentioned previously, it manifestly originates in the absorption of γ rays which emanate from the earth's surface. I conclude that *the γ rays from the earth's surface and the uppermost lower layers of the earth excite in zinc containers an ionization of about 3 ions cm^{-3} sec^{-1}.*

Already at altitudes of 2,000 m, a marked increase in radiation appears. In both thick-walled apparatuses, at 3,000–4,000 m the increase reaches the amount of 4 ions, at 4,000–5,200 m the amount of 16 to 18 ions. For the thin-walled apparatus 3 the increase appears much earlier and more strongly, if one reduces the values to normal atmospheric pressure.

What is the cause of this increase in penetrating radiation with altitude, which has been observed several times and simultaneously in all three apparatuses? If one restricts oneself to the point of view that only the well-known radioactive substances in the earth's crust and in the atmosphere emit radiation with the character of γ rays and produce an ionization in a closed container, then great difficulties face any explanation. . . .

Also, the variations in the radiation found by Pacini and Gockel on the land and sea and by me in a balloon above the ground cause great difficulties for an explanation of the penetrating radiation based exclusively on radioactive theory. I have observed such variations repeatedly in the middle of the night in a quiescent atmosphere. Because of the lack of any meteorological alteration, there is no reason for attributing the variations to changes in the distribution of radioactive substances in the atmosphere.

The results of the preceding observations may most easily be explained on the assumption that a radiation of very great penetrating power impinges on our atmosphere from above, and still evokes in the lowest layers a part of the ionization observed in closed vessels. The intensity of this radiation seems to be subject to oscillations in time which are still recognizable in hour-long read-off intervals. Because I did not find any decrease in radiation either at night or during a solar eclipse, one can hardly view the sun as the cause of this hypothetical radiation, at least as long as one thinks only of a direct γ ray with straight line propagation. . . .

The hitherto existing investigations have shown that the penetrating radiation observed in closed vessels has a very complex origin. A part of the radiation originates in the earth's surface and in the uppermost layers of the earth and is altered relatively little. A second portion, influenced by meteorological factors, originates in radioactive substances in the air, most significantly RaC. My balloon observations seem to prove that there exists still a third component in the total radiation. This component increases with altitude and also produces noteworthy intensity variations on the ground. The greatest attention will have to be paid to this component in any further researches.

34 / Discovery of Pluto

Throughout the second half of the nineteenth century, new members of the solar system were found with increased proficiency. There was Neptune's moon Triton and new moons for Uranus and Saturn, and in 1892 Edward Barnard at Lick Observatory revealed a fifth satellite for Jupiter, the first addition to Jupiter's set of moons in nearly three centuries. It was the last satellite to be found without the aid of photography. Even earlier, in 1877, Asaph Hall from the U.S. Naval Observatory in Washington, D.C., discovered the solar system's most oddball satellites—the two tiny moons of Mars, Deimos and Phobos, which rapidly circle their planet in low orbit. "The peculiar appearance of these two moons to an inhabitant of Mars is evident on the slightest consideration," reported Hall to England's Royal Astronomical Society. "On account of the rapid motion of the inner moon it will rise in the west and set in the east, and, meeting and passing the outer moon, it will go through all its phases in about eleven hours."[53]

Jupiter's great red spot became highly visible in 1878 and quickly became a must-see for anyone owning a telescope. (Smaller spots were noticed earlier; one discovered in 1665 vanished and reappeared some nine times over a forty-eight-year span.) And the mystery of Saturn's rings was at last solved. Entering a prize competition in 1857, the noted English theorist James Clerk Maxwell, then a young professor, crafted an elegant mathematical proof that the rings had to consist of "an indefinite number of unconnected parti-cles" in order to exist over time.[54] Observations eventually proved him right. In 1895 James Keeler, then director of the Allegheny Observatory in Pittsburgh, determined that the velocity of the inner ring was decidedly faster than that of the outer ring, which would not be the case if the ring were solid.[55]

Mars gained a special allure. A specialist in planetary observa-tions at the Brera Observatory in Milan, Giovanni Schiaparelli pro-duced a detailed map of Mars during the planet's close approach in 1877, giving special attention to the *canali*, dark channels crossing the planet's brighter orange surface. This sparked nearly a century of controversy on the true nature of the red planet (see Chapter 60). With *canali* being translated as "canals" in English, speculation arose whether the canals had been constructed by intelligent beings. This idea was especially promoted by Percival Lowell, who continued the examination of Mars during its next close approach in 1894, using an observatory he privately established near Flagstaff, Arizona.

Along with his belief in Martian-built canals, Lowell had another cause célèbre that he pursued with a passion: the search for a "Planet X" beyond Neptune. Analyzing discrepancies in the motions of Uranus and Neptune, similar to the calculations that Urbain Jean Joseph Le Verrier and John Adams had carried out in the 1840s to find Neptune (see Chapter 20), Lowell came up with a predicted location for the missing planet—out in the farthest realm of the comets. "The Perseids and the Lyrids go out to meet the unknown planet, which circles at a distance of about forty-five astro-nomical units from the Sun," Lowell surmised.[56]

Lowell died in 1916, unsuccessful in his effort, but his nephew eventually took over the observatory and by 1929 set up new equip-ment for the search. Clyde Tombaugh, a twenty-two-year-old ama-teur astronomer from Kansas, was hired to help carry it out. Unlike the searches for Uranus and Neptune, photography played a major role in the hunt. Photographs of a region were taken days apart and their images compared for signs of an object that had shifted. With a blink comparator, the separate images were swiftly alternated; in this way, the fixed stars remained in place, while any moving bodies, such as a planet or asteroid, appeared to move back and forth.

After a shaky start, Tombaugh set up a systematic schedule, photographing a different constellation of the zodiac every month. After five months, he arrived at Gemini. Comparing two images in the blink comparator that he had photographed in late January 1930 near the star δ Geminorum, Tombaugh saw a tiny dot jump back and forth. Its faintness and lack of a disk suggested it was located in the far reaches of the solar system. Rushing to the office of the observatory director, Vesto Slipher, Tombaugh declared: "I have found your Planet X."[57] Further observing over the next few weeks confirmed the find, and a discovery notice was sent out on March 13, the seventy-fifth anniversary of Lowell's birth. For the naming, classical tradition prevailed. The new planet became Pluto, brother of Jupiter and Neptune, as well as god of the underworld. It didn't go unnoticed that the first two letters—PL—also honored Percival Lowell, who initiated the search.

Pluto was once thought to be more massive than the Earth but current measurements peg its mass as 1/500 that of the Earth and its diameter at only 1,400 miles, just slightly larger than an asteroid. Its orbit crosses Neptune's and is more tilted with respect to the plane of the solar system. And Pluto's moon Charon, discovered in 1978, is about half its size. Given these statistics, some astronomers are ready to demote Pluto from planethood. By the 1990s other objects, smaller but similar to Pluto, were discovered out past Neptune, confirming the existence of the Kuiper belt, a vast region of leftover cometary material at the edge of the solar system.[58] Some like to think of Pluto not as a planet but as the largest object yet found in the Kuiper belt.

"The Discovery of a Solar System Body Apparently Trans-Neptunian." *Lowell Observatory Observation Circular*, March 13, 1930

The message sent last night (March 12) to Harvard Observatory for distribution to astronomers read as follows:

"Systematic search begun years ago supplementing Lowell's investigations for TransNeptunian planet has revealed object which since seven weeks has in rate of motion and path consistently conformed to TransNeptunian body at approximate distance he assigned. Fifteenth magnitude. Position March twelve days three hours GMT was seven seconds of time West from Delta Geminorum, agreeing with Lowell's predicted longitude."

$\Big($ For ease in finding object was referred to Delta Geminorum.
Position March 12.14 G.M.T. R.A. 7^h 15^m 50^s Dec. $22°$ $6'$ $49''$ $\Big)$

The finding of this object was a direct result of the search program set going in 1905 by Dr. Lowell in connection with his theoretical work on the dynamical evidence of a planet beyond Neptune. (See L. O. Memoirs, Vol. I, No. 1, "A Trans-Neptunian Planet," 1914.) The earlier searching work, laborious and uncertain because of the less efficient instrumental means, could be resumed much more effectively early last year with the very efficient new Lawrence Lowell telescope specially designed for this particular problem. Some weeks ago, on plates he made with this instrument, Mr. C. W. Tombaugh, assistant on the staff, using the Blink Comparator, found a very exceptional object, which since has been studied carefully. It has been photographed regularly by Astronomer Lampland with the 42-inch reflector, and also observed visually by Astronomer E. C. Slipher and the writer with the large refractor.

The new object was first recorded on the search plates of January 21 (1930), 23rd, and 29th, and since February 19 it has been followed closely. Besides the numerous plates of it with the new photographic telescope, the object has been recorded on more than a score of plates with the large reflector, by Lampland, who is measuring both series of plates for positions of the object. Its rate of motion he has measured for the available material at intervals between observations with results that appear to place the object outside Neptune's orbit at an indicated distance of about 40 to 43 astronomical units. During the period of more than 7 weeks the object has remained close to the ecliptic; the while it has passed from 12 days after opposition point to within about 20 days of its stationary point. Its rate of retrogression, March 10 to 11, was about $30''$ per day. In its apparent path and in its rate of motion it conforms closely to the expected behavior of a Trans-Neptunian body, at about Lowell's predicted distance. There has not been opportunity yet to complete measurements and accurate reductions of positions of the object requisite for use in the computation of the orbit, but it is realized that the orbital elements are much to be desired and this important work is in hand.

In brightness the object is only about 15th magnitude. Examination of it in the large refractor—but without very good seeing conditions—has not revealed certain indication of a planetary disk. Neither in brightness nor apparent size is the object comparable with Neptune. Preliminary attempts at comparative color tests photographically with large reflector and visually with refractor indicate it does not have the blue color of Neptune and Uranus, but hint rather that its color is yellowish, more like the inner planets. Such indications as we have of the object suggest low albedo and high density. Thus far our knowledge of it is based largely upon its observed path and its determined rates of motion. These with its position and distance appear to fit only those of an object beyond Neptune, and one apparently fulfilling Lowell's theoretical findings.

While it is thus too early to say much about this remarkable object and much caution and concern are felt—because of the necessary interpretations involved—in announcing its discovery before its status is fully demonstrated; yet it has appeared a clear duty to science to make its existence known in time to permit other astronomers to observe it while in favorable position before it falls too low in the evening sky for effective observation.

—V. M. Slipher

V

EINSTEINIAN COSMOS

Just as Isaac Newton had transformed the field of astronomy in the seventeenth and eighteenth centuries with his law of gravitation, so too did dual revolutions in physics drastically affect the course of astronomy in the twentieth century. Once the theories of quantum mechanics and relativity were in place, astronomers could at last begin to comprehend how stars generated enough energy to shine for billions of years, to perceive the intricate course of stellar evolution, to discover how the elements were constructed, and to contemplate the universe's very origin.

The man who stood at the epicenter of this twentieth-century transformation was Albert Einstein. His special theory of relativity, introduced in 1905, not only made physicists completely revise their conceptions of length, time, and mass, it provided the crucial link between mass and energy that helped astronomers understand the mystery of stellar power. The solution was elegantly summarized in his famous equation $E = mc^2$. And as the laws of nuclear physics were better understood, astronomers came to grasp how giant stars formed. Applying both the laws of quantum mechanics and special relativity to various states of stellar matter also allowed physicists to recognize (to their surprise and amazement) the existence of such compact stellar bodies as white dwarfs and neutron stars, types of celestial objects never before imagined. The neutron stars, it was suggested, were forged when massive stars spectacularly exploded as brilliant supernovae.

But that was only the beginning of Einstein's momentous influence on astronomical concerns in the twentieth century. In 1915 Einstein introduced his general theory of relativity, which radically amended Newton's law of gravitation. General relativity turned space-time into a tangible entity, whose geometric shape is determined by the matter within it; according to this new perspective, gravity arises when massive bodies, such as stars, indent space-time, and other objects are then attracted by following the space-time curvatures carved out by the star. Einstein was elevated to the pinnacle of scientific celebrity in 1919 when this novel vision of gravity was verified. Evidence was gathered during a solar eclipse showing that starlight does indeed bend around the Sun due to the Sun's space-time warping. Some theorists later took that bending to its ultimate limit and revealed that particularly bizarre entities might exist in the universe, what have come to be known as black holes.

General relativity also allowed cosmology, once the province of philosophy alone, to become a bona fide science. By applying the equations of general relativity to the cosmos at large, astrophysicists started predicting the ultimate geometric structure of the universe. Indeed, general relativity was able to explain the observational evidence gathered in the 1920s that the universe was not static but rather expanding outward. Whether the cosmos expands forever or eventually collapses, they found, depends on the total amount of mass-energy within the universe's space-time boundaries.

Knowledge of a cosmic expansion ultimately led to contemplation of the universe's creation. By taking the universe's ballooning growth and mentally putting it into reverse, astrophysicists came to picture the cosmos as originating out of a compact fireball, which had exploded in a Big Bang. Theorists then recognized that this primordial plasma offered a suitably hot environment in which to start constructing the lightest elements. Over the centuries, astronomical progress has largely been ruled by either new observational discoveries or new instruments. But in the early twentieth century new theories, particularly those introduced by Einstein, were the powerful engines for advancement of the field.

35 / Special Relativity and E = mc²

The special theory of relativity was not a sudden revelation to Albert Einstein, arrived at in a single eureka moment. It was the result of deep reflection over many years on the contradictions coming to light in electrodynamics.

Einstein was bothered by a paradox, which he describes in the introduction to his historic 1905 paper on relativity. Consider either a bar magnet moving through a fixed coil or a coil moving over a stationary bar magnet. According to the equations of electromagnetism, the description of what is happening is different for each case, yet the observed outcome is exactly the same—the flow of an electric current in the coil. Isaac Newton had long established that space was an empty vessel, and everything in the universe was either at rest or in motion within this fixed container (later thought to be filled with a motionless ether to transmit light waves). Einstein was disturbed that the laws of electromagnetism could not reveal which object—the coil or the magnet—was *really* moving in absolute space. Einstein's master stroke was finding an answer to this conundrum with the simplest assumption possible: by recognizing that space and time are not absolute. There is neither a universal clock nor a fixed rest frame shared by everyone in the universe. The luminiferous ether, as he put it, became "superfluous."

Not a favored student among his professors while at school because of his impatience with rote learning, Einstein found no academic post upon graduation and eventually found work in 1902 at the Swiss patent office in Bern. His work as a patent examiner turned out to be a blessing. Unencumbered by academic duties or pressures, Einstein was able to explore his ideas freely. By 1905, at the age of twenty-six, like a dormant desert flower that suddenly blooms, he burst forth with a historic series of papers published in the distinguished German journal *Annalen der Physik.* Inspired by the new quantum mechanics, he first proposed that light consists of discrete particles, what came to be known as photons (his Nobel Prize–winning idea). Second, he helped persuade the scientific community that atoms truly exist by explaining that the jittery dance of microscopic particles—Brownian motion—arose from the buffets of surrounding atoms. Lastly, he submitted a paper entitled "On the Electrodynamics of Moving Bodies" in which he revealed his "principle of relativity."

This paper is remarkable in that it makes few references to experimental measurements, includes a mere four footnotes, and cites no previous literature. Most of its mathematics can be understood by an astute high school graduate. It specifically proposed that *all* the laws of physics (both mechanics and electromagnetism) remain the same, whether an object is at rest or moving at a constant velocity. But for that to be true, the speed of light must also remain the same, whether measured on the Earth or aboard a speeding spacecraft. What resulted was a radically new outlook on how the universe works. When observers move either toward or away from one another, they will disagree on their measurements of length, time, and mass from their separate vantage points. Each believes the other experiences space shrinking, mass increasing, and time slowing down (most prominently as velocities approach the speed of light). Measurement becomes "relative," depending on the reference frame. The only thing the different travelers will agree on is the speed of light. It is the one universal constant.

Within a few months of publishing his theory, Einstein recognized another consequence of special relativity and quickly sent off a postscript to the *Annalen* entitled "Does the Inertia of a Body Depend Upon Its Energy-Content?" As he related to a close friend in a letter: "The relativity principle in connection with the basic Maxwellian equations [of electromagnetism] demands that the mass should be a direct measure of the energy contained in a body; light transfers mass. With radium there should be a noticeable diminution of mass. The idea is amusing and enticing; but whether the Almighty is laughing at it and is leading me up the garden path—that I cannot know."[1] We do not see Einstein's solution in its most famous form in this brief, initial paper. Rather, the concept is summarized in a sentence at the end, which relates a *loss* of energy, L, to a *reduction* in mass. The full equivalence of mass and energy, as stated in the celebrated equation $E = mc^2$, was described by Einstein in more detail in 1907.[2] While it was largely the ongoing revolution in atomic physics that allowed researchers to figure out how the stars generated their immense quantities of energy (see Chapter 43), $E = mc^2$ entered the scene to help confirm it.

"On the Electrodynamics of Moving Bodies."
Annalen der Physik, Volume 17 (1905)
by Albert Einstein

It is known that Maxwell's electrodynamics—as usually understood at the present time—when applied to moving bodies, leads to asymmetries which do not appear to be inherent in the phenomena. Take, for example, the reciprocal electrodynamic action of a magnet and a conductor. The observable phenomenon here depends only on the relative motion of the conductor and the magnet, whereas the customary view draws a sharp distinction between the two cases in which either the one or the other of these bodies is in motion. For if the magnet is in motion and the conductor at rest, there arises in the neighborhood of the magnet an electric field with a certain definite energy, producing a current at the places where parts of the conductor are situated. But if the magnet is stationary and the conductor in motion, no electric field arises in the neighborhood of the magnet. In the conductor, however, we find an electromotive force, to which in itself there is no corresponding energy, but which gives rise—assuming equality of relative motion in the two cases discussed—to electric currents of the same path and intensity as those produced by the electric forces in the former case.

Examples of this sort, together with the unsuccessful attempts to discover any motion of the earth relatively to the "light medium," suggest that the phenomena of electrodynamics as well as of mechanics possess no properties corresponding to the idea of absolute rest. They suggest rather that, as has already been shown to the first order of small quantities, the same laws of electrodynamics and optics will be valid for all frames of reference for which the equations of mechanics hold good. We will raise this conjecture (the purport of which will hereafter be called the "Principle of Relativity") to the status of a postulate, and also introduce another postulate, which is only apparently irreconcilable with the former, namely, that light is always propagated in empty space with a definite velocity c which is independent of the state of motion of the emitting body. These two postulates suffice for the attainment of a simple and consistent theory of the electrodynamics of moving bodies based on Maxwell's theory for stationary bodies. The introduction of a "luminiferous ether" will prove to be superfluous inasmuch as the view here to be developed will not require an "absolutely stationary space" provided with special properties, nor assign a velocity-vector to a point of the empty space in which electromagnetic processes take place. The theory to be developed is based—like all electrodynamics—on the kinematics of the rigid body, since the assertions of any such theory have to do with the relationships between rigid bodies (systems of co-ordinates), clocks, and electromagnetic processes. Insufficient consideration of this circumstance lies at the root of the difficulties which the electrodynamics of moving bodies at present encounter. . . .

[Einstein goes on to show that the length and mass of a moving object, when measured from a stationary frame, will differ from the length and mass of the object when measured at rest. Time within the moving frame will also appear different from the time passing in the stationary frame. The amounts of these differences depend on the relative motion between the two systems of reference. To summarize Einstein's arguments mathematically:

$$(1)\ \text{Mass}_{\text{moving}} = \frac{\text{Mass}_{\text{rest}}}{\sqrt{1 - \frac{v^2}{c^2}}}$$

(1) The mass of a body moving at speed v relative to an observer is larger than its mass when at rest relative to the observer. Note from this equation that as the velocity approaches the speed of light, the mass approaches infinity. "For velocities greater than that of light our deliberations become meaningless," wrote Einstein.[3]

$$(2)\ \text{Length}_{\text{moving}} = \text{Length}_{\text{rest}} \sqrt{1 - \frac{v^2}{c^2}}$$

(2) The length of an object moving at speed v relative to an observer is shorter than its length when at rest relative to the observer. "The greater the value of v, the greater the shortening," said Einstein. "For $v = c$ all moving objects—viewed from the 'stationary' system—shrivel up into plane figures."[4]

$$(3)\ \text{Time}_{\text{moving}} = \frac{\text{Time}_{\text{rest}}}{\sqrt{1 - \dfrac{v^2}{c^2}}}$$

(3) The interval of time between ticks of a clock in a reference frame moving at speed v relative to an observer is longer than it would be in the observer's frame. In other words, a clock in motion appears to run more slowly to an observer at rest. "Thence we conclude," wrote Einstein, "that a balance-clock at the equator must go more slowly, by a very small amount, than a precisely similar clock situated at one of the poles under otherwise identical conditions."[5]]

"Does the Inertia of a Body Depend Upon Its Energy-Content?" *Annalen der Physik*, Volume 17 (1905) by Albert Einstein

The results of the previous investigation* lead to a very interesting conclusion, which is here to be deduced.

I based that investigation on the Maxwell-Hertz equations for empty space, together with the Maxwellian expression for the electromagnetic energy of space, and in addition the principle that:—

The laws by which the states of physical systems alter are independent of the alternative, to which of two systems of coordinates, in uniform motion of parallel translation relatively to each other, these alterations of state are referred (principle of relativity).

With these principles as my basis I deduced *inter alia* the following result:—†

* The previous investigation to which he refers is "On the Electrodynamics of Moving Bodies," *Annalen der Physik* 17 (1905): 891–921.

† "The principle of the constancy of the velocity of light is of course contained in Maxwell's equations."

Let a system of plane waves of light, referred to the system of co-ordinates (x, y, z), possess the energy l; let the direction of the ray (the wave-normal) make an angle φ with the axis of x of the system. If we introduce a new system of co-ordinates (ξ, η, ζ) moving in uniform parallel translation with respect to the system (x, y, z), and having its origin of co-ordinates in motion along the axis of x with the velocity v, then this quantity of light—measured in the system (ξ, η, ζ)—possesses the energy

$$l^* = l \, \frac{1 - \dfrac{v}{c} \cos \phi}{\sqrt{1 - \dfrac{v^2}{c^2}}}$$

where c denotes the velocity of light. We shall make use of this result in what follows.

Let there be a stationary body in the system (x, y, z), and let its energy—referred to the system (x, y, z) be E_0. Let the energy of the body relative to the system (ξ, η, ζ) moving as above with the velocity v, be H_0.

Let this body send out, in a direction making an angle φ with the axis of x, plane waves of light, of energy $\frac{1}{2}L$ measured relatively to (x, y, z), and simultaneously an equal quantity of light in the opposite direction. Meanwhile the body remains at rest with respect to the system (x, y, z). The principle of energy must apply to this process, and in fact (by the principle of relativity) with respect to both systems of co-ordinates. If we call the energy of the body after the emission of light E_1 or H_1 respectively, measured relatively to the system (x, y, z) or (ξ, η, ζ) respectively, then by employing the relation given above we obtain

$$E_0 = E_1 + \tfrac{1}{2}L + \tfrac{1}{2}L,$$

$$H_0 = H_1 + \tfrac{1}{2}L \, \frac{1 - \dfrac{v}{c} \cos \phi}{\sqrt{1 - \dfrac{v^2}{c^2}}} + \tfrac{1}{2}L \, \frac{1 + \dfrac{v}{c} \cos \phi}{\sqrt{1 - \dfrac{v^2}{c^2}}}$$

$$= H_1 + \frac{L}{\sqrt{1 - \dfrac{v^2}{c^2}}}$$

By subtraction we obtain from these equations

$$H_0 - E_0 - (H_1 - E_1) = L \left\{ \frac{1}{\sqrt{1 - v^2/c^2}} - 1 \right\}$$

The two differences of the form H − E occurring in this expression have simple physical significations. H and E are energy values of the same body referred to two systems of co-ordinates which are in motion relatively to each other, the body being at rest in one of the two systems (system (*x, y, z*)). Thus it is clear that the difference H − E can differ from the kinetic energy K of the body, with respect to the other system (ξ, η, ζ), only by an additive constant C, which depends on the choice of the arbitrary additive constants of the energies H and E. Thus we may place

$$H_0 - E_0 = K_0 + C$$
$$H_1 - E_1 = K_1 + C$$

since C does not change during the emission of light. So we have

$$K_0 - K_1 = L \left\{ \frac{1}{\sqrt{1 - v^2/c^2}} - 1 \right\}$$

The kinetic energy of the body with respect to (ξ, η, ζ) diminishes as a result of the emission of light, and the amount of diminution is independent of the properties of the body. Moreover, the difference $K_0 - K_1$, like the kinetic energy of the electron depends on the velocity.

Neglecting magnitudes of fourth and higher orders we may place

$$K_0 - K_1 = \frac{1}{2} \frac{L}{c^2} v^2$$

From this equation it directly follows that:—

If a body gives off the energy L in the form of radiation, its mass diminishes by L/c². The fact that the energy withdrawn from the body becomes energy of radiation evidently makes no difference, so that we are led to the more general conclusion that:

The mass of a body is a measure of its energy-content; if the energy changes by L, the mass changes in the same sense by $L/9 \times 10^{20}$, the energy being measured in ergs, and the mass in grams.

It is not impossible that with bodies whose energy-content is variable to a high degree (e.g. with radium salts) the theory may be successfully put to the test.

If the theory corresponds to the facts, radiation conveys inertia between the emitting and absorbing bodies.

36 / General Relativity and the Solar Eclipse Test

Throughout the month of November 1915 Albert Einstein reported to the Prussian Academy of Sciences on his final progress toward a new theory of gravitation, one that would recast Newton's laws in the light of relativity. He had been struggling with the problem for nearly a decade. A breakthrough arrived in mid-month, when he was able to successfully explain a small displacement in Mercury's orbit, a nagging mystery to astronomers for decades (see Chapter 20).[6] Einstein later remarked that he had palpitations of the heart upon seeing this result: "I was beside myself with ecstasy for days."[7]

Complete triumph arrived on November 25, the day he presented his concluding paper. In this culminating talk he presented the decisive modifications that allowed him to secure a truly *general* theory of relativity. Special relativity was exactly that—special. It dealt only with a specific type of motion: objects moving at a constant velocity (see Chapter 35). With general relativity, Einstein extended the theory to handle all types of movement: things that are speeding up, slowing down, or changing direction. And what he discovered by working within this new universal framework was a unique way to picture gravity.

Written in the deceptively simple notation of tensor calculus—shorthand for a larger set of more complex equations—the general theory of relativity displays a mathematical elegance:

$$R_{uv} - \tfrac{1}{2}\, g_{uv}\, R = T_{uv}$$

On the left side of the equation are quantities that describe the gravitational field as a geometry of space-time. On the right side is a representation of mass-energy and how it is distributed. The equal sign sets up an intimate relationship between these two entities. With general relativity, Einstein offered that space and time can be viewed as joining up and forming a palpable object known as space-time, a sort of boundless rubber sheet. Masses, such as a star or planet, indent this flexible mat, curving space-time. From this perspective, planets circle the Sun not because they are held by invisible tendrils of force, as Newton had us think, but because they are caught in the natural hollow formed by the Sun in four-dimensional space-time. The more massive the object, the deeper the depression. Mercury's orbit is proof of this effect. The point of Mercury's orbit that is closest to the Sun—its perihelion—shifts around over time due to the combined gravitational tugs of the other planets. But there is an added shift—an extra 43 arcseconds per century— due to Mercury's proximity to the Sun; it has more of a space-time "dip" to contend with.

Einstein himself suggested another test to confirm the space-time curvatures proposed in his theory: to photograph a field of stars at night, then for comparison photograph those same stars when they pass near the Sun's limb during a solar eclipse. A stellar beam of light passing right by the Sun would be gravitationally attracted to the Sun and get bent. Moreover, the attraction would be twice the bending calculated from Newton's laws alone.[8] The extra contribution, according to Einstein, comes from the warping of space-time near the Sun. Einstein calculated that a ray of starlight just grazing the Sun would get deflected by 1.7 arcseconds.

Three solar eclipse expeditions were carried out prior to 1919 to detect the light bending but were unsuccessful due to either bad weather or war. The results of a fourth, an American effort, were plagued by data comparison problems and so were never published. For Einstein, this was a fortuitous turn of events. The American results went against him, and some of the other expeditions were carried out when his theory, not yet fully developed, was predicting a smaller deflection. Attention was thus focused on British astronomers, who as victors in World War I had the necessary funding to conduct a test during a very favorable solar eclipse in 1919, one occurring in a section of the sky with an exceptional patch of bright stars. To be in the path of this eclipse, Arthur Eddington and his assistant E. T. Cottingham journeyed to the tiny isle of Principe off the coast of West Africa, a trip arranged by the astronomer royal Frank Watson Dyson. To improve their chances for a clear view, two other astronomers traveled to the village of Sobral in northern Brazil.

On the day of the eclipse, May 29, Eddington and his colleague took sixteen photographs, all but two ultimately useless because of intervening clouds. "We have no time to snatch a glance at [the Sun]," wrote Eddington of his adventure. "We are conscious only of the weird half-light of the landscape and the hush of nature, broken by the calls of the observers, and beat of the metronome ticking out the 302 seconds of totality."[9] At Sobral, they had two instruments and better weather. With their astrographic telescope, sixteen photographs were taken; eight were taken with a 4-inch scope.

General relativity had not been immediately embraced by physicists outside Germany. World War I, for one, restricted Einstein's paper from being widely circulated, and scientists trained in classical physics were leery of relativity's radically new outlook. Eddington was an exception and actively championed Einstein's ideas, despite British prejudice against German science, a feverish aftermath of the war. Contrary to the standard story, widely circulated in many textbooks, the 1919 solar eclipse results were not clear-cut. The best results supporting Einstein came from the Sobral 4-inch telescope; from its plates, the British astronomers determined a starlight deflection of 1.98 arcseconds. The poorer images from Principe suggested a bending of 1.61 arcseconds (give or take 20 percent). These were the results that Eddington and Dyson stressed in their reports, which were widely hailed in newspaper headlines worldwide, turning the name Einstein into a synonym for genius. Because of various technical problems with the instrument, they selectively downplayed the larger data set from the Sobral astrographic telescope, which displayed a deflection of 0.93 arcseconds and favored Newton.

Eddington admitted he was unscientifically rooting for Einstein, but the results held up over time. With the introduction of new astronomical techniques, light deflection experiments are now performed with exquisite precision. Using globe-spanning networks of radio telescopes, astronomers can monitor how the separation of close pairs of quasars changes as their radio signals pass close to the Sun. The accuracy in this type of test is nearly a thousand times better than Eddington's first crude try, and the results match Einstein's theory with near perfection. In the 1960s Irwin Shapiro, then with MIT's Lincoln Laboratory, devised a new test altogether. Radar signals were transmitted to both Venus and Mercury and reflected back to the Earth as the planets were about to pass behind the Sun.[10] Shapiro figured that the signal's excursion would take a bit longer than normal, because the Sun's warp in space-time adds a tad more distance to the journey. He was right; the measured delay in the Venus round trip—1/5,000 of a second—was within 0.1 percent of that predicted by general relativity.

"A Determination of the Deflection of Light by the Sun's Gravitational Field, from Observations Made at the Total Eclipse of May 29, 1919." *Philosophical Transactions of the Royal Society of London, Series A*, Volume 220 (1920) by Frank W. Dyson, Arthur S. Eddington, and Charles Davidson

I. Purpose of the Expeditions

The purpose of the expeditions was to determine what effect, if any, is produced by a gravitational field on the path of a ray of light traversing it. Apart from possible surprises, there appeared to be three alternatives, which it was especially desired to discriminate between—

(1) The path is uninfluenced by gravitation.
(2) The energy or mass of light is subject to gravitation in the same way as ordinary matter. If the law of gravitation is strictly the Newtonian law, this leads to an apparent displacement of a star close to the sun's limb amounting to $0''.87$ [arcsecond] outwards.

(3) The course of a ray of light is in accordance with Einstein's generalized relativity theory. This leads to an apparent displacement of a star at the limb amounting to 1".75 outwards.

In either of the last two cases the displacement is inversely proportional to the distance of the star from the sun's center, the displacement under (3) being just double the displacement under (2).

It may be noted that both (2) and (3) agree in supposing that light is subject to gravitation in precisely the same way as ordinary matter. The difference is that, whereas (2) assumes the Newtonian law, (3) assumes Einstein's new law of gravitation. The slight deviation from the Newtonian law, which on Einstein's theory causes an excess motion of perihelion of Mercury, becomes magnified as the speed increases, until for the limiting velocity of light it doubles the curvature of the path.

The displacement (2) was first suggested by Prof. Einstein in 1911, his argument being based on the Principle of Equivalence, viz., that a gravitational field is indistinguishable from a spurious field of force produced by an acceleration of the axes of reference.* But apart from the validity of the general Principle of Equivalence there were reasons for expecting that the electromagnetic energy of a beam of light would be subject to gravitation, especially when it was proved that the energy of radioactivity contained in uranium was subject to gravitation. In 1915, however, Einstein found that the general Principle of Equivalence necessitates a modification of the Newtonian law of gravitation, and that the new law leads to the displacement (3).

The only opportunity of observing these possible deflections is afforded by a ray of light from a star passing near the sun. (The maximum deflection by Jupiter is only 0".017.) Evidently, the observation must be made during a total eclipse of the sun.

Immediately after Einstein's first suggestion, the matter was taken up by Dr. E. Freundlich, who attempted to collect information from eclipse plates already taken; but he did not secure sufficient material. At ensuing eclipses plans were made by various observers for testing the effect, but they failed through cloud or other causes. After Einstein's second suggestion had appeared, the Lick Observatory expedition attempted to observe the effect at the eclipse of 1918. The final results are not yet published. Some account of a preliminary discussion has been given, but the eclipse

* "*Annalen der Physik* 35, p. 898."

was an unfavorable one, and from the information published the probable accidental error is large, so that the accuracy is insufficient to discriminate between the three alternatives.

The results of the observations here described appear to point quite definitely to the third alternative, and confirm Einstein's generalized relativity theory. As is well-known the theory is also confirmed by the motion of the perihelion of Mercury, which exceeds the Newtonian value by 43″ [arcseconds] per century—an amount practically identical with that deduced from Einstein's theory. . . . Whether or not changes are needed in other parts of the theory, it appears now to be established that Einstein's law of gravitation gives the true deviations from the Newtonian law both for the relatively slow-moving planet Mercury and for the fast-moving waves of light.

It seems clear that the effect here found must be attributed to the sun's gravitational field and not, for example, to refraction by coronal matter. . . .

II. Preparations for the Expeditions

In March, 1917, it was pointed out as the result of an examination of the photographs taken with the Greenwich astrographic telescope at the eclipse of 1905 that this instrument was suitable for the photography of the field of stars surrounding the sun in a total eclipse. Attention was also drawn to the importance of observing the eclipse of May 29, 1919, as this afforded a specially favorable opportunity owing to the unusual number of bright stars in the field, such as would not occur again for many years.

With weather conditions as good as those at Sfax [Tunisia] in the 1905 eclipse—and these were by no means perfect—it was anticipated that twelve stars would be shown. . . .

The track of the eclipse runs from North Brazil across the Atlantic, skirting the African coast near Cape Palmas, passing through the Island of Principe, then across Africa to the western shores of Lake Tanganyika. Enquiry as to the suitable sites and probable weather conditions was kindly made. . . .

Acting on this information the Joint Permanent Eclipse Committee at a meeting on November 10, 1917, decided, if possible, to send expeditions to Sobral in North Brazil, and to the island of Principe. Application was made to the Government Grant Committee for £100 for instruments and £1,000 for the expedition. . . .

III. The Expedition to Sobral
(*Observers*, Dr. A. C. D. Crommelin and Mr. C. Davidson.)

Sobral is the second town of the State of Ceara, in the north of Brazil. Its geographical co-ordinates are: longitude 2h. 47m. 25s. west; latitude 3° 41′ 33″ south; altitude 230 feet. Its climate is dry and though hot not unhealthy. . . .

The morning of the eclipse day was rather more cloudy than the average, and the proportion of cloud was estimated at 9/10 at the time of first contact, when the sun was invisible; it appeared a few seconds later showing a very small encroachment of the moon, and there were various short intervals of sunshine during the partial phase which enabled us to place the sun's image at its assigned position on the ground glass, and to give a final adjustment to the rates of the driving clocks. As totality approached, the proportion of cloud diminished, and a large clear space reached the sun about one minute before second contact. Warnings were given 58s., 22s. and 12s. before second contact by observing the length of the disappearing crescent on the ground glass. When the crescent disappeared the word "go" was called and a metronome was started by Dr. Leocadio [assisting from the local state ministry of agriculture], who called out every tenth beat during totality, and the exposure times were recorded in terms of these beats. It beat 320 times in 310 seconds; allowance has been made for this rate in the recorded times. The program arranged was carried out successfully, 19 plates being exposed in the astrographic telescope with alternate exposures of 5 and 10 seconds, and eight in the 4-inch camera with a uniform exposure of 28 seconds. The region round the sun was free from cloud, except for an interval of about a minute near the middle of totality when it was veiled by thin cloud, which prevented the photography of stars, though the inner corona remained visible to the eye and the plates exposed at this time show it and the large prominence excellently defined. The plates remained in their holders until development, which was carried out in convenient batches during the night hours of the following days, being completed by June 5. . . .

IV. The Expedition to Principe
(*Observers*, Prof. A. S. Eddington and Mr. E. T. Cottingham.)

. . . Principe is a small island belonging to Portugal, situated just north of the equator in the Gulf of Guinea, about 120 miles from the African coast. The extreme length and breadth are about 10 miles and 6 miles. Near the center mountains rise to a height of 2500 feet, which generally attract heavy masses of cloud. Except for a certain amount of virgin forest, the island is covered with cocoa plantations. The climate is very moist, but not unhealthy. The vegetation is luxuriant, and the scenery is extremely beautiful. We arrived near the end of the rainy season, but the *gravana,* a dry wind, set in about May 10, and from then onwards no rain fell except on the morning of the eclipse. . . .

The days preceding the eclipse were very cloudy. On the morning of May 29 there was a very heavy thunderstorm from about 10 A.M. to 11:30 A.M.—a remarkable occurrence at that time of year. The sun then appeared for a few minutes, but the clouds gathered again. About half-an-hour before totality the crescent sun was glimpsed occasionally, and by 1:55 it could be seen continuously through drifting cloud. The calculated time of totality was from 2h. 13m. 5s. to 2h. 18m. 7s. G.M.T. Exposures were made according to the prepared program, and 16 plates were obtained. Mr. Cottingham gave the exposures and attended to the driving mechanism, and Prof. Eddington changed the dark slides. It appears from the results that the cloud must have thinned considerably during the last third of totality, and some star images were shown on the later plates. The cloudier plates give very fine photographs of a remarkable prominence which was on the limb of the sun.

A few minutes after totality the sun was in a perfectly clear sky, but the clearance did not last long. It seems likely that the break-up of the clouds was due to the eclipse itself, as it was noticed that the sky usually cleared at sunset.

It had been intended to complete all the measurements of the photographs on the spot; but owing to a strike of the steamship company it was necessary to return by the first boat, if we were not to be marooned on the island for several months. By the intervention of the Administrator berths, commandeered by the Portuguese Government, were secured for us on the crowded steamer. We left Principe on June 12, and after transhipping at Lisbon, reached Liverpool on July 14. . . .

V. General Conclusions

In summarizing the results of the two expeditions, the greatest weight must be attached to those obtained with the 4-inch lens at Sobral. From the superiority of the images and the larger scale of the photographs it was recognized that these would prove to be much the most trustworthy. Further, the agreement of the results derived independently from the right ascensions and declinations, and the accordance of the residuals of the individual stars . . . provides a more satisfactory check on the results than was possible for the other instruments.

These plates gave

> From declinations . . . . . . 1″.94
> From right ascensions . . . 2″.06

The result from declinations is about twice the weight of that from right ascensions, so that the mean result is

$$1″.98$$

with a probable error of about ± 0″.12.

The Principe observations were generally interfered with by cloud. The unfavorable circumstances were perhaps partly compensated by the advantage of the extremely uniform temperature of the island. The deflection obtained was

$$1″.61$$

The probable error is about ± 0″.30, so that the result has much less weight than the preceding.

Both of these point to the full deflection 1″.75 of Einstein's generalized relativity theory, the Sobral results definitely, and the Principe results perhaps with some uncertainty. There remain the Sobral astrographic plates which gave the deflection

$$0″.93$$

discordant by an amount much beyond the limits of its accidental error. For the reasons already described at length not much weight is attached to this determination.*

---

* The reasons, omitted in this excerpt, included the astrographic images being blurred, perhaps due to the sun's heat on the mirror, and a change in the instrument when the comparison plates were later made.

Table 36.1: Radial Displacement of Individual Stars.

Star.	Calculation.	Observation.
	″	″
11	0.32	0.20
10	0.33	0.32
6	0.40	0.56
5	0.53	0.54
4	0.75	0.84
2	0.85	0.97
3	0.88	1.02

It has been assumed that the displacement is inversely proportional to the distance from the sun's center, since all theories agree on this, and indeed it seems clear from considerations of dimensions that a displacement, if due to gravitation, must follow this law. From the results with the 4-inch lens, some kind of test of the law is possible though it is necessarily only rough. The evidence is summarized in the table and diagram [see Table 36.1 and Figure 36.1], which show the radial displacement of the

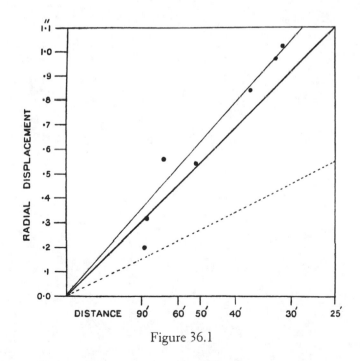

Figure 36.1

individual stars (mean from all the plates) plotted against the reciprocal of the distance from the center. The displacement according to Einstein's theory is indicated by the heavy line, according to the Newtonian law by the dotted line, and from these observations by the thin line.

Thus the results of the expeditions to Sobral and Principe can leave little doubt that a deflection of light takes place in the neighborhood of the sun and that it is of the amount demanded by Einstein's generalized theory of relativity, as attributable to the sun's gravitational field. But the observation is of such interest that it will probably be considered desirable to repeat it at future eclipses. The unusually favorable conditions of the 1919 eclipse will not recur, and it will be necessary to photograph fainter stars, and these will probably be at a greater distance from the sun. . . .

37 / Relativistic Models
of the Universe

Soon after the introduction of general relativity, Einstein and other theorists realized that its equations could also be used to determine the behavior of the universe at large—that it offered the means to move cosmology out of the realm of philosophy and turn it into a working science.

According to the theory as it was first introduced, the universe had to be in some kind of motion. But at the time it was generally believed among astronomers that the universe was static and unchanging. So in 1917 Einstein slightly altered his famous equation, adding a term λ that came to be called the cosmological constant. It was an added energy that permeated empty space and exerted a sort of outward "pressure" on it. This repulsive force—a kind of antigravity, actually—exactly balanced the inward gravitational attraction of all the matter in the universe, keeping it from moving. The universe remained immobile, "as required by the fact of the small velocities of the stars," wrote Einstein in his paper.[11] He was unaware that new astronomical observations were offering clues that the universe was quite different.

Where Einstein had a stationary universe filled with matter, the Dutch astronomer Willem de Sitter recognized that Einstein's theory also allowed for another solution: a universe that was stable and *empty*. It differed from Einstein's model in that it predicted that within such a space-time "the frequency of light-vibrations diminishes" (gets redder) with increasing distance from its source.[12] More up-to-date on the latest astronomical news, de Sitter believed that the spiral nebulae being sighted by astronomers in greater numbers were located outside the borders of the Milky Way and that their reported tendency to display redshifts (see Chapter 52) might be proof of his model (assuming that cosmic densities were so low that the universe could be considered essentially empty). The de Sitter universe also had the intriguing property that any bit of matter dropped inside its space-time would immediately fly off, another possible reason for the redshifts.

Astronomers, unfamiliar with relativity's complex mathematics and wary of its conclusions, were slow to design tests to distinguish between the de Sitter and Einstein models. But with the success of Arthur Eddington's 1919 solar eclipse expeditions in confirming general relativity (see Chapter 36), interest in these cosmological ideas grew throughout the 1920s. In 1922 the Russian mathematician Aleksandr Friedmann revealed another solution, the first non-stationary model of the cosmos that blended the best aspects of the de Sitter and Einstein universes: it allowed for a cosmos to be filled with matter, but at the same time it would be rapidly expanding outward. The universe changed over time. Moreover, depending on the amount of matter, the movement of space-time could be an expansion, a contraction, or even an oscillation between the two states. "We shall call this universe the *periodic world*," Friedmann wrote in his report to the *Zeitschrift für Physik*.[13] He even calculated an age for the universe—ten billion years—although he considered this estimate more a mathematical curiosity. He noted it might also be infinite. Unfortunately, Friedmann's work did not receive much attention (at least at first), and he had little chance to champion his ideas. In 1925, just a few years after completing his calculations, Friedmann became ill after conducting a record-breaking balloon ascent to make meteorological and medical observations and died while still in his thirties.

But the Belgian priest and astronomer Georges Lemaître, unaware of Friedmann's findings, independently reached the same conclusions in 1927. Published in an obscure Belgian journal, Lemaître's results went unnoticed until 1931, when Eddington, his former teacher, had the paper reprinted in the *Monthly Notices of the Royal Astronomical Society*. Unlike Friedmann, Lemaître directly linked his solution to astronomical observations, stressing in his conclusion that "the receding velocities of extragalactic nebulae are a cosmical effect of the expansion of the universe."[14]

"Cosmological Considerations on the General Theory of Relativity." *Sitzungsberichte der Preußischen Akademie der Wissenschaften zu Berlin* (1917) by Albert Einstein

. . . Regard the universe as a continuum which is *finite* (*closed*) *with respect to its spatial dimensions.* . . . We shall proceed to show that both the general postulate of relativity and the fact of the small stellar velocities are compatible with the hypothesis of a spatially finite universe; though certainly, in order to carry through this idea, we need a generalizing modification of the field equations of gravitation. . . .

. . . The system of equations allows a readily suggested extension which is compatible with the relativity postulate. . . . For on the left-hand side of [the] field equation we may add the fundamental tensor g_{uv}, multiplied by a universal constant, $-\lambda$, at present unknown, without destroying the general covariance. . . . We write

$$G_{uv} - \lambda g_{uv} = -K(T_{uv} - \tfrac{1}{2}g_{uv}T)$$

This field equation, with λ sufficiently small, is in any case also compatible with the facts of experience derived from the solar system. It also satisfies laws of conservation of momentum and energy. . . .

Thus the theoretical view of the actual universe, if it is in correspondence with our reasoning, is the following. The curvature of space is variable in time and place, according to the distribution of matter, but we may roughly approximate to it by means of a spherical space. At any rate, this view is logically consistent, and from the standpoint of the general theory of relativity lies nearest at hand; whether, from the standpoint of present astronomical knowledge, it is tenable, will not here be discussed. In order to arrive at this consistent view, we admittedly had to introduce an extension of the field equations of gravitation which is not justified by our actual knowledge of gravitation. It is to be emphasized, however, that a positive curvature of space is given by our results, even if the supplementary term is not introduced. That term is necessary only for the purpose of making possible a quasi-static distribution of matter, as required by the fact of the small velocities of the stars.

"On Einstein's Theory of Gravitation, and Its Astronomical Consequences. Third Paper." *Monthly Notices of the Royal Astronomical Society*, Volume 78 (November 1917) by Willem de Sitter

. . . Einstein's solution of the equations implies the existence of a "world-matter" which fills the whole universe. . . . It is, however, also possible to satisfy the equations without this hypothetical world-matter. . . .

In the [de Sitter] system. . . . the frequency of light-vibrations diminishes with increasing distance from the origin of co-ordinates. The lines in the spectra of very distant stars or nebulae must therefore be systematically displaced towards the red, giving rise to a spurious positive radial velocity.

. . . [F]or objects at very large distances we should expect a greater number of large or very large radial velocities. Spiral nebulae most probably are amongst the most distant objects we know. Recently a number of radial velocities of these nebulae have been determined. The observations are still very uncertain, and conclusions drawn from them are liable to be premature. Of the following three nebulae, the velocities have been determined by more than one observer:

Andromeda	(3 observers)	−311 km./sec.
N.G.C. 1068	(3 ")	+925 "
N.G.C. 4594	(2 ")	+1185 "

These velocities are very large indeed, compared with the usual velocities of stars in our neighborhood.

. . . The mean of the three observed radial velocities stated above is +600 km./sec. . . . Of course this result, derived from only three nebulae, has practically no value. If, however, continued observation should confirm

the fact that the spiral nebulae have systematically positive radial velocities, this would certainly be an indication to adopt the hypothesis B [de Sitter universe] in preference to A [Einstein universe]. . . .

"A Homogeneous Universe of Constant Mass and Increasing Radius Accounting for the Radial Velocity of Extra-Galactic Nebulae." *Monthly Notices of the Royal Astronomical Society,* Volume 91 (March 1931) by Abbé Georges Lemaître

Introduction.

According to the theory of relativity, a homogeneous universe may
exist such that all positions in space are completely equivalent; there is no
center of gravity. The radius of space R is constant; space is elliptic, *i.e.* of
uniform positive curvature $1/R^2$; straight lines starting from a point come
back to their origin after having traveled a path of length πR; the volume of
space has a finite value $\pi^2 R^3$; straight lines are closed lines going through
the whole space without encountering any boundary.

Two solutions have been proposed. That of de Sitter ignores the existence of matter and supposes its density equal to zero. It leads to special
difficulties of interpretation which will be referred to later, but it is of
extreme interest as explaining quite naturally the observed receding velocities of extra-galactic nebulae, as a simple consequence of the properties of
the gravitational field without having to suppose that we are at a point of
the universe distinguished by special properties.

The other solution is that of Einstein. It pays attention to the evident
fact that the density of matter is not zero, and it leads to a relation between
this density and the radius of the universe. This relation forecasted the existence of masses enormously greater than any known at the time. These
have since been discovered, the distances and dimensions of extra-galactic
nebulae having become known. From Einstein's formulae and recent
observational data, the radius of the universe is found to be some hundred
times greater than the most distant objects which can be photographed by
our telescopes.

Each theory has its own advantages. One is in agreement with the observed radial velocities of nebulae, the other with the existence of matter, giving a satisfactory relation between the radius and the mass of the universe. It seems desirable to find an intermediate solution which could combine the advantages of both.

At first sight, such an intermediate solution does not appear to exist. A static gravitational field for a uniform distribution of matter without internal stress has only two solutions, that of Einstein and that of de Sitter. De Sitter's universe is empty, that of Einstein has been described as "containing as much matter as it can contain." It is remarkable that the theory can provide no mean between these two extremes.

The solution of the paradox is that de Sitter's solution does not really meet all the requirements of the problem. Space is homogeneous with constant positive curvature; space-time is also homogeneous, for all events are perfectly equivalent. But the partition of space-time into space and time disturbs the homogeneity. The co-ordinates used introduce a center. A particle at rest at the center of space describes a geodesic of the universe; a particle at rest otherwhere than at the center does not describe a geodesic. The co-ordinates chosen destroy the homogeneity and produce the paradoxical results which appear at the so-called "horizon" of the center. When we use co-ordinates and a corresponding partition of space and time of such a kind as to preserve the homogeneity of the universe, the field is found to be no longer static; the universe becomes of the same form as that of Einstein, with a radius no longer constant but varying with the time according to a particular law.

In order to find a solution combining the advantages of those of Einstein and de Sitter, we are led to consider an Einstein universe where the radius of space or of the universe is allowed to vary in an arbitrary way.

Einstein Universe of Variable Radius. Field Equations. Conservation of Energy.

As in Einstein's solution, we liken the universe to a rarefied gas whose molecules are the extra-galactic nebulae. We suppose them so numerous that a volume small in comparison with the universe as a whole contains enough nebulae to allow us to speak of the density of matter. We ignore the possible influence of local condensations. Furthermore, we suppose that the nebulae are uniformly distributed so that the density does not depend on position. When the radius of the universe varies in an arbitrary way, the density, uniform in space, varies with time. . . .

[Omitted here are the derivations of the field equations for Lemaître's model of the universe.]

Conclusion.

We have found a solution such that

(1) The mass of the universe is a constant. . . .
(2) The radius of the universe increases without limit from an asymptotic value R_0 for $t = -\infty$.
(3) The receding velocities of extragalactic nebulae are a cosmical effect of the expansion of the universe. . . .

This solution combines the advantages of the Einstein and de Sitter solutions. . . . It remains to find the cause of the expansion of the universe. . . .

38 / Big Bang Versus Steady State

Astronomy's understanding of the structure and behavior of the universe underwent its most dramatic revision in the 1920s. Edwin Hubble confirmed, once and for all, that the Milky Way was but one of a multitude of other galaxies spread throughout the vast gulfs of space. He later enhanced the discovery in 1929 by proving that the very fabric of space-time was expanding, with galaxies continually riding the wave outward (see Chapters 51 and 52). Theorists such as Aleksandr Friedmann and Georges Lemaître, working on solutions to the equations of general relativity, already accounted for this motion in the light of Einstein's new law of gravitation (see Chapter 37). With theory and observation working hand in hand, astronomers could at last contemplate the universe's very creation, the unique moment when it all began. No longer was our cosmic origin a matter of metaphysics; it was a scientific theory that could be tested.

Soon after he introduced his model of an expanding universe, Lemaître was the first to contemplate in a scientific manner what that beginning might have been like. He mentally put the expansion of space-time into reverse and imagined the galaxies moving ever closer to one another, until they ultimately merged and formed a compact fireball of dazzling radiance. Boldly picturing the cosmos at earlier and earlier moments, Lemaître suggested that the universe emerged from a "primeval atom." Today's stars and galaxies, he surmised, were constructed from the fragments blasted outward from this original superatom. "The evolution of the world can be compared to a display of fireworks that has just ended: some few red wisps, ashes, and smoke," wrote Lemaître. "Standing on a well-chilled cinder, we see the slow fading of the suns, and we try to recall the vanished brilliance of the origin of the worlds."[15] From this poetic scenario arose the vision of the Big Bang, the cosmological model that shapes and directs the thoughts of cosmologists today as strongly as Ptolemy's crystalline spheres influenced natural philosophers in the Middle Ages.

For many, though, the contemplation of a singular moment of creation was philosophically distasteful. "The notion of a beginning of the present order of Nature is repugnant to me," said the British theorist Arthur Eddington. "By sweeping it far enough away from the sphere of our current physical problems, we fancy we have got rid of it. It is only when some of us are so misguided as to try to get back billions of years into the past that we find the sweepings all piled up like a high wall and forming a boundary—a beginning of time—which we cannot climb over."[16] There were scientific hurdles, too: estimates of the universe's age based on early (and incorrect) measurements of its rate of expansion were initially suggesting that the universe was younger than the stars, a paradox that posed a dilemma to Big Bang cosmologists for a while.

The notion of an evolving universe faced other challenges as well before it could be fully accepted. The most notable was the steady-state model of the universe. A group of young scientists at Cambridge University in the 1940s, contemplating a universe expanding and the density of matter thinning out, was concerned that the physical laws of nature would also change over time, making it impossible to compute anything about the universe's future or past. Mathematician Hermann Bondi and astrophysicist Thomas Gold (later joined by astrophysicist Fred Hoyle) figured this dire fate could be avoided if the density of matter did not change over time.[17] They conceded that the universe was eternally expanding (the observational evidence couldn't be denied), but it was an expansion with neither a beginning nor an end. It was in a "steady state." From wherever one viewed the universe, it always looked the same, because matter was continually and spontaneously being created to fill in the gaps opened up by the cosmic expansion. Galaxies were endlessly forming out of the new material to replace those that receded beyond our view, which meant the universe of the past would look very much like the universe of today. The Cambridge group estimated that to keep this process going, only one atom of hydrogen needed to be created each hour in roughly every cubic mile of intergalactic space.

For many years the steady-state universe was a potent competitor to the Big Bang theory. Ironically it was Hoyle, the ardent steady-stater, who gave his rivals a name for their cosmological model. During a British radio series on cosmology in 1949, Hoyle offhand-edly and derisively described the explosive version of creation as the "big bang idea."[18] The adjective stuck and turned into a noun. The rivalry between the two models inspired astronomers throughout the 1950s and into the 1960s to seek the observational evidence to decide the universe's true nature. The Big Bang was triumphant in 1964 with the discovery that the universe was awash in a sea of microwave radiation, the remnant echo of its thunderous concep-tion (see Chapter 63).

"The Beginning of the World from the Point of View of Quantum Theory." *Nature*, Volumes 127 (May 9, 1931) and 128 (October 24, 1931) by Georges Lemaître

First Paper

Sir Arthur Eddington states that, philosophically, the notion of a beginning of the present order of Nature is repugnant to him.* I would rather be inclined to think that the present state of quantum theory sug-gests a beginning of the world very different from the present order of Nature. Thermodynamical principles from the point of view of quantum theory may be stated as follows: (1) Energy of constant total amount is dis-tributed in discrete quanta. (2) The number of distinct quanta is ever

* "*Nature*, March 21, 1931, p. 447."

increasing. If we go back in the course of time we must find fewer and fewer quanta, until we find all the energy of the universe packed in a few or even in a unique quantum.

Now, in atomic processes, the notions of space and time are no more than statistical notions; they fade out when applied to individual phenomena involving but a small number of quanta. If the world has begun with a single quantum, the notions of space and time would altogether fail to have any meaning at the beginning; they would only begin to have a sensible meaning when the original quantum had been divided into a sufficient number of quanta. If this suggestion is correct, the beginning of the world happened a little before the beginning of space and time. I think that such a beginning of the world is far enough from the present order of Nature to be not at all repugnant.

It may be difficult to follow up the idea in detail as we are not yet able to count the quantum packets in every case. For example, it may be that an atomic nucleus must be counted as a unique quantum, the atomic number acting as a kind of quantum number. If the future development of quantum theory happens to turn in that direction, we could conceive the beginning of the universe in the form of a unique atom, the atomic weight of which is the total mass of the universe. This highly unstable atom would divide in smaller and smaller atoms by a kind of super-radioactive process. Some remnant of this process might, according to Sir James Jeans's idea, foster the heat of the stars until our low atomic number atoms allowed life to be possible.*

Clearly the initial quantum could not conceal in itself the whole course of evolution; but, according to the principle of indeterminacy, that is not necessary. Our world is now understood to be a world where something really happens; the whole story of the world need not have been written down in the first quantum like a song on the disc of a phonograph. The whole matter of the world must have been present at the beginning, but the story it has to tell may be written step by step.

Second Paper

... If I had to ask a question of the infallible oracle ... , I think I should choose this: "Has the universe ever been at rest, or did the expansion start from the beginning?" But, I think, I would ask the oracle not to

* Jeans was a British mathematician and astrophysicist.

give the answer, in order that a subsequent generation would not be deprived of the pleasure of searching for and of finding the solution.

If the total time of evolution did not exceed, say, ten times the age of the earth, it is quite possible to have a variation of the radius of the universe going on, expanding from zero to the actual value. I would picture the evolution as follows: At the origin, all the mass of the universe would exist in the form of a unique atom; the radius of the universe, although not strictly zero, being relatively very small. The whole universe would be produced by the disintegration of this primeval atom. It can be shown that the radius of space must increase. Some fragments retain their products of disintegration and form clusters of stars or individual stars of any mass. When the stars are formed, the process of formation of the extragalactic nebulae out of a gaseous material, proposed by Sir James Jeans, could be retained for the star-gas filling the space. The numerical test works out equally well for this case.

Whether this is wild imagination or physical hypothesis cannot be said at present, but we may hope that the question will not wait too long to be solved. . . .

"The Steady-State Theory of the Expanding Universe."
Monthly Notices of the Royal Astronomical Society,
Volume 108 (1948)
by Hermann Bondi and Thomas Gold

. . . Any interdependence of physical laws and large-scale structure of the universe might lead to a fundamental difficulty in interpreting observations of light emitted by distant objects. For if the universe, as seen from those objects, presented a different appearance, then we should not be justified in assuming familiar processes to be responsible for the emission of the light which we analyze. This difficulty is partly removed by the "cosmological principle." According to this principle all large-scale averages of quantities derived from astronomical observations (i.e. determination of the mean density of space, average size of galaxies, ratio of condensed to uncondensed matter, etc.) would tend statistically to a similar value independent of the positions of the observer, as the range of the observation is increased; provided only that the observations from different places are carried out at equivalent times. This principle would mean that there is nothing outstanding about any place in the universe, and that those differences which do exist are only of local significance; that seen on a large scale the universe is homogeneous.

This principle is widely recognized, and the observations of distant nebulae have contributed much evidence in its favor. An analysis of these observations indicates that the region surveyed is large enough to show us a fair sample of the universe, and this sample is homogeneous. . . .

We shall proceed quite differently at this point. As the physical laws cannot be assumed to be independent of the structure of the universe, and as conversely the structure of the universe depends upon the physical laws, it follows that there may be a stable position. We shall pursue this possibility that the universe is in such a stable, self-perpetuating state, without making any assumptions regarding the particular features which lead to this stability. We regard the reasons for pursuing this possibility as very compelling, for it is only in such a universe that there is any basis for the assumption that the laws of physics are constant; and without such an assumption our knowledge, derived virtually at one instant of time, must be quite inadequate for an interpretation of the universe and the dependence of its laws on its structure, and hence inadequate for any extrapolation into the future or the past.

Our course is therefore defined not only by the usual cosmological principle but by that extension of it which is obtained on assuming the universe to be not only homogeneous but also unchanging on the large scale. This combination of the usual cosmological principle and the stationary postulate we shall call the *perfect cosmological principle,* and all our arguments will be based on it. The universe is postulated to be homogeneous and stationary in its large-scale appearance as well as in its physical laws.

We do not claim that this principle must be true, but we say that if it does not hold, one's choice of the variability of the physical laws becomes so wide that cosmology is no longer a science. . . .

For the perfect cosmological principle to apply, one might at first sight expect that the universe would have to be static, i.e. to possess no consistent large-scale motion. This, however, would conflict with the observations of distant galaxies, and it would also conflict with the thermodynamic state which we observe. For such a static universe would be very different indeed from the universe we know. A static universe would clearly reach thermodynamical equilibrium after some time. An infinitely old universe would certainly be in this state. There would be complete equilibrium between matter and radiation, and (apart possibly from some slight variations due to gravitational potentials) everything would be at one and the same temperature. There would be no evolution, no distinguishing features, no recognizable direction of time. That our universe is not of this type is clear not only from astronomical observations but from local physics and indeed from our very existence. Accordingly there must be large-scale motions in our universe. The perfect cosmological principle permits only two types of motion, viz. large-scale expansion with a velocity proportional to distance, and its reverse, large-scale contraction.

In a contracting universe there would be even more radiation compared with matter than in a static universe. Therefore we reject this possibility and confine our attention to an expanding universe.

The observations of distant galaxies, which are now capable of a more rigorous interpretation by means of the perfect cosmological principle, inform us of the motion of expansion. This motion in which the velocity is proportional to the distance (apart from a statistical scatter) is well known to be of the only type compatible with homogeneity; but the compatibility with the hypothesis of a stationary property requires investigation. If we considered that the principle of hydrodynamic continuity were valid over large regions and with perfect accuracy then it would follow that the mean density of matter was decreasing, and this would contradict the perfect cosmological principle. It is clear that an expanding universe can only be stationary if matter is continuously created within it. The required rate of creation, which follows simply from the mean density and the rate of expansion, can be estimated as at most one particle of proton mass per liter per 10^9 years. . . .

We can now examine the requirements which the perfect cosmological principle places on the evolution of stars and galaxies. The mean ratio of condensed to uncondensed matter has to stay constant, and for this reason new galaxies have to be formed as older ones move away from each other. . . . In opposition to most other theories we should hence expect to find much diversity in the appearance of galaxies, as they will be of greatly different ages. . . . Furthermore the age distribution of galaxies in any volume will be independent of the time of observation, and it will hence be the same for distant galaxies as for near ones. . . .

39 / White Dwarf Stars

By the early decades of the twentieth century, astronomers had come to recognize a wide range of stellar sizes and types—from the large, white-hot O and B stars to the smaller and cooler M dwarf stars. What they didn't anticipate was a star the size of the Earth.

The first clue toward this revelation emerged between 1834 and 1844 when Friedrich Wilhelm Bessel, who also measured the first distance to a star (see Chapter 19), noticed that the bright star Sirius had a wavelike motion as it journeyed through the heavens. He reasoned that Sirius had an unseen companion that was gravitationally tugging on it. In 1862 the American telescope maker Alvan Clark Jr., while testing a new refractor, finally saw this dim companion. From its orbital movements, astronomers were able to determine that Sirius B weighed a solar mass, even though its light output was less than a hundredth of our Sun's. At the time they just figured it was a sunlike star cooling off at the end of its life.

That assessment would dramatically change in 1915 when Walter Adams at the Mount Wilson Observatory in California at last secured a spectrum of the faint light emanating from Sirius B, a difficult task due to the overwhelming brightness of the primary star. Even though the companion star was very dim, Adams was surprised to see that it displayed the spectral features of an intensely hot A star—at 25,000 K far hotter than our Sun. Adams knew that it wasn't impossible for a dim star to be so hot; just the year before he had noticed that a very faint star orbiting the star o Eridani was also an A-type star. Princeton astronomer Henry Norris Russell, in fact, had noticed the same star four years earlier in 1910 when constructing his famous diagram, which plotted stellar luminosities and types (see Chapter 28). The general rule is: the hotter the star, the brighter it is. But Russell saw o Eridani B enigmatically standing alone in a corner of his chart—white-hot yet somehow dim. "I was flabbergasted," recalled Russell. "I was really baffled trying to make out what it meant."[19]

Soon after Adams published his spectral findings, theorists, such as Arthur Eddington in Great Britain, recognized the meaning: If a star is both whiter and hotter than our Sun, it must be emitting more light over each square inch of its surface. But since Sirius B was so faint, that could only mean it had less surface area than our Sun—in other words, it was far smaller, roughly the size of the Earth. Such stars came to be called white dwarf stars.

Especially perplexing was the white dwarf's astounding density. Since a Sun's worth of mass is being squeezed into a tiny volume, a white dwarf star had to be incredibly compact. As Eddington wrote on this realization: "The message of the companion of Sirius, when decoded, ran: 'I am composed of material three thousand times denser than anything you've come across; a ton of my material would be a little nugget you could put in a matchbox.' What reply can one make to such a message? The reply which most of us made in 1914 was—'Shut up. Don't talk nonsense.'"[20]

By 1926 the British theorist Ralph Fowler realized that the laws of quantum mechanics, just then being developed, revealed the secret to a white dwarf star's curious bulk, the densest material then known in the universe. In fact, he was one of the first to apply the new physics to an astronomical problem. With a solar mass crushed into an Earth-sized space, Fowler figured that temperatures inside the dwarf become so extreme that all its electrons and atomic nuclei, like droves of little marbles, are packed into the smallest volume possible, creating an ultradense material impossible to assemble on Earth. A certain pressure exerted by the electrons (known as a degeneracy pressure) resists further compaction. Only the statistical laws of quantum mechanics, newly developed by Enrico Fermi and Paul Dirac, could explain it.

Astronomers later learned that the white dwarf is the end stage for a star of small to moderate mass; it is the luminous stellar core left behind after a star has run out of fuel and released its gaseous envelope into space. Radiating the energy left over from its fiery past, the white dwarf, like a dying ember, slowly cools down and fades away.

The white dwarf Sirius B offered another test of Einstein's general theory of relativity: the gravitational redshift. Since a white dwarf is so dense, its gravity is far stronger, causing the star's light waves to stretch out—get redder—as they "climb out" of the star's deep gravitational well. In 1925 Adams claimed he saw the redshift predicted by Eddington from Einstein's theory.[21] A more accurate assessment was made by the American astronomers Jesse Greenstein, J. Beverly Oke, and Harry L. Shipman in 1971. Analyzing data taken in the 1960s, they calculated a redshift in good agreement with that predicted to arise from a dense white dwarf star just 7,000 miles wide.[22]

"An A-Type Star of Very Low Luminosity." *Publications of the Astronomical Society of the Pacific*, Volume 26 (1914) by Walter S. Adams

It has been suggested by Hertzsprung that there is no such range in absolute brightness among the A-type stars as among those of types F to M, and, in fact, it is doubtful whether hitherto any certain case of a very faint A-type star has been found. A recent observation of the ninth-magnitude companion of o *Eridani* shows, however, that this star must be considered as such. The companion is at a distance of 83″ from the principal star and shares in its immense proper-motion of 4″.08 annually. Its parallax, therefore, may be assumed to be that of the bright star which is 0″.17. This would make the absolute magnitude of the companion 10.3, the Sun being taken as 5.5. The spectrum of the star is A_0.

"The Spectrum of the Companion of Sirius." *Publications of the Astronomical Society of the Pacific*, Volume 27 (1915) by Walter S. Adams

We have made several attempts during the past two years to secure a spectrum of the companion of *Sirius*. Its position is favorable, the distance, according to Professor Barnard's recent measures, being more than 10″ in a position angle of about 70°. The great mass of the star, equal to that of the Sun and about one-half that of *Sirius,* and its low luminosity, one one-hundredth part of that of the Sun and one ten-thousandth part of that of *Sirius,* make the character of its spectrum a matter of exceptional interest.

Most of the spectrum photographs have been taken at the 80-foot focus of the 60-inch reflector with the Cassegrain combination of mirrors. At this focus the distance of the companion from *Sirius* is 1.2^{mm} [1.2 millimeters on the plate]. The rays from *Sirius,* due to the supports of the auxiliary mirrors, are very prominent, but form angles of about 45° with the line joining *Sirius* with the companion, and so do not reach the slit unless the images begin to blur badly. The main difficulty in securing satisfactory photographs is, of course, the strong general illumination of the field and the presence of subsidiary rays which contribute more or less light to the slit as the seeing varies. During the exposures *Sirius* has been kept on the black metal screen in which is cut the opening forming the star window, while the companion is held in a position slightly to one side of the center of this window. Accordingly it is possible to compare on the photographs the spectrum of the point at which the companion is maintained with the spectrum due to the general illumination of *Sirius.* The exposure times given have been those normal for a star of 8.5 magnitude.

Two or three photographs obtained in this way showed a decided maximum in the spectrum at the point at which the companion was kept during the exposure. Still there was no distinct line of separation from the general spectrum due to *Sirius.* A photograph taken on October 18th under exceptionally good conditions of seeing does show such a demarcation, however, there being a narrow spectrum corresponding to the point on the slit at which the companion was held, which is separated by a distinct break from the intense spectrum of *Sirius* near the edge of the star window. It is difficult to avoid the conclusion that this is the spectrum of the companion. There was no ray from *Sirius* near this point of the slit and during the entire exposure the companion was well visible and accurate guiding was easily maintained.

The line spectrum of the companion is identical with that of *Sirius* in all respects so far as can be judged from a close comparison of the spectra, but there appears to be a slight tendency for the continuous spectrum of the companion to fade off more rapidly in the violet region. The suggestion has been made by several astronomers that at least a portion of the light of the companion is due to light reflected from *Sirius*. It is, however, by no means necessary to have recourse to this explanation, since in the case of the companion of o$_2$ *Eridani,* where there can be no question of reflected light, we know of a similar case of a star of very low intrinsic brightness which has a spectrum of type A$_0$.

Direct photographs taken by Dr. Van Maanen with and without the use of a yellow color screen agree with the spectrographic results in indicating that the companion of *Sirius* has a color index not appreciably different from that of the principal star.

"On Dense Matter." *Monthly Notices of the Royal Astronomical Society,* Volume 87 (December 1926) by Ralph H. Fowler

The accepted density of matter in stars such as the companion of Sirius is of the order of 10^5 gm./c.c. This large density has already given rise to most interesting theoretical considerations, largely due to Eddington. We recognize now that matter can exist in such a dense state if it has sufficient *energy,* so that the electrons are not bound in their ordinary atomic orbits of atomic dimensions but are in the main free—with sufficient energy to escape from any nucleus they may be near. The density of such "energetic" matter is then only limited *a priori* by the "sizes" of electrons and atomic nuclei. The "volumes" of these are perhaps 10^{-14} of the volume of the corresponding atoms, so that densities up to 10^{14} times that of terrestrial materials may not be impossible. Since the greatest stellar densities are of an altogether lower order of magnitude, the limitations imposed by the "sizes" of the nuclei and electrons can be ignored in discussions of stellar densities, and the structural particles of stellar matter can be treated as massive charged points.

Eddington has recently pointed out a difficulty in the theory of such matter.* Assuming it to behave more or less like a perfect gas, modified by its electrostatic forces and the sizes of such atomic structures as remain undissolved, there is a perfectly definite relation between the energy and the temperature, which depends on the density only to a minor degree. This assumption even here is not so unreasonable as appears at first sight. But even without it we naturally expect a perfectly definite relation between energy and temperature, in which there is a close correlation between large energies and large temperatures, small energies and small temperatures. The emission of energy by the star will proceed in the usual way at a rate depending on the surface temperature, and the internal temperatures must provide the gradient necessary to drive the radiation out. So long as the star contains matter at a high *temperature*, radiation of energy must presumably go on. But then, according to Eddington, there may come a time when a very curious state of affairs is set up. The stellar material will have radiated so much energy that it has less energy than the same matter in normal atoms expanded at the absolute zero of temperature. If part of it were removed from the star, and the pressure taken off, what could it do?

The present note is devoted to a further consideration of this paradox. It is clear that the crucial point is the connection between the energy and the temperature. In a sense the temperature measures the "looseness" of the system, the number of possible configurations which it can assume, and therefore its radiation. These depend directly on the temperature, and only on the energy in so far as the energy determines the temperature. The excessive densities involved suggest that the most exact form of statistical mechanics must be used to discuss the relationship between the energy, temperature, and density of the material. This is a form suggested by the properties of atoms and the new quantum mechanics, which has been already applied to simple gases by Fermi and Dirac. It may be accepted now as certain that classical statistical mechanics is not applicable at extreme densities, even to ideal material composed of extensionless mass-points, and that the form used here is fairly certainly the correct substitute. Its essential feature is a principle of exclusion which prevents two mass-

* "Eddington, *The Internal Constitution of the Stars*, Cambridge University Press (1926)."

points ever occupying exactly the same cell of extension h^3 of the six-dimensional phase-space of the mass-points. When this form of statistical mechanics is adopted, it at once appears that the suggested difficulty resolves itself, and there is really no difficulty at all. The apparent difficulty was due to the use of a wrong correlation between energy and temperature, suggested by classical statistical mechanics. When the correct relation is substituted, it is found that the limiting state of such dense stellar matter is one in which the *energy* is still, as it must be, excessively great, but the *temperature* is zero! Since the temperature determines the radiation, radiation stops when the dense matter has still ample energy to expand and form normal matter if the pressure happened to be removed. As the dense matter radiates its energy away, the number of its possible configurations rapidly falls, and therewith the temperature. The absolutely final state is one in which there is only one possible configuration left. Temperature then ceases to have any meaning, for the star is strictly analogous to one gigantic molecule in its lowest quantum state. We may call the temperature then zero.

Whether or not some such explanation may not be equally possible using other forms of statistical mechanics (perhaps the classical) I am not prepared to say. The new form used here seems for entirely independent reasons so satisfactory that its applicability need not be questioned. On application it clears up Eddington's question in a convincing manner, and I am content to leave the matter so. . . .

[Omitted here are Fowler's quantum mechanical calculations proving his statement above.]

40 / Beyond the White Dwarf

Within two decades of the discovery of the extremely dense white dwarf star, theorists working with the new laws of relativity and quantum mechanics were astonished to find that dying stars might face even stranger fates, if they had enough mass.

The first steps toward this realization were taken in 1930, during a sea voyage from India. Nineteen-year-old Subrahmanyan Chandrasekhar, while traveling to England to begin his graduate studies with Ralph Fowler at Cambridge University, explored the physics of white dwarfs and came to realize that velocities for some of the electrons in the dense stellar nugget would approach the speed of light. That meant it was necessary to apply the rules of special relativity to the star's behavior. Fowler had earlier shown (see Chapter 39) that the pressure from electrons, tightly packed in the compact star at a density of a ton per cubic inch, keeps a white dwarf intact. But could this go on forever? What happens, asked the young student from India, if a white dwarf is even more dense?

Chandrasekhar concluded that there is a critical limit to the mass of a white dwarf (now known to be 1.4 solar masses). If the dwarf is more massive, it collapses, overcome by the extreme pressure of gravity. In 1931 he published this result in a brief paper entitled "The Maximum Mass of Ideal White Dwarfs" in the *Astrophysical Journal*. The paper's abstract summarized it succinctly:

The theory of the *polytropic gas spheres* in conjunction with the equation of state of a *relativistically degenerate electron-gas* leads to a *unique value for the mass of a star* built on this model. This mass (= $0.91_{\odot}$ [solar mass]) is interpreted as representing the upper limit to the mass of an ideal white dwarf.[23]

But most astrophysicists at the time were skeptical and not interested in following up, although Lev Landau in the Soviet Union did reach the same conclusion independently, reporting that past the critical limit "there exists in the whole quantum theory no cause preventing the system from collapsing to a point."[24]

Chandrasekhar continued to pursue the problem and stressed his concern in a paper published in 1932.[25] "We may conclude," he said in the very last sentence, "that great progress in the analysis of stellar structure is not possible before we can answer the following fundamental question: *Given an enclosure containing electrons and atomic nuclei (total charge zero), what happens if we go on compressing the material indefinitely?*" What happens to the star? "It is necessary to emphasize one major result of the whole investigation," he later wrote in 1934, "namely, that it must be taken as well established that the life-history of a star of small mass must be essentially different from the life-history of a star of large mass. For a star of small mass the natural white-dwarf stage is an initial step towards complete extinction. A star of large mass . . . cannot pass into the white-dwarf stage, and one is left speculating on other possibilities."[26] The following year he reported "that when the central density is high enough . . . the configurations then would have such small radii they would cease to have any practical importance in astrophysics."[27] (See Figure 40.1.)

The great British theorist Arthur Eddington was not pleased to hear this and, during a discussion of Chandrasekhar's idea of drastic stellar collapse at a 1935 meeting of the Royal Astronomical Society in London, made the infamous declaration (often quoted) that "there should be a law of nature to prevent a star from behaving in this absurd way!"[28]

To be ridiculed by one of England's towering figures was a scientific humiliation and setback for the young investigator, and it took twenty years before the "Chandrasekhar limit" became a vital parameter in astrophysics textbooks.[29] Eddington was wrong. Nature did not provide a safety net against stellar collapse. In the end, Chandrasekhar opened the door for theorists to contemplate the existence of neutron stars and black holes.

"The Highly Collapsed Configurations of a Stellar Mass (Second Paper)." *Monthly Notices of the Royal Astronomical Society,* Volume 95 (1935) by Subrahmanyan Chandrasekhar

A study of the equilibrium of degenerate gas spheres has a twofold significance in the analysis of stellar structure, namely, in providing an approach to a proper theory of white dwarfs, and also, we shall see, in providing a certain limiting sequence of configurations to which all stars must tend eventually. . . . and this was the most important conclusion reached, these composite configurations have a *natural limit* [shown in Figure 40.1]. . . .

"Discussion by Arthur Eddington and Edward Milne." *The Observatory,* Volume 58 (1935)

Dr. Chandrasekhar read a paper describing the research which he has recently carried out, an account of which has already appeared in *The Observatory,* 57. 373, 1934, investigating the equilibrium of stellar configurations with degenerate cores. He takes the equation of state for degenerate matter in its exact form, that is to say, taking account of relativistic degeneracy. An important result of the work is that the life history of a star of small mass must be essentially different from that of a star of large mass. . . .

PROF. MILNE: I have had an opportunity of seeing Dr. Chandrasekhar's paper. We have both been working on the same problem. . . . In many ways the methods pursued and the results obtained are the same as

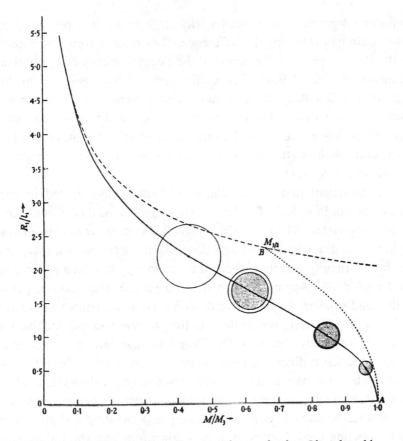

Figure 40.1: This graph summarizes the results that Chandrasekhar obtained in his eighteen-page paper, dense with calculations. It shows that once a white dwarf reaches a certain mass, it will face a drastic compression in size, its radius approaching zero.

Dr. Chandrasekhar's. I have pursued a cruder method of analysis, but I believe that my method gives more insight into the fundamental physical postulates underlying the work, takes account of our ignorance of the behavior of degenerate matter, and gives a more rational picture. A result common to our theory and Dr. Chandrasekhar's is that the more massive a star, the smaller its radius when completely collapsed. This has a bearing on the Russell diagram.

THE PRESIDENT: Fellows will wish to return their thanks to Dr. Chandrasekhar. I now invite Sir Arthur Eddington to speak on his paper "Relativistic Degeneracy."

SIR ARTHUR EDDINGTON: Dr. Chandrasekhar has been referring to degeneracy. There are two expressions commonly used in this connection, "ordinary" degeneracy and "relativistic" degeneracy, and perhaps I had better begin by explaining the difference. They refer to formulae expressing the electron pressure P in terms of the electron density σ. For ordinary degeneracy $P_e = K\sigma^{5/3}$. But it is generally supposed that this is only the limiting form at low densities of a more complicated relativistic formula, which shows P varying as something between $\sigma^{5/3}$ and $\sigma^{4/3}$, approximating to $\sigma^{4/3}$ at the highest densities. I do not know whether I shall escape from this meeting alive, but the point of my paper is that there is no such thing as relativistic degeneracy!

I would remark first that the relativistic formula has defeated the original intention of Prof. R. H. Fowler, who first applied the theory of degeneracy to astrophysics. When, in 1924, I suggested that owing to ionization we might have to deal with exceedingly dense matter in astronomy, I was troubled by a difficulty that there seemed to be no way in which a dense star could cool down. Apparently it had to go on radiating forever, getting smaller and smaller. Soon afterwards Fermi-Dirac statistics were discovered, and Prof. Fowler applied them to the problem and showed that they solved the difficulty; but now Dr. Chandrasekhar has revived it again. Fowler used the ordinary formula; Chandrasekhar, using the relativistic formula which has been accepted for the last five years, shows that a star of mass greater than a certain limit M remains a perfect gas and can never cool down. The star has to go on radiating and radiating and contracting and contracting until, I suppose, it gets down to a few km. radius, when gravity becomes strong enough to hold in the radiation, and the star can at last find peace.

Dr. Chandrasekhar had got this result before, but he has rubbed it in in his last paper; and, when discussing it with him, I felt driven to the conclusion that this was almost a *reductio ad absurdum* of the relativistic degeneracy formula. Various accidents may intervene to save the star, but I want more protection than that. I think there should be a law of Nature to prevent a star from behaving in this absurd way! . . .

41 / Supernovae and Neutron Stars

In the early 1930s Subrahmanyan Chandrasekhar spent several years trying to convince his colleagues in the British astrophysics community that if a star were massive enough it would not settle down as a white dwarf in its old age but rather undergo further stellar collapse, perhaps even to a singular point (see Chapter 40). Chandrasekhar did not speculate what other forms the star might take, but others soon did. In 1933 at a meeting of the American Physical Society, Walter Baade of the Mount Wilson Observatory and Fritz Zwicky of the California Institute of Technology introduced the idea that the star might end up as a relatively tiny ball of neutrons not much wider than a city, forged in a spectacular stellar explosion they had christened a "supernova."

Astronomers had long recognized that novae—"new stars"—occasionally appeared in the heavens. In the mid-nineteenth century, there were a number of quaint theories about their origin, including swarms of meteors colliding with one another, two stars in near collision, or even a star encountering a cloud of cosmic material and heated to superluminal brightness by the friction as it passed through.

By the early twentieth century, astronomers realized that a nova involved some kind of outburst on a star. Moreover, they began to notice that there were two kinds. There were the "common" novae that appeared up to thirty times a year in both the Milky Way and other galaxies (now known to occur when a white dwarf steals mass from a companion—matter that compresses on the dwarf and eventually ignites in a thermonuclear blast). And then there was a special class of novae, far more luminous and much rarer. Some called them "giant novae," others "exceptional novae." In his native German, Baade referred to them as "*Hauptnovae*" (chief novae). This was translated into English as "supernovae" during lectures by Zwicky and Baade in Pasadena.

More than providing a name, Zwicky and Baade offered a reason for the spectacular flare-up. Neutrons had just been discovered by particle physicists in 1932, and the Soviet physicist Lev Landau soon suggested that the compressed cores of massive stars might harbor extremely dense matter, "forming one gigantic nucleus."[30] Zwicky and Baade took the idea further by suggesting that under the most extreme conditions—during the explosion of a star—ordinary stars would completely transform themselves into naked spheres of neutrons. The negatively charged electrons and positively charged protons in the stellar core would be pressed inward to form a compact ball of neutral particles.

This proposal was considered wildly speculative, and only a handful of physicists proceeded to investigate a neutron star's possible structure. In 1937 George Gamow calculated that its density would be around 100 trillion grams per cubic centimeter, an entire Sun's worth of mass squeezed into a space only 10 miles wide.[31] J. Robert Oppenheimer, who went on to become the father of the atom bomb, also briefly dabbled in the subject, joining with two of his graduate students to ponder a neutron star's range of stable masses.[32] But for three decades, neutron stars remained theoretical fabrications, which astronomers figured would never be seen even if they did exist, due to their extremely small size. That changed when the first bona fide neutron star, beeping away as a pulsar, was at last discovered in 1967 (see Chapter 64).

"On Super-Novae" and "Cosmic Rays from Super-Novae."
Proceedings of the National Academy of Sciences,
Volume 20 (May 15, 1934)
by Walter Baade and Fritz Zwicky

First Paper

A. Common Novae.—The extensive investigations of extragalactic systems during recent years have brought to light the remarkable fact that there exist two well-defined types of new stars or novae which might be distinguished as *common novae* and *super-novae*. No intermediate objects have so far been observed.

Common novae seem to be a rather frequent phenomenon in certain stellar systems. Thus, according to Bailey, ten to twenty novae flash up every year in our own Milky Way.* A similar frequency (30 per year) has been found by Hubble in the well-known Andromeda nebula. A characteristic feature of these common novae is their absolute brightness (M) at maximum, which in the mean is −5.8 with a range of perhaps 3 to 4 mags. The maximum corresponds to 20,000 times the radiation of the sun. During maximum light the common novae therefore belong to the absolutely brightest stars in stellar systems. This is in full agreement with the fact that we have been able to discover this type of novae in other stellar systems near enough for us to reach stars of absolute magnitude − 5 with our present optical equipment.

B. Super-Novae.—The novae of the second group (super-novae) presented for a while a very curious puzzle because this type of new star was found, not only in the nearer systems, but apparently all over the accessible range of nebular distances. Moreover, these novae presented the new feature that at their maximum brightness they emit nearly as much light as the whole nebula in which they originate. Since the investigations of Hubble and others have revealed that the absolute total luminosities of extragalactic systems scatter with rather small dispersion around the mean value $M_{vis} = -14.7$, there is no doubt that we must attribute to this group of novae an individual maximum brightness of the order of $M_{vis} = -13$.

* Variable-star expert Solon I. Bailey of the Harvard College Observatory.

A typical specimen of these super-novae is the well-known bright nova which appeared near the center of the Andromeda nebula in 1885 and reached a maximum apparent brightness of $m = 7.5$. Since the distance modulus of the Andromeda nebula is

$$m - M = 22.2, \tag{1}$$

the absolute brightness of the nova at maximum was $M = -14.7$. An integration of the light-curve shows that practically the whole visible radiation is emitted during the 25 days of maximum brightness and that the total thus emitted is equivalent to 10^7 years of solar radiation of the present strength.

Finally, there exist good reasons for the assumption that at least one of the novae which have been observed in our Milky Way system belongs to the class of the super-novae. We refer to the abnormally bright nova of 1572 (Tycho Brahe's nova).

About the final state of super-novae practically nothing is known. The bright nova of 1885 in the Andromeda nebula has faded away and must now be fainter than absolute magnitude −2. Repeated attempts to identify the nova of 1572 with one of the faint stars near its former position have so far not been very convincing.

Regarding the initial states of super-novae only the following meager facts are known. First, super-novae occur not only in the blurred central parts of nebulae but also in the spiral arms, which in certain cases are clearly resolved into individual stars. Secondly, the super-nova of 1572 in its initial stage probably was not brighter than apparent magnitude 5 as otherwise it would be registered as such in the old catalogues, which, however, is not the case.

Super-novae are a much less frequent phenomenon than common novae. So far as the present observational evidence goes, their frequency is of the order of one super-nova per stellar system (nebula) [now called a galaxy] per several centuries.

We believe that on the basis of the available observations of super-novae the following assumptions are admissible:

(1) Super-novae represent a general type of phenomenon, and have appeared in all stellar systems (nebulae) at all times as far back as 10^9 years. To be conservative we shall assume for purposes of calculation that in every stellar system only one super-nova appears per thousand years.

(2) Super-novae, initially, are quite ordinary stars whose masses are not greater than 10^{33} gr. to 10^{35} gr.

(3) The super-nova of 1885 in Andromeda is a fair sample. We there-

fore base our calculations on the characteristics observed for this super-nova, namely:

(α) At maximum the visible radiation L_V emitted per second is equal to that of 6.3×10^7 suns. The radiation from our sun is

$$L_\odot = 3.78 \times 10^{33} \text{ ergs/sec.} \tag{2}$$

Therefore

$$L_V = 6.3 \times 10^7 L_\odot = 2.38 \times 10^{41} \text{ ergs/sec.} \tag{3}$$

The *total* visible radiation which was emitted by our super-nova represents an energy $E_V = 10^7$ years of $L_\odot$, that is

$$E_V = 1.19 \times 10^{48} \text{ ergs.} \tag{4}$$

(β) A common nova reaches maximum brightness in about two to three days. Indications are that a super-nova reaches maximum brightness during about the same interval.

[Omitted here are the authors' calculations of the total radiation from a supernova, which demonstrated that a star converts a considerable amount of mass into energy during the supernova event.]

. . . [I]t therefore becomes evident that *the phenomenon of a super-nova represents the transition of an ordinary star into a body of considerably smaller mass.* . . .

Second Paper

[In this second paper Baade and Zwicky presented their arguments that "cosmic rays are produced in the super-nova process." At the very end of the report they further speculated on the nature of the stellar remnant left behind after a supernova explosion.]

In addition, the new problem of developing a more detailed picture of the happenings in a super-nova now confronts us. With all reserve we advance the view that a super-nova represents the transition of an ordinary star into a *neutron star,* consisting mainly of neutrons. Such a star may possess a very small radius and an extremely high density. As neutrons can be packed much more closely than ordinary nuclei and electrons, the "gravitational packing" energy in a *cold* neutron star may become very large, and, under certain circumstances, may far exceed the ordinary nuclear packing fractions. A neutron star would therefore represent the most stable configuration of matter as such. . . .

42 / Black Holes

General relativity had profound effects in astrophysics, and not just as a tool in examining the behavior of the universe at large. It is also required to explain the most extreme cases of stellar evolution, when stars are massive enough at the end of their life that their inward gravitational attraction overwhelms all other forces, leading to the creation of what are now known as black holes.

An early version of the black hole idea was actually introduced in 1783, just as Great Britain was recovering from its war with colonial America. John Michell (who had already predicted the existence of double stars; see Chapter 13) presented a paper to the Royal Society of London suggesting that if a star got big enough, its gravitational pull would become so powerful that "all light emitted from such a body would be made to return towards it, by its own proper gravity."[33] This would happen, he figured, if the star were five hundred times wider than our Sun and just as dense throughout. Fifteen years later the French mathematician Pierre-Simon de Laplace arrived at a similar conclusion.[34] But no such stars exist (supergiant stars have far lower densities). It took the arrival of general relativity to describe a genuine black hole.

The first hint of this formidable stellar fate came in 1916, shortly after Einstein first introduced general relativity. The German astronomer Karl Schwarzschild, enamored with relativity, began to examine how its laws would operate in certain theoretical situations.

He wondered what would happen—gravitationally—if all the mass of an object, such as the Sun, were squeezed down to one point, a condition of zero volume and infinite density that mathematicians call a "singularity."[35] He discovered that around this hypothetical point a spherical region of space could be defined out of which nothing—no signal, not a glimmer of light nor bit of matter—could escape. The entire boundary—the point of no return—has come to be known as the event horizon, because no event occurring within its borders can be observed from the outside.

But Schwarzschild's initial relativistic calculations were simply an academic exercise. It wasn't until 1939 that researchers began to realize that the cosmos could well be generating these singularities. When studying the physics of neutron stars at the University of California at Berkeley (see Chapter 41), J. Robert Oppenheimer and his student George Volkoff concluded that if a neutron star were massive enough, it would have "to contract indefinitely, never reaching equilibrium."[36] Shortly afterward Oppenheimer and another graduate student, Hartland Snyder, went on to describe the nature of such "continued gravitational contraction," as they called it, establishing the first modern description of a black hole.

They began with a star that has depleted all its fuel. And to make the computation easier in that era of desktop calculators, they ignored certain pressures and the star's rotation. With the heat from its nuclear fires gone, the star's core becomes unable to support itself against the pull of its own gravity, and the stellar corpse begins to shrink. Oppenheimer and Snyder determined that if this core is weightier than a certain mass (now believed to be around two to three solar masses), the stellar remnant would not turn into a white dwarf star (our own Sun's fate) nor even settle down as a ball of neutrons, because once the material is squeezed to densities beyond 3 billion tons per cubic inch, the neutrons can no longer serve as an adequate brake against collapse. Degeneracy pressures no longer work. They calculated that the star would continue to contract indefinitely. The last light waves to flee before the "door" is irrevocably shut get so extended by the enormous pull of gravity (from visible, to infrared, to radio and beyond) that the rays become invisible, and the star vanishes from sight. Space-time is so warped around the collapsed star that it literally closes itself off from the rest of the universe. "Only its gravitational field persists," they reported.[37] But astronomers were not yet ready to believe that such bizarre objects could exist in the real world. Even Einstein wrote a paper attempting to prove they were impossible to form.[38]

Oppenheimer dropped the subject and never returned to it. Interest was not revived until the 1960s, when Princeton physicist John Archibald Wheeler began, once again, to ponder the fate of collapsed stars. Sure at first that some kind of force, yet to be considered, would step in to halt the contraction, he finally saw that nothing could prevent the collapse to a singularity. It was inevitable. Tired of awkwardly calling the bodies "gravitationally collapsed objects," Wheeler in 1967 adopted the catchy name "black hole," which was swiftly embraced.

The first evidence for a black hole arrived with the development of x-ray astronomy (see Chapter 61). One of the brightest sources spotted in the x-ray sky in the 1960s was Cygnus X-1, which optical and radio astronomers later pinpointed to a double-star system where a giant blue star (called HDE 226868) is coupled with an invisible companion orbiting around it every 5.6 days. By 1972 orbital measurements were suggesting that the unseen companion has a mass at least ten times greater than our Sun's—too massive for a neutron star and so a prime candidate to be a black hole.[39]

Two years later, theoretical physicist Stephen Hawking of Cambridge University gave the black hole a whole new look. By applying the laws of quantum mechnics to a black hole, Hawking startled the astrophysics community when he announced at a 1974 Oxford symposium that "quantum mechanical effects cause black holes to create and emit particles as if they were hot bodies. . . . This thermal emission leads to a slow decrease in the mass of the black hole and its eventual disappearance."[40] For a stellar-mass black hole (or larger), such a decay would take more than 10^{67} years, but atom-sized black holes (possibly forged in the Big Bang) would vanish in just billions of years.

"On Continued Gravitational Contraction."
Physical Review, Volume 56 (September 1, 1939)
by J. Robert Oppenheimer and Hartland Snyder

Abstract

When all thermonuclear sources of energy are exhausted a sufficiently heavy star will collapse. Unless fission due to rotation, the radiation of mass, or the blowing off of mass by radiation, reduce the star's mass to the order of that of the sun, this contraction will continue indefinitely. In the present paper we study the solutions of the gravitational field equations which describe this process. . . . [G]eneral and qualitative arguments are given on the behavior of the metrical tensor as the contraction progresses: the radius of the star approaches asymptotically its gravitational radius; light from the surface of the star is progressively reddened, and can escape over a progressively narrower range of angles. . . . The total time of collapse for an observer comoving with the stellar matter is finite, and for this idealized case and typical stellar masses, of the order of a day; an external observer sees the star asymptotically shrinking to its gravitational radius.

Recently it has been shown that the general relativistic field equations do not possess any static solutions for a spherical distribution of cold neutrons if the total mass of the neutrons is greater than $\sim 0.7_\odot$ [solar mass]. It seems of interest to investigate the behavior of nonstatic solutions of the field equations.

In this work we will be concerned with stars which have large masses, $> 0.7_\odot$, and which have used up their nuclear sources of energy. A star under these circumstances would collapse under the influence of its gravitational field and release energy. This energy could be divided into four parts: (1) kinetic energy of motion of the particles of the star, (2) radiation, (3) potential and kinetic energy of the outer layers of the star which could be blown away by the radiation, (4) rotational energy which could divide the star into two or more parts. If the mass of the original star were sufficiently small, or if enough of the star could be blown from the surface by radiation, or lost directly in radiation, or if the angular momentum of the star were great enough to split it into small fragments, then the remaining matter could form a stable static distribution, a white dwarf star. We consider the case where this cannot happen.

If then, for the late stages of contraction, we can neglect the gravitational effect of any escaping radiation or matter, and may still neglect the deviations from spherical symmetry produced by rotation. . . . we should now expect that since the pressure of the stellar matter is insufficient to support it against its own gravitational attraction, the star will contract, and its boundary r_b will necessarily approach the gravitational radius r_0. Near the surface of the star, where the pressure must in any case be low, we should expect to have a local observer see matter falling inward with a velocity very close to that of light; to a distant observer this motion will be slowed up by a factor $(1 - r_0/r_b)$. All energy emitted outward from the surface of the star will be reduced very much in escaping, by the Doppler effect from the receding source, by the large gravitational redshift . . . and by the gravitational deflection of light which will prevent the escape of radiation except through a cone about the outward normal of progressively shrinking aperture as the star contracts. The star thus tends to close itself off from any communication with a distant observer; only its gravitational field persists. . . . Although it takes, from the point of view of a distant observer, an infinite time for this asymptotic isolation to be established, for an observer comoving with the stellar matter this time is finite and may be quite short. . . .

. . . We expect that this behavior will be realized by all collapsing stars which cannot end in a stable stationary state. . . .

[Omitted here are the authors' solutions to the field equations demonstrating their points above.]

43 / Source of Stellar Power

I n the course of the great industrial revolution of the nineteenth
century, scientists came to understand the nature of heat and
energy in far more detail. And with that increased awareness of
thermodynamics emerged astronomy's great dilemma: what was the
source of the Sun's immense, long-lasting power? In 1848 Julius R.
Mayer, a German physician also involved in physics, calculated that
if the Sun were one huge lump of coal, it would have burned up
within a few thousand years. So Mayer suggested that the Sun might
be refueled by a continuous stream of meteors from interplanetary
space, kindled by their plunge into the Sun. But others figured this
wouldn't work; the added mass needed to keep the Sun burning
would have altered its gravitational pull enough to shorten the
length of the year by weeks within a few thousand years, an effect
clearly not occurring. More than that, the Earth would also have
been under bombardment and turned blazing hot from the impacts.

More promising was an idea championed in the 1850s by both
the German physicist Hermann von Helmholtz and William
Thomson (later known as Lord Kelvin after acquiring a peerage)
that the Sun could be deriving its thermal and radiant heat from
gravitational contraction, a formidable force when dealing with
celestial-sized masses. They surmised that as gravity pulls the solar
material inward, compressing and heating it up, some of the resultant energy continually flows into space. It was calculated that the

Sun—nearly a million miles wide—had to shrink only a couple of hundred feet a year to account for its luminosity, a rate of collapse that would be hardly noticed over the span of recorded history. In this way gravitational power could keep the Sun shining for a few dozen million years.

This reasonable deduction faltered, however, once evidence from the fossil and geologic records, such as rates of erosion and sedimentation, disclosed that the Earth might be at least a billion years old.* This paradox of an Earth older than the Sun was not adequately resolved until the first decades of the twentieth century, when the discovery of radioactivity, the introduction of quantum mechanics, and the realization that hydrogen was the Sun's primary constituent at last allowed physicists to consider atomic transformations that could keep the Sun burning for billions of years.

There were several foreshadowings to this solution. In 1873 the British astronomer Norman Lockyer suggested that within the fiery Sun some elements were broken up into smaller constituents and somehow produced energy by combining. "An interesting physical speculation connected with this working hypothesis is the effect on the period of duration of a star's heat which would be brought about by assuming that the original atoms of which a star is composed are possessed with the increased potential energy of combination which this hypothesis endows them with," reported Lockyer in a Bakerian lecture to the Royal Society of London. "From the earliest phase of a star's life the dissipation of energy would, as it were, bring into play a new supply of heat, and so prolong the star's light."[41]

In 1919 the French physicist Jean Perrin, noted for his elegant experiments on Brownian motion that demonstrated the existence of atoms, proposed that the condensation of hydrogen into heavier atoms could account for the Sun's power. The next year Francis Aston at the Cavendish Laboratory in Great Britain confirmed that one helium atom weighed slightly less—roughly 1 percent less—than four hydrogen atoms. That lost mass, transformed into pure energy according to Einstein's famous equation $E = mc^2$, could provide "a colossally larger reserve than the energy of gravitation to

---

* Meteorites, the rocky fragments left over after the solar system coalesced out of a nebulous cloud of interstellar gas, now tell us that the Sun and its attendant planets formed about 4.6 billion years ago.

which Helmholtz and Kelvin have thought they could attribute the origin of solar heat," Perrin later asserted.[42] British astrophysicist Arthur Eddington agreed. "To my mind the *existence* of helium is the best evidence we could desire of the possibility of the *formation* of helium," he wrote in his 1927 book *Stars and Atoms*. "I am aware that many critics consider the conditions in the stars not sufficiently extreme to bring about the transmutation—the stars are not hot enough. The critics lay themselves open to an obvious retort; we tell them to go and find a hotter place." But there was still a formidable roadblock to that scenario: figuring out how four hydrogen nuclei (four protons, actually) could randomly bump into one another with enough energy to overcome their electrical repulsion and fuse. According to classical physics, the probability of that happening was zero.

That obstacle was overcome in 1928 when the Russian-American theorist George Gamow demonstrated that the new rules of quantum mechanics did allow two like-charged particles to "tunnel" through their electromagnetic barrier and occasionally approach one another.[43] The following year British astronomer Robert Atkinson and Austrian physicist Fritz Houtermans, then working together in Göttingen, Germany, and guided by Gamow, showed that there is a small but real chance that some of the protons fiercely moving within the Sun's core, where temperatures reach over 10 million K, can get close enough to overcome their repulsion and, with the assistance of some other light element, get glued together by the strong nuclear force.[44] It was the first attempt at a theory for generating nuclear energy within stars.

Through the 1930s physicists learned of other atomic particles—the neutron and the positron (the electron's antimatter mate)—which allowed researchers to figure out the most feasible nuclear pathways that would transmute hydrogen into helium. In the United States in 1938 Charles Critchfield (Gamow's graduate student) and Hans Bethe worked out the proton-proton chain, now known to be the dominant process for stars similar to our Sun or less massive.[45] First, two protons join up and are transformed into a deuteron (H^2, a proton-neutron pair). In less than a second, the deuteron adds on another proton, to create a light form of helium (He^3). In a final step that can take up to a million years, two of these light helium nuclei collide to form a helium nucleus with two protons and two neutrons, releasing two protons back into the stellar plasma. Each second about four million tons of mass are transformed into pure energy within the Sun in this way. This has been going on for five billion years and will continue for about five billion more.

Meanwhile, in Germany, Carl von Weizsäcker (and Bethe independently) recognized another important reaction route, one in which protons are fused using carbon and nitrogen as catalysts (with oxygen later recognized to assist as well). This CNO cycle is now known to be the most important process in stars more massive than our Sun.[46] The decade of research into the problem of stellar energy culminated in 1939 when Bethe published his landmark paper "Energy Production in Stars," which analyzed a host of possible nuclear reactions, calculated the rates of these reactions, decided the proton-proton chain and CNO cycle were the most important, and estimated a central temperature for the Sun that is within 20 percent of the current value (nearly 16 million Kelvins). A young and up-and-coming nuclear physicist, Bethe had been introduced to the astrophysical stellar energy problem only the previous year and completed this paper in six months.

Within it, Bethe also spotlighted the difficulty (given the physics then known at the time) in generating any elements beyond helium in stars, which spurred interest in seeing if they could be created in the Big Bang (see Chapter 44). It took more than a decade of further work before astronomers more fully understood how the heavier elements were created within stars (see Chapter 46).

"Energy Production in Stars." *Physical Review,* Volume 55 (March 1, 1939) by Hans A. Bethe

Abstract

It is shown that the *most important source of energy in ordinary stars is the reactions of carbon and nitrogen with protons.* These reactions form a cycle in which the original nucleus is reproduced, viz. $C^{12} + H = N^{13}$, $N^{13} = C^{13} + e^{+}$, $C^{13} + H = N^{14}$, $N^{14} + H = O^{15}$, $O^{15} = N^{15} + e^{+}$, $N^{15} + H = C^{12} + He^{4}$. Thus carbon and nitrogen merely serve as catalysts for the combination of four protons (and two electrons) into an α-particle [helium nucleus of two protons and two neutrons].

The carbon-nitrogen reactions are unique in their cyclical character. For all nuclei lighter than carbon, reaction with protons will lead to the emission of an α-particle so that the original nucleus is permanently destroyed. For all nuclei heavier than fluorine, only radiative capture of the protons occurs, also destroying the original nucleus. Oxygen and fluorine reactions mostly lead back to nitrogen. Besides, these heavier nuclei react much more slowly than C and N and are therefore unimportant for the energy production.

The agreement of the carbon-nitrogen reactions with observational data is excellent. In order to give the correct energy evolution in the sun, the central temperature of the sun would have to be 18.5 million degrees while integration of the Eddington equations gives 19. For the brilliant star Y Cygni the corresponding figures are 30 and 32. This good agreement holds for all bright stars of the main sequence, but, of course, not for giants.

For fainter stars, with lower central temperatures, the reaction $H + H = D + e^+$ and the reactions following it, are believed to be mainly responsible for the energy production.

It is shown further that *no elements heavier than* He^4 *can be built up in ordinary stars.* This is due to the fact, mentioned above, that all elements up to boron are disintegrated by proton bombardment (α-emission!) rather than built up (by radiative capture). The instability of Be^8 reduces the formation of heavier elements still further. The production of neutrons in stars is likewise negligible. The heavier elements found in stars must therefore have existed already when the star was formed.

Finally, the suggested mechanism of energy production is used to draw conclusions about astrophysical problems, such as the mass-luminosity relation, the stability against temperature changes, and stellar evolution.

Introduction

The progress of nuclear physics in the last few years makes it possible to decide rather definitely which processes can and which cannot occur in the interior of stars. Such decisions will be attempted in the present paper, the discussion being restricted primarily to main sequence stars. The results will be at variance with some current hypotheses.

The first main result is that, under present conditions, no elements heavier than helium can be built up to any appreciable extent. Therefore we must assume that the heavier elements were built up *before* the stars reached their present state of temperature and density. No attempt will be made at speculations about this previous state of stellar matter.

The energy production of stars is then due entirely to the combination of four protons and two electrons into an α-particle. This simplifies the discussion of stellar evolution inasmuch as the amount of heavy matter, and therefore the opacity, does not change with time.

The combination of four protons and two electrons can occur essentially only in two ways. The first mechanism starts with the combination of two protons to form a deuteron [denoted by either D or H^2] with positron [e^+] emission, viz.

$$H + H = D + e^+ \tag{1}$$

The deuteron is then transformed into He^4 by further capture of protons; these captures occur very rapidly compared with process (1). The second mechanism uses carbon and nitrogen as catalysts, according to the chain reaction*

$$
\begin{aligned}
C^{12} + H &= N^{13} + \gamma \qquad N^{13} = C^{13} + e^+ \\
C^{13} + H &= N^{14} + \gamma \\
N^{14} + H &= O^{15} + \gamma \qquad O^{15} = N^{15} + e^+ \\
N^{15} + H &= C^{12} + He^4
\end{aligned}
\tag{2}
$$

The catalyst C^{12} is reproduced in all cases except about one in 10,000, therefore the abundance of carbon and nitrogen remains practically unchanged (in comparison with the change of the number of protons). The two reactions (1) and (2) are about equally probable at a temperature of $16 \cdot 10^6$ degrees which is close to the central temperature of the sun ($19 \cdot 10^6$ degrees). At lower temperatures (1) will predominate, at higher temperatures, (2).

No reaction other than (1) or (2) will give an appreciable contribution to the energy production at temperatures around $20 \cdot 10^6$ degrees such as are found in the interior of ordinary stars. The lighter elements (Li, Be, B) would "burn" in a very short time and are not replaced as is carbon in the cycle (2), whereas the heavier elements (O, F, etc.) react too slowly. Helium, which is abundant, does not react with protons because the prod-

---

* The symbol γ represents the emission of a gamma ray and e^+ a positron, or anti-electron.

uct, Li⁵, does not exist; in fact, the energy evolution in stars can be used as a strong additional argument against the existence of He⁵ and Li⁵.

Reaction (2) is sufficient to explain the energy production in very luminous stars of the main sequence as Y Cygni (although there are difficulties because of the quick exhaustion of the energy supply in such stars which would occur on any theory). Neither of the reactions (1) or (2) is capable of accounting for the energy production in giants. . . .

The Reactions Following Proton Combination

. . . All elements lighter than carbon, with the exception of H^1 and He^4, have an exceedingly short life in the interior of stars. Such elements can therefore only be present to the extent to which they are continuously produced in nuclear reactions from elements of longer life. This is in accord with the small abundance of all these elements both in stars and on earth.

Of the two more stable nuclei, He^4 is too inert to play an important role. It combines neither with a proton nor with another α-particle since the product would in both cases be an unstable nucleus. The only way in which He^4 can react at all, is by triple collisions. . . .

As the only primary reaction between elements lighter than carbon, there remains therefore the reaction between two protons [Equation 1 from above is restated here],

$$H^1 + H^1 = H^2 + e^+ \tag{1}$$

According to Critchfield and Bethe, this process gives an energy evolution of 2.2 ergs/g sec. under "standard stellar conditions" ($2 \cdot 10^7$ degrees, $\rho = 80$, hydrogen content 35 percent).* The reaction rate under these conditions is $2.5 \cdot 10^{-19}$ sec.$^{-1}$, corresponding to a mean life of $1.2 \cdot 10^{11}$ years for the hydrogen in the sun. This lifetime is about 70 times the age of the universe as obtained from the redshift of nebulae.

According to the foregoing, any building up of elements out of hydrogen will have to start with reaction (1). The deuteron will capture another proton,

$$H^2 + H^1 = He^3 \tag{17}$$

* *"Physical Review*, volume 54, 248."

This reaction follows almost instantaneously upon (1), with a delay of only 2 sec. There is, therefore, always statistical equilibrium between protons and deuterons, the concentration (by number of atoms) being in the ratio of the respective lifetimes. . . .

The further development of the He^3 produced according to (17) depends on the question of the stability of Li^4 and the relative stability of H^3 and He^3. . . .

[For the remainder of his twenty-three-page paper, Bethe calculates the energy production of various nuclear reactions, displays the difficulty in building up heavier elements, and discusses the effect of his conclusions on the mass-luminosity relation and stellar evolution.]

Stellar evolution

. . . Very few stars will actually be found near the end of their lives even if the age of the stars is comparable with their total lifespan. . . . In reality, the lifespan of all stars, except the most brilliant ones, is long compared with the age of the universe as deduced from the redshift. . . . E.g., for the sun, only one percent of the total mass transforms from hydrogen into helium every 10^9 years so that there would be only 2 percent He in the sun now, provided there was none "in the beginning." The prospective future life of the sun should according to this be $12 \cdot 10^9$ years.

It seems to us that this comparative youth of the stars is one important reason for the existence of a *mass*-luminosity relation—if the chemical composition, and especially the hydrogen content, could vary absolutely at random we should find a greater variability of the luminosity for a given mass.

It is very interesting to ask what will happen to a star when its hydrogen is almost exhausted. Then, obviously, the energy production can no longer keep pace with the requirements of equilibrium so that the star will begin to contract. . . . Gravitational attraction will then supply a large part of the energy. The contraction will continue until a new equilibrium is reached. For "light" stars . . . the electron gas in the star will become degenerate and a white dwarf will result. In the white dwarf state, the necessary energy production is extremely small so that such a star will have an almost unlimited life. This evolution was already suggested by Strömgren.*

* Swedish-born astronomer Bengt Strömgren.

For heavy stars, it seems that the contraction can only stop when a neutron core is formed. The difficulties encountered with such a core may not be insuperable in our case because most of the hydrogen has already been transformed into heavier and more stable elements so that the energy evolution at the surface of the core will be by gravitation rather than by nuclear reactions. However, these questions obviously require much further investigation.

44 / Creating Elements in the Big Bang

Before the atomic age, scientists had generally assumed that the elements always were and would always be. But the revelations emerging from atomic physics laboratories in the first half of the twentieth century made the idea of constructing the elements more and more attractive.

Initially, the most plausible factory for their manufacture was the hot interiors of stars. As the British astronomer Arthur Eddington remarked in 1920 upon seeing Ernest Rutherford and his expert staff at Cambridge University transforming elements by bombarding materials with atomic particles, "What is possible in the Cavendish Laboratory may not be too difficult in the sun."[47] But in the years that followed, any theoretical scheme to fabricate elements more massive than helium within stars was stymied by a formidable nuclear roadblock: no theorist could get past helium-4 (He^4), the common isotope of helium whose nucleus consists of two protons and two neutrons. Any atom with five nuclear particles decays extremely quickly, and without a sizable supply of such atoms, it seemed impossible to proceed from helium to lithium (its stable forms having three protons and three or four neutrons), from lithium to beryllium, from beryllium to boron, from boron to carbon, and so on through the periodic table. In his landmark 1939 paper on energy production in stars, Hans Bethe had to declare that "no elements heavier than He^4 can be built up in ordinary stars" (see Chapter 43).[48]

Facing what looked like an insurmountable barrier to stellar nucleosynthesis, others began to look for other fiery environments to create the elements, and the discovery by Edwin Hubble and Milton Humason in 1929 that the universe was expanding provided a new candidate. Georges Lemaître, a Belgian priest-astronomer, mentally put this expansion into reverse and imagined the universe exploding from a "primeval atom," a state of extremely hot and highly compressed matter (see Chapter 38). The Russian-American physicist George Gamow recognized that Lemaître's cosmic atom provided an alternate locale for cooking all the chemical elements in one fell swoop, allowing elements beyond helium to form. In the mid-1940s he and Ralph Alpher, his graduate student at George Washington University, attempted to demonstrate how it could be done.

By the 1930s, based on meteoritic, terrestrial, and astronomical data, geochemists had constructed a graph that displayed the relative amounts of each element found throughout the universe. Together hydrogen and helium account for 98 percent of the universe's visible matter, and all the remaining elements constitute a mere 2 percent, generally becoming increasingly rare as the weight of the atom increases (see Figure 44.1). This was the road map that Alpher and Gamow tried to duplicate. They were aided by the timely release of a government survey on neutron-capture rates. The information had been gathered to see what materials might best absorb neutrons in nuclear reactors, but it provided vital clues on element construction: Alpher recognized that the materials on the list that captured neutrons readily were the rarest in the universe (the added particles quickly convert these nuclei into other elements); conversely, substances slow to capture neutrons (thus avoiding conversion) were the most plentiful.

Using these guidelines, they visualized the early universe as a hot and highly compressed stew of neutrons, which Alpher dubbed *ylem* (pronounced "I-lem"), a derivation of an ancient Greek word meaning the basic substance out of which all matter evolved.[49] According to their scheme, as the temperature of the universal ylem dropped, some of the neutrons decayed into protons by the emission of electrons (beta decay), and these protons promptly began to stick to available neutrons, first forming nuclei of deuterium (H^2), then tritium (H^3), and finally helium and the heavier elements. They imagined matter being built up by a succession of neutron captures and beta decays. All the major reactions were essentially complete in less than half an hour, brought to an end once the short-lived free neutrons decayed away. In this way, they were able to duplicate, fairly roughly, the cosmic abundance chart.

Alpher and Gamow reported their first results in a short, one-page synopsis now as famous for its byline as for its content. Gamow, a merry prankster, listed the paper's authors as Alpher, Bethe, and Gamow, even though Hans Bethe never participated in the work. Gamow couldn't resist the pun on the first three letters of the Greek alphabet: alpha, beta, and gamma ($\alpha\beta\gamma$). That the paper chanced to be published on April Fool's Day only added to the fun.

For several years afterward, Alpher teamed up with Robert Herman (both were then employees at the Applied Physics Laboratory of Johns Hopkins University in Maryland) to carry out further primordial nucleosynthesis calculations.[50] In the end, though, they came to realize that the cosmic expansion both dispersed and cooled down the hot ylem before the heavier elements had any real chance of forming. Moreover, astronomers were beginning to notice that young stars were more enriched with heavier elements than older stars, which suggested the new stars were inheriting elements cooked up by previous generations of stars. This spurred other researchers to once again consider stellar nucleosynthesis (see Chapter 46).

But the idea of a burst of elemental cooking that took but minutes in the early universe, introduced in the $\alpha\beta\gamma$-paper, still prevails in creating the lighter elements. Starting with a primordial soup composed solely of neutrons, though, was an oversimplification. Physicists now deal with a mix of protons, neutrons, electrons, and neutrinos. Presently they calculate that nearly one-quarter of this primordial mass was converted into helium (He^4) in the first three minutes of the universe's life; three-quarters remained as hydrogen nuclei (protons); and a tiny smattering of lithium, helium-3, and deuterium accounted for the rest. Manufacture of the heavier elements had to await the births (and deaths) of the first stars.

Currently every measurement of the key light elements is close—remarkably close—to the chemical abundances predicted from Big Bang nucleosynthesis calculations. The existence of deuterium is a very strong argument in itself for a Big Bang to have occurred. Nuclear processing invariably causes deuterium to be destroyed in a star—never created. Yet deuterium is seen throughout the universe. Only a Big Bang can adequately explain its presence.

"The Origin of Chemical Elements." *Physical Review,* Volume 73 (April 1, 1948) by Ralph A. Alpher, Hans Bethe, and George Gamow

As pointed out by one of us,* various nuclear species must have originated not as the result of an equilibrium corresponding to a certain temperature and density, but rather as a consequence of a continuous building-up process arrested by a rapid expansion and cooling of the primordial matter. According to this picture, we must imagine the early stage of matter as a highly compressed neutron gas (overheated neutral nuclear fluid) which started decaying into protons and electrons when the gas pressure fell down as the result of universal expansion. The radiative capture of the still remaining neutrons by the newly formed protons must have led first to the formation of deuterium nuclei, and the subsequent neutron captures resulted in the building up of heavier and heavier nuclei. It must be remembered that, due to the comparatively short time allowed for this process, the building up of heavier nuclei must have proceeded just above the upper fringe of the stable elements (short-lived Fermi elements), and the present frequency distribution of various atomic species was attained only somewhat later as the result of adjustment of their electric charges by β-decay [emission of an electron].

Thus the observed slope of the abundance curve must not be related to the temperature of the original neutron gas, but rather to the time period permitted by the expansion process. Also, the individual abundances of various nuclear species must depend not so much on their intrinsic stabilities (mass defects) as on the values of their neutron capture cross-sections. . . .

We may remark at first that the building-up process was apparently completed when the temperature of the neutron gas was still rather high, since otherwise the observed abundances would have been strongly

* "G. Gamow. *Physical Review* 70 (1946): 572."

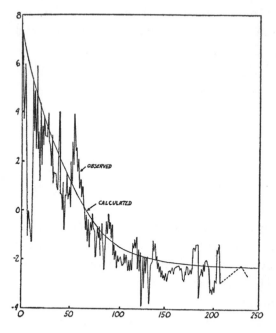

Figure 44.1: Log of relative abundance both observed and calculated.

affected by the resonances in the region of the slow neutrons. According to Hughes, the neutron capture cross sections of various elements (for neutron energies of about 1 MeV) increase exponentially with atomic number halfway up the periodic system, remaining approximately constant for heavier elements.*

Using these cross sections, one finds . . . that the relative abundances of various nuclear species decrease rapidly for the lighter elements and remain approximately constant for the elements heavier than silver. . . .

More detailed studies . . . leading to the observed abundance curve and discussion of further consequences will be published by one of us (R. A. Alpher) in due course.†

* Donald J. Hughes was a physicist at the Argonne National Laboratory in Illinois.

† R. A. Alpher, "A Neutron-Capture Theory of the Formation and Relative Abundance of the Elements," *Physical Review* 74 (1948): 1577–89.

45 / Cosmic Microwave Background Predicted

Working as a consultant at the Applied Physics Laboratory of Johns Hopkins University in Maryland in the late 1940s, the Russian-American theorist George Gamow came to work closely with Ralph Alpher and Robert Herman, two young employees at the lab who eagerly joined Gamow's crusade to study the physics of the Big Bang model of the cosmos. The trio were particularly interested in seeing if the elements were generated all at once out of the primordial plasma of the universe's fiery birth (see Chapter 44).

Along the way, Alpher and Herman came to predict that the present-day universe should be bathed in a uniform wash of radiation. The flood of highly energetic photons released in the aftermath of the Big Bang, they figured, should cool down with the expansion of the universe and appear today as centimeters-long microwaves. This momentous prediction had a curious debut. It was tucked away in a short *Nature* note, written to correct some errors that Gamow had made in a paper published two weeks earlier on the universe's evolution. In the very last sentence Alpher and Herman reported that the present-day microwave background should register a temperature of 5 K, five degrees on the Kelvin scale above absolute zero. (Today, it is measured at 3 K.) Over the next few years, the two young physicists went on to develop a detailed evolution of the newborn universe, work described as "the first thoroughly modern analysis of the early history of the universe."[51]

No one at the time did anything about their fascinating forecast of a cosmic microwave background or seemed to take note, despite its usefulness as a tool for deciding between the steady-state and Big Bang models of the universe then being debated (see Chapter 38). Perhaps it was because cosmology was still a young discipline skeptically regarded by mainstream astronomers, and radio astronomy, also in its infancy in 1948, had other pressing concerns. Gamow, Alpher, and Herman never pushed to make a search for the remnant echo of the primeval blast, and so the prediction fell into obscurity. Most astronomers forgot it altogether. The idea did not resurface until the 1960s and came to be verified serendipitously (see Chapter 63).

"Evolution of the Universe." *Nature*, Volume 162 (November 13, 1948) by Ralph A. Alpher and Robert Herman

In checking the results presented by Gamow in his recent article on "The Evolution of the Universe" [*Nature* of October 30, p. 680], we found that his expression for matter-density suffers from the following errors: (1) an error of not taking into account the magnetic moments in Eq. (7) for the capture cross-section, (2) an error in estimating the value of α by integrating the equations for deuteron formation (the use of an electronic analogue computer leads to $\alpha = 1$), and (3) an arithmetical error in evaluating ρ_0 from Eq. (9). In addition, the coefficient in Eq. (3) is 1.52 rather than 2.14. Correcting for these errors, we find

$$\rho_{\text{mat.}} = \frac{4.83 \times 10^{-4}}{t^{3/2}}$$

The condensation-mass obtained from this corrected density comes out not much different from Gamow's original estimate. However, the intersection point $\rho_{\text{mat.}} = \rho_{\text{rad.}}$ occurs at $t = 8.6 \times 10^{17}$ sec. $\cong 3 \times 10^{10}$ years (that is, about ten times the present age of the universe). This indicates that, in finding the intersection, one should not neglect the curvature term in the general equation of the expanding universe. In other words, the formation of condensations must have taken place when the expansion was becoming linear with time.

Accordingly, we have integrated analytically the exact expression:

$$\frac{dl}{dt} = \left[\frac{8\pi G}{3} \left(\frac{aT^4}{c^2} + \rho_{mat} \right) l^2 - \frac{c^2 l_0^2}{R_0^2} \right]^{1/2}$$

with $T \propto 1/l$ and $R_0 = 1.9 \times 10^9 \sqrt{-1}$ light-years. The integrated values of $\rho_{mat.}$ and $\rho_{rad.}$ intersect at a reasonable time, namely, 3.5×10^{14} sec. $\cong 10^7$ years, and the masses and radii of condensations at this time become, according to the Jeans' criterion, $M_c = 3.8 \times 10^7$ sun masses, and $R_c = 1.1 \times 10^3$ light-years. The temperature of the gas at the time of condensation was 600° K., and the temperature in the universe at the present time is found to be about 5° K. . . .

46 / Creating Elements in the Stars

O ne of the triumphs of twentieth-century astrophysics was determining the source of stellar power and demonstrating how the elements are generated as the stars progress through their various stages of nuclear burning. This knowledge was gained not suddenly but over many years, as astronomers gathered data on stellar compositions and physicists carried out accelerator experiments to understand the exact nuclear pathways in building up the elements.

Physicists in the 1930s, such as Hans Bethe and Carl von Weizsäcker, were initially able to show how hydrogen is transformed into helium within the sun, releasing enormous amounts of energy (see Chapter 43), but they couldn't adequately resolve how to fuse and build elements beyond helium. This difficulty spurred others, particularly George Gamow, Ralph Alpher, and Robert Herman, to see if the elements could have been brewed during the blazing first minutes of the Big Bang (see Chapter 44). Big Bang nucleosynthesis was eventually quite successful in explaining the abundances of helium, deuterium, and lithium seen throughout the universe but was ineffective in concocting heavier atoms. This dilemma turned the spotlight back on stars in the search for the origin of the elements.

A breakthrough in this venture occurred in 1951. Up until then, physicists knew that the fusing of two helium nuclei was a dead end:

the resulting nucleus of beryllium disintegrates in less than a trillionth of a second, putting an end to any further element construction. But Edwin Salpeter of Cornell University (and independently Ernst Öpik, director of the Armagh Observatory in northern Ireland) figured out if *three* helium nuclei come together in just the right way, they can form a stable nucleus of carbon — thus bypassing the nuclear roadblock.[52] From there British astrophysicist Fred Hoyle soon proceeded to describe carbon burning and oxygen burning in stars.[53] Even earlier, in 1946, he had already demonstrated from statistical arguments that heavy elements could be constructed and spewed into space during a supernova explosion.[54]

All the while astronomers were gathering definitive evidence of stellar nucleosynthesis from spectroscopic studies of various stars. Some stars, for example, were found to contain technetium, a rare radioactive element with a lifetime of less than a million years, which indicated it had to have been made in the star and not the Big Bang.[55] And in nuclear physics laboratories, physicists were measuring exactly how neutrons and protons can be captured by various atomic nuclei, transforming themselves into other elements.

Around 1953 Hoyle began collaborating with William Fowler, a physicist with the Kellogg Radiation Laboratory at the California Institute of Technology, and Geoffrey and Margaret Burbidge, a young British husband-and-wife team newly interested in the problem of stellar nucleosynthesis. Together, first in England and later in Pasadena, they laid out a comprehensive theory of nucleosynthesis in the stars. Their paper was more than one hundred pages in length and unique for opening with quotations from Shakespeare's *King Lear* and *Julius Caesar*. This seminal publication has attained a stature that transcends ordinary citations in that astronomers simply refer to it, like some chemical formula, as B^2FH (pronounced B-squared FH, from the initials of the authors' surnames). Within this formidable paper, the four researchers elucidate a variety of routes to chemical synthesis, including the direct fusing of atoms, the addition of protons and alpha particles (helium nuclei) to existing elements, or a nucleus's capture, either slowly (the s-process in red giant stars) or rapidly (the r-process in supernovae), of additional neutrons. They demonstrated how each route occurs at a specified time and stellar condition. Their quest benefited from both the release of a formerly classified government list of neutron-capture rates and the publication of the most up-to-date table of cosmic abundances by chemists Hans Suess and Harold Urey. Astrophysicist Alastair G. W. Cameron, then working in Canada, independently reached similar conclusions, although he received little notice at the time, having published in a less available journal.[56]

The B²FH paper served as the framework for all subsequent work in this arena and ultimately helped astronomers confirm the fate of various types of stars (see Chapter 47). Hoyle, a committed advocate of steady-state cosmology (see Chapter 38), had hoped to have all elements beyond hydrogen created by stars, and this agenda can be seen in the B²FH paper. But in time the evidence became overwhelming that Big Bang nucleosynthesis is required to explain the presence of deuterium and the bulk of the helium in the universe.

"Synthesis of the Elements in Stars." Reviews of Modern Physics, Volume 29 (October 1957) by E. Margaret Burbidge, Geoffrey R. Burbidge, William A. Fowler, and Fred Hoyle

"It is the stars, The stars above us, govern our conditions";

—(King Lear, Act IV, Scene 3)

but perhaps
"The fault, dear Brutus, is not in our stars, But in ourselves."

—(Julius Caesar, Act 1, Scene 2)

I. Introduction

A. Element Abundances and Nuclear Structure

Man inhabits a universe composed of a great variety of elements and their isotopes. . . . Ninety elements are found terrestrially and one more, technetium, is found in stars; only promethium has not been found in nature. Some 272 stable and 55 naturally radioactive isotopes occur on the earth. In addition, man has been able to produce artificially the neutron, technetium, promethium, and ten transuranic elements. The number of radioactive isotopes he has produced now numbers 871 and this number is gradually increasing.

Each isotopic form of an element contains a nucleus with its own characteristic nuclear properties which are different from those of all other nuclei. Thus the total of known nuclear species is almost 1200, with some 327 of this number known to occur in nature. In spite of this, the situation is not as complex as it might seem. Research in "classical" nuclear physics since 1932 has shown that all nuclei consist of two fundamental building blocks. These are the proton and the neutron which are called nucleons in this context. As long as energies below the meson production threshold are not exceeded, all "prompt" nuclear processes can be described as the shuffling and reshuffling of protons and neutrons into the variety of nucleonic packs called nuclei. Only in the slow beta-decay processes is there any interchange between protons and neutrons at low energies, and even there, as in the prompt reactions, the number of nucleons remains constant. Only at very high energies can nucleons be produced or annihilated. Prompt nuclear processes plus the slow beta reactions make it possible in principle to transmute any one type of nuclear material into any other even at low energies of interaction.

With this relatively simple picture of the structure and interactions of the nuclei of the elements in mind, it is natural to attempt to explain their origin by a synthesis or buildup starting with one or the other or both of the fundamental building blocks. The following question can be asked: What has been the history of the matter, on which we can make observations, which produced the elements and isotopes of that matter in the abundance distribution which observation yields? This history is hidden in the abundance distribution of the elements. To attempt to understand the sequence of events leading to the formation of the elements it is necessary to study the so-called universal or cosmic abundance curve. . . .

It seems probable that the elements all evolved from hydrogen, since the proton is stable while the neutron is not. Moreover, hydrogen is the most abundant element, and helium, which is the immediate product of hydrogen burning by the *pp* [proton-proton] chain and the CN [carbon-nitrogen] cycle, is the next most abundant element. The packing-fraction curve shows that the greatest stability is reached at iron and nickel. However, it seems probable that iron and nickel comprise less than 1% of the

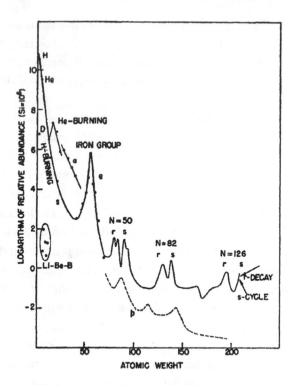

Figure 46.1: "Schematic curve of atomic abundances as a function of atomic weight based on the data of Suess and Urey. . . ."*

total mass of the galaxy. It is clear that although nuclei are tending to evolve to the configurations of greatest stability, they are still a long way from reaching this situation.

It has been generally stated that the atomic abundance curve has an exponential decline to A ~ 100 and is approximately constant thereafter. Although this is very roughly true it ignores many details which are important clues to our understanding of element synthesis. These details are shown schematically in Figure [46.1] and are outlined in the left-hand column of Table [46.1]. . . .

B. Four Theories of the Origin of the Elements

Any completely satisfactory theory of element formation must explain in quantitative detail all of the features of the atomic abundance curve. Of the theories so far developed, three assume that the elements were built in a primordial state of the universe. . . .

* "H. E. Suess and H. C. Urey, *Reviews of Modern Physics* 28 (1956): 53–74."

Table 46.1: Features of the Abundance Curve

Feature	Cause
Exponential decrease from hydrogen to $A \sim 100$	Increasing rarity of synthesis for increasing A, reflecting that stellar evolution to advanced stages necessary to build high A is not common.
Fairly abrupt change to small slope for $A > 100$	Constant σ (n, γ) in s process. Cycling in r process.
Rarity of D, Li, Be, B as compared with their neighbors H, He, C, N, O	Inefficient production, also consumed in stellar interiors even at relatively low temperatures.
High abundance of alpha-particle nuclei such as O^{16}, Ne^{20} ... Ca^{40}, Ti^{48} relative to their neighbors	He burning and α process more productive than H burning and s process in this region.
Strongly-marked peak in abundance curve centered on Fe^{56}	e process; stellar evolution to advanced stage where maximum energy is released (Fe^{56} lies near minimum of packing-fraction curve).
Double peaks $\begin{cases} A = 80, 130, 196 \\ A = 90, 138, 208 \end{cases}$	Neutron capture in r process (magic N = 50, 82, 126 for progenitors). Neutron capture in s process (magic N = 50, 82, 126 for stable nuclei).
Rarity of proton-rich heavy nuclei	Not produced in main line of r or s process; produced in rare p process.

Each of these theories possesses some attractive features, but none succeeds in meeting all of the requirements. It is our view that these are mainly satisfied by the fourth theory in which it is proposed that the stars are the seat of origin of the elements. In contrast with the other theories which demand matter in a particular primordial state for which we have no evidence, this latter theory is intimately related to the known fact that nuclear transformations are currently taking place inside stars. This is a strong argument, since the primordial theories depend on very special initial conditions for the universe. Another general argument in favor of the stellar theory is as follows.

It is required that the elements, however they were formed, are distributed on a cosmic scale. Stars do this by ejecting material, the most efficient mechanisms being probably the explosive ejection of material in supernovae, the less energetic but more frequent novae, and the less rapid and less violent ejection from stars in the giant stages of evolution and from planetary nebulae. Primordial theories certainly distribute material on a cosmic scale but a difficulty is that the distribution ought to have been spatially uniform and independent of time once the initial phases of the universe were past. This disagrees with observation. There are certainly differences in composition between stars of different ages, and also stars at particular evolutionary stages have abnormalities such as the presence of technetium in the S-type stars and Cf^{254} in supernovae. . . .

It is not known for certain at the present time whether all of the atomic species heavier than hydrogen have been produced in stars without the necessity of element synthesis in a primordial explosive stage of the universe. Without attempting to give a definite answer to this problem we intend in this paper to restrict ourselves to element synthesis in stars and to lay the groundwork for future experimental, observational, and theoretical work which may ultimately provide conclusive evidence for the origin of the elements in stars. However, from the standpoint of the nuclear physics alone it is clear that our conclusions will be equally valid for a primordial synthesis in which the initial and later evolving conditions of temperature and density are similar to those found in the interiors of stars.

C. General Features of Stellar Synthesis

Except at catastrophic phases a star possesses a self-governing mechanism in which the temperature is adjusted so that the outflow of energy through the star is balanced by nuclear energy generation. The temperature required to give this adjustment depends on the particular nuclear fuel available. Hydrogen requires a lower temperature than helium; helium requires a lower temperature than carbon, and so on, the increasing temperature sequence ending at iron since energy generation by fusion processes ends here. If hydrogen is present the temperature is adjusted to hydrogen as a fuel, and is comparatively low. But if hydrogen becomes exhausted as stellar evolution proceeds, the temperature rises until helium becomes effective as a fuel. When helium becomes exhausted the temperature rises still further until the next nuclear fuel comes into operation, and so on. The automatic temperature rise is brought about in each case by the conversion of gravitational energy into thermal energy.

In this way, one set of reactions after another is brought into operation, the sequence always being accompanied by rising temperature. Since penetrations of Coulomb barriers occur more readily as the temperature rises it can be anticipated that the sequence will be one in which reactions take place between nuclei with greater and greater nuclear charges. As it becomes possible to penetrate larger and larger barriers the nuclei will evolve towards configurations of greater and greater stability, so that heavier and heavier nuclei will be synthesized until iron is reached. Thus there must be a progressive conversion of light nuclei into heavier ones as the temperature rises.

There are a number of complicating factors which are superposed on these general trends. These include the following.

The details of the rising temperature and the barrier effects of nuclear reactions at low temperatures must be considered.

The temperature is not everywhere the same inside a star, so that the nuclear evolution is most advanced in the central regions and least or not at all advanced near the surface. Thus the composition of the star cannot be expected to be uniform throughout. A stellar explosion does not accordingly lead to the ejection of material of one definite composition, but instead a whole range of compositions may be expected.

Mixing within a star, whereby the central material is mixed outward, or the outer material inward, produces special effects.

Material ejected from one star may subsequently become condensed in another star. This again produces special nuclear effects.

All of these complications show that the stellar theory cannot be simple, and this may be a point in favor of the theory, since the abundance curve which we are trying to explain is also not simple. Our view is that the elements have evolved, and are evolving, by a whole series of processes. These are marked in the schematic abundance curve, Fig. [46.1], as H burning, He burning, α, e, r, s, and p processes. . . .

II. Physical Processes Involved in Stellar Synthesis, Their Place of Occurrence, and the Time-Scales Associated with Them

A. Modes of Element Synthesis

. . . It appears that in order to explain all of the features of the abundance curve, at least eight different types of synthesizing processes are demanded, if we believe that only hydrogen is primeval. . . .

(i) Hydrogen Burning

Hydrogen burning is responsible for the majority of the energy production in the stars. By hydrogen burning in element synthesis we shall mean the cycles which synthesize helium from hydrogen and which synthesize the isotopes of carbon, nitrogen, oxygen, fluorine, neon, and sodium which are not produced by processes (*ii*) and (*iii*). . . .

(ii) Helium Burning

These processes are responsible for the synthesis of carbon from helium, and by further α-particle addition for the production of O^{16}, Ne^{20}, and perhaps Mg^{24}. . . .

(iii) α Process

These processes include the reactions in which α particles are successively added to Ne^{20} to synthesize the four-structure nuclei Mg^{24}, Si^{28}, Si^{32}, [Ar^{36}], Ca^{40}, and probably Ca^{44} and Ti^{48}. . . . The source of the α particles is different in the α process than in helium burning.

(iv) e Process

This is the so-called equilibrium process . . . in which under conditions of very high temperature and density the elements comprising the iron peak in the abundance curve (vanadium, chromium, manganese, iron, cobalt, and nickel) are synthesized. . . .

(v) s Process

This is the process of neutron capture with the emission of gamma radiation (n, γ) which takes place on a long time-scale, ranging from ~100 years to ~10^5 years for each neutron capture. The neutron captures occur at a *slow* (s) rate compared to the intervening beta decays. This mode of synthesis is responsible for the production of the majority of the isotopes in the range $23 \leq A \leq 46$ (excluding those synthesized predominantly by the α process), and for a considerable proportion of the isotopes in the range $63 \leq A \leq 209$. . . . The s process produces the abundance peaks at A = 90, 138, and 208.

(vi) r Process

This is the process of neutron capture on a very short time-scale, ~0.01–10 sec for the beta-decay processes interspersed between the neutron captures. The neutron captures occur at a *rapid* (r) rate compared to the beta decays. This mode of synthesis is responsible for production of a large number of isotopes in the range $70 \leq A \leq 209$, and also for synthesis of uranium and thorium. This process may also be responsible for some light element synthesis, e.g., S^{36}, Ca^{46}, Ca^{48}, and perhaps Ti^{47}, Ti^{49}, and Ti^{50}. . . . The *r* process produces the abundance peaks at A = 80, 130, and 194.

(vii) p Process

This is the process of proton capture with the emission of gamma radiation (*p*, *γ*), or the emission of a neutron following gamma-ray absorption (*γ*, *n*), which is responsible for the synthesis of a number of proton-rich isotopes having low abundances as compared with the nearby normal and neutron-rich isotopes. . . .

(viii) x Process

This process is responsible for the synthesis of deuterium, lithium, beryllium, and boron. More than one type of process may be demanded here (described collectively as the *x* process), but the characteristic of all of these elements is that they are very unstable at the temperatures of stellar interiors, so that it appears probable that they have been produced in regions of low density and temperature. . . .

[After this introduction to their 103-page paper, the authors meticulously detail the various nuclear pathways involved in each process leading to element synthesis and show how this generates the cosmic abundances observed.]

Conclusion

It is impossible in a short space to summarize the advantages and disadvantages of a theory with as many facets as this. However, it may be reasonable to conclude as follows. The basic reason why a theory of stellar origin appears to offer a promising method of synthesizing the elements is that the changing structure of stars during their evolution offers a succession of conditions under which many different types of nuclear processes can occur. Thus the internal temperature can range from a few million degrees, at which the *pp* chain first operates, to temperatures between 10^9 and 10^{10} degrees when supernova explosions occur. The central density can also range over factors of about a million. Also the time-scales range between billions of years, which are the normal lifetimes of stars of solar mass or less on the main sequence, and times of the order of days, minutes, and seconds, which are characteristic of the rise to explosion. On the other hand, the theory of primeval synthesis demands that all the varying conditions occur in the first few minutes, and it appears highly improbable that it can reproduce the abundances of those isotopes which are built on a long time-scale in a stellar synthesis theory. . . .

47 / A Star's Life Cycle

As soon as astronomers recognized the existence of giant and dwarf stars in the first decades of the twentieth century (see Chapter 28), it seemed natural to assume that stars began as giants and then contracted to white-hot stars. But the story of a star's life cycle turned out to be quite different from this initial guess. The first clues arrived as astronomers and physicists were spiritedly debating the exact mechanisms that provided atomic energy for powering the Sun and stars (see Chapters 43 and 46). Simultaneously they were also considering how thermonuclear burning might explain the evolution of stars, particularly the development of giant stars.

As early as 1938, both Russian-American theorist George Gamow and Estonian astronomer Ernst Öpik, then at Tartu University, proposed a shell model for a star's burning.[57] Öpik's paper, over a hundred pages in length, fleshed it out extensively. Thermonuclear fusion begins at the star's center and then moves outward. This would explain the creation of a giant star. Once the hydrogen in the star's center has been converted into helium, the core contracts and heats up, releasing energy that causes the outer envelope of the star to expand into the vast giant structure. Over the next two decades, this nascent idea would be more thoroughly developed by a host of other theorists, working hand in hand with astronomers observing sequences of stars in clusters, to explain the full history of a star based on its mass.[58]

For a star like our Sun, when the hydrogen in its core is completely converted to helium, its central furnace will flame out and the core will contract. The gravitational energy released will help push its outer envelope outward to create a giant star. Nuclear burning then takes place in a shell of hydrogen around the inert helium core. With continued compression and heating the helium finally ignites and begins to fuse into carbon and oxygen. Our Sun, a rather small star, will stop fabricating elements at this stage. Shedding its glowing red outer envelope over time, creating a planetary nebula, the Sun will eventually shrink and become a white dwarf star, what is essentially the luminous remains of its blazing stellar core.

For stars more massive than the Sun, the fusion process continues. The carbon core gets surrounded by a helium-burning shell, with a hydrogen-burning shell farther out. With its nested shells, the center of the red giant begins to resemble the structure of an onion. The carbon and oxygen eventually fuse into neon and magnesium. These, in turn, serve as the raw materials for the construction of such elements as silicon, sulfur, argon, and calcium. Meanwhile, free neutrons can be slowly captured by the nuclei to seed the star with even heavier elements. If the star is massive enough, its core will continue fusing until iron is formed, which is the end of the line. The fusion of iron requires more energy than it releases. At this point, the core collapses to form a neutron star (or if heavy enough, a black hole). The firestorm of neutrinos released in the formation of the neutron star powers a shock wave that works its way through the star's outer atmosphere, allowing the elements there to rapidly absorb neutrons and create more elements, including those beyond iron. The ensuing explosion spreads these elemental ashes through space, until they condense once again into new stars.

From "Stellar Structure, Source of Energy, and Evolution." *Publications de l'Observatoire astronomique de l'Université de Tartu*, Volume 30 (1938)
by Ernst Öpik

. . . A core devoid of hydrogen, thus presumably devoid of subatomic sources of energy, is doomed to collapse on a "Kelvin" time scale, i.e., with gravitation as the source of energy; high densities can be attained, and a super-dense core may be formed. The hydrogen-containing envelope cannot be sucked into the core as long as traces of hydrogen are present, because the corresponding immense increase of temperature and density would lead to an instantaneous release of the whole store of subatomic energy, sufficient to disperse all the envelope into space. Actually no such catastrophe happens, the contraction of the core being a gradual one; instead of blowing up, the envelope gradually expands and adjusts itself to such low values of the effective density and temperature that the release of subatomic energy remains more or less normal (it may be even less than

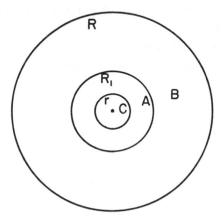

Figure 47.1: Öpik's model for the structure of a giant star. Hydrogen has been exhausted in the inner core C of radius r, which has contracted. Hydrogen continues to be converted into helium in the shell A with radius R_1. The outer envelope B of the star with radius R has expanded in response to the energy released by the contracting core.

the "normal," as the gravitational energy of the core supplies now a large fraction of the star's needs). In spite of the high gravitational force exerted by the core, the transition from the superdense core to the envelope of "normal" density and temperature is made possible by the peculiar distribution of the energy sources, and the smaller molecular weight of the envelope; the presence of subatomic energy sources suddenly beginning to work outside the core creates radiation pressure that "blows away" the matter of the shell, leaving a small density of matter just sufficient for the subatomic sources to work. The conditions are similar to those in the mathematical point-source model, except that here the subatomic source of energy is not concentrated in exactly one point, and that an additional point-source of energy and a considerable point-mass complicate the problem. . . .

A typical giant structure results, consisting of a vast extended envelope of low density in radiative or adiabatic equilibrium, an intermediate zone in adiabatic (convectional) equilibrium, of a density about the central density of main sequence stars, containing active sources of subatomic energy, and a contracting superdense core of zero hydrogen content and no subatomic energy. The intermediate zone, with active atomic synthesis, is supposed to contain a decreased amount of hydrogen and to get in this way definitely separated from the outer envelope; if not, the whole outer mass except the core may be stirred by convection currents (as in the purely adiabatic model), and the outer radius becomes little sensitive to eventual changes in the luminosity (corresponding to the changing mass of the core which must increase with the progress of time from the exhausted material of the shell, and decrease as the result of energy losses), in which case an apparently "main sequence" star with a superdense core may result. . . .

VI

THE MILKY WAY
AND BEYOND

By 1900 astronomy began to differentiate into separate fronts: on one hand, there was the study of the stars—their evolution, variety of types, and states of death; on the other hand, there was growing interest in discerning the overall structure of the universe (then largely defined as the Milky Way) and understanding the role of the nebulae within it. In 1845 William Parsons, the third earl of Rosse, had detected a spiraling structure in the M51 nebula, and for the rest of that century astronomers found other spiraling clouds in the nighttime sky. Around 1898 at the Lick Observatory in California, James Keeler began to systematically photograph them, estimating there were 120,000 objects potentially accessible to him with his 36-inch reflecting telescope.

Even though they looked more like oval blobs, these distinctive patches came to be called "spiral nebulae" to differentiate them from the irregular nebulae found within the plane of the Milky Way. In the eighteenth century, Immanuel Kant had described them as "higher universes" that "constitute . . . a still more immense system."[1] But (with a few exceptions) most astronomers concluded that the spiraling nebulae were more likely solar systems in the making within the Milky Way itself. In 1889 the British astronomer William Huggins wrote that the Andromeda nebula "shows a planetary system at a somewhat advanced stage of evolution; already several planets have been thrown off, and the central gaseous mass has condensed to a moderate size as compared with the dimensions it must have possessed before any planets had been formed."[2] Many others came to believe this as well, especially when reports were issued that the spiraling nebulae seemed to be changing and rotating. At the start of the twentieth century Agnes C. Clerke, a noted historian of astronomy, concluded that "the question whether nebulae are external galaxies hardly any longer needs discussion. . . . No competent thinker with the whole of the available evidence before him can now, it is safe to say, maintain any single nebula to be a star system of coordinate rank with the Milky Way."[3] To her and others, our galaxy defined the borders of the known universe.

This conclusion abruptly changed with new developments in instrumentation and the opening of new arenas for observation. As early as 1717 Isaac Newton had recognized that atmospheric turbulence affected a telescope's performance and suggested it could be remedied by moving to "a most serene and quiet air, such as may perhaps be found on the tops of the highest mountains above the grosser clouds."[4] His advice was applied in 1888, when Lick Observatory began operations atop Mount Hamilton, south of San Francisco. This was followed by the Mount Wilson Observatory near Pasadena, California, where new records were set in the size of reflecting telescopes: first a 60-inch reflector in 1908, followed by a 100-inch in 1917. With such instruments, funded by benefactors made rich in a growing industrialized nation, the universe was redefined. First, Harlow Shapley determined that our solar system did not reside in the center of the Milky Way but rather was positioned off to the side of the flat disk of stars. Several years later, Edwin Hubble solved the mystery of the spiral nebulae, proving they were indeed separate galaxies arrayed outward in space to great depths. The Milky Way was not alone. From that point on, the study of galaxies and the information they could provide on the evolution of the universe became premier interests within the field of astronomy.

48 / Cepheids: The Cosmic Standard Candles

Working at the Harvard College Observatory in 1907, Henrietta Leavitt examined photographs of the Magellanic clouds taken at Harvard's southern station in Peru and produced a catalog of 1,777 variable stars. For a special group of variables (later dubbed Cepheids after one of the first discovered, δ Cephei), she determined the period over which their light regularly waxed and waned. Looking over the results for these sixteen stars situated in the Small Magellanic Cloud, she discovered an interesting feature: "It is worthy of notice," she stated, "that . . . the brighter variables have the longer periods."[5] Wary that her sample was too small, she added nine more Cepheids to her list over the next four years before establishing the distinct mathematical relationship between a Cepheid's period and its luminosity reported in her classic 1912 Harvard observatory paper. (The star's magnitude steadily decreases with the logarithm of the period; see Figure 48.2.)

Since these variable stars were about the same distance from the Earth — all being situated in the Magellanic cloud — Leavitt keenly reasoned that "their periods are apparently associated with their actual emission of light."[6] A specific period was truly linked with a set level of brightness. This characteristic made Cepheid variables superb beacons. In revealing this association between a star's period and its luminosity, Leavitt provided astronomers with one of their handiest yardsticks for measuring distances throughout the Milky

Way galaxy and beyond. Once they were calibrated, Cepheids became astronomy's most useful standard candles. Observers were soon able to pick out a far-off Cepheid, note its period, and then infer its true *absolute* brightness (the luminosity you would observe if you were essentially right near the star). The distance to the Cepheid then followed: by measuring the Cepheid's *apparent* brightness in the sky (a much fainter magnitude), astronomers could figure out how far away it must be to appear that dim. This relatively simple relationship has been the keystone to measuring distances in the universe.

It had been long believed that Cepheids were eclipsing binary stars. But soon after Leavitt's 1912 report, both Henry Plummer, royal astronomer of Ireland, and Harvard's Harlow Shapley correctly perceived that these variables were stars that were pulsating, their atmospheres regularly ballooning in and out.[7] Several years later, Shapley expertly used the period-luminosity rule to discover our Sun's true place within the Milky Way galaxy (see Chapter 49).

"Periods of 25 Variable Stars in the Small Magellanic Cloud." *Harvard College Observatory Circular No. 173* (March 3, 1912) by Henrietta Leavitt

The following statement regarding the periods of 25 variable stars in the Small Magellanic Cloud has been prepared by Miss Leavitt.

A Catalogue of 1777 variable stars in the two Magellanic Clouds is given in H. A. 60, No. 4.* The measurement and discussion of these objects present problems of unusual difficulty, on account of the large area covered by the two regions, the extremely crowded distribution of the stars contained in them, the faintness of the variables, and the shortness of their periods. As many of them never become brighter than the fifteenth magnitude, while very few exceed the thirteenth magnitude at maximum, long

---

* Henrietta S. Leavitt, "1777 Variables in the Magellanic Clouds," *Annals of the Astronomical Observatory of Harvard College*, volume 60, number 4, 1908, pp. 104–7.

Table 48.1: Periods of Variable Stars in the Small Magellanic Cloud

H.	Max.	Min.	Epoch	Period	Res. M	Res. m	H.	Max.	Min.	Epoch	Period	Res. M	Res.
			d.	d.						d.	d.		
1505	14.8	16.1	0.02	1.25336	−0.6	−0.5	1400	14.1	14.8	4.0	6.650	+0.2	−0.3
1436	14.8	16.4	0.02	1.6637	−0.3	+0.1	*1355*	14.0	14.8	4.8	7.483	+0.2	−0.2
1446	14.8	16.4	1.38	1.7620	−0.3	+0.1	1374	13.9	15.2	6.0	8.397	+0.2	−0.3
1506	15.1	16.3	1.08	1.87502	+0.1	+0.1	818	13.6	14.7	4.0	10.336	0.0	0.0
1413	14.7	15.6	0.35	2.17352	−0.2	−0.5	*1610*	13.4	14.6	11.0	11.645	0.0	0.0
1460	14.4	15.7	0.00	2.913	−0.3	−0.1	*1365*	13.8	14.8	9.6	12.417	+0.4	+0.2
1422	14.7	15.9	0.6	3.501	+0.2	+0.2	*1351*	13.4	14.4	4.0	13.08	+0.1	−0.1
842	14.6	16.1	2.61	4.2897	+0.3	+0.6	827	13.4	14.3	11.6	13.47	+0.1	−0.2
1425	14.3	15.3	2.8	4.547	0.0	−0.1	822	13.0	14.6	13.0	16.75	−0.1	+0.3
1742	14.3	15.5	0.95	4.9866	+0.1	+0.2	823	12.2	14.1	2.9	31.94	−0.3	+0.4
1646	14.4	15.4	4.30	5.311	+0.3	+0.1	824	11.4	12.8	4.	65.8	−0.4	−0.2
1649	14.3	15.2	5.05	5.323	+0.2	−0.1	821	11.2	12.1	97.	127.0	−0.1	−0.4
1492	13.8	14.8	0.6	6.2926	−0.2	−0.4							

exposures are necessary, and the number of available photographs is small. The determination of absolute magnitudes for widely separated sequences of comparison stars of this degree of faintness may not be satisfactorily completed for some time to come. With the adoption of an absolute scale of magnitudes for stars in the North Polar Sequence, however, the way is open for such a determination.

Fifty-nine of the variables in the Small Magellanic Cloud were measured in 1904, using a provisional scale of magnitudes, and the periods of seventeen of them were published in H.A. 60, No. 4, Table VI. They resemble the variables found in globular clusters, diminishing slowly in brightness, remaining near minimum for the greater part of the time, and increasing very rapidly to a brief maximum. Table [48.1] gives all the periods which have been determined thus far, 25 in number, arranged in the order of their length. The first five columns contain the Harvard Number, the brightness at maximum and at minimum as read from the light curve, the epoch expressed in days following J.D. 2,410,000, and the length of the period expressed in days. The Harvard Numbers in the first column are placed in italics, when the period has not been published hitherto. A remarkable relation between the brightness of these variables and the length of their periods will be noticed. In H.A. 60, No. 4, attention was called to the fact that the brighter variables have the longer periods, but at that time it was felt that the number was too small to warrant the drawing of general conclusions. The periods of 8 additional variables which have been determined since that time, however, conform to the same law.

The relation is shown graphically in Figure [48.1], in which the abscissas are equal to the periods, expressed in days, and the ordinates are equal to the corresponding magnitudes at maxima and at minima. The two resulting curves, one for maxima and one for minima, are surprisingly smooth, and of remarkable form. In Figure [48.2] the abscissas are equal to the logarithms of the periods, and the ordinates to the corresponding magnitudes, as in Figure [48.1]. A straight line can readily be drawn among each of the two series of points corresponding to maxima and minima, thus showing that there is a simple relation between the brightness of the variables and their periods. The logarithm of the period increases by about 0.48 for each increase of one magnitude in brightness. The residuals of the maximum and minimum of each star from the lines in Figure [48.2] are given in the sixth and seventh columns of Table [48.1]. It is possible that the deviations from a straight line may become smaller when an absolute scale of magnitudes is used, and they may even indicate the corrections that need to be applied to the provisional scale. It should be noticed that the average range, for bright and faint variables alike, is about 1.2 magnitudes. Since the variables are probably at nearly the same distance from the Earth, their periods are apparently associated with their actual emission of light, as determined by their mass, density, and surface brightness.

The faintness of the variables in the Magellanic Clouds seems to preclude the study of their spectra, with our present facilities. A number of brighter variables have similar light curves, as UY Cygni, and should repay careful study. The class of spectrum ought to be determined for as many such objects as possible. It is to be hoped, also, that the parallaxes of some variables of this type may be measured. Two fundamental questions upon

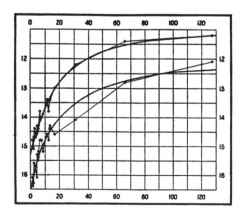

Figure 48.1

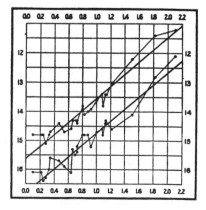

Figure 48.2

which light may be thrown by such inquiries are whether there are definite limits to the mass of variable stars of the cluster type, and if the spectra of such variables having long periods differ from those of variables whose periods are short.

The facts known with regard to these 25 variables suggest many other questions with regard to distribution, relations to star clusters and nebulae, differences in the forms of the light curves, and the extreme range of the length of the periods. It is hoped that a systematic study of the light changes of all the variables, nearly two thousand in number, in the two Magellanic Clouds may soon be undertaken at this Observatory.

49 / Sun's Place in the Milky Way

According to historians Michael Hoskin and David Dewhirst, Harlow Shapley liked to think of himself as a latter-day Copernicus. In the late 1910s he banished humanity to the sidelines of the Milky Way galaxy.

After working briefly as a crime reporter in Kansas and Missouri, Shapley enrolled in the University of Missouri to enter its school of journalism. Finding the school not yet open, he offhandedly took up astronomy and ended up graduating with honors. After receiving his doctorate at Princeton, he became a staff member in 1914 at the Mount Wilson Observatory in southern California. There he had access to what was then the most powerful telescope in the world, a 60-inch reflector. At the suggestion of a colleague, he used this instrument to identify variable stars called Cepheids in globular clusters—highly spherical, compact systems of stars—dispersed around the Milky Way.

Astronomers had already noticed that the largest fraction of these clusters, about one-third, were curiously concentrated in the direction of the constellation Sagittarius. The Swedish astronomer Karl Bohlin in 1909 suggested that the center of the galaxy was in that direction with the clusters huddled around it, but no one at the time took the idea seriously.[8] It was long assumed that the Sun was positioned in the center of the galaxy. Even Shapley rejected Bohlin's hypothesis at first, until his data forced him to radically alter his opinion.

At the start of his project, Shapley took swift advantage of an intriguing property of Cepheids newly revealed by Henrietta Leavitt at Harvard (see Chapter 48): that the period of a Cepheid's variation was directly related to its luminosity. The brighter the Cepheid, the longer its period. By knowing the distance to just one Cepheid, an astronomer could calibrate it as a standard candle and gauge the distance to other Cepheids by measuring their periods. Shapley obtained the needed calibration and then studiously determined the periods (and hence the distances) of dozens of Cepheids in the nearest globular clusters — a formidable task, as the stars were very faint. For clusters farther out, he used the brightest stars as distance markers, and when the stars themselves were no longer useful, he judged distance by the apparent size of the cluster on the sky. It was yeoman's work. "The work on clusters goes on monotonously — monotonous so far as labor is concerned," he wrote a colleague in 1917. "But the results are continual pleasure."[9] He published this growing body of data in an extended series of papers. It was the twelfth article that officially announced his historic conclusion.

Once he had determined the distances to sixty-nine globular clusters and mapped their distribution in space, Shapley clearly saw that our Sun was not situated in the galaxy's center but off to one side of the flat disk of stars. There was other evidence to support this view; as he pointed out in his paper, astronomers had been long aware that stars appeared more plentiful in the direction of Sagittarius. "The remarkable one-sidedness of the Milky Way has been little considered heretofore in works on stellar distribution," he noted. At the same time, Shapley also estimated that the total width of the Milky Way was some 100,000 parsecs (300,000 light-years), many times greater than previously assumed. He presumed (wrongly) that such a large size meant that the mysterious spiral nebulae were not "island universes" (as some were then arguing; see Chapter 51) but "dependents of the Galaxy."

Shapley's new vision of the Milky Way was not accepted right away. Around the same time, the noted Dutch astronomer Jacobus C. Kapteyn and his associate Pieter van Rhijn had just updated a competing model, the "Kapteyn universe," which kept the Sun near the center.[10] Their Milky Way was also much smaller, around 40,000 light-years wide. But further observations confirmed Shapley's finding. In the course of his comprehensive surveys of the distribution and motion of stars in the Milky Way, Kapteyn had noticed that stars in the galaxy preferred to move in two opposite directions.[11] In 1925 Bertil Lindblad, a Swedish expert on stellar dynamics, proposed that a disk-shaped galaxy rotating about the center of Shapley's system of globular clusters could explain this particular mode of "star streaming."[12] Two years later Jan Oort at the Leiden Observatory in Holland verified it; stars farther out from the galactic center were traveling more slowly than those closer in (as expected from orbital dynamics).[13]

The size of Shapley's Milky Way was eventually adjusted downward to 100,000 light-years, once the existence of interstellar dust was recognized, which affects distance estimates (see Chapter 50), but even then the Milky Way remained far larger than astronomers has previously assumed. "I have always admired the way in which Shapley . . . [ended] up with a picture of the Galaxy that just about smashed up all the old school's ideas about galactic dimensions . . . for these distances seemed to be fantastically large, and the 'old boys' did not take them sitting down," recalled Mount Wilson astronomer Walter Baade.[14]

"Studies Based on the Colors and Magnitudes in Stellar Clusters. Twelfth Paper: Remarks on the Arrangement of the Sidereal Universe."
Astrophysical Journal, Volume 49 (June 1919)
by Harlow Shapley

I. The General Galactic System

1. *Introduction.*—A fairly definite conception of the arrangement of the sidereal system evolves naturally from the observational work discussed in the preceding *Contributions.** We find, in short, that globular clusters, though extensive and massive structures, are but subordinate items in the immensely greater organization which is dimly outlined by their positions. From the new point of view our galactic universe appears as

---

* *Contributions from the Mount Wilson Solar Observatory.*

a single, enormous, all-comprehending unit, the extent and form of which seem to be indicated through the dimensions of the widely extended assemblage of globular clusters. The fundamental nature of the galactic plane, in the dynamical structure of all that we now recognize as the sidereal universe, is manifested by the distribution of clusters in space. Near this plane lie the celestial objects that we customarily study. The open clusters, the diffused and planetary nebulae, the naked-eye stars, most variables, the objects that define and compose the star streams—all of these appear to be far within a relatively narrow equatorial region of the greater galactic system, a region in which globular clusters are not found. The Orion nebula and even the Magellanic clouds are miniature organizations in this general scheme, and undoubtedly are dependents of the Galaxy.

The adoption of such an arrangement of sidereal objects leaves us with no evidence of a plurality of stellar "universes." Even the remotest of recorded globular clusters do not seem to be independent organizations. The hypothesis that spiral nebulae are separate galactic systems now meets with further difficulties. So long as the high velocities of nebulae were unapproached by the motions of other objects and the maximum luminosity attainable by stars was beyond estimate, and so long as the diameter of the galactic system was thought to be only a thousand light-years or so, we had a fairly plausible case for the "island universe" hypothesis. But now we must consider radial velocities of several hundred kilometers a second as quite possible for objects in our own system; we must assume a moderate upper limit of luminosity, perhaps even for the most massive of novae; and any external "universe" must now be compared with a galactic system probably more than three hundred thousand light-years in diameter. As seen from the center of the galactic system, globular clusters would be distributed in the sky much as the spirals are when observed from the earth.

It is probable that the further accumulation of observations will modify to some extent the views outlined above and discussed more fully in the following pages. The present data may in some cases be susceptible of alternative interpretation, or possibly the conclusions may be questioned in the belief that the material is insufficient. But the greater part of the hypothesis proposed is merely the most direct and simple reading of recent observations.

2. *Outline of interpretation.*—The suggested plan of the galactic system may be concretely formulated through the following series of propositions. . . .

A. The globular clusters are a part of the galactic system and knowledge of their distances seems at present to afford the best way to fathom the system.

B. The system of globular clusters, which is coincident in general, if not in detail, with the sidereal arrangement as a whole, appears to be somewhat ellipsoidal. The longest axis of the ellipsoid lies in the galactic plane and passes the sun at a distance of approximately three thousand parsecs. Its nearest point is in galactic longitude 240°, nearly coincident with the direction of the center assigned to the local system of stars. See . . . Fig. [49.1].

C. The center of the sidereal system is distant from the earth some twenty thousand parsecs in the direction of the constellation Sagittarius; it lies in the galactic plane, which dynamically and statistically appears to be the symmetrical plane of the entire sidereal universe as now known. As seen from the sun the thinnest part of the Milky Way lies in Gemini, Taurus, and Auriga—a region rich in bright open clusters close to the galactic plane.

D. The axes of the system in the galactic plane and perpendicular to it may not differ greatly; but the gravitationally important equatorial segment, which apparently contains most of the stars, is at least thirty times as extended in the plane as at right angles thereto.

E. The equatorial region appears to be uninhabitable by compact systems, such as globular clusters, notwithstanding the greater abundance there of stellar material.

F. The stars in the neighborhood of the sun (practically all that go into our catalogues of spectrum, position, and motion) appear to compose (1) a large, open, moving subordinate group, and (2) a part of the surrounding and interpenetrating star fields of the equatorial segment of the greater galactic system. The center of the local system is in the direction of the constellation Carina, nearly at right angles to the direction of the center of the general galactic system, but less than one two-hundredths as far away. The plane of symmetry and condensation of the local cluster is inclined to the galactic plane about 12°; the center of the cluster is north of the galactic plane, and the sun is north of both planes.

G. The volume of space occupied by stars brighter than the sixth apparent magnitude, some of which, being absolutely very bright, are extremely distant as compared with the majority of naked-eye stars, is at most only a hundred-thousandth of the volume occupied by the other parts of the galactic system.

3. *Relation of present interpretation to earlier hypotheses.*—In order to show where the earlier working hypotheses stand with respect to the interpretation now offered, it may be of interest to note the development, during the course of this work on clusters and variable stars, of the ideas concerning the relation of globular clusters to the galactic organization. Until the last year or so most students of stellar problems believed rather vaguely that the sun was not far from the center of the universe, and that the radius of the galactic system was of the order of 1,000 parsecs [around 3,000 light-years]. From the earlier observational data Seeliger and Newcomb derived a fairly central position for the sun.* Hertzsprung in 1906 estimated the "Dimensionen" of the visible Milky Way system to be of the order of 2,000 parsecs, and some years later Walkey,† from consideration of extensive distributional data, estimated a distance of about seventeen hundred parsecs for the galactic main stream. In 1914, referring to the apparently lens-shaped sidereal system, Eddington wrote, "There is little evidence as to the sun's position with respect to the perimeter of the lens; all that we can say is that it is not markedly eccentric"; and the diameter of the whole system (possibly excluding the peripheral ring of galactic clouds) was placed at some two or three thousand parsecs, with emphasis on the uncertainty. For a later computation Eddington assumed the distance of the Milky Way to be 2,000 parsecs. . . .

. . . The conviction [later] grew that the galactic system had an extent of at least 15,000 parsecs along its plane. This left little occasion for the direct comparison with globular clusters, the diameters of which were found by further study to be of the order of 150 parsecs. As a consequence, their relation to the general system was quite uncertain until the present determination of parallaxes and the discussion of the distribution in space indicated the position of globular clusters in the arrangement of sidereal objects and suggested that the actual diameter of the galactic system is of the order of 100,000 parsecs.

4. *The plane of symmetry and the equatorial segment.*—In the figures and discussion of the seventh paper of this series the dependence of globular clusters upon a larger symmetrical organization is definitely shown. Apparently there is no occasion to doubt the identity of the plane of

---

* German astronomer Hugo von Seeliger and Simon Newcomb of the U.S. Naval Observatory.

† British astronomer Oliver Rowland Walkey.

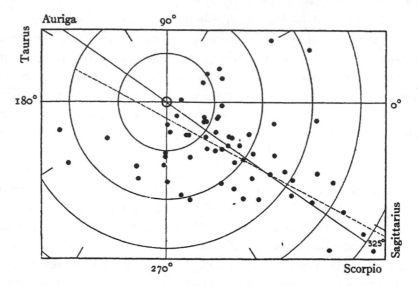

Figure 49.1: "The system of globular clusters projected on the plane of the Galaxy. The galactic longitude is indicated for every 30°. The 'local system' is completely within the smallest circle, which has a radius of 1,000 parsecs. The larger circles, which are also heliocentric, have radii increasing by intervals of 10,000 parsecs. The dotted line indicates the suggested major axis of the system, and the cross the adopted center. The dots are about five times the actual diameters of the clusters on this scale. Nine clusters more distant from the plane than 15,000 parsecs are not included in the diagram."

symmetry in this system of globular clusters with the galactic plane defined by stellar condensation and the Milky Way. . . .

That the equatorial segment is populated by stars throughout its whole extent seems very probable; both the arrangement of the clusters and the appearance of the Milky Way agglomerations support this view. . . .

5. *The Milky Way and its asymmetry; regions of maximum star density.*—According to the present view of the galactic system the phenomenon of the Milky Way is largely an optical one. Although the existence of local and occasionally very extensive condensations of Milky Way stars is not denied, the conception of a narrow encircling ring is abandoned. The Milky Way girdle is chiefly a matter of star depth, and its long recognized weakness between longitudes 90° and 180° is now taken to be a reflection of the eccentric position of the sun.

... We estimate provisionally that the limit of the Galaxy is three times greater in longitude 325° than in the opposite direction. This does not require an impossible difference of stellar density in the two directions, even if there is a considerable condensation toward the center. A star of a given absolute luminosity situated in the galactic plane would appear less than two and a half magnitudes fainter at the boundary of the system beyond the center than at the opposite point, which is nearest the sun. The remarkable one-sidedness of the Milky Way has been little considered heretofore in works on stellar distribution. Nort, in studying the Harvard map, has made an important beginning by showing that the star density is four or five times greater in the direction of the southern star clouds than in some of the shallower galactic regions of the north. . . .*

The possibly ellipsoidal form of the system of globular clusters is indicated in Fig. [49.1], which gives a projection on the galactic plane of the 60 clusters. . . . If the elongation be accepted as a real characteristic of the stars also, it is evident that the apparently densest star regions, depending on the faintness of the stars involved in the estimate, may lie in a longitude differing considerably from that of the center. The general direction of the galactic center is clearly toward the dense star clouds of Sagittarius and Scorpio; but the adopted galactic longitude, 325°, and the corresponding equatorial co-ordinates of the center, α [right ascension] = $17^h.5$, δ [declination] = $-30°$, are necessarily approximate. . . .

---

* The reference is to Dutch astronomer Henri Nort.

50 / Dark Nebulae and Interstellar Matter

A stronomers long believed that the vast expanses between the stars were a pristine emptiness. That space could be murky — that space might be "dirty" — was hard to imagine.

In the course of his extensive surveys over the nighttime sky, the noted eighteenth-century British astronomer William Herschel could not help but notice dark regions that appeared to be devoid of stars. His younger sister Caroline, his tireless assistant, heard him exclaim one night in his native German after a long silence, *"Hier ist wahrhaftig ein Loch im Himmel!* [Here is truly a hole in the heavens]," at the sight of a black chasm in the Ophiuchus constellation.[15] He surmised that as bright nebulae condensed into stars, they swept out these starless cavities, revealing deeper depths beyond.

This became the popular view among astronomers, but murmurs of dissent began to rise by the mid-nineteenth century. In Italy Giovanni Schiaparelli talked of dark nebulae wandering through the depths of space, and spectroscopist Angelo Secchi at the Collegio Romano called Herschel's dark-hole hypothesis "quite improbable . . . It is much more probable that the blackness results from a dark nebulosity projected on a bright background which intercepts its rays."[16] But few listened, since the sheer immensity of such dark structures — far bigger than glowing nebulae — was difficult to accept. That changed when Edward E. Barnard gathered exquisite photographic evidence, first at the Lick Observatory in California in the 1890s and then at the University of Chicago's Yerkes Observatory in the early 1900s, that the coal-black patches were true opaque bodies. Barnard, who revered Herschel from his early childhood reading, originally thought all voids were actual holes but gradually changed his mind in the course of his surveys, especially when examining black lanes meandering through the Taurus constellation in 1907. He thought of them as "dead nebulae." (Astronomers now know they are actually the incubators for the birth of new stars.)

In 1919 Barnard published a monumental paper cataloging 182 dark markings in the sky.[17] Unwilling to abandon Herschel altogether, though, he cautiously added (incorrectly) that some of the dark zones were "doubtless only vacancies."[18] Around the same time at an observatory in Germany, Max Wolf approached the problem more quantitatively, conducting star counts in both dark and bright regions. The dropoffs that occurred as he neared the dark voids matched the decreases expected due to intervening clouds of matter. He concluded that the dark nebulae were composed of dust.[19]

Yet even as astronomers came to recognize the existence of separate and distinct dark nebulae, they held on to the view that interstellar space was largely empty. Learning that space overall harbored matter was a slow revelation. As early as 1720, Edmond Halley wondered whether cosmic matter might darken the light of distant stars, and a century later the German astronomer F. G. W. Struve produced a mathematical model of interstellar absorption. It wasn't until 1904, though, that there was firmer evidence. At the Potsdam Astrophysical Observatory in Germany, Johannes Franz Hartmann observed the double star δ Orionis, one of the glittering jewels in Orion's belt, and noticed spectral absorption lines produced by calcium atoms caught by chance in front of the star system. It was proof that a gas was present somewhere over the vast distance between the Earth and the star. Later, other astronomers found evidence for additional interstellar atoms and molecules, such as sodium, iron, CN, and CH. But since such atoms pose no obstacle to viewing, astronomers still assumed that space was essentially transparent; many of their measurements in astronomy—from the brightness of stars to the distance of selected objects—depended on space having no *obscuring* matter.

Attitudes dramatically changed when Arthur Eddington, the most influential astrophysicist in his day, announced in a 1926

Royal Society lecture in Great Britain that interstellar matter was assuredly spread throughout the galaxy like a thin haze, being particularly dense in the dark nebulae. Barnard's superb photographs, Wolf's star counts, and the increasing spectroscopic evidence also led him to conclude that dust—very fine, solid particles—accompanied the gas, as others already suspected. "I have great reluctance (which is perhaps a prejudice) to admit meteoric matters of this kind in interstellar regions," he confessed, "but I cannot suggest an alternative."[20]

Final confirmation arrived in 1930 when Robert Trumpler at the Lick Observatory beautifully demonstrated how the light from distant stars is absorbed by dust in the Milky Way. After a decade of work on galactic star clusters—so named because they are located in the disk of our galaxy—the Swiss-born astronomer became quite familiar with the typical size, color, and brightness of these rich groupings. In the process, he could estimate the distance of a far-off cluster by judging how much smaller the cluster appeared when compared with a nearby cluster. In doing so, he noticed that the stars in the distant cluster were fainter than distance alone would warrant. The distant clusters were also noticeably redder than their counterparts closer to the Earth. It all made sense when he assumed that the disk of the Milky Way was suffused with a subtle dustiness, which dimmed and reddened the starlight much the way the Earth's dust-filled atmosphere turns the Sun a deep red-orange at sunset.

Trumpler's findings forced galactic mapmakers back to the drawing board, reducing Harlow Shapley's calculation of the Milky Way's size from 300,000 light-years to 100,000 (see Chapter 49). The intervening matter made faraway stars appear fainter and hence more distant than originally estimated. Interstellar gases constitute 5–10 percent of the matter in the Milky Way; microscopic dust grains—many fabricated in the outer envelopes of red giant stars— contribute roughly another 1 percent.

"On a Nebulous Groundwork in the Constellation Taurus."
Astrophysical Journal, Volume 25 (1907)
by Edward E. Barnard

I have elsewhere at various times called attention to the connection of nebulosities with some of the vacant regions of the sky. The finest example of this remarkable and suggestive peculiarity is shown in the great nebula of ρ *Ophiuchi*. In connection with some of these vacant regions I have remarked on the singular fact that in some cases—especially in the regions of θ and ρ *Ophiuchi*—these vacancies are vacancies only in the apparent absence of stars, for they are really often filled with a luminous veiling in which darker perforations occur.

The extraordinary vacant lanes among the Milky Way stars, in *Ophiuchus* and elsewhere, have often suggested that they are not only devoid of stars, but that they are darker than the immediate sky. In some cases there has been a suspicion that this was a matter of contrast, and that, if the remaining stars were removed, the lanes would also disappear. While this might be true in some cases, there are others where the appearance is strictly conclusive that the vacancies are not only due to the absence of stars, but that the channels are in a bed-work or nebulous substratum, and that, if the stars were removed, the lanes would still exist.

It will be seen that much importance depends upon whether these lanes are subjective—due to the scarcity of stars alone—or whether they reveal to us a nebulous substratum in certain parts of the sky.

In some of my early photographs north and east of the *Pleiades* the plates showed the existence of peculiar lanes far to the east of the cluster. Opportunity did not offer itself until the past winter to investigate their peculiarities by photography.

In the first part of January of this year I made several long exposures which covered the region in question. The result is very striking, and I believe of great importance; for the plates show that these lanes are undoubtedly in a substratum of some kind, as well as among the stars themselves.

The dying-out of nebulae—since it does not seem any longer necessary to use these vast bodies of gaseous matter for the making of suns—is

N

REGION OF VACANCIES IN *TAURUS*

Figure 50.1: Dark nebulae photographed by E. E. Barnard in 1907 (courtesy of Philip Myers).

a probability fully as warranted as the belief and certainty that the stars must die out. What would be the condition of a nebula that no longer emitted light, is a question; but as this light in all probability is not the product of heat or combustion in the ordinary sense, it is likely that we should simply have a dark nebula which would not be visible in the blackness of space unless its presence were made known by its absorption of the light of the stars beyond it—if this absorption were sufficient to be effective.

We have rather looked upon the nebulae as transparent bodies, like the comets; but there are no observations to warrant this idea, since in no case do we know that a nebula is on this side of the stars or beyond. True, there are cases where a star is palpably involved and seen through at least a portion of the nebula. There is nothing, however, to show whether the light of such a star has not been very greatly reduced by the interposition of the nebulous matter, and whether a much smaller star would not have been entirely invisible through the veiling of nebulosity.

This idea of the absorption of the light of the stars by a dead nebula or other absorbing matter has been used by some astronomers as an explanation of the dark or starless regions of the sky. Though this has not in general appealed to me as the true explanation—an apparently simpler one being that there are perhaps no stars at these places—there is yet considerable to commend it in some of the photographs. . . .

The connection of nebulosities with vacancies, and the apparent mingling of the outer portion of the nebula with the darkness of the sky, as if that darkness were something really tangible, as suggested in the case of the nebula of ρ *Ophiuchi,* is an extremely important feature, from which I believe there will some day develop facts of the greatest importance in explaining the real structure of the heavens. It would therefore be a very valuable work to locate all these regions and to secure the best long-exposure photographs of them. In this way a consistent study of their peculiarities may be made by those interested in their nature. Of late years I have endeavored to find as many of these places as possible. This has resulted in the development of the extraordinary regions of *Ophiuchus* and *Scorpio,* where these singular features are perhaps best shown. . . .

The region here shown is extraordinary. The narrow vacant lanes are as singular examples of the peculiarities I have mentioned as any that I know of, and they show perhaps even better the fact that the lanes actually exist in the sky independent of the stars. . . .

"Preliminary Results on the Distances, Dimensions, and Space Distribution of Open Star Clusters." *Lick Observatory Bulletin,* Volume 14, Number 420 (1930) by Robert J. Trumpler

Although the observations of magnitudes and spectral types in open star clusters of the Milky Way undertaken by the writer are still far from being complete, it seemed of interest to utilize the data at present available for a preliminary investigation of the distances and diameters of these clusters and for a study of their space distribution. . . .

[Omitted here are Trumpler's determination of the cluster distances from magnitudes and spectral types and his classification of the clusters based on appearance.]

Absorption of Light in the Milky Way System

Our method of deriving cluster distances from magnitudes and spectral types was based on [a] formula which expresses the law that the apparent brightness of a star diminishes with the square of the distance from the observer. If interstellar space is not perfectly transparent this law does not hold; the apparent brightness decreases more rapidly, our distance results are too large, and the error increases with the distance of the cluster. The linear diameters [of the clusters] computed with these distance results are then also too large, and the error also progresses with distance. . . .

[Here Trumpler derives formulas that correct both the distances and diameters of the clusters due to absorption by material uniformly distributed throughout the galactic system. He then applies the formulas to his cluster data and finds that the distant clusters are no longer larger than the nearer clusters (a phenomenon otherwise difficult to explain).]

. . . The apparent increase in the linear diameters of open star clusters is fully removed by the assumption of an absorption of light within our Milky Way system to the amount of 0.79 magnitudes per 1,000 parsecs for photographically effective rays.

If the light of the stars in the more distant clusters has been dimmed by the passage through an absorbing material, it seems *a priori* likely that such absorption is selective, and varies with the wavelength of the light and thus changes the color of the stars. Since the color of a star depends on its temperature a change of color by absorption can only be detected if its temperature is observable by some other means; *e.g.,* from the spectral types which measure the temperature by means of the ionization and excitation of the atoms in the stellar atmosphere. Spectral type estimates based on comparison of intensities of spectral lines are not affected by a general absorption. For the brighter and nearer stars which are little affected by absorption the relation between color indices and spectral types has been well established, but differs slightly for giant and dwarf stars. From this relation we can find the normal color index corresponding to a given spectral type. The difference between this and the color index actually observed is called the color excess of the star. The existence of large discrepancies between observed color indices and spectral types in open star clusters has been known for some time, but to test our hypothesis of an absorption of light in space it is necessary to show that the average color excesses in various clusters depend on their distance. The spectral types in open clusters determined with the 1-prism slit spectrograph and the slitless quartz spectrograph made it possible to compute the average color excess for a number of clusters in which color indices have been measured by various observers. Leaving out the few nearest clusters which are of little interest for our purpose and using only color indices determined by direct comparison with the North Polar sequence, seven clusters ranging widely in distance were available. Table [50.1] gives for each of these the observed distance r' (from magnitude and spectral types), the finally adopted distance which is corrected for absorption, the average color excess of the cluster stars, [and] the number of stars used. . . .

In all clusters we find a positive excess, and the latter is largest for the three most distant clusters. It must be kept in mind, of course, that color indices are subject to systematic errors of observation and especially that errors in the zero point of the photographic and visual magnitude scales may occur, due to changes in the observing conditions between the exposures on the cluster and on the North Pole. It is, however, out of [the] question that errors on the order of $0^m.5$–$0^m.7$ could result from this source, especially if the North Polar comparisons have been repeated on different days. Wallenquist, aware of the large discrepancy between his color indices

Table 50.1: Color Excess of Open Clusters

Cluster	Observed dist. r' (parsecs)	Adopted corrected distance	Color excess	Number of stars
NGC 2682 Messier 67	690	740	$+0^m.26$	81
1647	800	610	+0.17	33
			+0.19	6
2099 Messier 37	1450	820	+0.05	25
1960 Messier 36	1650	980	+0.05	40
6705 Messier 11	2200	1340	+0.65	46
7654 Messier 52	2400	1360	+0.49	43
663	3500	2170	+0.71	41

of NGC 663 and the spectral types of the same stars, and unable to account for it by observational errors, concludes: "The most probable explanation is, perhaps, the assumption of selective absorbing clouds within (and in the surroundings of) the cluster NGC 663."* That the effect is not due to an error in method is well illustrated by the fact that Wallenquist observed three clusters by the same method and instrument and that only the two more distant ones show a large color excess; the same is true for the two open clusters investigated by Shapley. In a former publication, I drew attention to the large excess of Shapley's color observations in the cluster Messier 11 and, averse to the idea of a general selective absorption in our stellar system, took rather a skeptical attitude concerning the correctness of Shapley's results until these were confirmed by Wallenquist's observations of two more distant clusters.† We are thus led to the assumption of a general absorption of light in our stellar system by two quite independent sets of observations; by the study of the linear diameters of star clusters as well as by color-index observations in such clusters. That the absorption is not caused merely by an absorbing cloud involved in the cluster itself . . . is shown by the increase in the color indices with the distance and by a similar increase in the apparent linear diameters.

. . . The hypothesis [is] that the absorbing medium . . . is very much concentrated toward the galactic plane. . . . Two-thirds of all open clusters lie within 100 parsecs of their plane of symmetry, and it is not improbable that the absorbing material has a similar distribution, thinning out very rapidly at greater distances from the galactic plane and forming so to speak a thin sheet (perhaps 200–300 parsecs thick) extending along the galactic plane to distances of at least 2,000 and perhaps 4,000 or 6,000 parsecs. . . .

* He is referring to the Swedish astronomer Åke Wallenquist.
† *"Lick Observatory Bulletin* 12 (1925): 12."

51 / Discovery of Other Galaxies

T he universe as we know it was revealed to astronomers on New Year's Day 1925. The man responsible, Edwin Hubble, was not present. His historic paper, primly entitled "Cepheids in Spiral Nebulae," was read to the thirty-third annual meeting of the American Astronomical Society in Washington, D.C., by Henry Norris Russell of Princeton. Reluctant to divulge his findings too soon, Hubble had to be persuaded to release a preliminary report at the conference. The announcement changed the face of the universe—the Milky Way was suddenly humbled, becoming just one of a multitude of galaxies residing in the vast gulfs of space. Hints of the Milky Way's true place in the universe had been cropping up for years, but Hubble's observations provided the evidence that at last convinced the community at large.

Establishing that certain nebulae were sister galaxies of the Milky Way had a long and tumultuous history. In the eighteenth century the great British astronomer William Herschel discovered hundreds of nebulae scattered over the celestial sphere and, like Immanuel Kant earlier (see Chapter 21), first thought of them as island universes, separate congregations of stars. But over time support for this view waned. The work of William Huggins and other spectroscopists in the nineteenth century showed that many nebulae were gaseous in nature (see Chapter 25), leading to the popular view that *all* nebulae were members of the Milky Way, possibly other solar systems or new star clusters in the making. "A good many converging lines of evidence," wrote historian Arthur Berry in 1898, "indicate the probability that [the celestial] bodies should be regarded as belonging to a single system."[21] In 1885, for example, there was a dramatic flaring in the Andromeda nebula. This nova, at the height of its brilliance, rose to the seventh magnitude. If Andromeda were an external universe, far beyond the Milky Way's borders, that nova had to be shining with the energy of some fifty million suns. Unaware as yet that a star could explode as a supernova, astronomers considered such an output preposterous. Nineteenth-century historian Agnes Clerke remarked that it would "have been on a scale of magnitude such as the imagination recoils from contemplating."[22] Ten years later, another bright nova was spotted in NGC 5253, a spiral nebula in the constellation Centaurus.* To Clerke, the idea that nebulae were other universes had "ceased to exist."[23]

Despite these declarations, a few lone astronomers began gathering evidence that Andromeda and other spiral nebulae were remote galaxies of stars after all. In 1917 at Mount Wilson Observatory George Ritchey, closely examining a photograph, spotted a previously unseen star in NGC 6946, but one far dimmer than the 1885 flare-up in Andromeda.[24] Heber Curtis at California's Lick Observatory found other cases of faint novae in spiral nebulae, one in NGC 4257 and two others in NGC 4321.[25] This suggested there were *two* types of flaring: the bright novae of 1885 and 1895 were exceptions rather than the rule and reopened the possibility that the spiral nebulae were indeed islands of stars. Curtis had been conducting a photographic survey of nebulae and was already convinced. In 1914, in an in-house Lick report, he remarked that the spirals are "inconceivably distant, galaxies of stars or separate stellar universes so remote that an entire galaxy becomes but an unresolved haze of light."[26] He even found the reason that spiral nebulae tended to crowd around the poles of the Milky Way and were not seen in its plane. He photographed some spiral nebulae edge-on and noticed dark lanes of obscuring matter. Such dusty material in the disk of

* NGC stands for New General Catalogue, published by J. L. E. Dreyer in 1888 by request of the Royal Astronomical Society. It extended the general catalogue published by John Herschel in 1864 and remains a standard for referencing deep-sky objects.

our own galaxy would of course, reasoned Curtis, hide our view of the distant universe along that direction.

In 1920, under the sponsorship of the National Academy of Sciences in Washington, D.C., Curtis squared off with Harvard's Harlow Shapley, chief proponent of the view that the Milky Way was the universe's sole galaxy, in what has come to be known in astronomy as the "Great Debate." Shapley had one, very compelling rebuttal to Curtis's arguments. At Mount Wilson, the Dutch astronomer Adriaan van Maanen was examining photographs of spiral nebulae taken at different times and was claiming to see the nebulae rotate, an effect impossible to see unless the nebulae were small, and hence fairly close, within the Milky Way itself. The debate was a draw.

In 1922 Ernst Öpik at the Dorpat Observatory in Estonia determined that the Andromeda nebula was some 1.5 million light-years distant, by assuming that its mass and luminosity were comparable to those of the Milky Way.[27] But his paper received little notice. It was Hubble who provided the conclusive evidence when he directly determined the distance to Cepheid variable stars within Andromeda (M31) and the M33 spiral in Triangulum. It was the type of distance measurement in which astronomers had the most confidence.

A Rhodes scholar trained in law, Hubble had returned to graduate school and his favorite college subject, astronomy, in 1914 at the age of twenty-four and five years later became a member of the Mount Wilson Observatory staff, which gave him access to the 100-inch Hooker reflector. He was part of a select group in California that for several decades dominated astronomy's discoveries in the far universe because of its employment of the world's largest and best telescopes. In the fall of 1923, Hubble began a study of Andromeda, spotting two ordinary novae and a third faint star. Perusing the library of plates on Andromeda, going back to 1909, he came to realize that the third stellar object was a *variable* star. With more than sixty plates on hand, he plotted the star regularly rising and falling in intensity. Six nights at the telescope in February 1924 confirmed it: the star's brightness rose rapidly that week, just like a Cepheid. Its period was thirty-one days, which, according to the period-luminosity relation established by Henrietta Leavitt and calibrated by Shapley, meant it was an extremely luminous star (see Chapter 48). But Hubble determined that this Cepheid, being only of eighteenth magnitude, had to be almost a million light-years away (285,000 parsecs) to appear so faint.* His estimate of Andromeda's distance fell short of today's value of 2 million light-years, but it still put the nebula far beyond the confines of the Milky Way.† Hubble eventually found other Cepheid variables, in both M31 and M33, that clinched his conclusion. He wrote Shapley, who recognized the overwhelming evidence and quickly conceded that he had been wrong about Andromeda. But van Maanen (whose rotation data were later found to be erroneous) resisted, which kept Hubble from making an immediate announcement. Shapley and others, though, convinced Hubble to have a paper summarizing his findings read at the 1925 astronomy meeting. There is no direct mention of external galaxies or a cosmological shake-up in his report; as was his style, Hubble kept his announcement low-key. Only in the American Astronomical Society's report of the meeting was it noted that Hubble had brought "confirmation to the so-called island universe theory."[28]

Hubble devoted the rest of his professional life to the realm of the nebulae. He soon introduced a classification scheme to distinguish the various types.[29] There are the E or elliptical galaxies, the smooth spheroidal bulges that run from round to oval in shape; the S or spiral galaxies, which consist of a central bulge in a range of sizes surrounded by a spiraling disk that can be tightly coiled or spread out wide; and lastly the irregular galaxies, such as the chaotic Magellanic clouds.

"Cepheids in Spiral Nebulae." *Publications of the American Astronomical Society*, Volume 5 (1925) by Edwin P. Hubble

Messier 31 and 33, the only spirals that can be seen with the naked eye, have recently been made the subject of detailed investigations with the

---

* 1 parsec = 3.26 light-years (see Chapter 19).

† Because Hubble was not aware that there were two classes of Cepheids, with different period-luminosity relationships, his distance was off by a factor of about two. The error would be discovered in the 1950s (see Chapter 53). Öpik's 1922 estimated distance to M31 turned out to be more accurate in the end.

100-inch and 60-inch reflectors of the Mount Wilson Observatory. Novae are a common phenomenon in M31, and Duncan* has reported three variables within the area covered by M33.[30] With these exceptions there seems to have been no definite evidence of actual stars involved in spirals. Under good observing conditions, however, the outer regions of both spirals are resolved into dense swarms of images in no way differing from those of ordinary stars. A survey of the plates made with the blink-comparator has revealed many variables among the stars, a large proportion of which show the characteristic light-curve of the Cepheids.†

Up to the present time some 47 variables, including Duncan's three, and one true nova have been found in M33. For M31, the numbers are 36 variables and 46 novae, including the 22 novae previously discovered by Mount Wilson observers. Periods and photographic magnitudes have been determined for 22 Cepheids in M33 and 12 in M31. Others of the variables are probably Cepheids, judging from their sharp rise and slow decline, but some are definitely not of this type. One in particular, Duncan's No. 2 in M33, has been brightening fairly steadily with only minor fluctuations since about 1906. It has now reached the 15th magnitude and has a spectrum of the bright line B type.

For the determinations of periods and normal curves of the Cepheids, 65 plates are available for M33, and 130 for M31. The latter object is too large for the area of good definition on one plate, so attention has been concentrated on three regions: around BD +41° 151, BD +40° 145, and a region some 45′ along the major axis south preceding the nucleus.

Photographic magnitudes have been determined from twelve comparisons with selected areas No. 21 and 45, made with the 100-inch using exposures from 30 to 40 minutes. This procedure seemed preferable to the much longer exposures required for direct polar comparisons with the 60-inch. It involves, however, a considerable extrapolation based on scales determined from the faintest magnitudes available for the selected areas.

Tables [51.1] and [51.2] give the data for the Cepheids in M33 and M31 respectively. No magnitudes fainter than 19.5 are recorded, because of the uncertainty involved in their precise determinations. The now familiar period-luminosity relation is conspicuously present.

For more detailed investigation of the relation, the magnitudes at max-

---

* John C. Duncan, director of the Wellesley College observatory, who first spotted a variable star in M31 in 1922.

† A blink comparator allows a viewer to quickly alternate between two photographic plates taken of the same field at different times. The blinking proceeds so rapidly that a changing or moving object will stand out, while those that remain fixed appear still.

Table 51.1: Cepheids in M33

Var. No.	Period in Days	Log. P	Photographic Max.	Magnitudes Min.
30	46.0	1.66	18.35	19.25
3	41.6	1.62	18.45	19.4
36	38.2	1.58	18.45	19.1
31	37.3	1.57	18.30	19.2
29	37.2	1.57	18.55	19.15
20	35.95	1.56	18.50	19.2
18	35.5	1.55	18.45	19.15
35	31.5	1.50	18.55	19.35
42	31.1	1.49	18.65	19.35
44	30.2	1.48	18.70	
40	26.0	1.41	19.00	
17	23.6	1.37	18.80	
11	23.4	1.37	18.85	
22	21.75	1.34	19.00	
12	21.2	1.33	18.80	
27	21.05	1.32	18.85	
43	20.8	1.32	18.95	
33	20.8	1.32	18.75	
10	19.6	1.29	18.80	
41	19.15	1.28	18.75	
37	18.05	1.26	18.95	
15	17.65	1.25	19.05	

ima have been plotted against the logarithm of the period in days. This procedure is necessary, not only because of the uncertainties in the fainter magnitudes, but also because most of the fainter variables at minimum are below the limiting magnitude of the plates. It assumes that there is no relation between period and range, for otherwise a systematic error in the slope of the period-luminosity curve is introduced. Among the brighter Cepheids of M33 the assumption appears to be allowable, for the ranges show a very small dispersion about the mean value of 0.8 magnitude. The average range and the dispersion are somewhat larger in M31, but the data are too limited for a complete investigation.

The curve for M33 appears to be very definite. The average deviation

Table 51.2: Cepheids in M31

Var. No.	Period in Days	Log. P	Photographic Magnitude Max.
5	50.17	1.70	18.4
7	45.04	1.65	18.15
16	41.14	1.61	18.6
9	38	1.58	18.3
1	31.41	1.50	18.2
12	22.03	1.34	19.0
13	22	1.34	19.0
10	21.5	1.33	18.75
2	20.10	1.30	18.5
17	18.77	1.28	18.55
18	18.54	1.27	18.9
14	18	1.26	19.1

is about 0.1 magnitude, although a considerable systematic error is allowable in the slope. For M31 the slope is very closely the same but the dispersion is much greater, averaging about 0.2 magnitude. This is probably greater than the accidental errors of measurement.

Shapley's period-luminosity curve for Cepheids, as given in his study of globular clusters, is constructed on a basis of visual magnitudes.[31] It can be reduced to photographic magnitudes by means of his relation between period and color-index, given in the same paper, and the result represents his original data. The slope is of the order of that for the spirals, but is not precisely the same. In comparing the two, greater weight must be given the brighter portion of the curve for the spirals, because of the greater reliability of the magnitude determinations. When this is done, the resulting values of $M - m$ are −21.8 and −21.9 for M31 and M33, respectively. These must be corrected by half the average ranges of the Cepheids in the two spirals, and the final values are then on the order of −22.3 for both nebulae. The corresponding distance is about 285,000 parsecs. The greatest uncertainty is probably in the zero point of Shapley's curve.

The results rest on three major assumptions: (1) The variables are actually connected with the spirals. (2) There is no serious amount of absorption due to amorphous nebulosity in the spirals. (3) The nature of Cepheid variation is uniform throughout the observable portion of the universe. As for the first, besides the weighty arguments based on analogy and probability, it may be mentioned that no Cepheids have been found on the several plates of the neighboring selected areas No. 21 and 45, on a special series of plates centered on BD +35° 207, just midway between the two spirals, nor in ten other fields well distributed in galactic latitude, for which six or more long exposures are available. The second assumption is very strongly supported by the small dispersion in the period-luminosity curve for M33. In M31, in spite of the somewhat larger dispersion, there is no evidence of an absorption effect to be measured in magnitudes.

These two spirals are not unique. Variables have also been found in M81, M101 and NGC 2403, although as yet sufficient plates have not been accumulated to determine the nature of their variation.

52 / Expansion of the Universe

B y the end of the nineteenth century, it was generally viewed
that the width and breadth of the Milky Way defined the borders of the universe. The one outstanding problem was determining, once and for all, the exact nature of the spiral nebulae in
this system. Improved instrumentation in the early twentieth century at last allowed Edwin Hubble to confirm that the nebulae were
separate "island universes," sister galaxies to the Milky Way (see
Chapter 51). Their movements, however, led to a more startling revelation: that the universe was not only far bigger than previously
imagined but also expanding, carrying the galaxies outward over
time.

The story of this discovery began before spiral nebulae were even
generally accepted as galaxies. First in 1910 and then over the fall
and winter of 1912 at the Lowell Observatory in Arizona, Vesto
Slipher obtained spectra of the Andromeda nebula, initially hoping
to uncover clues to the origin of our Sun and planets. Many at the
time believed the spiral nebulae might be other solar systems in the
making. It was painstaking work and required exposing the photographic plate for dozens of hours at the telescope before the nebula's faint spectral features could be interpreted. The nebula's
velocity, for example, could be pegged by noting the shift in its spectral lines (see Chapter 26). Expecting to find Andromeda moving at
about 20 kilometers per second, like other stars in our galaxy,

Slipher was surprised to find it was rushing toward the Earth at an astounding speed of 300 kilometers per second, the greatest velocity then measured for a celestial object in the universe.[32] Over the next year, he determined the velocities of fourteen more spirals; three years later he had an additional ten. Some (like Andromeda) were approaching us, but the majority were moving away (their light waves stretched out or "redshifted"), some at velocities surpassing 1,000 kilometers per second. With this evidence in hand, Slipher concluded that the Milky Way was a spiral nebula, drifting among other galaxies just like itself.[33] His finding added fuel to the ongoing debate on the island universe theory, still unresolved at the time.

Several astronomers began examining whether there was a pattern to the motions of the spiral nebulae—a difficult task, as the speeds measured were entangled with other velocities, such as the Earth's orbital motion and the Sun's movement through the galaxy. As early as 1918 Carl Wirtz in Germany began subtracting out these extra factors and saw that the nebulae speeds were still enormous. Moreover, he recognized that the nebulae were generally moving outward in all directions. Roughly gauging a nebula's distance by its size and luminosity, he and others noticed that the more distant a nebula, the faster it was receding.[34] What was missing to confirm this trend were reliable distance measurements.

A resolution swiftly arrived as soon as Edwin Hubble used Cepheid variable stars as distance markers and proved that spiral nebulae were galaxies. With the assistance of Milton Humason (a former Mount Wilson mule driver turned gifted observer), he then secured the definitive law for the recession of the galaxies, finding the direct relationship between a galaxy's velocity and its distance. In doing this, Hubble established the very foundation of modern-day cosmology. Using the largest telescope in their day—Mount Wilson Observatory's 100-inch reflector in California—the two collaborators were able to gather the key evidence that convinced the astronomical community that the universe was expanding in a specific way. While Humason focused on getting the redshift data to figure out the galaxies' velocities, Hubble determined their distances. By the time he published his historic discovery paper in 1929, Hubble had twenty-four galaxies analyzed out to a distance of 2 million parsecs (~6 million light-years). From that initial sample (which included Slipher's velocity data without acknowledgment), it appeared that for every million parsecs outward, the velocity of a galaxy increased by 500 kilometers per second. In his paper he referred to this factor as *K*, a term introduced by others in earlier analyses. Astronomers later changed it to H_0 and called it the Hubble constant.

Hubble had specifically designed his observations to test for an expansion. He knew it was a contentious issue and didn't publish for more than a year to make sure he had taken care of every possible criticism. He and Humason continued their survey outward and by 1931 were finding galaxies receding at velocities of 12,000 kilometers per second at a distance of 32 million parsecs. Yet they remained cautious in their explanations: "The interpretation of the redshifts as velocities of recession is controversial," wrote Humason that year. "For the present we prefer to speak of these velocities as *apparent*."[35] Einstein's theory of general relativity could account for the receding galaxies, but several relativistic models of the universe were in play and some explained the redshifts without the need for an expansion. As Hubble pointed out in his 1929 paper, the cosmology of Willem de Sitter could explain the redshifts as "an apparent slowing down of atomic vibrations" with increasing distance (see Chapter 37). Hubble kept his options open and spent years gathering data to compare the various models.

Others were not so hesitant, and by shifting the rate of expansion into reverse to the moment when it all began, astronomers were able to extrapolate an age for the universe. Using Hubble's initial constant of 500, they concluded it was 2 billion years old. That was less than the age of the Earth based on geologic evidence, which aroused suspicions that the redshifts might not be proof of a cosmic expansion after all. But over time, the Hubble constant was lowered with improved measurements, extending the age of the universe to between 10 and 20 billion years. For several decades, estimates of the Hubble constant ranged between 50 and 100. Each camp— those defending a low Hubble constant and those fighting for the higher number—fiercely argued that its methods were better calibrated and thus more precise. It is fitting that in the late 1990s the space telescope named after Hubble enabled a team of astronomers led by Wendy Freedman of the Carnegie Observatories (Hubble's professional home) to make the most accurate measurement to date, bringing it to the middle of the long-disputed range.[36]

"Nebulae." *Proceedings of the American Philosophical Society*, Volume 56 (1917) by Vesto M. Slipher

In addition to the planets and comets of our solar system and the countless stars of our stellar system there appear on the sky many cloud-like masses— the nebulae. These for a long time have been generally regarded as presenting an early stage in the evolution of the stars and of our solar system, and they have been carefully studied and something like 10,000 of them catalogued.

Keeler's classical investigation of the nebulae with the Crossley reflec- tor by photographic means revealed unknown nebulae in great numbers.* He estimated that such plates as his if they were made to cover the whole sky would contain at least 120,000 nebulae, an estimate which later obser- vations show to be considerably too small. He made also the surprising dis- covery that more than half of all nebulae are spiral in form; and he expressed the opinion that the spiral nebulae might prove to be of particu- lar interest in questions concerning cosmogony.

I wish to give at this time a brief account of a spectrographic investiga- tion of the spiral nebulae which I have been conducting at the Lowell Observatory since 1912. Observations had been previously made, notably by Fath at the Lick and Mount Wilson Observatories, which yielded valu- able information on the character of the spectra of the spiral nebulae.† These objects have since been found to be possessed of extraordinary motions and it is the observation of these that will be discussed here. . . .

Spiral nebulae are intrinsically very faint. The amount of their light admitted by the narrow slit of the spectrograph is only a small fraction of the whole and when it is dispersed by the prism it forms a continuous spec- trum of extreme weakness. The faintness of these spectra has discouraged their investigation until recent years. It will be only emphasizing the fact that their faintness still imposes a very serious obstacle to their spectro- graphic study when it is pointed out, for example, that an excellent spec- trogram of the Virgo spiral NGC 4594 secured with the great Mount Wilson reflector by Pease was exposed eighty hours. . . .‡

* Lick Observatory astronomer James Keeler.
† Carleton College astronomer Edward Fath.
‡ Francis Pease, instrument maker at Mount Wilson Observatory, who went on to design its historic 100-inch telescope.

. . . I have secured between forty and fifty spectrograms of 25 spiral nebulae. The exposures are long—generally from twenty to forty hours. It is usual to continue the exposure through several nights but occasionally it may run into weeks owing to unfavorable weather or the telescope's use in other work. Besides the exposures cannot be continued in the presence of bright moonlight and this seriously retards the accumulation of observations. . . .

The plates are measured under the Hartmann spectrocomparator in which one optically superposes the nebular plate of unknown velocity upon one of a like dark-line spectrum of known velocity, used as standard. A micrometer screw, which shifts one plate relatively to the other, is read when the dark lines of the nebula and the standard spectrum coincide; and again when the comparison lines of the two plates coincide. The difference of the two screw readings with the known dispersion of the spectrum gives the velocity of the nebula. By this method weak lines and groups of lines can be utilized that otherwise would not be available because of faintness or uncertainty of wavelength.

In table [52.1] are given the velocities for the twenty-five spiral nebulae thus far observed. In the first column is the New General Catalogue number of the nebula and in the second the velocity. The plus sign denotes the nebula is receding, the minus sign that it is approaching. . . .

Table 52.1: Radial Velocities of Twenty-five Spiral Nebulae

Nebula	Vel. km [sec^{-1}]	Nebula	Vel. km [sec^{-1}]
NGC 221	− 300	NGC 4526	+ 580
224	− 300	4565	+ 1,100
598	− 260	4594	+ 1,100
1023	+ 300	4649	+ 1,090
1068	+1,100	4736	+ 290
2683	+ 400	4826	+ 150
3031	− 30	5005	+ 900
3115	+ 600	5055	+ 450
3379	+ 780	5194	+ 270
3521	+ 730	5236	+ 500
3623	+ 800	5866	+ 650
3627	+ 650	7331	+ 500
4258	+ 500		

Referring to the table of velocities again: the average velocity 570 km [sec^{-1}] is about thirty times the average velocity of the stars. And it is so much greater than that known of any other class of celestial bodies as to set the spiral nebulae aside in a class to themselves. Their distribution over the sky likewise shows them to be unique—they shun the Milky Way and cluster about its poles.

The mean of the velocities with regard to sign is positive, implying the nebulae are receding with a velocity of nearly 500 km [sec^{-1}]. This might suggest that the spiral nebulae are scattering but their distribution on the sky is not in accord with this since they are inclined to cluster. . . .

As noted before the majority of the nebulae here discussed have positive velocities, and they are located in the region of sky near right ascension twelve hours which is rich in spiral nebulae. In the opposite point of the sky some of the spiral nebulae have negative velocities, *i.e.*, are approaching us; and it is to be expected that when more are observed there, still others will be found to have approaching motion. It is unfortunate that the twenty-five observed objects are not more uniformly distributed over the sky as then the case could be better dealt with. It calls to mind the radial velocities of the stars which, in the sky about Orion, are receding and in the opposite part of the sky are approaching. This arrangement of the star velocities is due to the motion of the solar system relative to the stars. . . .

We may in like manner determine our motion relative to the spiral nebulae, when sufficient material becomes available. A preliminary solution of the material at present available indicates that we are moving in the direction of right-ascension 22 hours and declination −22° with a velocity of about 700 km [sec^{-1}]. While the number of nebulae is small and their distribution poor this result may still be considered as indicating that we have some such drift through space. For us to have such motion and the stars not show it means that our whole stellar system moves and carries us with it. It has for a long time been suggested that the spiral nebulae are stellar systems seen at great distances. This is the so-called "island universe" theory, which regards our stellar system and the Milky Way as a great spiral nebula which we see from within. This theory, it seems to me, gains favor in the present observations.

It is beyond the scope of this paper to discuss the different theories of the spiral nebulae in the face of these and other observed facts. However, it seems that, if our solar system evolved from a nebula as we have long believed, that nebula was probably not one of the class of spirals here dealt with.

"A Relation Between Distance and Radial Velocity Among Extra-Galactic Nebulae." *Proceedings of the National Academy of Sciences*, Volume 15 (March 15, 1929) by Edwin Hubble

Determinations of the motion of the sun with respect to the extragalactic nebulae have involved a K term of several hundred kilometers which appears to be variable. Explanations of this paradox have been sought in a correlation between apparent radial velocities and distances, but so far the results have not been convincing. The present paper is a re-examination of the question, based on only those nebular distances which are believed to be fairly reliable.

Distances of extragalactic nebulae depend ultimately upon the application of absolute-luminosity criteria to involved stars whose types can be recognized. These include, among others, Cepheid variables, novae, and blue stars involved in emission nebulosity. Numerical values depend upon the zero point of the period-luminosity relation among Cepheids, the other criteria merely check the order of the distances. This method is restricted to the few nebulae which are well resolved by existing instruments. A study of these nebulae, together with those in which any stars at all can be recognized, indicates the probability of an approximately uniform upper limit to the absolute luminosity of stars, in the late type spirals and irregular nebulae at least, of the order of M (photographic) $= -6.3$. The apparent luminosities of the brightest stars in such nebulae are thus criteria which, although rough and to be applied with caution, furnish reasonable estimates of the distances of all extragalactic systems in which even a few stars can be detected. . . .

Radial velocities of 46 extragalactic nebulae are now available, but individual distances are estimated for only 24. For one other, NGC 3521, an estimate could probably be made, but no photographs are available at Mount Wilson. The data are given in table [52.2]. The first seven distances are the most reliable, depending, except for M32 [NGC 221 in the table] the companion of M31 [NGC 224], upon extensive investigations of many stars involved. The next thirteen distances, depending upon the criterion of a uniform upper limit of stellar luminosity, are subject to considerable

Table 52.2: Nebulae Whose Distances Have Been Estimated from Stars
Involved or from Mean Luminosities in a Cluster

Object	m_s	r	v
S. Mag.		0.032	+ 170
L. Mag.		0.034	+ 290
NGC 6822		0.214	− 130
598		0.263	− 70
221		0.275	− 185
224		0.275	− 220
5457	17.0	0.45	+ 200
4736	17.3	0.5	+ 290
5194	17.3	0.5	+ 270
4449	17.8	0.63	+ 200
4214	18.3	0.8	+ 300
3031	18.5	0.9	− 30
3627	18.5	0.9	+ 650
4826	18.5	0.9	+ 150
5236	18.5	0.9	+ 500
1068	18.7	1.0	+ 920
5055	19.0	1.1	+ 450
7331	19.0	1.1	+ 500
4258	19.5	1.4	+ 500
4151	20.0	1.7	+ 960
4382		2.0	+ 500
4472		2.0	+ 850
4486		2.0	+ 800
4649		2.0	+ 1090

m_s = photographic magnitude of brightest stars involved
r = distance in units of 10^6 parsecs
v = measured velocities in km/sec

probable errors but are believed to be the most reasonable values at present available. The last four objects appear to be in the Virgo Cluster. The distance assigned to the cluster, 2×10^6 parsecs, is derived from the distribution of nebular luminosities, together with luminosities of stars in some of the later-type spirals, and differs somewhat from the Harvard estimate of ten million light years [about 3×10^6 parsecs].

The data in the table indicate a linear correlation between distances and velocities, whether the latter are used directly or corrected for solar motion, according to the older solutions. This suggests a new solution for the solar motion in which the distances are introduced as coefficients of the *K* term, i.e., the velocities are assumed to vary directly with the distances, and hence *K* represents the velocity at unit distance due to this effect. . . .

Two solutions have been made, one using the 24 nebulae individually, the other combining them into 9 groups according to proximity in direction and in distance. The results are

	24 Objects	9 Groups
K	+465 ± 50	+513 ± 60 km/sec per 10^6 parsecs

For such scanty material, so poorly distributed, the results are fairly definite. Differences between the two solutions are due largely to the four Virgo nebulae, which, being the most distant objects and all sharing the peculiar motion of the cluster, unduly influence the value of *K*. . . . New data on more distant objects will be required to reduce the effect of such peculiar motion. Meanwhile round numbers, intermediate between the two solutions, will represent the probable order of the values. . . .

. . . In order to exhibit the results in a graphical form [figure 52.1], the solar motion has been eliminated from the observed velocities and the remainders, the distance terms plus the residuals, have been plotted against the distances. The run of the residuals is about as smooth as can be expected, and in general the form of the solutions appears to be adequate. . . .

The results establish a roughly linear relation between velocities and distances among nebulae for which velocities have been previously published, and the relation appears to dominate the distribution of velocities. In order to investigate the matter on a much larger scale, Mr. Humason at Mount Wilson has initiated a program of determining velocities of the most distant nebulae that can be observed with confidence. These, naturally, are the brightest nebulae in clusters of nebulae. The first definite result, $v = +3779$ km/sec for NGC 7619, is thoroughly consistent with the present conclusions. Corrected for the solar motion, this velocity is +3910, which, with $K = 500$, corresponds to a distance of 7.8×10^6 parsecs. Since the apparent magnitude is 11.8, the absolute magnitude at such a distance is −17.65, which is of the right order for the brightest nebulae in a cluster. A preliminary distance, derived independently from the cluster of which this nebula appears to be a member, is of the order of 7×10^6 parsecs.

New data to be expected in the near future may modify the significance of the present investigation or, if confirmatory, will lead to a solution having many times the weight. For this reason it is thought premature to dis-

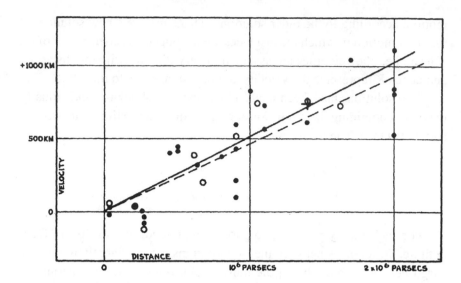

Figure 52.1: "Velocity-Distance Relation Among Extragalactic Nebulae Radial velocities, corrected for solar motion, are plotted against distances estimated from involved stars and mean luminosities of nebulae in a cluster. The black discs and full line represent the solution for solar motion using the nebulae individually; the circles and broken line represent the solution combining the nebulae into groups; the cross represents the mean velocity corresponding to the mean distance of 22 nebulae whose distances could not be estimated individually."

cuss in detail the obvious consequences of the present results. For example, if the solar motion with respect to the clusters represents the rotation of the galactic system, this motion could be subtracted from the results for the nebulae and the remainder would represent the motion of the galactic system with respect to the extragalactic nebulae.

The outstanding feature, however, is the possibility that the velocity-distance relation may represent the de Sitter effect, and hence that numerical data may be introduced into discussions of the general curvature of space. In the de Sitter cosmology, displacements of the spectra arise from two sources, an apparent slowing down of atomic vibrations and a general tendency of material particles to scatter. The latter involves an acceleration and hence introduces the element of time. The relative importance of these two effects should determine the form of the relation between distances and observed velocities; and in this connection it may be emphasized that the linear relation found in the present discussion is a first approximation representing a restricted range in distance.

53 / Stellar Populations and Resizing the Universe

In 1952, while attending the International Astronomical Union meeting in Rome, Walter Baade of the Mount Wilson Observatory rose to announce a new calibration for Cepheid variable stars. This singular moment "electrified the entire astronomical world," as one leading astronomer at the time described it.[37] Baade's years of work on stellar populations—recognizing that there were two distinct groupings of stars within galaxies and clusters—ultimately altered the cosmic distance scale, doubling the size of the universe overnight. Galaxies that were once thought to be 50 million light-years distant were suddenly found to be 100 million light-years away.

It all began with an intriguing yet troubling deviation. When the Andromeda nebula was recognized as a distinct galaxy, it seemed to share so many features with the Milky Way: the same disk of stars, the same system of globular clusters arranged in a halo around it, the same types of variable stars. Yet, based on Hubble's initial distance calculations in the 1920s, all these objects in Andromeda were fainter than those in the Milky Way. And Andromeda itself was smaller than our galaxy, which bothered astronomers who believed in the Copernican rule—that it is unlikely that we occupy a privileged place in the universe.

Alternative measurements were hinting that Andromeda was actually farther away than 900,000 light-years, as Hubble first estimated, but his use of Cepheid variable stars to gauge the distance was considered the most reliable method. "These discrepancies interested me exceedingly at that time," recalled Baade.[38] Because of the blackouts during World War II in the Los Angeles area, Baade used Mount Wilson's 100-inch telescope to great advantage on this problem. While other observatory staff members were involved in wartime work, Baade, a German citizen, was named an enemy alien and restricted to Pasadena, allowing him nearly unlimited telescope time. Over this period he developed special techniques for keeping his photographic plates in focus over the course of hours-long exposures and minimized temperature changes within the dome that affected the shape of the telescope mirror. He also used a special photographic film sensitive to the red bandwidth, newly developed for wartime reconnaissance. Using this new emulsion, he was able to resolve the stars within the bulge of the Andromeda galaxy.

In this way he discovered that stars within a galaxy were largely distributed according to their color: luminous blue giants and stars like our Sun (what he labeled as Type I population stars) tended to reside in the disk of a spiral galaxy, while older and redder stars (Type II population stars) were the primary stellar populations in a spiral galaxy's central bulge, as well as in elliptical galaxies and globular clusters. This finding would have profound effects on theoretical work on the evolution of both stars and galaxies. In 1962, gathering a wealth of evidence on stellar compositions and velocities, O. J. Eggen, Donald Lynden-Bell, and Allan Sandage wrote a classic paper indicating that our galaxy formed from the collapse of a cloud of gas, the oldest stars forming first in the initial halo, the youngest in or near the plane of the disk.[39] It showed that Baade's stellar populations were a reflection of galaxy formation.

After the opening in 1948 of the 200-inch telescope on Palomar mountain in California, Baade was able to follow up on his findings. With the added telescope power, he discovered that there were also two populations of Cepheid variable stars. Population I Cepheids, the kind that Hubble used to determine his distances to Andromeda and other galaxies, were actually more luminous than the Population II Cepheids that Shapley used to determine his distances to the globular clusters surrounding the Milky Way (and the type Hubble thought he was seeing). Recognizing this difference immediately doubled the estimated distance to Andromeda and brought its properties more in line with the Milky Way's. The Andromeda galaxy was indeed farther away than previously assumed (as some suspected), which in turn made the luminosities and sizes of its celestial objects similar to those in our own galaxy. Moreover, Baade's finding adjusted the distances to all the galaxies yet measured, essentially doubling the size and age of the universe. This allowed the Big-Bang model of the universe's creation, which had been under a cloud, to be revived, as Baade's readjustment got rid of the paradox of having an Earth older than the universe (see Chapter 52).

"The Resolution of Messier 32, NGC 205, and the Central Region of the Andromeda Nebula." *Astrophysical Journal*, Volume 100 (1944) by Walter Baade

In contrast to the majority of the nebulae within the local group of galaxies which are easily resolved into stars on photographs with our present instruments, the two companions of the Andromeda nebula—Messier 32 and NGC 205—and the central region of the Andromeda nebula itself have always presented an entirely nebulous appearance. Since there is no reason to doubt the stellar composition of these unresolved nebulae—the high frequency with which novae occur in the central region of the Andromeda nebula could hardly be explained otherwise—we must conclude that the luminosities of their brightest stars are abnormally low . . . compared with . . . the brightest stars in our own galaxy and for the resolved members of the local group. Although these data contain the first clear indication that in dealing with galaxies we have to distinguish two different types of stellar populations, the peculiar characteristics of the stars in unresolved nebulae remained, in view of the vague data available, a matter of speculation; and, since all former attempts to force a resolution of these nebulae had ended in failure, the problem was considered one of those which had to be put aside until the new 200-inch telescope should come into operation.

It was therefore quite a surprise when plates of the Andromeda nebula, taken at the 100-inch reflector in the fall of 1942, revealed for the first time unmistakable signs of incipient resolution in the hitherto apparently amorphous central region—signs which left no doubt that a comparatively small additional gain in limiting magnitude, of perhaps 0.3–0.5 mag., would bring out the brightest stars in large numbers.

How to obtain these few additional tenths in limiting magnitude was another question. Certainly there was little hope for any further gain from the blue-sensitive plates hitherto used, because the limit set by the sky fog, even under the most favorable conditions, had been reached. However, the possibility of success with red-sensitive plates remained. . . .

The minimum exposure times required . . . turned out to be 4 hours. Exposures of this length with a large reflector present a number of problems if critical definition is the prime requisite. That only nights with exceptionally fine definition, together with a practically perfect state of the mirror, would do hardly needs mention. Fortunately, these conditions are easily met on Mount Wilson during the fall months when the Andromeda region is in opposition. But real difficulties were presented by changes of focus during the relatively long exposures. . . . It seemed best to use only nights on which the focus-changes at the 100-inch are very small if not entirely negligible. Such conditions are not infrequently met on Mount Wilson during the fall, when, owing to a temperature inversion, the temperature stays practically constant all night. Neither was it difficult in the present case to select the proper nights. Since in the fall the Andromeda region culminates around midnight, a careful watch of the state of the mirror and of the temperature in the early evening hours permits a fair prediction of the focus-changes during the latter part of the night. Eventual small changes in focus during the exposure can then be inferred from changes in the coma of the guiding star. . . .

The plates of the Andromeda nebula, of Messier 32, and of NGC 205, taken in this manner at the 100-inch reflector during the fall months of 1943, led to the expected results. All three systems were resolved into stars. . . .

[Omitted here are Baade's descriptions of the photographic plates of Messier 32, NGC 205, and the inner region of the Andromeda nebula and the conditions under which they were taken.]

The main facts presented in the preceding descriptions can be summarized in the following four statements:

1. By using red-sensitive plates we have recorded the brightest stars in the hitherto unresolved members of the local group of galaxies.

2. The apparent magnitudes of the brightest stars are closely the same in all three systems, a result which was to be expected because the three nebulae form a triple system.

3. At the upper limit of stellar luminosity, stars appear at once in great numbers in these systems. . . .

4. With our present instruments early-type nebulae can be resolved on red-sensitive plates if their distance modulus does not exceed that of the Andromeda group. . . .

With these data at hand we are in the position to draw an important conclusion regarding the Hertzsprung-Russell diagram of the stars in early-type nebulae. . . . It has been known for some time that the highly luminous stars of the main branch (O- and B-type stars), together with the supergiants of types F–M, are absent in these systems; in fact, their absence was the reason why up to now the early-type nebulae have proved to be unresolvable. But neither are the brightest stars which we find in them the common giants of the ordinary H-R diagram, because as a group they are nearly 3 mag. brighter. . . . It is significant that the same situation is known to exist in the globular clusters. . . . Similarly, there is perfect agreement in the color indices of the brightest stars in early-type nebulae and globular clusters. . . . We conclude, therefore, that, within the present uncertainties, absolute magnitude and color index of the brightest stars in early-type nebulae are the same as those of the brightest stars in globular clusters.

[Omitted here are Baade's technical arguments that show how Cepheid variable stars in globular clusters differ from stars found in our solar neighborhood (that is, in the disk of a galaxy). He goes on to state that the same should hold true for early-type nebulae, whose stellar populations are "similar, if not identical" to globular clusters. By 1952 his follow-up on this (see next paper) affected all previous Cepheid distance measurements.]

But we can advance a third argument which explains at the same time why the globular clusters happen to be the prototypes of this peculiar type of stellar population which we will call type II in distinction from populations defined by the ordinary H-R diagram—type I. This is the fact that, as far as the present evidence goes, globular clusters are always associated with stellar populations of type II. . . .

Although the evidence presented in the preceding discussion is still very fragmentary, there can be no doubt that, in dealing with galaxies, we have to distinguish two types of stellar populations, one which is represented by the ordinary H-R diagram (type I), the other by the H-R diagram of the globular clusters (type II). Characteristic of the first type are highly luminous O- and B-type stars and open clusters; of the second, globular clusters and short-period Cepheids. Early-type nebulae (E–Sa) [elliptical galaxies and spirals with large central bulges] seem to have populations of pure type II. Both types coexist, although differentiated by their spatial arrangement, in the intermediate spirals like the Andromeda nebula and our own galaxy. In the late-type spirals and in most of the irregular nebulae the highly luminous stars of type I are the most conspicuous feature. It would probably be wrong, however, to conclude that we are dealing with populations of pure type I, because the occurrence of globular clusters in these late-type systems, for instance, in the Magellanic Clouds, indicates that a population of type II is present too. Altogether it seems that, whereas stars of the second type may occur alone in a galaxy, those of type I occur only in association with type II.

In conclusion it should be pointed out that these same two types of stars were recognized in our own galaxy by Oort as early as 1926.* Oort showed that the high-velocity stars of our galaxy (our type II) are of a kind quite different from the slow-moving stars (type I) which predominate in the solar neighborhood. Since his conclusions are based on entirely different material and since they supplement those derived in the present paper, they are worth recalling. . . .

"A Revision of the Extra-Galactic Distance Scale."
Transactions of the International Astronomical Union, Volume 8 (1952)
by Walter Baade†

In his opening remarks Dr. Baade pointed out that although in the past instrumental opportunities for the study of extragalactic problems had been

* Jan Oort, *Groningen Publications*, 40:6 (1926).
† As recorded by British astrophysicist Fred Hoyle, secretary of the IAU session on extragalactic nebulae.

extremely limited, there was now hope that several large telescopes would soon become available. In particular, Dr. Baade referred to the progress that had been made with the new reflector at the Lick Observatory. . . .

Dr. Baade then went on to describe several results of great cosmological significance. He pointed out that, in the course of his work on the two stellar populations in M31 [Andromeda galaxy], it had become more and more clear that either the zero-point of the classical cepheids or the zero-point of the cluster variables must be in error. Data obtained recently— Sandage's color-magnitude diagram of [globular cluster] M3—supported the view that the error lay with the zero-point of the classical cepheids, not with the cluster variables.* Moreover, the error must be such that our previous estimates of extragalactic distances—not distances within our own Galaxy—were too small by as much as a factor of 2. Many notable implications followed immediately from the corrected distances: the globular clusters in M31 and in our own Galaxy now come out to have closely similar luminosities; and our Galaxy may now come out to be somewhat smaller than M31. Above all, Hubble's characteristic time scale for the Universe must now be increased from about 1.8×10^9 years to about 3.6×10^9 years†. . . .

. . . Prof. [Harlow] Shapley . . . asked Dr. Baade if he could describe in a little more detail how he (Dr. Baade) had arrived at the error in the zero-point of the classical cepheids. Dr. Baade offered [this argument] in support of his conclusion:

Argument—According to the present zero-points we should expect to find the cluster-type variables of the Andromeda nebula at m_{pg} [photographic magnitude] = 22.4 since the distance modulus of this system, derived from classical cepheids, is $m - M = 22.4$.‡ The very first exposures of M31, taken at the 200-inch telescope, showed at once that something was wrong. Tests had shown that we reach with this instrument, using the *f*/3.7 correcting lens, stars of $m_{pg} = 22.4$ in an exposure of 30 min. Hence we should just reach in such an exposure the cluster-type variables in M31, at least in their maximum phases. Actually we reach only the brightest

---

* Astronomer Allan Sandage of the Carnegie Observatories.

† The Hubble constant was further adjusted over the following decades, leading to an age of the universe of around 14 billion years.

‡ Distance modulus is the difference between a star or galaxy's apparent magnitude (the magnitude you see directly in the sky) and its absolute magnitude (the object's true brightness). This term is used as a means of measuring the object's distance from the Earth.

stars of population II in M31 with such an exposure. Since, according to the latest color-magnitude diagrams of globular clusters, the brightest stars of the population II are photographically about 1.5 mag. brighter than the cluster-type variables we must conclude that the latter are to be found in M31 at $m_{pg} = 23.9 \pm$ [a lower brightness], and not at $m_{pg} = 22.4$ as predicted on the basis of our present zero-points.

We have also convincing proof that the brightest stars of population II in M31 are properly identified because when they emerge above the plate limit the globular clusters of M31 begin to be resolved into stars. . . .

Dr. Dufay pointed out that the change in the scale of extragalactic distance had an interesting application so far as the sizes of gaseous emission nebulae were concerned.* Formerly there had been systematic differences of size between such nebulae in the Galaxy and both the Magellanic Clouds and NGC 6822. The changed distance scale now equalized the sizes. . . .

* French astronomer Jean Dufay.

54 / Mapping the Milky Way's Spiral Arms

With William Parsons, the third earl of Rosse, revealing in 1845 that some nebulae in the celestial sky displayed intriguing spiraling patterns (see Chapter 22), it wasn't long before others began to imagine the Milky Way itself having a similar structure. Within seven years Stephen Alexander, a professor at what is now Princeton University, published a paper in the *Astronomical Journal* surmising that "the Milky Way and the stars within it together constitute a spiral with several (it may be *four*) branches, and a central (probably spheroidal) cluster."[40] And in 1869 an English astronomer named Richard Proctor described our galaxy as a kind of coiling ring.[41] These ideas were greatly popularized when the Dutch journalist and amateur astronomer Cornelis Easton in the early 1900s independently concluded that the Milky Way was composed of "dark spaces surrounded by luminous streams" of stars and published a beautiful drawing of a spiraling Milky Way seen face-on.[42]

But these reports were more speculation than science. Astronomers realized they at last had a chance at mapping the Milky Way's true structure after Mount Wilson astronomer Walter Baade in the 1940s recognized a unique feature about spiral galaxies: that old red stars tend to huddle in the galaxy's central bulge and young blue-white giant stars and bright gaseous nebulae generally line up along the spiral arms, like the luminous lights along an airport runway (see Chapter 53).

The first to apply this newfound knowledge to the Milky Way were William Morgan at the Yerkes Observatory in Williams Bay, Wisconsin, and two of his student assistants, Stewart Sharpless and Donald Osterbrock. With painstaking care, they determined the distances to dozens of blue giant stars and bright nebulae of ionized hydrogen (known as H II regions) within the solar neighborhood, which enabled them to trace segments of a few spiral arms (later labeled the Orion, Perseus, and Sagittarius arms). Morgan announced these findings at a 1951 meeting of the American Astronomical Society, presenting a model of the spiral arms that used cotton balls to depict the positions of the luminous nebulae. The results were greeted with a rare, emotional ovation that included clapping of hands and stomping of feet.

Morgan's view was far from complete, because it is difficult for optical telescopes to peer deeply into the dust- and gas-filled plane of the Milky Way. Within two years, the spiraling pattern was confirmed and extended with the use of a new instrument available to astronomers—the radio telescope, which could penetrate the dust and haze by tuning into a radio frequency emitted by hydrogen atoms (see Chapter 57). Some of the data for the first radio maps of the Milky Way's spiral structure were obtained at Leiden Observatory in the Netherlands using a 25-foot dish that was turned by hand, using two small cranks, every two and a half minutes. Other observations were carried out in Australia with a 40-foot antenna. To analyze the reams of data, Leiden astronomy students were herded into a lecture room, where they spent a week translating the chart-paper recordings into radio intensities over the sky.[43]

Why are there spiral arms at all? The answer quickly arrived once astronomers realized that the arms are not permanent but continuously changing. In the 1960s, building on an idea introduced by Swedish astronomer Bertil Lindblad forty years earlier, C.-C. Lin, of the Massachusetts Institute of Technology, and Frank Shu, then with the Harvard College Observatory, demonstrated how spiral arms simply mark the position of a density wave, a spiral-shaped region of compression that slowly rotates through the flat disk of the galaxy. As the disk's gas, which travels faster, passes through this compression wave, it gets squeezed in the cosmic traffic jam, huge clouds form, and within several million years new bright stars turn on to illuminate the spiral structure.[44]

"Some Features of Galactic Structure in the Neighborhood of the Sun." *The Astronomical Journal*, Volume 57 (1952) by William W. Morgan, Stewart Sharpless, and Donald Osterbrock

The distribution in space of the nearer regions of ionized hydrogen [H II] has been investigated by spectroscopic parallaxes determined with the 40-inch Yerkes refractor. The regions north of −10° declination occur in two long, narrow belts similar to the spiral arms observed by [Walter] Baade in the Andromeda nebula. The nearer arm extends from galactic longitude 40° to 190° and passes at its nearest point about 300 parsecs distant from the sun in a direction opposite to that of the galactic center. The observed length of the arm is about 3,000 parsecs; its width is of the order of 250 parsecs. Among the constituents of this arm are the nebulosities in the neighborhood of P Cygn, the North American nebula, the ξ Persei nebulosity, the Orion nebula and loop, and the H II regions near λ Orionis and S Monocerotis.

A second arm can be traced from galactic longitude 70° to 140°. This arm is parallel to the first and is situated at a distance of about 2,000 parsecs from it in the anti-center direction. There is some evidence for another arm located at a distance of around 1,500 parsecs in the direction toward the galactic center. This is defined by the series of condensations of O and B stars from galactic longitude 253° in Carina to 345°, the small cloud in Sagittarius. The data are so fragmentary, however, that more observations from the southern hemisphere will be necessary before a definite conclusion can be reached.

Both arms are inclined with respect to the normal to a radius vector by approximately 25°; when this tilt is combined with the known direction of galactic rotation the arms are found to be trailing.

The dimensions of the H II regions are similar to those observed by Baade in the Andromeda nebula; the width of the arms is also similar, as is the frequency of H II regions along the arms.

The structure described above is also shown by the blue giants, O–A_5 stars having M_{vis} brighter than −4.0 mag. The great aggregates of early-type stars, Perseus double cluster, P Cygni region, Orion, are condensations in the arms similar to the condensations observed by Hubble in the Andromeda nebula.

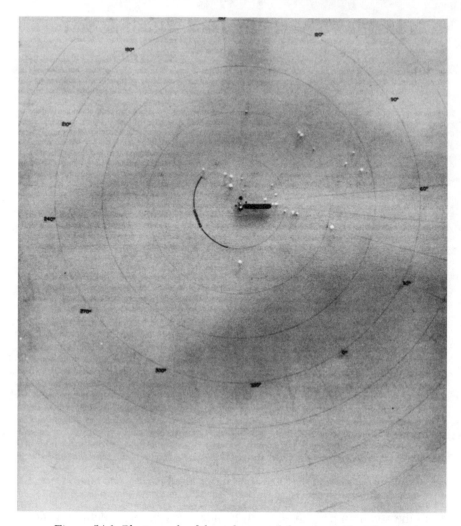

Figure 54.1: Photograph of the galaxy model presented by Morgan, Sharpless, and Osterbrock. With the sun at the center, it displays the regions of ionized hydrogen (white dots), which line up along spiral arms. The dark spots mark obscuring clouds of dust, including the long Great Rift. One arm passes through the sun's neighborhood; a second runs parallel farther out. The one H II region below the sun marks a third arm in the direction of the galactic center, which is located toward the bottom of the picture.

"The Spiral Structure of the Outer Part of the Galactic System Derived from the Hydrogen Emission at 21 cm Wavelength." *Bulletin of the Astronomical Institutes of the Netherlands,* Volume 12 (May 14, 1954) by Hendrik C. van de Hulst, C. Alex Muller, and Jan H. Oort

Interpretation of Results

. . . It is evident that the hydrogen is concentrated in relatively narrow lanes separated by regions of much smaller density. We shall see . . . that in the latter regions the density is probably negligible. The long stretches of hydrogen are evidently to be identified with "spiral" arms.

The first successful identification of such arms in the Galactic System has been made by W. W. Morgan and his collaborators at the Yerkes Observatory from the distribution of regions of ionized hydrogen [see paper above]. This investigation was later extended through a study of the space distribution of O associations. . . .

It is convenient to attach names to the various arms that can be distinguished. . . . In order to provide a provisional means for referring to the various arms already discovered, especially those that are also observable in optical wavelengths, we propose to name the principal arms after conspicuous associations contained in them. After consultation with W. W. Morgan the following designations are proposed: For the arm passing through the sun: Orion arm; for that passing through h and χ Persei: Perseus arm; and for the first arm encountered when proceeding in the direction of the center: Sagittarius arm. . . .

The Perseus arm is most clearly seen from longitude 50° to 115°. It contains, as one of its most conspicuous features, the rich association of early-type supergiants surrounding the double cluster in Perseus. In longitudes 120° to 130° it is probably present in comparable strength, but more difficult to separate from the Orion arm, due to the smaller differential rotation, combined, probably, with the effect of the branching of the Orion arm. . . . The first signs of such separation appear again at $\ell = 170°$, where the distant maximum is of considerable strength. . . . This powerful outer arm may be a continuation of the Perseus arm. . . .

Perhaps the most striking feature of Figure [54.2] is the strong outer

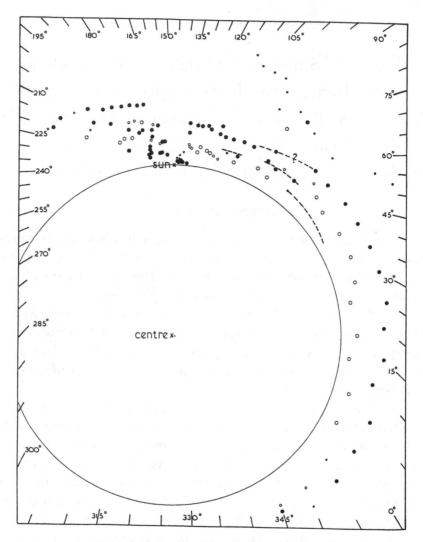

Figure 54.2: The solid dots mark the regions of high hydrogen density (the spiral arms); the open circles indicate where hydrogen densities are low.

arm seen from $\ell = 55°$ down to $\ell = 345°$. Over the entire range between $0°$ and $55°$ the arm is separated by a deep and wide minimum from the more inward parts of the system. We shall provisionally refer to this arm as the "distant arm." As we see from the figure, there may well be a continuous transition between this and the Perseus arm. If this is so, the latter would be practically circular over the whole interval from $350°$ to $125°$. . . .

Between $\ell = 60°$ and $120°$ the line profiles clearly show the existence of a third, more distant arm. . . .

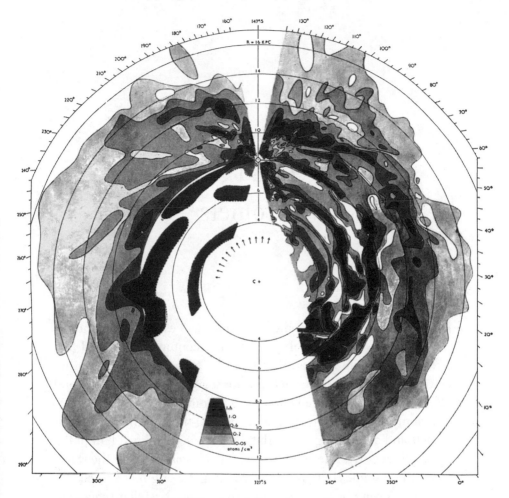

Figure 54.3: By 1958 Dutch and Australian radio astronomers combined their data on the distribution of neutral hydrogen in the plane of our galaxy to produce this classic image—the Leiden-Sydney map—of the Milky Way's spiraling arms. The circles are centered at C+, the galactic center, and spaced 2 kiloparsecs apart. The Sun is denoted by ☉. The numbers around the figure are galactic longitudes. (Image from Oort, Kerr, and Westerhout, "The Galactic System as a Spiral Nebula," *Monthly Notices of the Royal Astronomical Society* 118 [1958]: 379–89.)

55 / Source and Composition of Comets

The modern era of cometary studies was essentially launched when Tycho Brahe in the sixteenth century determined that comets were not earthly atmospheric phenomena but rather celestial visitors traveling through the solar system. After Edmond Halley's successful prediction of a comet's return in 1758, astronomers spent many decades gathering data on the motions of comets and computing their orbital paths. Over the first half of the twentieth century, astronomers began to closely examine the statistics of these studies in hope of determining the origin of comets.

Comets were once thought to arrive from interstellar space, but the orbital paths followed by comets did not support that idea. By 1948 Adrianus J. J. van Woerkom showed how comets can be ejected from the solar system—flung out to far distances—by gravitationally interacting with Jupiter and the other gas giants. Drawing on this work in 1950, Jan Oort, then director of the Leiden Observatory in the Netherlands, proposed that comets emanated from a vast and distant reservoir composed of the primordial fragments kicked out during the solar system's youth. This huge storage area is now known as the Oort cloud. Oort's initial calculations figured this spherical shell extended from 25,000 AU (around 2 trillion miles, 1 astronomical unit being equal to 93 million miles, the Sun-Earth distance) out to 150,000 AU (halfway to the nearest stars) and contained up to 200 billion comets. When disturbed by a gravitational perturbation, such as a passing star, one is occasionally dislodged and sent inward toward the Sun. While Ernst Öpik had a similar idea in 1932, Oort crucially demonstrated how such a cloud could account for the observational data.[45]

Oort's paper was a major advance for the field, and within months it was joined by another revolutionary treatise on comets written by Fred Whipple of the Harvard College Observatory. At that time comets were thought to be composed of dust and rock. But after studying the spectroscopic and dynamical evidence then available, Whipple proposed that a comet is best described as an icy conglomerate—dust particles caught in a matrix of water ice and frozen gases, which sublimate as the comet approaches the Sun and gets heated. Whipple described it as his "dirty snowball" model. A flyby of Halley's comet in 1986 by the European Space Agency's Giotto spacecraft and the Soviet Union's Vega probe directly revealed that Whipple's description was correct.

"The Structure of the Cloud of Comets Surrounding the Solar System, and a Hypothesis Concerning Its Origin."
Bulletin of the Astronomical Institutes of the Netherlands,
Volume 11 (January 13, 1950)
by Jan H. Oort

Abstract

The combined effects of the stars and of Jupiter appear to determine the main statistical features of the orbits of comets.

From a score of well-observed original orbits it is shown that the "new" long-period comets generally come from regions between about 50,000 and 150,000 A.U. distance. The sun must be surrounded by a general cloud of comets with a radius of this order, containing about 10^{11} comets of observable size; the total mass of the cloud is estimated to be of the order of $\frac{1}{10}$ to $\frac{1}{100}$ of that of the earth. Through the action of the stars fresh comets are continually being carried from this cloud into the vicinity of the sun.

The article indicates how three facts concerning the long-period comets, which hitherto were not well understood, namely the random distribution of orbital planes and of perihelia, and the preponderance of nearly-parabolic orbits, may be considered as necessary consequences of the perturbations acting on the comets.

The theoretical distribution curve . . . following from the conception of the large cloud of comets . . . is shown to agree with the observed distribution . . . , except for an excess of observed "new" comets. The latter is taken to indicate that comets coming for the first time near the sun develop more extensive luminous envelopes than older comets. . . .

Sketch of the Problem

Among the so-called long-period comets there are 22 for which, largely by the work of Elis Strömgren, accurate calculations have been made of the orbits followed when they were still far outside the orbits of the major planets. Approximate calculations of the original orbits . . . are available for 8 other comets with well-determined osculating orbits. . . .

. . . We may conclude that a sensible fraction of the long-period comets must have come from a region of space extending from a distance [of] . . . 20,000 to distances of at least 150,000 A.U. [0.3 to 2.3 light-years] from the sun; that is, almost to the nearest star. This does not mean that they are interstellar. They belong very definitely to the solar system, because they share accurately the sun's motion. Yet, the prevalence of these very large major axes has led several astronomers to investigate the question whether the comets could not be of interstellar origin. It is evident that they cannot *directly* come from interstellar space, for in that case there would have to be many more outspoken hyperbolic orbits than nearly parabolic ones. So far, no comet has been found for which the eccentricity exceeds 1 by an amount large enough to be considered as real. It is conceivable, however, that comets would be caught from an interstellar field by the action of the major planets, and would then move for a long time in orbits of large dimensions, so that the number of comets caught would gradually become far larger than the number of hyperbolic comets passing through the solar system. This suggestion has recently been studied by Dr. van Woerkom. He concludes that this possibility must be ruled out. . . . There would again be a large preponderance of hyperbolic comets, which is contradicted by observations. . . .

There is no reasonable escape, I believe, from the conclusion that the comets have always belonged to the solar system. They must then form a huge cloud, extending, according to the numbers cited above, to distances of at least 150,000 A.U., and possibly still further. It is not necessary at this point to enter upon the question how this cloud has originated.* It might conceivably be considered as part of the remnants of a disrupted planet. An alternative hypothesis, repeatedly put forward, according to which comets would be formed by eruptions from Jupiter and the other planets, does not appear to be likely.

Accepting this existence of a huge cloud of comets we are still faced with a difficulty that has been put into full light by van Woerkom's study. Jupiter, and to a lesser extent the other planets, exert a diffusing action on the long-period comets. According to van Woerkom's calculations the small perturbations by Jupiter suffered by an observable comet during its passage through the "inner" part of the planetary system will on the average change the reciprocal major axis by about 0.0005; positive and negative changes are equally probable. By these perturbations the long-period comets will gradually disappear, partly into interstellar space, partly into the families of short-period comets. In addition, the comets may gradually diminish in brightness through the sun's action, or be dissolved. It is evident from van Woerkom's study that within one or two million years after their first perihelion passage practically all long-period comets will have disappeared. As it is highly improbable that the comets we observe have only originated within the last two million years we are led to conclude that comets already existing outside the region where they are subject to the perturbing action of sun and planets are continually being brought into this region. . . .

If we assume that at the start the velocity distribution of the comets in the huge cloud surrounding the planetary system was a random distribution, there must have been comets, even in the outer parts of the cloud, whose velocities were so nearly directed towards the sun that these comets would eventually pass through the "observable region" (i.e., the region within about 2 A.U. from the sun). Even if the radius of the cloud was 150,000 A.U. all the comets which could come into the vicinity of the earth would have done so within roughly 20 million years. All these comets will diffuse into space or be disintegrated. No new comets would come in after this period unless they were made to do so by some perturbation. Van Woerkom's discussions make it clear that perturbations by *planets* cannot be effective in bringing comets into the observable region: their influence on the major axis and the period is always much more important than on

---

* Astronomers today generally view comets as frozen remnants of the same nebula that condensed to form the Sun and its planetary system.

the perihelion distance. Their perturbations will diffuse the comets out of the long-period range long before they have caused a change of any importance in the perihelion distance.

Two alternative types of perturbations offer themselves, namely resistance by an interplanetary medium, and influence of passing stars. It seems extremely unlikely that the former mechanism could have an observable influence on the perihelia of comets. For a general influence of this kind to be effective a density of interplanetary gas would be required that is quite inadmissible on dynamical grounds. Moreover, a resisting medium would in the first place tend to decrease the major axes, while for the nearly parabolic comets the perihelion distances would appear to be practically unaffected, so that it could never solve our problem.

The purpose of the present paper is to investigate the second possibility, the action of passing stars.

[In his extensive nineteen-page paper, Oort proceeds to calculate the density, shape, and velocity distribution of his proposed cloud of comets. He then investigates the number of comets required in the cloud to explain the new comets observed passing through our solar system. He concludes that "the total number of comets in the cloud is . . . found to be $1.9 \cdot 10^{11}$."]

There are no good estimates of the average mass of a comet, except that it must probably be larger than about 10^{14}, and smaller than 10^{20} grams. A plausible estimate is perhaps about 10^{16} g. . . . With such an average mass the total mass of the cloud of comets would be 10^{27}, or about $\frac{1}{10}$ of the earth's mass. This estimate is uncertain by one or two factors of 10. . . .

The enormous size of the cloud of comets presents an interesting problem in itself. It seems most unlikely that in the regions between 50,000 and 200,000 A.U. from the sun, where probably the general gas density will never have been much higher than the average density in interstellar space, bodies as large as the comets could have been built up by condensation or accretion. . . . It appears far more probable that instead of having originated in these far away regions, comets were born among the planets. . . .

It seems a reasonable hypothesis to assume that the comets originated together with the minor planets, and that those fragments whose orbits deviated so much from circles between the orbits of Mars and Jupiter that they became subject to large perturbations by the planets, were diffused away by these perturbations, and that, as a consequence of the added effect of the perturbations by stars, part of these fragments gave rise to the formation of the large cloud of comets which we observe today.

. . . The mechanism proposed would give rise to a density distribution showing exactly those characteristics that are exhibited by the cloud of comets. A cloud so formed would necessarily extend to the limit set by the dissolving action of the stars, that is, to about 200,000 A.U. The inner limit should lie near 25,000 A.U., where the perturbing action of the stars is no longer strong enough to have shifted bodies from the elongated orbits into orbits that remain outside the region of the large planets; the cloud would therefore contain practically no members with mean distances less than 25,000 A.U. . . .

"A Comet Model. I. The Acceleration of Comet Encke."
Astrophysical Journal, Volume 111 (1950)
by Fred L. Whipple

Abstract

A new comet model is presented that resolves the chief problems of abnormal cometary motions and accounts for a number of other cometary phenomena. The nucleus is visualized as a conglomerate of ices, such as H_2O, NH_3, CH_4, CO_2 or CO, $(C_2N_2?)$, and other possible materials volatile at room temperature, combined in a conglomerate with meteoric materials, all initially at extremely low temperatures ($<50°$ K). Vaporization of the ices by externally applied solar radiation leaves an outer matrix of non-volatile insulating meteoric material. . . .

Introduction

. . . The lifetime of a short-period comet must lie generally in the range of from 3,000 (one hundred comets being lost at a rate of three per century) to possibly 60,000 years. Probably more important is the number of small perihelion passages that can be weathered by a comet—of the order of several hundreds, at least, for perihelion distances as small as 0.5–1.0 A.U. For considerably greater perihelion distances the number probably increases to several thousand, thus permitting comets with periods up to

10^6 years to persist throughout all or most of the past history of the solid earth.

Even though parts of the preceding discussion are somewhat conjectural, we must certainly accept the conclusion that individual short-period comets cannot exist indefinitely in their present orbits and also that they must previously have existed at great distances from the sun, where their temperature throughout remained at extremely low values. . . . In the present discussion I propose to investigate the possibility that the molecules responsible for most of the light of comets near perihelion arise primarily from gases long frozen in the nuclei of comets. Furthermore, I propose that these primitive gases constitute an important, if not a predominant, fraction of the mass of a "new" or undisintegrated comet.

On the basis of these assumptions, a model comet nucleus then consists of a matrix of meteoric material with little structural strength, mixed together with the frozen gases—a true conglomerate. Since no meteorites are known certainly to arise from cometary debris, we know very little about the physical structure of the meteoric material except that the pieces seem generally to be small. Hence we assume that the larger pieces are perhaps a few centimeters in radius and the smallest are perhaps molecular. As a convenience in terminology, the term "ices" will be used in referring to substances with melting points below about 300°C and "meteoric material" to substances with higher melting points.

Our only chemical knowledge of the meteoric material comes from the spectra of meteors, which tell us that *Fe, Ca, Mn, Mg, Cr, Si, Ni, Al,* and *Na,* at least, are present. Physically the meteoric material is strong enough to withstand some shock in the atmosphere, but more than 3 per cent of the Harvard photographic meteors are observed to break into two or more pieces. A much larger percentage show flares in brightness, an indirect evidence of breaking. The high altitude of the disappearance of the photographic Giacobinid meteors of October 9, 1946 . . . suggests that those meteoric bodies may have been unusually fragile or porous. It is difficult to defend the hypothesis that, as a whole, the bodies producing photographic meteors possess great physical rigidity or strength.

A careful determination of the relative abundances of the primitive ices in the nucleus of a comet and their physical properties will require an exhaustive study of the theory of cometary spectra and related phenomena, including evolutionary hypotheses. Only a few comments will be made here. The observed gases *CH, CH$^+$, CH$_2$, CO, NH, NH$_2$, OH,* and *OH$^+$* can be accounted for by four possible parent-molecules of great stability, viz.,

CH_4, CO_2, NH_3, and H_2O. Photodissociation appears capable of producing the various radicals from these parent-molecules. . . .

As our model comet nucleus approaches perihelion, the solar radiation will vaporize the ices near the surface. Meteoric material below some limiting size will blow away because of the low gravitational attraction of the nucleus and will begin the formation of a meteor stream. Some of the larger or denser particles may be removed by shocks, but the largest particles or matrix will remain on the surface, to produce an insulating layer. After a short time (probably in the geologic past for all known comets) the loss of gas will be reduced materially by the insulation so provided. . . .

The weakening of the upper layers of the icy core by selective vaporization of the ices may be expected to produce cometary activity of considerable intensity, especially near the sun. The surface gravities of cometary nuclei are certainly extremely low; hence surprisingly weak structures can persist over rather large areas of the nucleus. At irregular intervals collapses must occur. The heated meteoric material will then fall into the ices and produce rapid vaporization. The dust and smaller particles held in the upper layers will be shaken out and blown away, so that insulation produced by this material will be much reduced. Solar heat, consequently, will be much more effective in vaporizing the ices in the pit until equilibrium is again established. Such "cave-ins," might spread over appreciable areas. Other effects might occur if "pockets" of an ice with low melting points exist within an ice of higher melting point. Phenomena of mildly explosive, jet, or cracking types may occur, forcing out pieces of material much larger than those carried normally by the outgoing gas. Hence the type of nuclear activity that is observed for large comets with small perihelion distances would be expected from this type of comet model.

If the primitive ices constitute a large percentage of the total mass, the comet truly disintegrates with time. Its actual substance vaporizes; the surface gravity decreases; and, finally, all activity ceases as the last of its ice reservoir is exhausted. The observed sequence of phenomena in dying comets is entirely consistent with this picture. In the later stages, only a very small nucleus of the largest meteoric fragments remains. . . .

The period of rotation of a comet with a single spheroidal nucleus would generally remain constant with age, so that the comet might dissipate slowly and uniformly. If, however, the nucleus were multiple or irregular in shape, the vaporization of ices could materially affect the rotation. Suppose, for example, a part of the surface were nearly in a plane passing through the center of gravity of the nucleus, while the remaining surface were generally smooth and approximately oval in shape. Meteoric material would fall from the vertical surface, exposing it to the full action of sunlight. Hence the excess of gas evolved from this surface would exert a force moment on the nucleus as a whole.

The effect of the resulting rotation, depending upon the initial circumstances, might easily produce rotational instability, permitting the sun's tidal action to complete the splitting of the nucleus. If the larger parts of the separated nucleus were unstable, the comet might disappear quickly. On the other hand, the pieces might be large enough to persist for a long period of time as individual comets. In fact, the phenomenon of splitting has occurred for several comets and has been followed by disappearance in some cases, but not in others. Either possibility may be expected on the basis of the present comet model, depending upon the mass, shape, and rotation of the nucleus. . . .

[Whipple then examines heat transfer in his icy comet nucleus and discusses how his model can be used to demonstrate possible mechanisms for the observed acceleration of Comet Encke, then a problem in cometary studies.]

VII

NEW EYES,
NEW UNIVERSE

For most of astronomy's history, observers gathered solely the visible light rays emanating from the heavens—first with their eyes and later with the lenses and mirrors of telescopic instruments. And what was detected was largely a serene universe. Even with Edwin Hubble's disquieting discovery of an expanding cosmos, where billions of other galaxies rush away from one another at tremendous speeds, the stars and elegant spiraling galaxies still maintained a graceful and dignified composure. Not until the mid-twentieth century did astronomers become fully aware that the universe could reveal a vastly different portrait of itself in other regions of the electromagnetic spectrum.

The roots of this transition go back to the late nineteenth century, when the German physicist Heinrich Hertz first generated radio waves in his laboratory; seven years later, Wilhelm Röntgen discovered X rays. These and other findings proved that there were both longer waves of electromagnetic radiation (a single radio wave can stretch out for miles) and shorter waves (x-ray wavelengths measure less than a millionth of an inch across). As soon as observers were able to effectively explore these and other parts of the spectrum, a new and golden era for astronomy arose. "In all of history," a U.S. report on the state of astronomy and astrophysics in the 1980s stated, "there have been only two periods in which our view of the universe has been revolutionized within a single human lifetime. The first occurred three and a half centuries ago at the time of Galileo; the second is now under way."[1]

Although nineteenth-century astronomers had dabbled with infrared sensors (see Chapter 65), a more extensive spectral examination of the cosmos did not begin until later. It was initiated in the 1930s when a physicist at the Bell Telephone Laboratories in New Jersey, Karl Jansky, serendipitously discovered radio waves arriving from the center of the Milky Way, hinting at processes in our galaxy's center not revealed in visible light. Over the succeeding decades, radio telescopes were erected around the world, examining corners of the cosmos once hidden to our eyes. In the farthest reaches of the universe, radio astronomers discovered quasars, young galaxies disgorging the energy of a trillion suns from their centers. Closer in within the Milky Way, they detected pulsars, 10-mile-wide neutron stars spinning rapidly and emitting periodic beeps in the process. Meanwhile, in the dust- and gas-filled sectors of our galaxy, the distinctive radio tones of a host of molecules were heard for the first time, pointing astronomers to the birthplaces of new stars. And permeating all of space was the leftover microwave radiation from the Big Bang itself.

Our cosmic vision widened even more with the dawn of the space age. The many probes sent throughout the solar system changed long-held concepts about the planets; it introduced us to an Earthlike Mars, a hellish Venus, and a volcanic Jovian moon. X-ray satellites, gamma-ray telescopes, and infrared and ultraviolet observatories at last rose above the Earth's obscuring blanket of air, allowing astronomers to closely examine galaxies vigorously interacting with their neighbors—swapping mass, colliding, and triggering stupendous bursts of star formation in one another. Within our own galaxy were seen a host of highly energetic x-ray sources, some providing the first evidence for black holes. Infrared telescopes examined a whirlpool of matter in the heart of our galaxy, swirling around what seems to be a supermassive black hole. The beauty of the universe of old has been retained, but the new eyes available to astronomers now reveal it to be suffused with titanic energies and explosive behaviors.

Twentieth-century astronomers also realized that collecting electromagnetic radiation is not the only means of studying the universe. Special instruments were developed to capture ghostlike neutrino particles originating from the core of the Sun. These underground observatories not only revealed new properties in particle physics but also unexpectedly detected neutrinos emanating from the spectacular explosion of a star. Meanwhile, other observers discerned indirect evidence for gravity waves—vibrations in space-time predicted by Einstein to be generated by moving matter—in the motions of two neutron stars circling one another. By the end of the twentieth century, neutrino and gravity-wave "telescopes" were being developed and enhanced to extend astronomy's reach into the universe.

56 / Radio Astronomy

Within a few years of Heinrich Hertz's discovery of radio waves in 1888, researchers wondered whether such waves could be emanating from the heavens. The prolific inventor Thomas Edison proposed stringing a loop of telephone wire around a field of iron ore, figuring radio waves from the Sun would induce a current. British physicist Oliver Lodge, a pioneer in radio telegraphy, carried out a search for solar radio waves from Liverpool, England, in the 1890s, but electrical interference from the industrial town overwhelmed any sources from space. Afterward, physicists incorrectly assumed from the new quantum theories coming into vogue that radio emissions from the Sun and stars would be feeble, and so further measurements were not pursued.

It took a bit of serendipity to reestablish interest. In the early 1930s Karl Jansky, a radio physicist at the Bell Telephone Laboratories in Holmdel, New Jersey, built a steerable antenna to study the sources of interference—atmospheric static—that could play havoc with transatlantic radio-telephone service. Known around the lab as the "merry-go-round," this aerial was 30 meters long and rotated on four wheels taken off a Model T Ford (see Figure 56.1). With it, Jansky came to recognize three types of interference: intermittent static from local thunderstorms; weaker static due to distant storms; and, finally, a low steady hiss emanating from a direction that regularly moved around the sky. Initially he linked it with the Sun, but after a year of monitoring the strange signal, he came to realize it was arriving from the direction of the Sagittarius constellation—the center of our galaxy. He offered up several conjectures as to the source. This collective radio cry, he speculated, might be coming from the masses of stars in that region. He also suggested "some sort of thermal agitation of charged particles," which turned out to be closer to the truth. The long radio waves of 14.6 meters (20.5 megahertz frequency) that he detected are now known to be generated as energetic charged particles race through interstellar space.

Bell Labs publicized where radio listeners could tune in to hear this "hiss of the universe," which a reporter likened to "steam escaping from a radiator."[2] Jansky wanted to expand his search with a 100-foot-wide dish antenna, a proposal that his employer rejected. The company assigned him to other engineering tasks, which brought his nascent radio astronomy career to an end. Frail in health, he died in 1950 at the age of forty-four.

No one in academia immediately followed up, but Jansky's work did not go unnoticed. Radio engineer and avid ham-radio operator Grote Reber, inspired by Jansky's journal articles, built an antenna in the backyard of his home in Wheaton, Illinois. Specifically designed for celestial observations, the 31-foot-wide dish was constructed out of sheets of galvanized iron that were screwed onto wooden rafters cut to form a shallow parabola (a design that became radio astronomy's trademark; see Figure 56.3). With this huge saucer, Reber confirmed Jansky's discovery that celestial radio waves were most intense along the plane of the Milky Way; he published the results in 1940, the first paper on radio astronomy that the *Astrophysical Journal* had ever received. Only the intervention of a farsighted editor kept it from being rejected.[3] Collecting a shorter radio wavelength (1.87 meters; 160 megahertz frequency) than Jansky did, Reber thought he was observing radiation from hot interstellar gas. By 1944 Reber completed the first map showing the distribution of radio waves across the sky. It displayed a strong peak at the galactic center, with secondary peaks in the direction of Cygnus and Cassiopeia.

The astronomical community, then more at home with lenses and mirrors, did not rush to embrace this new method of observing, but that attitude changed with World War II. James S. Hey, while working with the British Army Operational Research Group analyzing the jamming of radar sets in 1942, came to discover intense radio emission from the Sun.* After the war, many physicists and engineers newly trained in radar transformed themselves into radio astronomers. In 1946, Hey and his team went on to discover a discrete and variable radio source in the sky, which came to be labeled Cygnus A for its location in that constellation.[4] (In the 1950s others would link the signal in Cygnus, as well as others in the Centaurus and Virgo constellations, to specific galaxies, establishing a new celestial category—the active radio galaxy.[5]) Meanwhile, Martin Ryle in England and J. L. Pawsey in Australia began to develop radio interferometry, techniques that combined the signals from two or more radio telescopes for better resolution. In this way the Australians John Bolton and Gordon Stanley in 1949 were the first to link a small discrete radio source with a visible object—the famous Crab nebula, the remnant of a star that exploded in 1054.[6]

"Directional Studies of Atmospherics at High Frequencies." *Proceedings of the Institute of Radio Engineers*, Volume 20 (December 1932); "Electrical Disturbances Apparently of Extraterrestrial Origin," Volume 21 (October 1933); "A Note on the Source of Interstellar Interference," Volume 23 (October 1935) by Karl G. Jansky

First Paper

For some time various investigators have made records of one type or another of the direction of arrival of static on the long wavelengths. . . . Very little work, however, has been done on the direction of arrival of short and very shortwave static. . . .

Since the middle of August, 1931, records have been taken at Holmdel,

---

* Hey's work was at first classified and not made public right away. Reber independently detected the radio Sun and mentions it, almost as an afterthought, in his 1944 *Astrophysical Journal* paper excerpted below.

Figure 56.1

N.J., of the direction of arrival and the intensity of static on 14.6 meters. . . . The rotating antenna, a photograph of which is shown in Figure [56.1], is a Bruce type broadside receiving array two wavelengths long made of ¾-inch brass pipe. The array was designed to operate on a wavelength of 14.5 meters. As shown in the photograph it is mounted on a wooden framework which in turn is mounted on a set of four wheels and a central pivot. The structure is connected by a chain drive to a small synchronous motor geared down so that the array makes a complete rotation once every twenty minutes. . . .

From the data obtained it is found that three distinct groups of static are recorded. The first group is composed of the static received from local thunderstorms and storm centers. Static in this group is nearly always of the crash type. It is very intermittent, but the crashes often have very high peak voltages. The second group is composed of very steady weak static coming probably by Heaviside layer refractions from thunderstorms some distance away. The third group is composed of a very steady hiss type static the origin of which is not yet known. . . . It is readily distinguished from ordinary static and probably does not originate in the thunderstorm areas. The direction of arrival of this static changes gradually throughout the day going almost completely around the compass in twenty-four hours. . . .

During the latter part of December and the first part of January the direction of arrival of this static coincided, for most of the daylight hours, with the direction of the sun from the receiver. However, during January and February the direction has gradually shifted so that now (March 1) it precedes in time the direction of the sun by as much as an hour. . . .

The fact that the direction of arrival changes almost 360 degrees during twenty-four hours and that the shift in the position of the curve observed during the three months over which data [have] been taken corresponds to the change in latitude of the sun affords definite indication that the source of this static is somehow associated with the position of the sun. . . .

Second Paper

During the progress of a series of studies that were being made at Holmdel, N.J., on the direction of arrival of atmospherics at high frequencies, records were obtained that showed the presence of weak but steady electromagnetic waves of an unknown origin. The first indications of these waves were obtained on records taken during the summer and fall of 1931. However, a comprehensive study of them was not begun until January 1932. The first complete records obtained showed the surprising fact that the horizontal component of the direction of arrival of these waves changed nearly 360 degrees in 24 hours, and at that time this horizontal component was approximately the same as the azimuth of the sun. These facts led to the assumption that the source of these waves was somehow associated with the sun.

Records of these waves have now been taken at frequent intervals for a period of more than a year. The data obtained from these records, contrary to the first indications, are not consistent with the suppositions made above relative to the source of the waves, but indicate that the direction of the phenomenon remains fixed in space, that is to say, its right ascension and declination [celestial coordinates] remain constant. . . .

If, now, the horizontal component of the direction of arrival is plotted against the time of day a curve similar to one of those of Figure [56.2] is obtained. . . . The figure shows curves for eleven different days spaced approximately one month apart during the year 1932. There is no curve for the month of November. These curves were obtained by averaging the data taken over several consecutive days so as to eliminate the errors made in measuring the records. The day assigned to a given curve is the middle day of the group over which the data for that curve were obtained. . . .

This figure shows: first, that the horizontal component of the direction of arrival changes nearly 360 degrees in 24 hours, and, then that there is a

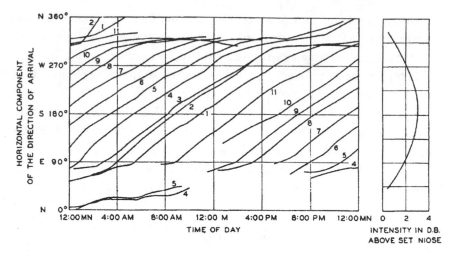

Figure 56.2: "Direction of arrival of waves of extraterrestrial origin."

1. Jan. 21, 1932	5. May 8, 1932	8. Aug. 21, 1932
2. Feb. 24, 1932	6. June 11, 1932	9. Sept. 17, 1932
3. March 4, 1932	7. July 15, 1932	10. Oct. 8, 1932
4. April 9, 1932		11. Dec. 4, 1932

uniformly progressive shift of the curves to the left from month to month which at the end of one sidereal year brings the curve back to its initial position. These facts show that the waves come, not from the sun, but from a direction which remains constant throughout the year. . . .

It may very well be that the waves that reach the receiver instead of coming from a single point fixed in space originate in the earth's atmosphere, but are secondary radiations caused by some primary rays of unknown character, coming from a source or sources fixed in space, and striking the earth's atmosphere. If this is so the disturbance measured by the receiver is probably the summation of very many waves of various intensities coming from secondary sources in the earth's atmosphere that are scattered over a considerable area. . . .

On the other hand it may be that the waves that reach the receiver are the primary waves themselves coming from a great many sources scattered throughout the heavens. In this case the direction measured would be the direction of the center of activity. . . .

The apparent direction of arrival of the waves has not as yet been definitely associated with any region fixed in space; however, there are two such regions that should be seriously considered. The point on the celestial sphere of right ascension 18 hours and declination −10 degrees, the direction from which the waves seem to come, is very near the point where the line drawn from the sun through the center of the huge galaxy of stars and nebulae of which the sun is a member would strike the celestial sphere. The coördinates of that point are approximately right ascension of 17 hours, 30 minutes, declination −30 degrees (in the Milky Way in the direction of Sagittarius). It is also very near that point in space towards which the solar system is moving with respect to the stars. The coördinates of this point are right ascension 18 hours and declination +28 degrees. Whether or not the actual direction of arrival of the primary rays coincides with either of these directions cannot be determined definitely until some method of accurately measuring their declination is devised and the measurements made. . . .

Third Paper

. . . Since the publication of the above papers further consideration of the data has led to some very interesting conclusions and speculations. . . . [I]t is discovered that when the peaks [of the radio signal] are broad, the antenna is so located in space that it sweeps along the Milky Way and the maximum response is obtained when it points in the direction of the center of the Milky Way. . . .

If we consider the belief now held by astronomers that the Milky Way is a large galaxy of stars having the same general shape as a huge discus or grindstone with the solar system, and therefore the earth, located at some distance from the center and almost in the galactic plane, then the phenomena described above would seem to indicate that the disturbances recorded are due to radiations emanating from the stars themselves. The various heights and widths of the peaks obtained on the record would then be explained in the following manner.

If the axis of rotation of the antenna were perpendicular to the plane of the Milky Way the antenna would rotate so that it always pointed at some part of the Milky Way and therefore would always receive some energy. This energy should reach a maximum value when the antenna points in the direction of the center of the Milky Way System, for the greatest number of stars would then be included within the angle of reception of the antenna. As the antenna rotates the number of stars included within this angle would very gradually decrease until the antenna points in just the opposite direction when the number of stars within the angle would be a minimum. As the antenna rotates further the number of stars within the angle would again increase until the maximum was again reached, etc. Thus the energy received at such a time would show a gradual decrease and increase with one maximum and one minimum for a single rotation of the antenna. . . .

A more detailed analysis of the data has shown that every time the antenna points toward some part of the Milky Way the record shows an increase in the energy received, and also every time the record shows an increase of energy received the antenna is found to be pointed towards some part of the Milky Way.

As said before, the most obvious explanation of these phenomena is one that assumes that the stars themselves are sending out these radiations and that the direction of arrival at the receiving location, instead of being confined to a single direction as was formerly intimated, include all directions, a greater indication being obtained for those directions confined to the Milky Way because of the greater star density there.

Another plausible explanation is one based on an hypothesis previously suggested, that the waves which reach the antenna are secondary radiations caused by some form of bombardment of the atmosphere by high speed particles which are shot off by the stars.

Upon examining the characteristics of these radiations for further clues as to their source, one is immediately struck by the similarity between the sounds they produce in the receiver headset and that produced by the thermal agitation of electric charge. In fact the similarity is so exact that it leads one to speculate as to whether or not the radiations might be caused by some sort of thermal agitation of charged particles. Such particles are found not only in the stars, but also in the very considerable amount of interstellar matter that is distributed throughout the Milky Way, which matter, according to [Arthur] Eddington has an effective temperature of 15,000 degrees centigrade. If the radiations come from such particles one would expect the response obtained to depend upon the directional characteristic and gain of the antenna and the way it is pointed relative to the Milky Way, an expectation which agrees with the observed facts. . . .

"Cosmic Static." *Astrophysical Journal*, Volume 100 (November 1944) by Grote Reber

Experiments on the measurement of electromagnetic energy at radio wavelengths arriving from the sky have been conducted at Wheaton, Illinois, for a number of years. Preliminary results have already been published. Dur-

Figure 56.3

ing the year 1943 [a] new and improved apparatus was put into operation and considerably better data were obtained.

The electromagnetic energy is captured by the mirror shown in Figure [56.3] and is directed to the mouth of the drum at the focal point of the mirror. Within the drum are a pair of cone antennae. These convert the electromagnetic energy into alternating current. . . . The mirror of Figure [56.3] is mounted on an east-west axis so that it may be pointed to any angle of declination between the limits of −32.5° and +90° along the north-south meridian. . . . The mirror is set to point at the desired declination; and then, as the earth rotates, the mirror sweeps out a band in the sky along this particular declination. . . . If no cosmic static is intercepted, the recorder will draw a straight line on the chart. If cosmic static is encountered, the pen will move up. . . .

. . . About two hundred charts were obtained in 1943. The final results plotted on a flattened globe are shown in Figure [56.4], *a* and *b,* for the two hemispheres of the sky. . . . The points at declination −48° in Figure [56.4*a*] are beyond the normal range of the collector machine. They were obtained as the result of an accident when the machine was run off the end of the track in a heavy snowstorm. It lodged with the mirror nearly vertical and the drum somewhat out of focus and resting in the service tower. These

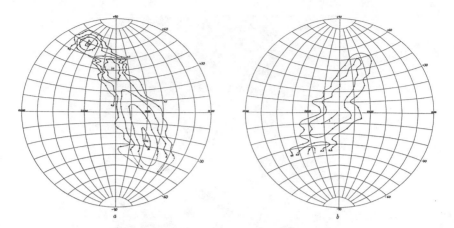

Figure 56.4: "[First radio map of the celestial sky.] Constant intensity lines in terms of 10^{-22} watt/sq. cm./cir. deg./M.C. band."

points show the general trend of the phenomena but are not of a high order of accuracy and, consequently, are connected with dashed lines.

Too little is known about the cause of this phenomenon to read a great deal from Figure [56.4]. However, it is suggested that this disturbance is in some way connected with the amount of material in space. Since the wavelength is long (1.87 meters), the absorption caused by dust is small. Therefore, the intensity is roughly indicative of the amount of material between us and the edge of the Milky Way. On this basis the various maxima point to the directions of projections from the Milky Way. These projections may be similar to the arms often photographed in other spiral nebulae. In the case of the Milky Way this general picture would call for the center toward Sagittarius, and arms in the directions of Cygnus, Cassiopeiae, and Canis Major. A minimum occurs in Perseus, indicating that we are nearest the edge of the galaxy in that direction. The maximum in lower Puppis is possibly a general rise toward the center. This region from Puppis to Scorpius is out of reach at the latitude of Wheaton. . . .

. . . The sun had the rather surprising center intensity of 10×10^{-22} watts/sq. cm., cir. deg., M.C. band. In spite of the apparent great strength from the sun, this source must be greatly discounted when explaining the origin of cosmic static. If it were the source and the Milky Way were made of average stars like the sun, a very large area in Sagittarius would have a visible intensity equal to that of the sun. Since this is not the case, some other cause must be found to make up the difference of 20 or 30 mag. . . .

57 / Interstellar Hydrogen

Although their country was under German military occupation, researchers in the Netherlands during World War II managed to smuggle in copies of the *Astrophysical Journal* from America. In it, they read of Grote Reber's first mappings of the radio sky (see Chapter 56). A young Dutch graduate student at the Leiden Observatory, Hendrik van de Hulst, was advised to examine how interstellar matter might be traced through its radio emissions—a radical idea at the time since radio astronomy was as yet far from established as a vital tool in the field.

The recommendation was a profitable one. Van de Hulst proceeded to undertake one of the first extensive analyses of the new observations. In his historic 1945 paper on the subject, he predicted the characteristics of a number of radio emissions within the galaxy and the universe. But his most notable contribution came at the end of the paper. In the course of his calculations, van de Hulst came to see that single atoms of neutral hydrogen dispersed through interstellar space would be emitting a specific type of signal, an unvarying radio hum. This would happen, he reasoned from atomic theory, when the lone spinning electron in a hydrogen atom occasionally flips over like a top (in his paper he calls it "spontaneous reversal of the spin") to a lower energy state. This generates an electromagnetic wave 21 centimeters in length (a radio frequency of 1,420 megahertz). Any one hydrogen atom performs this flip, on average, once every eleven million years or so, but the supply of hydrogen atoms in interstellar space is so enormous that van de Hulst figured there were more than enough atoms to produce a continuous, detectable drone.

It took time to develop the necessary equipment to look for van de Hulst's predicted signal, still considered quite speculative. The Dutch looked for years, but an American team on March 25, 1951, was the first to record hydrogen's monotonic tone. At Harvard University graduate student Harold Ewen and his advisor Edward Purcell picked up the radio waves with a large horn antenna built of plywood and copper foil that jutted out from their physics laboratory window (which created a convenient target for undergraduates lobbing snowballs). Within a few months, the 21 cm waves were also detected in the Netherlands and Australia. The Dutch were successful after Ewen and Purcell instructed radio astronomers there how to build Harvard's unique receiver, which employed a novel switched-frequency mode of detection.

This success was a turning point for the infant field of radio astronomy. Previous to that, radio astronomers were considered more as engineers than as celestial observers. Traditional astronomers, who dealt solely with the optical portion of the electromagnetic spectrum, believed that nothing new would be learned from celestial radio emissions. But the 21 cm line quickly became astronomy's most efficient surveyor of our galaxy's construction. Because radio waves can travel through interstellar clouds and dust with relative ease (unlike visible light), radio astronomers were soon able to tune into the hydrogen signal and map the Milky Way's structure. Several bands were traced nearly all the way around the galaxy, confirming that the Milky Way has several spiraling arms, perhaps four in all, that wrap themselves around the galactic center like coiled streamers (see Chapter 54).

"Origin of the Radio Waves from Space."
Nederlands tijdschrift voor natuurkunde, Volume 11 (1945)
by Hendrik C. van de Hulst

Although the existence of radio waves of extraterrestrial origin has already been known for a decade, astronomers have not yet paid much attention to them. This is partly due to the crude data thus far furnished by the observations; not much more has been established than the order of magnitude of the intensity and the rough directional dependence of the radiation. Thus, little can be expected from a detailed discussion of the observational facts.

Neither is the existence of these waves especially interesting from a purely theoretical viewpoint. The production of radio waves is not at all an essential feature of the physical conditions in the interstellar gas. The amount of kinetic energy which is converted into radiation energy is negligible; it is like a small leak in the overall energy budget which begins with the ionization of interstellar atoms by starlight. From purely theoretical considerations of the production of these radio waves, we cannot expect any new insights into the physical state of the interstellar gas, mainly characterized by its density, degree of ionization and distribution of electron velocities.

It is nevertheless the possibility of direct observation of these waves which makes this subject attractive. For twenty centuries astronomers have drawn all their knowledge from observations in the rather narrow frequency range around the visual frequencies. They have designed powerful instruments and for these observations have spared themselves no trouble. The Earth's atmosphere, however, leaves open another frequency range, near the radio frequencies, for astronomical observations. The first measurements have now been made in this range, but the observational technique is still in its infancy. . . .

The long wavelength does, however, entail one difficulty. Without telescopes of enormous aperture the radio waves will never furnish a detailed picture of the heavens. For the time being we must be quite content if a resolving power of about one degree can be realized.* The Sun, the Milky

---

* Today, arrays of radio telescopes and the electronic linking of radio telescopes around the globe enable radio astronomers to obtain resolutions matching and even surpassing those of ground-based optical telescopes.

Way, and the brightest extragalactic nebulae would then be measurable objects. In the Milky Way, the variation of intensity with longitude and latitude as well as other individual details, especially the distribution of H I and H II regions in the interstellar gas, could be investigated. There is the further possibility that the very distant extragalactic nebulae form a measurable diffuse background which would be of particular cosmological importance.

[At this point, van de Hulst theoretically calculates the overall radio spectrum expected to be emitted by the stars, dust, and gas within the Milky Way and compares it with the observations carried out by Grote Reber and Karl Jansky. He then considers whether there might also be more narrow spectral lines to be sought.]

Are There Also Discrete Spectral Lines?

We have established that on average the bound-bound transitions contribute an imperceptible amount to the continuous spectrum. The energy liberated in these transitions, however, is emitted in discrete spectral lines, and it is conceivable that within the rather narrow lines the intensity could be appreciably higher than in the continuum. . . .

All of these lines, just as the free-free continuum, would only be formed in H II regions. But there still remains a completely different possibility. The ground level of hydrogen is split by *hyperfine structure* into two levels with a separation of 0.047 cm^{-1}. The spins of the electron and the proton are pointed in the same direction in one state and are opposite in the other state. A quantum of wavelength 21.2 cm is emitted due to a spontaneous flip of the spin.* Such a transition is, of course, forbidden. On the other hand, the ground state, on account of the extreme dilution of radiation in interstellar space, is preferred above all other states (including the free states) by a factor [of] 10^{14} compared with that in thermodynamic equilibrium. In the H I regions, where the ionized state is completely absent, practically all of the atoms are in the ground state. We presume that these atoms are about equally divided between the two sublevels.

It would be especially interesting if this line were observable. . . . This possibility does not appear hopeless, even when we consider that the sensitivity of today's receiver installations must be improved by still another factor of 100. . . . Until a rigid calculation is made, the existence of this line remains speculative.

---

* The actual value for the wavelength is 21.1 cm.

"Radiation from Galactic Hydrogen at 1,420 Mc./sec."
Nature, Volume 168 (September 1, 1951)
by Harold I. Ewen and Edward M. Purcell

The ground-state of the hydrogen atom is a hyperfine doublet the splitting of which, determined by the method of atomic beams, is 1,420.405 Mc/sec [megacycles per second = megahertz]. . . . The possibility of detecting this transition in the spectrum of galactic radiation, first suggested by H. C. van de Hulst, has remained one of the challenging problems of radio-astronomy. In interstellar regions not too near hot stars, hydrogen atoms are relatively abundant, there being, according to the usual estimate, about one atom per cm^3. Most of these atoms should be in the ground-state. The detectability of the hyperfine transition hinges on the question whether the temperature which characterizes the distribution of population over the hyperfine doublet—which for want of a better name we shall call the hydrogen "spin temperature"—is lower than, equal to, or greater than the temperature which characterizes the background radiation field in this part of the galactic radio spectrum. If the spin temperature is lower than the temperature of the radiation field, the hyperfine line ought to appear in absorption; if it is higher, one would expect a "bright" line; while if the temperatures are the same no line could be detected. The total intensity within the line, per unit band-width, should depend only on the difference between these temperatures, providing the source is thick enough to be opaque.

We can now report success in observing this line. A microwave radiometer, built especially for the purpose, consists mainly of a double superheterodyne receiver with pass band of 17 kc [kilocycles], the band being shifted back and forth through 75 kc thirty times per second. The conventional phase-sensitive detector and narrow (0.016 c/s) filter then enable the radiometer to record the apparent radio temperature *difference* between two spectral bands 75 kc apart. These bands are slowly swept in frequency through the region of interest. The overall noise figure of the receiver, measured by the glow-discharge method, is 11 db [decibels], and the mean output fluctuation at the recorder corresponds to a temperature change of 3.5°. The antenna is a pyramidal horn of about 12° half-power beam-width. It is rigidly mounted at declination −5°; scanning is effected by the earth's rotation.

The line was first detected on March 25, 1951. It appeared in emission with a width of about 80 kc, and was most intense in the direction 18 hr right ascension. Many subsequent observations have established the following facts. At declination −5° the line is detectable, by our equipment, over a period of about six hours, during which the apparent temperature at the center of the line rises to a maximum of 25° above background and then subsides into the background. The source appears to be an extended one approximately centered about the galactic plane. The frequency of the center of the line, which was measured with an accuracy of ±5 kc, was displaced some 150 kc above the laboratory value, and this shift varied during an observing period. Both the shift and its variation are reasonably well accounted for by the earth's orbital motion and the motion of the solar system toward Hercules. The period of reception shifts two hours per month, in solar time, as it ought to.

Some conclusions can already be drawn from these results. Extrapolation of radio temperature data for somewhat lower frequencies suggests that the background radiation temperature near the 21-cm line is not more than 10° K. Then the hydrogen spin temperature is not more than 35° K, if the source is "thick." But we can calculate the opacity of the source on the assumption of a spin temperature of 35° K and 1 atom/cm^3, using only the observed line-width and the matrix element of the transition in question, and we obtained 900 light-years for the absorption-length. As this is much smaller than galactic dimensions, we conclude that the temperature observed corresponds indeed to the spin temperature at the source. To the extent that "self-absorption" contributes to the observed line-width, the true absorption length at the frequency of the center may be *less* than that computed. Further evidence for relatively high opacity is the absence of large frequency-shifts, which would be expected to arise from galactic rotation were the opacity-thickness comparable to the size of the galaxy. . . .

We have made rough theoretical estimates of the efficacy of various processes through which energy is exchanged between the hydrogen hyperfine levels and the other thermal reservoirs in the interstellar matter plus radiation complex. Of these we find exchange with the radiation field (involving spontaneous emission) and exchange with gas-kinetic energy of the hydrogen atoms (via H–H collisions) much the most important, with the latter process probably dominant. This is consistent with the observation, and if correct implies that the gas-kinetic temperature of the hydrogen exceeds, but not greatly, the spin temperature. The estimated spin relaxation time for these processes is of the order of 10^5 years.

58 / Molecules in Space

In 1955 physicist Charles Townes, who had just invented the maser and was well known for his work on the spectroscopy of molecules, was invited to an international symposium on radio astronomy convening in England. Just a few years earlier, radio astronomers had discovered a unique radio signal emitted by hydrogen atoms in interstellar space (see Chapter 57), and conferees were interested in hearing from Townes whether it might be possible to detect other cosmic substances by tuning into their specific radio emissions. In his presentation Townes named several promising candidates, including such molecules as carbon monoxide (CO, the stuff of car exhaust), ammonia (NH_3), water (H_2O), and the hydroxyl radical OH, the oxygen-hydrogen combination that distinguishes all alcohols.[7] But radio astronomers at the time were wary of using valuable telescope time to search for substances that, according to conventional wisdom, were rare and unimportant to celestial processes. Only one other researcher, Soviet astrophysicist Iosif Shklovsky, had even published on the subject.[8] Optical astronomers had already recognized a few species, such as carbon-hydrogen (CH), carbon-nitrogen (CN), and CH^+ (an ionized CH), when wisps of gas containing these molecules were caught in front of bright stars, but theorists were convinced that such molecules were just dust-grain residue and quickly destroyed by ultraviolet and cosmic rays in space.

Despite these warnings Alan Barrett, a student of Townes's at Columbia University in the 1950s, took on the challenge of finding OH molecules in space. At his first job at the Naval Research Laboratory in Washington, D.C., he and Harvard astronomer Arthur E. Lilley carried out an unsuccessful search using a 50-foot dish perched on the roof of the laboratory. Barrett persisted and was at last victorious in 1963, after he had moved to the Massachusetts Institute of Technology. By then Townes's Columbia laboratory had better pinpointed OH's most prominent frequency—1,667 megahertz—which hastened the discovery. Using a new digital receiver designed by Sander Weinreb, Barrett and his colleagues saw OH absorb energy at that exact frequency in the supernova remnant Cassiopeia A and later in dust clouds situated in front of the galactic center.

Still unsure of the importance of celestial molecules, though, astronomers were slow to follow up, but eventually other groups widened the search. In the 1960s Townes transferred to the University of California at Berkeley specifically to work in radio astronomy. Using a dish at Berkeley's Hat Creek Radio Observatory, his group in 1968 recorded the radio cries of ammonia and water vapor.[9] Soon after, Lewis Snyder and David Buhl at the National Radio Astronomy Observatory in Green Bank, West Virginia, detected a signal from the embalming fluid formaldehyde (H_2CO).[10] A race quickly ensued to snare the next new molecules. By 1973, nearly thirty cosmic molecules were identified in space; by the end of the twentieth century, the total was more than a hundred—from ethyl alcohol (C_2H_5OH) and hydrogen cyanide (HCN) to methane (CH_4) and nitrous oxide or laughing gas (N_2O). Astronomers had not realized that dust grains in a cloud protect the molecules from being destroyed by radiation and also serve as sites for their construction.

An unexpected revelation in this growing field occurred in 1965 when Harold Weaver and colleagues at Berkeley tuned to the distinctive frequency of OH to survey interstellar clouds and came across an enormously intense beam of radio emission. "To emphasize the surprising nature of the observation," the Berkeley researchers reported in the journal Nature, "we shall speak of this unidentified line as arising from 'mysterium,'" an unknown substance.[11] Within a year, it was determined that mysterium was actually a cloud of hydroxyl molecules that had coalesced to form a gigantic natural maser, the microwave equivalent of the laser.[12] Under certain conditions, molecules in a cloud can absorb energy from their surroundings and release it in unison, producing an intense and pure beam of radio waves. Cosmic masers, now known to be generated by a variety of different molecules, are often found in clumps of gas surrounding newborn stars. By the mid-1970s astronomers began to detect such masers in far-off galaxies and are now using them to directly measure the distances to galaxies.[13]

Overturning conventional wisdom, the new field of astrochemistry proved a boon to the study of interstellar clouds and star formation. In the densest clumps of gas in the galaxy, hydrogen atoms tend to join up to form H_2 molecules, which are effectively silent in the radio regime. But traces of other molecules are sprinkled in with the H_2 and emit radiation profusely. By tracking these signals across the celestial sky, radio astronomers discovered a whole new class of objects: the giant molecular clouds, which can be up to a thousand light-years wide and contain enough hydrogen to form a million suns. Bright nebulae, such as the Orion nebula, are luminous blisters situated on the sides of these huge invisible clouds. Most of the stars in our galaxy are created in molecular clouds, where one generation of stars kindles a succession of new generations that surge through the cloud.

"Radio Observations of OH in the Interstellar Medium."
Nature, Volume 200 (November 30, 1963)
by Sander Weinreb, Alan H. Barrett,
M. Littleton Meeks, and John C. Henry

In this article we wish to report the detection of 18-cm absorption lines of the hydroxyl (OH) radical in the radio absorption spectrum of Cassiopeia A, thereby providing positive evidence for the existence of OH in the interstellar medium. The microwave transitions of OH in the ground-state . . . arise from two Λ-type doublet-levels, each of which is split by hyperfine interactions with the hydrogen nucleus, so that four transitions result. The two strongest lines have been previously measured in the laboratory at 1,667.34 $\pm$ 0.03 Mc/s and 1,665.46 $\pm$ 0.10 Mc/s with relative

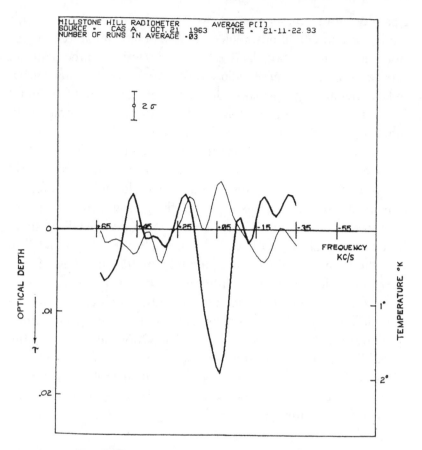

Figure 58.1: "Observed 1,667 Mc/s OH absorption spectrum in Cassiopeia A. The heavy line shows 8,000 sec of data taken with the antenna beam directed at Cassiopeia A, and the light line shows 6,000 sec of data taken with the beam displaced slightly from Cassiopeia A. . . ."

intensities of 9 and 5, respectively; these results are in agreement with theory. The suggestion that these lines might be detected in the radio spectrum of the interstellar medium has been made by [Iosif] Shklovsky and [Charles] Townes. A previous search by Barrett and [Arthur E.] Lilley, in 1956, was unsuccessful, primarily because the laboratory measurements of the frequencies had not been made. A recent search for OH emission also yielded negative results.

Our observations were conducted on 10 days between October 15 and October 29, 1963, using the 84-ft. parabolic antenna of the Millstone Hill Observatory of Lincoln Laboratory, Massachusetts Institute of Technology, and the spectral-line autocorrelation radiometer designed by Weinreb. The receiver uses digital techniques to determine the autocorrelation function of the received signal. The resulting autocorrelation function is then coupled directly into a digital computer that performs a Fourier transformation and displays the resulting spectrum on a cathode-ray tube or a precision *x-y* plotter. During one integration time-interval of 2,000 sec, a 100-kc/s portion of the spectrum is determined with a frequency resolution of 7.5 kc/s. The ability to see immediately a calibrated visual display of the measured spectrum and average this result with others greatly facilitated the conduct of the experiment and eliminated almost all post-observation data handling. The system noise temperature was 420° K, of which 110° K was due to Cassiopeia A. System tests were performed by observing the hydrogen line.

The results obtained during the first evening of our observations showed strong evidence of the 1,667 Mc/s line in Cassiopeia A; the signal is visible after 2,000 sec of integration. We decided that positive identification of OH absorption lines of Cassiopeia A would be secured before proceeding to observations of other regions. . . . A typical record showing the 1,667 Mc/s line . . . is shown in Figure [58.1]. . . . The evidence that we are indeed detecting interstellar OH in these observations may be summarized as follows:

(1) Lines at both 1,667 Mc/s and 1,665 Mc/s have been detected with frequencies and intensity ratios that are in good agreement with the expected values.

(2) The OH absorption spectra at both frequencies show general agreement with the H absorption spectra.

(3) The absorption lines disappear when the antenna is positioned off Cassiopeia A by one degree in both azimuth and elevation.

(4) The lines shifted 20 kc/s between October 17 and October 29; this is the shift expected from the orbital velocity of the Earth during this time-interval.

A quantity of immediate astrophysical interest which follows from our observations is the abundance ratio of OH to H. . . .

[Omitted here is the theoretical discussion of the calculation with its equations.]

. . . The OH/H abundance ratio can be calculated from the results of the H absorption on Cassiopeia A, and gives typical ratios of 1×10^{-7}. . . .

59 / Van Allen Radiation Belts

On October 4, 1957, the space age dawned in a dramatic fashion with the Soviet Union's launch of Sputnik I, the first artificial Earth satellite. Sputnik II quickly followed a month later. By February 1, 1958, competitive Americans had their own satellite in space, Explorer 1 (called "earth satellite 1958 α" in official reports), designed and built by scientists from the University of Iowa led by James Van Allen. These endeavors were part of the International Geophysical Year, a global program set up for scientists to carry out a series of coordinated observations of various geophysical phenomena from July 1957 to December 1958.

Explorer I was equipped with a Geiger counter, which Van Allen intended to use to measure the intensity of cosmic rays arriving from space before they plunged into the atmosphere. Unlike the Sputniks, Explorer 1 had a highly elliptical orbit, and this led to its puzzling finding. With no flight recorder on board, data could be transmitted only intermittently, when the satellite was within range of a tracking station. From a low point in its orbit, Explorer 1 reported the expected number of energetic particles. But when it rose to its high altitude of about 1,600 miles, no particles were counted.

With Explorer 2 failing to orbit, Explorer 3 (earth satellite 1958 γ, launched on March 26) solved the mystery. This time the satellite carried a tape recorder, and Van Allen's team saw that the Geiger counter was actually saturating at high altitudes from high counting rates, causing a discharge to zero. What the Explorers had uncovered was the first conclusive evidence for radiation belts surrounding the Earth, where high-energy electrons and protons are geomagnetically trapped. Norwegian physicists Kristian Birkeland and Carl Störmer had anticipated the existence of such a reservoir in the early part of the twentieth century, to explain how auroras and geomagnetic disturbances occurred.[14]

The first Explorers revealed a doughnut-shaped region of radiation, situated about an Earth radius above our planet's equator. A rocket flight the following year detected another belt about 2.5 radii out. In honor of their discoverer, they are now known as the Van Allen belts.

"Observation of High Intensity Radiation by Satellites 1958 Alpha and Gamma." *Jet Propulsion*, Volume 28 (September 1958) by James A. Van Allen, George H. Ludwig, Ernest C. Ray, and Carl E. McIlwain

This is a preliminary report of results obtained concerning radiation intensities measured with a single geiger tube carried by the artificial earth satellites 1958 α and 1958 γ.

The counting rate of the counter in 1958 α was transmitted continuously, and the data were recorded only when the satellite was quite near one of the 16 receiving stations distributed over the earth.

The data collected by 1958 γ were also telemetered continuously. In addition, a small magnetic tape recorder stored the data obtained during each entire orbit. Then, as the satellite passed near one of the receiving stations, a radio command from the ground caused these data to be read out.

A preliminary study of the data obtained from 1958 α and several interrogations of 1958 γ has been carried out, with the following results.

Reasonable cosmic ray counting rates have been obtained for altitudes below about 1000 km. In particular, we have obtained a plot of omnidirectional intensity vs. height in the vicinity of California for the first two weeks in February. This curve, extrapolated down to altitudes previously reached by rockets, agrees with earlier data.

At altitudes greater than about 1100 km, very high counting rates were obtained. This conclusion is the result of a somewhat lengthy analysis. Geiger tube output rates up to about 140/sec have actually been observed. In addition, periods have been found during which the geiger tube put out less than 128 pulses in 15 min. (We have a scaling factor of 128.) [Certain] considerations cause us to conclude that this is not due to equipment malfunction, but is caused by a blanking of the geiger tube by an intense radiation field. We estimate that if the geiger tube had had zero dead time, it would on these occasions have been producing at least 35,000 counts/sec.

We surmise that the radiation we have found is closely related to the soft radiation previously detected during rocket flights in the auroral zone.

The radiation intensity necessary just to blank the geiger tube is equivalent to 60 mr/hr. In this connection the recommended permissible dose for human beings is 0.3 r/week. The present radiation is 0.3 r in 5 hr or less.

Several geophysical effects of this radiation seem possible. It is very likely closely related to aurorae and geomagnetic storms. In addition, a rough calculation suggests that the radiation may be sufficiently intense to contribute important heating to the upper atmosphere. It will be important to investigate the amount of atmospheric ionization, light and radio noise which would be produced, under various assumptions as to the nature of the radiation.

Instrumentation for 1958 α and 1958 γ

The instrumentation for 1958 α consisted essentially of a single Geiger Mueller tube, a scaling circuit for reducing the number of pulses to be worked with, and telemetry systems for transmitting the scaler output to the ground receiving stations. The system contained in 1958 γ was identical, with the addition of a miniature tape recorder for storing the data for the duration of each orbit and a command system to cause the telemetry of the stored information over a ground receiving station.

[Omitted here are detailed descriptions of the instruments and their operation.]

Summary of Preliminary Observations

Table [59.1] is a list of the stations receiving data and reporting them to us. The stations labeled JPL are operated under the auspices of the Jet

Propulsion Laboratory at Pasadena, Calif. Those labeled NRL are operated by the Naval Research Laboratory in Washington, D.C. Data were obtained from 1958 α only when it was reasonably near one of these stations, since it had no provision for storing data for a later readout. We have already analyzed most of the data from the JPL stations, and some of that from the NRL stations as well. This work is continuing.

A small magnetic tape recorder in 1958 γ stored the cosmic ray information for an entire orbit, and then played it into a transmitter on command from the ground. Data from nine of these orbits have been reduced in a preliminary way. We already have on hand many more of these passes, and are reducing the data from them in a routine way. . . .

Table 59.1: Receiving Stations

Blossom Point, Md.	NRL
Fort Stewart, Ga.	NRL
Antigua, Br. W. Ind.	NRL
Havana, Cuba	NRL
San Diego, Calif.	NRL
Quito, Ecua.	NRL
Lima, Peru	NRL
Antofagasta, Chile	NRL
Santiago, Chile	NRL
Woomera, Aus.	NRL
Patrick Air Force Base, Fla.	JPL
Earthquake Valley, Calif.	JPL
Singapore	JPL
Ibadan, Nigeria	JPL
Temple City, Calif.	JPL
Pasadena, Calif.	JPL

. . . The passes fall into two classes. In the first case, one obtains a counting rate of about 30/sec, a roughly reasonable value. In the second case, the telemetered signal fails to show a single scaler output pulse during the approximately 2 min of clean signal. This represents an input rate to the scaler of less than about 0.1/sec. There are, in addition, a few cases showing a strong change in counting rate during the pass.

For reasons discussed [below], we believe that the extremely low output rate of the scaler is caused by very intense radiation which "jams" the

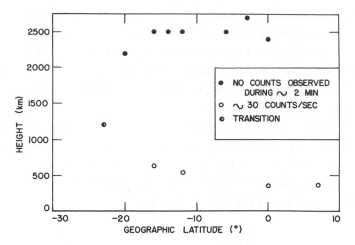

Figure 59.1: "Positions in altitude vs. latitude for telemetry of data from 1958 α over South America."

geiger tube so that it puts out pulses of such low height that they are below the threshold of the counting circuits. Laboratory tests show that this first happens for the present equipment when the radiation reaches such an intensity that a counter of the same effective dimensions and efficiency as the present geiger counter but with a zero dead time would produce 35,000 counts/sec.

Figure [59.1] is a plot of height vs. geographic latitude in the vicinity of 75 W longtitude. The positions of 1958 α during reception of its telemetering signal by various of the NRL stations are marked. A code designates the kind of information received. It is at once evident that the extremely low counting rates observed all occur at a high altitude, while the more or less normal rates occur at a low altitude. . . . Transitional cases occur at intermediate altitudes. . . .

These data already suggest a picture of the geophysical phenomenon being measured. The data from 1958 γ are much more explicit. Figure [59.2] is a plot of the scaler output as a function of time as given by the tape recorder readout for the pass ending near San Diego on March 28, 1748 UT [universal time]. Since the tape recorder can only record one scaler output pulse each second the maximum indication on the tape recorder output corresponds to 128 counts/sec for the geiger tube output rate. (Our scaling factor is 128 in this case.) It is evident from the figure that reasonable counting rates occur near the two ends of the pass. These ends correspond to the most northern latitudes and the lowest heights above the earth. The section where the counting rate indication is zero corresponds to a portion of the

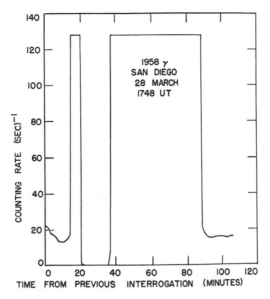

Figure 59.2: "A sample of the results of a tape recorder readout near San Diego."

magnetic tape where no tuning fork pulses were missing, and hence no scaler output pulses occurred. This condition lasted 15 min, and 128 pulses were fed to the scaler during this time. This is an average counting rate for the interval of 0.14/sec, to be compared with the usual cosmic ray rate for a geiger tube of this sort of about 50/sec. The counter goes through the transition from putting out essentially no counts to putting out a great many very quickly, and we presume that most of the 128 counts observed during this 15 min interval occurred near the ends of the interval. There is, of course, no real evidence for this.

As discussed in detail in the next section, we believe that if we had had a detector with zero dead time, and a storage mechanism of unlimited capacity, Fig. [59.2] would begin where it does now and at about 13 min would have begun rising rapidly to a peak near 25 min at which point the counting rate would have been greater than 35,000 counts/sec. After this time, the rate would gradually have subsided, returning finally to about the value actually recorded near the end of the pass. . . .

Interpretation of Observed Data

We now propose to justify our claim that when essentially no scaler output pulses occur, the apparatus is, in fact, exposed to very intense radiation.

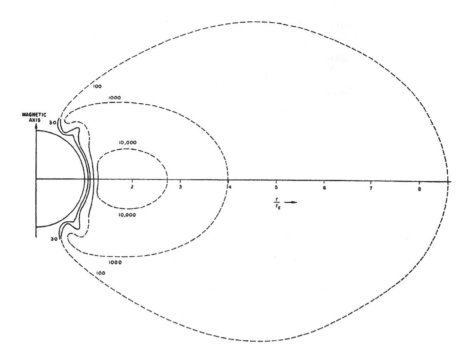

Figure 59.3: First cross-section of the Van Allen belt from a later paper. The solid-line contours specify the counting rate of the detector aboard Satellite 1958 ε. These lines are extended as dashed lines in a speculative way. The numbers are the detector's counting rates; the linear scale is in units of the Earth's radius— 6,371 kilometers. (James A. Van Allen, Carl E. McIlwain, and George H. Ludwig, "Radiation Observations with Satellite 1958 ε," *Journal of Geophysical Research* 64 [March 1959]: 271–86.)

Three possibilities are immediately evident. The apparatus may have some simple malfunction. This possibility can immediately be rejected except for the scalers, geiger tubes, and geiger tube voltage supplies, since the subsequent treatment of the information is completely different in the 1958 α and 1958 γ. Some effect of temperature seems the only reasonable possibility here. The temperature of the geiger tube was measured in 1958 γ and telemetered on the continuously operating transmitter. The observed temperatures range from zero to about 15 C. . . . The operating range of the circuitry is −15 to 85 C. In addition, the frequencies of the continuously telemetering channels which carried the cosmic ray information are significantly temperature sensitive. These showed that no extreme temperatures occurred at the location of the corresponding sub-carrier generators.

Another possibility might be that the satellite passed through regions which very few cosmic rays could reach. This is extremely unlikely. A magnetic field of the order of one gauss extending over thousands of kilometers and remaining unbelievably free of local irregularities would be required to exclude a sufficient fraction of the cosmic radiation.

The possibility that we firmly believe is correct is that the geiger tube encountered such intense radiation that dead time effects reduced the counting rate essentially to zero. . . .

We have little concrete evidence concerning the nature of this radiation. Apparently, however, it is not electromagnetic. It makes its effects felt through the 1.5 g/cm^2 of absorber which constitute the hull of the satellite and the walls of the counter. Photons with such energy should then be seen down to the lowest altitudes our equipment reaches. The radiation can presumably be either protons or electrons. . . .

Implications

Any reasonable identification of this radiation strongly suggests several geophysical consequences. It is unlikely that the particles have several Bev of energy each. Then in order to reach such low heights through the geomagnetic field they must at least initially be associated with plasmas which seriously perturb the magnetic field at an earth radius or so. We presume that this plasma is closely related to geomagnetic storms and aurorae. . . .

60 / Geology of Mars

P lanetary studies, so long eclipsed by discoveries in the far universe in the first half of the twentieth century, experienced a renaissance in the 1960s with the arrival of the space age. The many spacecraft sent on missions to the planets transmitted back a veritable treasure chest of details, information that was out of reach by telescope alone. In 1959 the Soviet Luna 3 captured the first view of the backside of the Moon; by the end of the twentieth century, spacecraft had journeyed to all the planets (except Pluto) and twelve astronauts had walked on the Moon, while other probes roamed over Mars, parachuted onto Venus, landed on an asteroid, and plunged into Jupiter's dense atmosphere.[15] The ability to journey robotically through the solar system altered planetary astronomy in a dramatic fashion. With the Pioneer and Voyager grand tours of the gas giants in the 1970s and 1980s, "all the existing information about Jupiter and Saturn, so slowly and painstakingly gathered in three and a half centuries of telescopic astronomy, was made obsolete almost in one stroke," said science historian Albert Van Helden.[16] And a series of Mariner spacecraft launched by the United States dashed, once and for all, the many exotic imaginings about the red planet, Mars.

Fascination with Mars arose soon after the invention of the telescope. By the mid-1600s astronomers were discerning markings on the planet's surface. Later, bright patches around its poles were seen to wax and wane with the Martian seasons. In 1784 the British astronomer William Herschel reported that Mars "is not without a considerable atmosphere . . . so that its inhabitants probably enjoy a situation in many respects similar to ours."[17]

In September 1877, during a favorable opposition, when Earth and Mars were at their closest on the same side of the Sun, the Italian astronomer Giovanni Schiaparelli was able to view the red planet in far greater detail than previously and noticed numerous dark streaks crossing Mars' reddish ochre regions, then known as "continents." He called these markings *canali* in his native language (best translated as "channels"). After initial doubts, other astronomers came to see the lines as well and began to offer a number of explanations. Some believed they were just optical illusions—the eye creating patterns out of irregular forms—while others suggested they were either chains of mountains or vast rifts created as the primordial Mars cooled down.

The most controversial theory was championed by the American astronomer Percival Lowell. In 1894 the wealthy Lowell established a private observatory in Flagstaff, Arizona, for the express purpose of observing Mars and its "canals," as he called them. That year he observed that Mars's south polar cap had disappeared, and he went on to claim that the canals, in turn, had become darker. This led him to speculate that the dark bands were actually waterways, lined on either side by banks of vegetation. The reddish continents, he surmised, were deserts, while the more greenish areas were vast marshes of plant life. To Lowell, the canals were irrigation works built by advanced beings, who were directing their scarce resources over the surface of the planet for cultivation. He wrote in 1895:

A mind of no mean order would seem to have presided over the system we see—a mind certainly of considerably more comprehensiveness than that which presides over the various departments of our own public works. Party politics, at all events, have had no part in them; for the system is planet wide. Quite possibly, such Martian folk are possessed of inventions of which we have not dreamed, and with them electrophones and kinetoscopes are things of a bygone past, preserved with veneration in museums as relics of the clumsy contrivances of the simple childhood of the race.

Certainly what we see hints at the existence of beings who
are in advance of, not behind us, in the journey of life.[18]

These were ideas that captivated the public and, due to Lowell's
persistent advocacy, even caused some other astronomers to seriously consider the possibility.

That Mars was in reality a barren world with no visible signs of
life was proven with a series of Mariner missions launched by the
National Aeronautics and Space Administration (NASA). Mariner
4, the first spacecraft to successfully fly by Mars, revealed on July 15,
1965, both a cratered surface similar to the Moon's and an atmosphere thinner than expected. Further photographs taken by
Mariners 6 and 7 in the summer of 1969 reinforced this bleak image
and led to the conclusion that craters were the dominant landform
on Mars. That verdict, however, had to be drastically revised with
the next successful mission to Mars, Mariner 9, the first spacecraft
to orbit the red planet. Arriving on November 14, 1971, and operating for almost a year, its planetwide photographic survey displayed
gigantic volcanoes, an enormous canyon as long as the continental
United States, and most surprisingly ancient riverbeds with tributar-
ies and erosion patterns that appeared to have been carved by catastrophic flooding episodes in Mars's distant past. Water is present on
Mars. Mariner 9 and later missions found the Martian polar caps are
composed of both frozen carbon dioxide ("dry ice") and frozen
water, which suggests that liquid water flowing in Mars's past was
possible.

The periodic darkenings observed by Lowell were determined to
be the result of deposits shifting around during the planet's many
dust storms. Mars, concluded Mariner scientists in their report, "has
a geological style of its own different from that of either the Earth or
Moon, and . . . represents a body intermediate." Mars became the
planet that was still alien yet also strangely familiar.

"Preliminary Mariner 9 Report on the Geology of Mars."
Icarus, Volume 17 (1972)
by John F. McCauley, Michael H. Carr, James A. Cutts,
William K. Hartmann, Harold Masursky, Daniel J. Milton,
Robert P. Sharp, and D. E. Wilhelms

Abstract

Mariner 9 pictures indicate that the surface of Mars has been shaped by impact, volcanic, tectonic, erosional and depositional activity. The moon-like cratered terrain, identified as the dominant surface unit from the Mariner 6 and 7 flyby data, has proven to be less typical of Mars than previously believed, although extensive in the mid- and high-latitude regions of the southern hemisphere. Martian craters are highly modified but their size-frequency distribution and morphology suggest that most were formed by impact. Circular basins encompassed by rugged terrain and filled with smooth plains material are recognized. These structures, like the craters, are more modified than corresponding features on the Moon and they exercise a less dominant influence on the regional geology. Smooth plains with few visible craters fill the large basins and the floors of larger craters; they also occupy large parts of the northern hemisphere where the plains lap against higher landforms. The middle northern latitudes of Mars from 90 to 150° longitude contain at least four large shield volcanoes each of which is about twice as massive as the largest on Earth. Steep-sided domes with summit craters and large, fresh-appearing volcanic craters with smooth rims are also present in this region. Multiple flow structures, ridges with lobate flanks, chain craters, and sinuous rilles occur in all regions, suggesting widespread volcanism. Evidence for tectonic activity postdating formation of the cratered terrain and some of the plains units is abundant in the equatorial area from 0 to 120° longitude. Some regions exhibit a complex semiradial array of graben that suggest doming and stretching of the surface. Others contain intensely faulted terrain with broader, deeper graben separated by a complex mosaic of flat-topped blocks. An east-west-trending canyon system about 100–200 km wide and about 2,500 km long extends through the Coprates-Eos region. The canyons have gullied walls indicative of extensive headward erosion since their initial formation.

Regionally depressed areas called chaotic terrain consist of intricately broken and jumbled blocks and appear to result from breaking up and slumping of older geologic units. Compressional features have not been identified in any of the pictures analyzed to date. Plumose light and dark surface markings can be explained by eolian transport. Mariner 9 has thus revealed that Mars is a complex planet with its own distinctive geologic history and that it is less primitive than the Moon.

Introduction

Mars, as revealed to Mariner 9 after the great dust storm of 1971, proved to be geologically far more heterogeneous than previously suspected. Certain regions of the planet have been shaped principally by impact, others by volcanism. Tectonism, erosion, and deposition appear to dominate other parts of the surface. The dominant geological processes that have shaped the surface of Mars have varied not only from place to place but also from time to time throughout the planet's history.

The equatorial region displays most of the distinctive geologic features of Mars although some features appear to be restricted to the high-latitude and polar regions described in a companion paper.[19] This region (30° south latitude to 30° north latitude) [is] an area of about 8×10^7 km^2 or roughly 10 times the area of the conterminous United States of America. . . .

Cratering and Circular Basins

Mariner 4, 6, and 7 photographs suggested that most of Mars was highly cratered, like the Moon's southern highlands and far side. Although Mariner 9 has revealed extensive relatively uncratered regions, cratering still appears to have been the dominant geologic process in approximately 40% of the equatorial region and in much of the area outside the map. Martian craters grossly resemble lunar craters of comparable size and most of the differences between them can be attributed to greater degradation. . . . We believe that, as on the Moon, the majority of the craters are of impact origin as indicated by their morphologies and the slope of their size-frequency distributions. An unknown but probably small fraction is almost certainly of volcanic origin, but it is very difficult to distinguish individual volcanic craters from impact craters when both are severely degraded.

Circular basins complete the series of craters at the upper end of the size range. Some of these appear to have multiple rims, although the outer rims are not as well developed as around the fresher lunar basins. Four basins have been identified in the equatorial region. The largest Martian basin is Hellas, about 2,000 km in diameter. . . . The second largest basin is here named the "Libya basin" for the telescopic feature that coincides approximately with the most conspicuous segment of its main ring. . . .

The two smaller basins are named for the features Edom, near 340° W and the equator, and lapygia, near 305° W and 15° S. The Edom basin has a conspicuous outer scarp and both basins have inner and outer rings that appear as gentle highs. Conspicuous radial fractures and flat-bottomed troughs, partly filled by cratered plains, in lineated terrain southwest of the lapygia basin, are much like the radial structures around lunar basins, though less conspicuous and extensive. These smaller basins are comparable in size to the Crisium basin on the Moon.

Volcanism

One of the most significant Mariner 9 results is the recognition of the major role of volcanism in the formation of the surface of Mars. The most striking of the volcanic features are the four enormous shield volcanoes of the Tharsis-Amazonis-Elysium region each of which is at least twice as massive as the largest comparable features on Earth. South Spot and Middle Spot have simple, circular, summit calderas whereas Nix Olympica [now called Olympus Mons], and North Spot have composite calderas consisting of several intersecting craters with floors at different levels. The floors are generally smooth and very sparsely cratered. . . . The Martian volcanic shields have no lunar analogs but are strikingly similar to some terrestrial volcanoes, for example, Fernandina in the Galapagos Islands. . . .

Volcanic domes also occur mainly in the Tharsis region. These generally have a central caldera one-half to one-third the diameter of the dome. The flanks of the domes are convex upward and appear smooth in the low-resolution A-frame pictures. In the high resolution B-frames numerous radial channels are visible, some terminating in shallow depressions. The domes and the channels terminate abruptly against the surrounding plains. . . .

The striking edifices radially symmetric about a central vent are not the only probable volcanic features observed on Mars. Some of the plains, apparently featureless in A-frame pictures, appear in B-frames to have low ridges and hills, or finely lobate escarpments suggestive of flow fronts. These fronts strongly resemble those of the flows in Mare Ibrium on the Moon. . . .

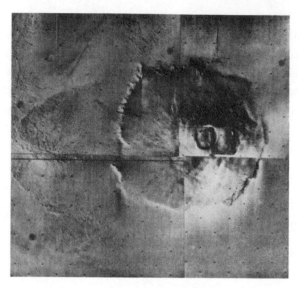

Figure 60.1: "Volcanic shield of Nix Olympica [Olympus Mons]. The shield, approximately 600 km across, slopes gently away from the central crater and is surrounded by a peripheral escarpment. The central crater is multiple, consisting of several intersecting circular depressions with level floors. Smooth plains in lower right corner of picture. Grooved terrain surrounding Nix Olympica at left of picture. A-frame mosaic centered at 18° N and 137° W."

Canyons

The discovery of the equatorial canyon system of the Coprates region was another startling Mariner 9 result. The canyon system consists of a series of roughly parallel steep-walled, linear depressions that range from 1 to 3 or more km in depth with an average width of about 100–150 km and an overall length of more than 2,500 km. In the Melas Lacus region the main canyon widens to about 250 km. Here two large roughly rectilinear troughs border the main canyon so that the total width is on the order of 500 km.

The canyon walls are rarely smooth, but exhibit a variety of re-entrants, ranging from alcoves with gently curved broad outlines to a complex branching system of ravines with steep gradients. In addition, large gullies, some of which have a dendritic tributary pattern, extend backward from the canyon walls for as much as 150 km. Hummocky terrain which appears to be landslide debris occurs at the base of some re-entrants, particularly in the open alcoves. The gullies and the ravines resemble in form those cut by running water on Earth. Many gullies show an orthogonal pattern suggesting structural control. Whatever originally formed the canyons, the present shape appears to be the result of slope retreat effected by processes the same as or closely analogous to those that widen terrestrial canyons: landsliding, debris flow, artesian sapping, and possibly erosion by running water. . . .

Channels

Many of the channels . . . are remarkably similar to fluvial channels on Earth. These sinuous multichannel features with their discontinuous marginal terraces contain teardrop-shaped islands and channel bars must have been carved by running fluids. The current working hypothesis is that the fluid was water, although the possibility of erosion by fluidized solid-gas systems must also be considered. Integrated drainage systems like those of established river systems on Earth, composed of small channels that successively join to form wide channels of higher order and that culminate in a master channel, are markedly absent. The Martian channel systems, most of which show little change in character from head to mouth, resemble features produced by episodic floods on Earth—ephemeral channels on desert fans or, on a scale more nearly approaching the Martian channels, the channels cut by catastrophic draining of ice-dammed lakes or the melting of ice by subglacial volcanism. Braiding, evident in many channels, is indicative of a strongly varying flow regime, in which the stream during the waning stages of a flood is unable to transport the sediment carried during peak flow.

A source must be found for the enormous quantities of water (or other fluid) that carved the channels, some of which are over 200 km wide and 1,500 km long. Theoretical considerations[20] suggest that the Martian atmosphere could never have produced the volumes of water necessary, although an adequate quantity of water episodically released from the polar caps has been conjectured.[21] The source of the fluids could be lithospheric rather than atmospheric. Of particular significance is the apparent relation between the braided channel deposits and the chaotic terrain in the Margaritifer Sinus region. Some of the largest channels appear to originate in patches of chaotic terrain and flow northward into the Chryse region. . . .

Surface Markings and Eolian Activity

As indicated from Earth-based observations, Mars does not have homogeneous tones and colors. Apart from the polar caps, clouds and dust

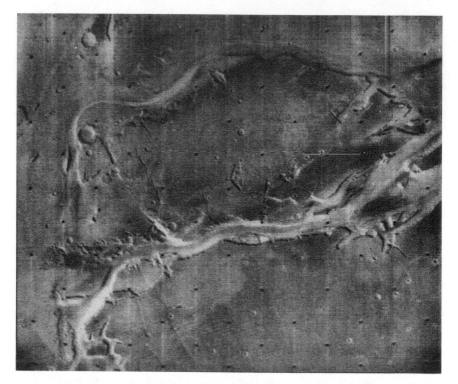

Figure 60.2: "Martian channels photographed by Mariner 9."

storms, many tonal differences or so-called albedo markings are observed that must relate to regional variations of topography, composition, or texture of the materials on the surface. . . .

Dominating many Mariner 9 photographs are conspicuous streaks or plumes which originate at a crater, ridge, or scarp and extend up to hundreds of kilometers across the surface of the planet. . . . Light and dark plumes are found, and although both exhibit large variations in morphology, they have characteristic differences. . . . Most dark plumes develop from within the floor of a crater and extend outside it; most light plumes develop tangentially to crater rims. Dark plumes in most places contrast sharply with their surroundings; light plumes are usually of low contrast and have diffuse edges. . . .

The morphology and distribution of the plumes reinforce earlier ideas that wind moves surficial materials on Mars. We can at present only speculate on the details of the formative mechanism, but most of them must result from deposition, and scouring and sorting of surface materials in the lee of topographic obstructions to the prevailing wind. One series of

Mariner 9 photographs shows that a dark area apparently formed behind a bright dust cloud. Dark plumes with high contrasts and sharp boundaries may be the scoured areas that are depleted of the fine particles. The surface now exposed may consist of coarser sand, gravel, or even outcrops of bedrock any one of which account for the lower albedo. . . .

Topographic forms, as opposed to albedo markings, probably produced by eolian deposition have been recognized in only a few places. The most spectacular example is the field of transverse dune ridges covering more than 2,000 km^2 on the floor of a large crater at 48° S latitude and 330° longitude. . . . Other dune fields are found in the south polar area. Wind erosion has probably affected much of the surface. . . . Outside the polar region, probable wind-eroded forms are less striking. However, everywhere on Mars, ridges typically have concave slopes meeting at a sharp crest suggestive of sculpturing by the wind. The general softness of crater rims in contrast with those on the Moon further suggests eolian activity. The lineaments of the grooved terrain and less extensive patches of similar appearance elsewhere may be fractures etched out by wind scouring. . . .

Major Conclusions

Mariner 9 has shown that Mars is geologically far more heterogeneous than previously suspected from the earlier flyby missions. The analyses to date indicate convincingly that it has a geological style of its own different from that of either the Earth or Moon, and suggest that it represents a body intermediate in its evolutionary sequence somewhere between the Earth and Moon. It is now tempting to consider Mars as a planet that has partly made the transition from a relatively primitive impact-dominated (but not primordial) body like the Moon to an orogenically mobile, volcanically active, water-dominated planet like the Earth. Phobos and Deimos are clearly the most primitive solar system bodies closely investigated to date.

Like the Moon, Mars shows extensively cratered regions as well as numerous large circular basins. The basins, some of which are larger than any lunar basin, seem to exercise less control on the regional topography and distribution of the volcanic units than they do on the Moon.

The crater and basin terrains and the plains material that fills depressions in it are reminiscent of the Moon. But over much of the planet, the later parts of the Martian geological record are punctuated by huge and spectacular tectonic and volcanic features, not moonlike and only partly earthlike, that have destroyed or covered its earlier crater and basin aspect.

Extensive tectonic activity has occurred in huge regions of Mars. Much of this can be ascribed to circumferential tension in the upper parts of the lithosphere and to local doming. No shear or compressional features have been identified to date.

Volcanism has also played an important role in shaping the surface of Mars and probably has contributed in great part to its tenuous atmosphere. Martian volcanism is dramatically more varied and may span a larger part of the planet's history than lunar volcanic activity. Preliminary crater frequency studies point to the possibility that the major shield volcanos, calderas, plains, and other volcanic features could be relatively young.

Erosion and sedimentation, neither of them related to impact cratering, [have] occurred on a planetwide scale and surface modification processes are more widespread than previously envisioned. Erosion channels and depositional features abound, commonly related geographically and probably genetically to terrain that has collapsed chaotically. Extensive transport of materials has occurred in these channels, and moving fluids probably in episodic surges seem to be the only possible mechanism by which this can be explained, that is consistent with the present observational data.

Mars has clearly undergone a different proportionate mix of major surface-shaping processes than the Earth; the interplay between impact, volcanism, tectonism, and various erosion and sedimentation processes is clearly distinctive. Elucidation of these relations certainly will be the major fruit of the Mariner 1971 mission and should contribute significantly to a better understanding of the Earth.

Most markings observed by Mariner 9 seem to be surficial and of probable eolian origin. They are partly controlled by topographic features such as craters and scarps and appear to be excellent indicators of both past and recent wind regimes.

61 / Extrasolar X-Ray Sources

Soon after World War II scientists in the United States used surplus V-2 rockets, captured from the Germans, to loft instruments into space high above our x-ray-absorbing atmosphere. In this way, Herbert Friedman and colleagues at the U.S. Naval Research Laboratory in 1949 first detected X rays from the Sun.[22] By 1956 detectors aboard a "rockoon," a balloon-launched rocket, registered an x-ray background that appeared quite different from the radiation generated in the Earth's atmosphere by cosmic rays. Friedman suspected it was extraterrestrial, but confirmation had to await the development of more sensitive detectors capable of scanning large regions of the sky.[23]

That moment arrived on June 18, 1962, when Riccardo Giacconi, Herbert Gursky, and Frank Paolini of American Science and Engineering, Inc., and Massachusetts Institute of Technology physicist Bruno Rossi mounted geiger-counter detectors onto a small Aerobee rocket and launched it from the White Sands Missile Range in southern New Mexico for a short (350-second) flight. The official mission was to search for solar-induced x-ray fluorescence from the lunar surface, to help determine the Moon's composition. But Giacconi and Rossi had long suspected that extrasolar X rays would be radiating from such objects as supernova remnants, and they used the U.S. Air Force–funded test as an opportunity to look around. They were not disappointed; during the brief venture, the rocket-borne instrument not only detected a diffuse x-ray background but also noticed a huge flux of x-ray radiation emanating from the general vicinity of the galactic center. Later the source was better pinpointed to the Scorpius constellation. Sco X-1, as the mysterious source came to be labeled, blazed with an x-ray intensity far beyond what anyone expected based on the Sun's relatively low output—a hundred million times stronger than normal stellar sources. A few years later other astronomers only added to the mystery when they identified the visible object releasing the x-ray torrent. It turned out to be a faint and variable blue star.[24]

The first dedicated x-ray satellite, Uhuru, launched in 1970, found more than three hundred discrete sources like Sco X-1 within the Milky Way galaxy and discovered that many of these "x-ray stars" were highly variable. An explanation finally arrived in 1971, when the Uhuru team, led by Giacconi, studied one particular source, Cen X-3 in the southern constellation Centaurus, and found that its X rays were turning on and off every 4.8 seconds. Cen X-3 was the first known x-ray pulsar.[25] Moreover, every two days for a dozen hours the pulses virtually disappeared, evidence that led them to realize that Cen X-3 was actually part of a double-star system and was periodically eclipsed as it was orbiting a companion.[26] Harvey Tananbaum, who first noticed the cutoff, recalled that he and his colleagues did not immediately associate it with a binary star system. "Most of us in the early days of x-ray astronomy were physicists by training," he said, "[and] didn't have a sufficient background in astronomy."[27] The greatest number of bright x-ray sources in the galaxy turned out to be compact neutron stars in binary systems. Here the neutron star draws gas away from its companion, wrapping itself in an accretion disk and ultimately funneling the material toward its two magnetic poles. Most of the X rays are emitted as the gas swirls within the disk, accelerated to near the speed of light by the neutron star's intense gravitational field.

The new field of x-ray astronomy fully blossomed with the development of special space-borne telescopes that could focus the extremely short x-ray wavelengths and form pictures of its sources, much like an optical telescope. The first of these was the Einstein Observatory, launched by NASA in 1978. The Chandra telescope, put into operation in 1999, has a sensitivity a hundred million times greater than the rocket-borne detector that first discovered Sco X-1.

Study of the universe in X rays has turned out to be crucial in investigating its most compact stellar objects—neutron stars and black holes (see Chapters 41 and 42)—as well as the origin of the immense energies emitted from active galaxies and quasars.

"Evidence for X Rays from Sources Outside the Solar System." *Physical Review Letters*, Volume 9 (December 1, 1962) by Riccardo Giacconi, Herbert Gursky, Frank R. Paolini, and Bruno B. Rossi

Data from an Aerobee rocket carrying a payload consisting of three large area Geiger counters have revealed a considerable flux of radiation in the night sky that has been identified as consisting of soft x rays.

The entrance aperture of each Geiger counter consisted of seven individual mica windows comprising 20 cm^2 of area placed into one face of the counter. Two of the counters had windows of about 0.2-mil mica, and one counter had windows of 1.0-mil mica. The sensitivity of these detectors for x rays was between 2 and 8 Å, falling sharply at the extremes due to the transmission of the filling gas and the opacity of the windows, respectively. The mica was coated with lampblack to prevent ultraviolet light transmission. The three detectors were disposed symmetrically around the longitudinal axis of the rocket, the normal to each detector making an angle of 55° to that axis. Thus, during flight, the normal to the detectors swept through the sky, at a rate determined by the rotation of the rocket, forming a cone of 55° with respect to the longitudinal axis. . . . Each Geiger counter was placed in a well formed by an anticoincidence scintillation counter designed to reduce the cosmic-ray background. The experiment was intended to study fluorescence x rays produced on the lunar surface by x rays from the sun and to explore the night sky for other possible sources. On the basis of the known flux of solar x rays, we had estimated a flux from the moon of about 0.1 to 1 photon cm^{-2} sec^{-1} in the region of sensitivity of the counter.

The rocket launching took place at the White Sands Missile Range, New Mexico, at 2359 MST on June 18, 1962. The moon was one day past

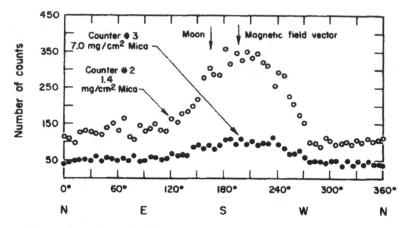

Figure 61.1: "Number of counts versus azimuth angle. The numbers represent counts accumulated in 350 seconds in each 6° angular interval."

full and was in the sky about 20° east of south and 35° above the horizon. The rocket reached a maximum altitude of 225 km and was above 80 km for a total of 350 seconds. The vehicle traveled almost due north for a distance of 120 km. Two of the Geiger counters functioned properly during the flight; the third counter apparently arced sporadically and was disregarded in the analysis. The optical aspect system functioned correctly. The rocket was spinning at 2.0 rps around the longitudinal axis. . . . Each complete rotation of the rocket was divided into sixty equal intervals, and the number of counts in each of these intervals was recorded separately.

The total data accumulated in this manner during the entire flight are shown in Fig. [61.1] for the operating Geiger counters. The observed region of the sky is shown in Fig. [61.2]. The counting rates show an altitude dependence on both the ascending and descending portions of the flight. . . . The rocket had begun tumbling during descent. The data in that portion of the flight are difficult to interpret and have not been included in the analysis.

The residual cosmic-ray background could not be determined directly. However, the strong angular dependence of the counting rate and the large difference between the counting rates of the counters provided with windows of different thickness clearly show that most of the recorded counts are due to a strongly anisotropic and very soft radiation. Thus, the possible existence of a small cosmic-ray effect is not an essential element in the discussion of the results.

The large peak that appears at about 195° in both counters shows that part of the recorded radiation is in the form of a well collimated beam. The fact that the counting rate does not go to zero on either side of the peak shows that this beam is superimposed on a diffuse background radiation. The background radiation itself is not isotropic, but appears to have a higher intensity in directions to the east of the peak than in the direction to the west of the peak, suggesting a secondary maximum centered around 60°. . . .

At the location where the measurements were obtained, the magnetic field has an inclination of 63° and a declination of 13° east of north. Thus, the field lines are at an azimuth of 193°, which is about the same as the azimuth of the observed radiation peak. This coincidence makes one wonder whether the radiation might not consist of charged particles spiraling along the field lines. . . . [But] the sharpness and azimuth of the observed peak requires that the pitch angles of these particles be very small. It is hard to find a reasonable source for particles with the required pitch angles at the location of our measurements. In particular, particles spilling out of the inner radiation belt ought to have a broad pitch-angle distribution. . . .

It is also clear that the radiation responsible for the asymmetry of the background cannot consist of charged particles. Thus, we conclude that the bulk of the observed radiation is not corpuscular, but electromagnetic in nature.

The counters were so constructed as to be insensitive to visible or ultraviolet light. The data themselves provide a definite test on this point since a strong visible light source, the moon, and two comparatively strong ultraviolet light sources, Virgo and presumably the moon, went through the field of view of the counters, and yet were not detected. Thus, if the radiation is electromagnetic, it must consist of soft x rays. . . .

[Omitted here is the authors' calculation of the wavelength and origin of the peak source, which they assume is a point source.]

. . . The peak appears to be due to a source emitting x rays of a 3 Å wavelength whose origin is about 10° above the horizon. The location of this source is shown as "source position" on the sky map in Fig. [61.2]. The measured flux from this source is 5.0 photons cm^{-2} sec^{-1}.

The diffuse character of the observed background radiation does not permit a positive determination of its nature and origin. However, the apparent absorption coefficient in mica and the altitude dependence is con-

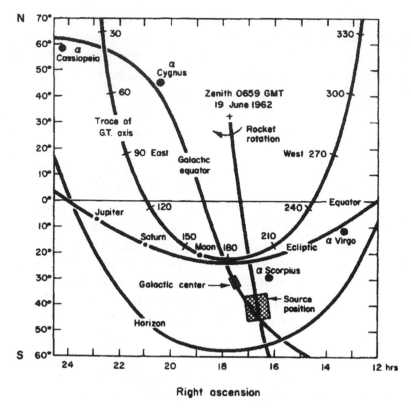

Figure 61.2: "Chart showing the portion of sky explored by the counters."

sistent with radiation of about the same wavelength as that responsible for the peak. Assuming the source lies close to the axis of the detectors, one obtains the intensity of the x-ray background as 1.7 photons cm^{-2} sec^{-1} sr^{-1} and of the secondary maximum (between 102° and 18°) as 0.6 photon cm^{-2} sec^{-1}. In addition, there seems to be a hard component to the background of about 0.5 cm^{-2} sec^{-1} sr^{-1} which does not show an altitude dependence and which is not eliminated by the anticoincidence.

The question arises whether the source of the observed x radiation could be associated with the earth's atmosphere and ascribed to some form of auroral activity. The rarity of occurrence of auroras of the magnitude required to account for the observed intensities at the latitude of the measurement makes this possibility very unlikely. . . .

From Fig. [61.2], showing the locations of the source as well as of the moon and planets, it is clear that the observed source does not coincide with any obvious scattering body belonging to our solar system. Further, the intensity of solar x radiation at the observed wavelength is much too low in this period of the solar cycle to account for the observed intensities of the peak or of the background on the basis of back-scattered solar radiation. It would thus appear that the radiation does not originate in our solar system.

From Fig. [61.2] we see that the main apparent source is in the vicinity of the galactic center at a G.T. azimuthal angle of about 195°. We also see that the trace of the G.T. axis lies close to the galactic equator for a value of the azimuthal angle near 40°, which is the region where the background radiation is recorded with greater intensity. This apparent maximum of the background radiation is the general region of the sky where two peculiar objects—Cassiopeia *A* and Cygnus *A*—are located. It is perhaps significant that both the center of the galaxy where the main apparent source of x rays lies, and the region of Cassiopeia *A* and Cygnus *A* where there appears to be a secondary x-ray source, are also regions of strong radio emission. Clark has pointed out that the probable mechanism for the production of the nonthermal component of the radio noise, namely, synchrotron radiation from cosmic electrons in the galactic magnetic fields, can also give rise to the x rays we observe.*

In the cosmic-ray air shower experiment presently being carried out in Bolivia, tentative evidence has been obtained for the existence of cosmic γ rays in the energy region of 10^{14} eV at a rate of 10^{-3}–10^{-4} of the charged cosmic-ray flux at the same energy with an indication of enhanced emission in the galactic plane. Clark has shown that cosmic electrons must be produced along with γ-rays by the decay of mesons that arise in the interactions of cosmic rays with interstellar matter. Since electrons at these energies lose their energy predominantly via synchrotron radiation in the galactic magnetic field, one should observe roughly the same total energy in synchrotron radiation at the earth as in γ-ray energy. For electrons of 2×10^{14} eV in a field of 3×10^{-6} gauss, the peak of the synchrotron emission is at 3 Å; in a stronger field this will happen at lower electron energies. It has been shown that x rays in this wavelength region are not appreciably absorbed over interstellar distances.

With this one experiment it is impossible to completely define the nature and origin of the radiation we have observed. Even though the statistical precision of the measurement is high, the numerical values for the

* MIT physicist George W. Clark.

derived quantities and angles are subject to large variation depending on the choice of assumptions. However, we believe that the data can best be explained by identifying the bulk of the radiation as soft x rays from sources outside the solar system. Synchrotron radiation by cosmic electrons is a possible mechanism for the production of these x rays. Ordinary stellar sources could also contribute a considerable fraction of the observed radiation.

62 / Quasars

T hrough the 1940s and 1950s astronomers began to gather evidence that galaxies were not necessarily the serene and graceful objects long imaged. By 1963 this lesson was conveyed with utmost drama when Maarten Schmidt realized, to his surprise, that an unusual radio-emitting "star" called 3C 273 was actually a distant galaxy radiating from its center the energy of a hundred normal galaxies. Schmidt had discovered the first quasar.

Radio-wave observations provided the first clues toward this discovery. As the new field of radio astronomy expanded after World War II, radio astronomers detected a number of sources across the sky. Once radio telescopes attained better resolutions, astronomers were able to link some of these sources of radio emission to specific visible objects, such as the Crab nebula and Cassiopeia A, the remnants of exploded stars (see Chapter 56). In 1954 a source known as Cygnus A was associated with an unusual galaxy, the first time that a radio signal was linked to an object outside the Milky Way. There were also several dozen pointlike sources. A decade earlier Carl Seyfert, then at the Mount Wilson Observatory, had studied spirals with particularly bright nuclei (now known as Seyfert galaxies), and theorist Thomas Gold suggested that such galaxies, if far enough away, might appear as discrete radio sources.[28] But, by and large, radio astronomers figured that most of these pointlike sources were likely invisible "radio stars" situated within our galaxy.

By the 1960s the positions of several of these compact radio sources were pinned down more accurately, allowing optical astronomers to photograph them in visible light. To all appearances they looked like blue stars, except that they exhibited spectral features unlike any star ever observed. Astronomers wondered whether unknown elements or unusual stellar conditions were involved. 3C 273 (the 273rd listing in the third Cambridge catalog of radio sources) proved to be the Rosetta Stone in breaking the code. In 1962, 3C 273 was occulted by the Moon, an eclipse that allowed Australian radio astronomers Cyril Hazard, M. Mackey, and A. Shimmins to peg its position on the sky to within one second of arc, precision enough to see that 3C 273 was a telescopically bright (thirteenth-magnitude) starlike object. At the end of that year Maarten Schmidt at the California Institute of Technology used the 200-inch telescope on Palomar mountain, then the world's largest reflector, to obtain an optical spectrum of this unusual blue star. For weeks Schmidt was baffled by his results. But then on February 5, 1963, sitting at his office desk with the optical spectrum of 3C 273 spread before him, along with a crucial near-infrared line obtained by Caltech astronomer J. Beverley Oke, he at last recognized a familiar pattern that had eluded him. He saw the well-known emission lines of hydrogen (the Balmer series), only the lines were in the wrong place. The spectral bands had been collectively shifted toward the red end of the spectrum by 16 percent. "Astronomers had never seen a *starlike* object in the sky with a large redshift," he explained. "It wasn't even imagined."[29]

Schmidt immediately grasped the implications of this enormous shift: 3C 273 was not an unusual star in our galaxy but rather an extraordinarily luminous object located nearly 2 billion light-years away, rushing into space at some 30,000 miles per second as it is carried outward—its light waves stretched—with the expansion of the universe. Excited by his discovery, Schmidt quickly conferred with his Caltech colleague Jesse Greenstein. They soon confirmed that the spectrum of another puzzling radio star, 3C 48, was redshifted even more, by 37 percent, which translates into a distance of about 3.5 billion light-years. Such starlike extragalactic objects were soon christened quasi-stellar radio sources, or quasars.

Astronomers avidly debated the origin of such huge galactic energies emanating from a compact region. At first it was suggested that a series of supernovae were exploding in a galaxy's nucleus, one explosion triggering another in a vast chain reaction. For many years a small but vocal group argued that the quasars were not situated at cosmological distances at all but instead were dense objects set closer in — the redshifts caused by the object's light waves stretching as they escaped a particularly intense gravitational field (an explanation that Schmidt considered in his discovery report but found problematic). Or perhaps, they said, a new principle of physics was at work.

By the end of the century, evidence from optical, x-ray, and radio telescopes was confirming that quasar activity arises from a supermassive black hole residing in the center of a young, gas-filled galaxy. The vast energies are released as matter spirals in toward the black hole and by the spinning hole itself acting as a powerful dynamo. Quasars are now observed back to a time when the universe was less than a billion years old.

"3C 273: A Star-Like Object with Large Red-Shift."
Nature, Volume 197 (March 16, 1963)
by Maarten Schmidt

The only objects seen on a 200-in. plate near the positions of the components of the radio source 3C 273 reported by Hazard, Mackey and Shimmins . . . are a star of about thirteenth magnitude and a faint wisp or jet.[30] The jet has a width of 1″–2″ and extends away from the star in position angle 43°. It is not visible within 11″ from the star and ends abruptly at 20″ from the star. The position of the star, kindly furnished by Dr. T. A. Matthews, is R.A. 12h 26m 33.35s ± 0.04s, Decl. + 2° 19′ 42.0″ ± 0.5″ (1950), or 1″ east of component B of the radio source.* The end of the jet is 1″ east of component A. The close correlation between the radio structure and the star with the jet is suggestive and intriguing.

Spectra of the star were taken with the prime-focus spectrograph at the 200-in. telescope with dispersions of 400 and 190 Å [angstrom] per mm. They show a number of broad emission features on a rather blue continuum. The most prominent features, which have widths around 50 Å, are, in

---

* Caltech radio astronomer Thomas A. Matthews.

order of strength, at 5632, 3239, 5792, 5032 Å. These and other weaker emission bands are listed in the first column of Table [62.1]. For three faint bands with widths of 100–200 Å the total range of wavelength is indicated.

The only explanation found for the spectrum involves a considerable redshift. A redshift $\Delta\lambda/\lambda_0$ of 0.158 allows identification of four emission bands as Balmer [hydrogen] lines, as indicated in Table [62.1]. Their relative strengths are in agreement with this explanation. Other identifications based on the above redshift involve the Mg II lines around 2798 Å, thus far only found in emission in the solar chromosphere, and a forbidden line of [O III] at 5007 Å. On this basis another [O III] line is expected at 4959 Å with a strength one-third of that of the line at 5007 Å. Its detectability in the spectrum would be marginal. A weak emission band suspected at 5705 Å, or 4927 Å reduced for redshift, does not fit the wavelength. No explanation is offered for the three very wide emission bands.

It thus appears that six emission bands with widths around 50 Å can be explained with a redshift of 0.158. The differences between the observed and the expected wavelengths amount to 6 Å at the most and can be entirely understood in terms of the uncertainty of the measured wavelengths. The present explanation is supported by observations of the infrared spectrum communicated by Oke[31] . . . and by the spectrum of another star-like object associated with the radio source 3C 48 discussed by Greenstein and Matthews in another communication.[32]

The unprecedented identification of the spectrum of an apparently stellar object in terms of a large redshift suggests either of the two following explanations.

Table 62.1: Wavelengths and Identifications

λ	$\lambda/1.158$	λ_0	
3239	2797	2798	Mg II
4595	3968	3970	Hε
4753	4104	4102	Hδ
5032	4345	4340	Hγ
5200–5415	4490–4675		
5632	4864	4861	Hβ
5792	5002	5007	[O III]
6005–6190	5186–5345		
6400–6510	5527–5622		

(1) The stellar object is a star with a large gravitational redshift. Its radius would then be of the order of 10 km. Preliminary considerations show that it would be extremely difficult, if not impossible, to account for the occurrence of permitted lines and a forbidden line with the same redshift with widths of only 1 or 2 percent of the wavelength.

(2) The stellar object is the nuclear region of a galaxy with a cosmological redshift of 0.158, corresponding to an apparent velocity of 47,400 km/sec. The distance would be around 500 megaparsecs, and the diameter of the nuclear region would have to be less than 1 kiloparsec. This nuclear region would be about 100 times brighter optically than the luminous galaxies which have been identified with radio sources thus far. If the optical jet and component A of the radio source are associated with the galaxy, they would be at a distance of 50 kiloparsecs, implying a timescale in excess of 10^5 years. The total energy radiated in the optical range at constant luminosity would be of the order of 10^{59} ergs.

Only the detection of an irrefutable proper motion or parallax would definitively establish 3C 273 as an object within our Galaxy. At the present time, however, the explanation in terms of an extragalactic origin seems most direct and least objectionable.

63 / Evidence for the Big Bang

Casual readers of the July 1, 1965, issue of the *Astrophysical Journal* could easily have missed the article. Its title was fairly prosaic ("A Measurement of Excess Antenna Temperature at 4080 Mc/s"), and it was tucked away at the very end of the journal. Yet the one-page report was conveying one of the most important discoveries in twentieth-century astronomy. In their eight short paragraphs plus a note, Bell Telephone Laboratory researchers Arno Penzias and Robert Wilson were announcing the first definitive evidence for the Big Bang. The excess energy they had detected with their radio antenna—a background of microwaves filling the celestial sky—unexpectedly turned out to be the remnant echo of the explosive creation of our universe.

After Edwin Hubble discovered in 1929 that the cosmos was expanding (see Chapter 52), attention was soon focused on its origin. The expansion implied a beginning, a unique moment when space and time first emerged. Some scientists were unsettled by this notion of an initial cosmic bang and so offered an alternative cosmological theory: a universe in steady-state expansion, with neither a beginning nor an end (see Chapter 38). As the galaxies receded from one another, matter was spontaneously generated to form new galaxies, which meant the universe of the past would look very much like the universe of today. This was in stark contrast to the predicted behavior of a Big Bang universe, where galaxies would age and evolve over the eons. Moreover, a primeval explosion would have released a flood of energetic photons that eventually cooled down with the expansion and would currently appear as a faint glow of microwave radiation.

Ralph Alpher and Robert Herman, while working with George Gamow in the 1940s on models of element production in the Big Bang, were the first to suggest the presence of a cosmic microwave background. They figured that the overall temperature of the waning cosmic fire would by now have dropped to within several degrees of absolute zero (see Chapter 45). But no one followed up on the prediction, even though it provided a clear-cut means of deciding between the two opposing theories of creation. Cosmological tests were not a high priority in radio astronomy at the time, and so their calculation was eventually forgotten.

The idea did not resurface until the 1960s, when Robert Dicke and P. James E. Peebles at Princeton University, as well as Yakov Zel'dovich in the Soviet Union, again reasoned that residual heat from the Big Bang must be permeating the universe. Peebles figured it was "as low as 3.5° K."[33] Dicke and several colleagues began constructing the equipment to measure this radiation, but in the process of setting up their antenna on a campus rooftop they learned that they had been scooped; by chance, two radio astronomers with Bell Labs had already been listening to the weak cosmic hiss.

In 1964 Penzias and Wilson had begun to calibrate a massive horn-shaped antenna, three stories high, located in Holmdel, New Jersey, not far from the site where Karl Jansky first detected celestial radio waves. They were converting the receiver, originally used for satellite communications, into a radio telescope to study our galaxy. During their initial tests, they consistently registered an excess signal, no matter where the instrument was pointed. They spent months investigating possible sources, from atmospheric radiation to electromagnetic noises emanating from nearby New York City. They even cleaned up pigeon droppings within the antenna to rule out biological interference.

Despairing that he and Wilson would ever locate the origin of the noise, Penzias chanced to mention the problem to a friend, who knew of the plan under way at Princeton to search for a cosmic microwave background. Penzias soon invited the Princeton group to visit the Holmdel installation, just a few dozen miles from the university, whereupon it was confirmed that Penzias and Wilson had indeed been listening to the faint reverberation of the Big Bang all along. Their initial report focused solely on all the possible sources contributing to their reception. They refer to a companion article by Dicke and his colleagues in the same journal issue for an explanation of the cosmological consequences of "the remaining unaccounted-for antenna temperature."

Others had actually detected the cosmic noise earlier. In 1941 the Canadian astronomer Andrew McKellar recognized that cyanogen (CN) molecules in space were being energized by a thermal background of about 2.3 K, but he failed to understand the implications of his find.[34] Others wrote off the extra heat as a systematic error in their instrumentation.

Since Penzias and Wilson's discovery, the cosmic microwave background has been measured from both the ground and space with ever finer precision. The Cosmic Background Explorer satellite (COBE), launched by NASA in 1989, detected a Big Bang afterglow of 2.7 degrees above absolute zero on the Kelvin scale, with an uncertainty of only 0.01 K.[35]

"A Measurement of Excess Antenna Temperature at 4080 Mc/s." *Astrophysical Journal*, Volume 142 (July 1, 1965) by Arno A. Penzias and Robert W. Wilson

Measurements of the effective zenith noise temperature of the 20-foot horn-reflector antenna at the Crawford Hill Laboratory, Holmdel, New Jersey, at 4080 Mc/s [million cycles per second] have yielded a value about 3.5° K higher than expected. This excess temperature is, within the limits of our observations, isotropic, unpolarized, and free from seasonal variations (July, 1964–April, 1965). A possible explanation for the observed excess noise temperature is the one given by Dicke, Peebles, Roll, and Wilkinson in a companion letter in this issue.[36]

The total antenna temperature measured at the zenith is 6.7° K of which 2.3° K is due to atmospheric absorption. The calculated contribution due to ohmic losses in the antenna and back-lobe response is 0.9° K.

The radiometer used in this investigation has been described elsewhere.[37] It employs a traveling-wave maser, a low-loss (0.027-db) comparison switch, and a liquid helium-cooled reference termination. Measurements were made by switching manually between the antenna input and the reference termination. The antenna, reference termination, and radiometer were well matched so that a round-trip return loss of more than 55 db existed throughout the measurement; thus errors in the measurement of the effective temperature due to impedance mismatch can be neglected. The estimated error in the measured value of the total antenna temperature is 0.3° K and comes largely from uncertainty in the absolute calibration of the reference termination.

The contribution to the antenna temperature due to atmospheric absorption was obtained by recording the variation in antenna temperature with elevation angle and employing the secant law. The result, 2.3° ± 0.3° K, is in good agreement with published values.

The contribution to the antenna temperature from ohmic losses is computed to be 0.8° ± 0.4° K. In this calculation we have divided the antenna into three parts: (1) two non-uniform tapers approximately 1 m in total length which transform between the 2⅛-inch round output waveguide and the 6-inch-square antenna throat opening; (2) a double-choke rotary joint located between these two tapers; (3) the antenna itself. Care was taken to clean and align joints between these parts so that they would not significantly increase the loss in the structure. Appropriate tests were made for leakage and loss in the rotary joint with negative results.

The possibility of losses in the antenna horn due to imperfections in its seams was eliminated by means of a taping test. Taping all the seams in the section near the throat and most of the others with aluminum tape caused no observable change in antenna temperature.

The back-lobe response to ground radiation is taken to be less than 0.1° K for two reasons: (1) Measurements of the response of the antenna to a small transmitter located on the ground in its vicinity indicate that the average back-lobe level is more than 30 db below isotropic response. The horn-reflector antenna was pointed to the zenith for these measurements, and complete rotations in azimuth were made with the transmitter in each of ten locations using horizontal and vertical transmitted polarization from each position. (2) Measurements on smaller horn-reflector antennas at these laboratories, using pulsed measuring sets on flat antenna ranges, have consistently shown a back-lobe level of 30 db below isotropic response. Our larger antenna would be expected to have an even lower back-lobe level.

From a combination of the above, we compute the remaining unaccounted-for antenna temperature to be $3.5° \pm 1.0°$ K at 4080 Mc/s. In connection with this result it should be noted that DeGrasse *el al.*[38] and Ohm[39] give total system temperatures at 5650 Mc/s and 2390 Mc/s, respectively. From these it is possible to infer upper limits to the background temperatures at these frequencies. These limits are, in both cases, of the same general magnitude as our value.

Note added in proof.—The highest frequency at which the background temperature of the sky had been measured previously was 404 Mc/s, where a minimum temperature of 16° K was observed. Combining this value with our result, we find that the average spectrum of the background radiation over this frequency range can be no steeper than $\lambda^{0.7}$. This clearly eliminates the possibility that the radiation we observe is due to radio sources of types known to exist, since in this event, the spectrum would have to be very much steeper.

64 / Pulsars

T he first neutron star was found in 1967, a discovery that came
as a complete surprise. No one had ever imagined that a
compact star—a mere dozen miles wide—would be emitting
clocklike radio pulses. "No event in radio astronomy seemed more
astonishing and more nearly approaching science fiction," said the
British radio-astronomy pioneer James S. Hey.[40]

That neutron stars might exist was not unforeseen. Theorists had
been contemplating their creation since the 1930s; they were pic-
tured as the compressed stellar cores left behind after supernova
explosions (see Chapter 41). But no one seriously thought neutron
stars would be detected, since calculations showed them to be so
small. In a 1966 paper on such "superdense stars," general relativist
John Archibald Wheeler figured the only opportunity to catch sight
of this "nuclear matter in bulk" would be immediately after a stellar
explosion (a rare event in our galaxy), when the remnant core was
still hot and fiercely emitting a host of electromagnetic radiations.[41]

But within a year this opinion was completely overturned by
British radio astronomers, who stumbled upon a neutron star by
accident. Their report in the journal *Nature*, terse and dense with
scientific data, offers few details on the serendipity that led to their
finding. A small platoon of students and technicians, led by Cambridge University radio astronomer Antony Hewish, had just completed the construction of a sprawling radio telescope near the

university: more than 2,000 dipole antennas, lined up like rows of corn and connected by dozens of miles of wire. While the celestial sky moved overhead, the telescope passively searched for fast variations in the intensities of radio sources, an indication that they were of small angular size, possibly quasars. The data continually registered on a strip-chart recorder, and it was Jocelyn Bell's job to analyze the long stream of paper—400 feet a week—for her doctoral thesis. Reviewing the output one day, she noticed "there was a little bit of what I call 'scruff,' which didn't look exactly like [man-made] interference and didn't look exactly like [quasar] scintillation. . . . I began to remember that I had seen some of this unclassifiable scruff before, and what's more, I had seen it from the same patch of sky."[42]

Eventually observing it with a higher-speed recording, Bell (later Bell Burnell) came to see that the scruff was actually a methodical succession of pulses spaced 1.3 seconds apart. The unprecedented precision caused Hewish and his group to briefly label the source LGM for "little green men," a jesting nod to the possibility that the regular pulsations might be coming from an extraterrestrial-built beacon. But within a few months, the team uncovered three more rhythmical signals in different regions of the sky. Inspired by the name of the recently discovered quasars (see Chapter 62), a British journalist dubbed them *pulsars*, for pulsating stars, a label that astronomers swiftly adopted. Discovery of the exotic pulsings had to await Hewish's new telescope, which, unlike previous radio telescopes, had been specifically designed to examine celestial radio waves over short time intervals.

In their discovery report Hewish, Bell, and their colleagues right away pointed out that the exceedingly brief duration of the *beep* itself—around a hundredth of a second—meant that the source could span no more than 5,000 kilometers (around the distance light can travel in a hundredth of a second, close to the width of the planet Mercury). This suggested the pulsar was either a white dwarf or a neutron star. They wondered whether the entire star was pulsating in and out, with the radiation then "likened to radio bursts from a solar flare occurring over the entire star during each cycle of the oscillation." Within a year, though, Cornell University theorist Thomas Gold developed the model that best explained a pulsar's behavior: the neutron star, a highly magnetized body that is rapidly spinning, transfers its rotational energy into electromagnetic energy that is beamed outward like a lighthouse beacon from its north and south magnetic poles. Depending on the pulsar's alignment with Earth, we observe either one or two blips of radio energy with each pulsar rotation.

Since neutron stars can spin quite fast, Gold predicted that radio astronomers should also detect pulsars with shorter periods than those first discovered. This was successfully confirmed when astronomers found extremely fast-spinning pulsars within the Vela and Crab nebulas, with periods of .089 and .033 second, respectively.[43] Since each nebula was a supernova remnant, these finds also validated the idea that neutron stars would be found at the sites of stellar explosions. In the concluding remarks of his paper, Gold further anticipated that a pulsar would slow down over time, as it depletes its rotational energy. This was first demonstrated with the Crab nebula pulsar, when its spin was measured to be diminishing ever so slightly with each passing year.[44]

"Observation of a Rapidly Pulsating Radio Source."
Nature, Volume 217 (February 24, 1968)
by Antony Hewish, S. Jocelyn Bell, John D. H. Pilkington, Paul Frederick Scott, and Robin Ashley Collins

In July 1967, a large radio telescope operating at a frequency of 81.5 MHz was brought into use at the Mullard Radio Astronomy Observatory. This instrument was designed to investigate the angular structure of compact radio sources by observing the scintillation caused by the irregular structure of the interplanetary medium. The initial survey includes the whole sky in the declination range $-08° < \delta < 44°$ and this area is scanned once a week. A large fraction of the sky is thus under regular surveillance. Soon after the instrument was brought into operation it was noticed that signals which appeared at first to be weak sporadic interference were repeatedly observed at a fixed declination and right ascension; this result showed that the source could not be terrestrial in origin.

Systematic investigations were started in November and high speed records showed that the signals, when present, consisted of a series of pulses each lasting ~0.3 s[econd] and with a repetition period of about 1.337 s which was soon found to be maintained with extreme accuracy. Further observations have shown that the true period is constant to better than 1 part in 10^7 although there is a systematic variation which can be ascribed to the orbital motion of the Earth. The impulsive nature of the recorded signals is caused by the periodic passage of a signal of descending frequency through the 1 MHz pass band of the receiver.

The remarkable nature of these signals at first suggested an origin in terms of man-made transmissions which might arise from deep space probes, planetary radar or the reflection of terrestrial signals from the Moon. None of there interpretations can, however, be accepted because the absence of any parallax shows that the source lies far outside the solar system. A preliminary search for further pulsating sources has already revealed the presence of three others having remarkably similar properties which suggests that this type of source may be relatively common at a low flux density. A tentative explanation of these unusual sources in terms of the stable oscillations of white dwarf or neutron stars is proposed.

Position and Flux Density

The aerial consists of a rectangular array containing 2,048 full-wave dipoles arranged in sixteen rows of 128 elements. Each row is 470 m long in an E.-W. direction and the N.-S. extent of the array is 45 m. Phase-scanning is employed to direct the reception pattern in declination and four receivers are used so that four different declinations may be observed simultaneously. Phase-switching receivers are employed and the two halves of the aerial are combined as an E.-W. interferometer. . . . For detailed studies of the pulsating source a time constant of 0.05 s was usually employed and the signals were displayed on a multi-channel "Rapidgraph" pen recorder with a time constant of 0.03 s. Accurate timing of the pulses was achieved by recording second pips derived from the *MSF* Rugby time transmissions.

A record obtained when the pulsating source was unusually strong is shown in Figure [64.1]. This clearly displays the regular periodicity and also the characteristic irregular variation of pulse amplitude. On this occasion the largest pulses approached a peak flux density (averaged over the 1 MHz pass band) of 20×10^{-26} W m^{-2} Hz^{-1}, although the mean flux density integrated over one minute only amounted to approximately 1.0×10^{-26} W m^{-2} Hz^{-1}. On a more typical occasion the integrated flux density would be several times smaller than this value. It is therefore not surprising that the

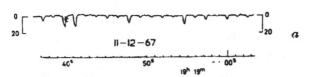

Figure 64.1: "A record of the pulsating radio source in strong signal conditions . . ."

source has not been detected in the past, for the integrated flux density falls well below the limit of previous surveys at meter wavelengths. . . .

Pulse Recurrence Frequency and Doppler Shift

By displaying the pulses and time pips from *MSF* Rugby on the same record the leading edge of a pulse of reasonable size may be timed to an accuracy of about 0.1 s. Observations over a period of 6 h[ours] taken with the tracking system mentioned earlier gave the period between pulses as $P_{obs} = 1.33733 \pm 0.00001$ s. . . . The true periodicity of the source, making allowance for the Doppler shift [due to Earth's motion] . . . is then

$$P_0 = 1.3372795 \pm 0.0000020 \text{ s}$$

. . . It is also interesting to note the possibility of detecting a variable Doppler shift caused by the motion of the source itself. Such an effect might arise if the source formed one component of a binary system, or if the signals were associated with a planet in orbit about some parent star. For the present . . . there is no evidence for an additional orbital motion comparable with that of the Earth.

The Nature of the Radio Source

The lack of any parallax greater than about 2′ places the source at a distance exceeding 10^3 A.U. The energy emitted by the source during a single pulse, integrated over 1 MHz at 81.5 MHz, therefore reaches a value which must exceed 10^{17} erg if the source radiates isotropically. It is also possible to derive an upper limit to the physical dimension of the source. The small instantaneous bandwidth of the signal (80 kHz) and the rate of sweep (-4.9 MHz s^{-1}) show that the duration of the emission at any given frequency does not exceed 0.016 s. The source size therefore cannot exceed 4.8×10^3 km. . . .

The positional accuracy so far obtained does not permit any serious attempt at optical identification. The search area, which lies close to the galactic plane, includes two twelfth magnitude stars and a large number of weaker objects. In the absence of further data, only the most tentative suggestion to account for these remarkable sources can be made.

The most significant feature to be accounted for is the extreme regularity of the pulses. This suggests an origin in terms of the pulsation of an entire star, rather than some more localized disturbance in a stellar atmosphere. In this connection it is interesting to note that it has already been suggested that the radial pulsation of neutron stars may play an important part in the history of supernovae and supernova remnants. . . .

If the radiation is to be associated with the radial pulsation of a white dwarf or neutron star there seem to be several mechanisms which could account for the radio emission. It has been suggested that radial pulsation would generate hydromagnetic shock fronts at the stellar surface which might be accompanied by bursts of X-rays and energetic electrons. The radiation might then be likened to radio bursts from a solar flare occurring over the entire star during each cycle of the oscillation. Such a model would be in fair agreement with the upper limit of $\sim 5 \times 10^3$ km for the dimension of the source. . . .

More observational evidence is clearly needed in order to gain a better understanding of this strange new class of radio source. If the suggested origin of the radiation is confirmed further study may be expected to throw valuable light on the behavior of compact stars and also on the properties of matter at high density.

"Rotating Neutron Stars as the Origin of the Pulsating Radio Sources." *Nature*, Volume 218 (1968) by Thomas Gold

Abstract—The constancy of frequency in the recently discovered pulsed radio sources can be accounted for by the rotation of a neutron star. Because of the strong magnetic fields and high rotation speeds, relativistic velocities will be set up in any plasma in the surrounding magnetosphere, leading to radiation in the pattern of a rotating beacon.

The case that neutron stars are responsible for the recently discovered pulsating radio sources appears to be a strong one. No other theoretically known astronomical object would possess such short and accurate period-icities as those observed, ranging from 1.33 to 0.25 s. Higher harmonics of a lower fundamental frequency that may be possessed by a white dwarf have been mentioned; but the detailed fine structure of several short pulses repeating in each repetition cycle makes any such explanation very unlikely. Since the distances are known approximately from interstellar dispersion of the different radio frequencies, it is clear that the emission per unit emitting volume must be very high; the size of the region emitting any one pulse can, after all, not be much larger than the distance light travels in the few milliseconds that represent the lengths of the individual pulses. No such concentrations of energy can be visualized except in the presence of an intense gravitational field.

The great precision of the constancy of the intrinsic period also suggests that we are dealing with a massive object, rather than merely with some plasma physical configuration. Accuracies of one part in 10^8 belong to the realm of celestial mechanics of massive objects, rather than to that of plasma physics.

It is a consequence of the virial theorem that the lowest mode of oscillation of a star must always have a period which is of the same order of magnitude as the period of the fastest rotation it may possess without rupture. The range of 1.5 s to 0.25 s represents periods that are all longer than the periods of the lowest modes of neutron stars. They would all be periods in which a neutron star could rotate without excessive flattening. . . . If the rotation period dictates the repetition rate, the fine structure of the observed pulses would represent directional beams rotating like a lighthouse beacon. . . .

There are as yet not really enough clues to identify the mechanism of radio emission. It could be a process deriving its energy from some source of internal energy of the star, and thus as difficult to analyze as solar activity. But there is another possibility, namely, that the emission derives its energy from the rotational energy of the star (very likely the principal remaining energy source), and is a result of relativistic effects in a co-rotating magnetosphere.

In the vicinity of a rotating star possessing a magnetic field there would normally be a co-rotating magnetosphere. Beyond some distance, external influences would dominate, and co-rotation would cease. In the case of a fast rotating neutron star with strong surface fields, the distance out to which co-rotation would be enforced may well be close to that at which co-rotation would imply motion at the speed of light. The mechanism by which the plasma will be restrained from reaching the velocity of light will be that of radiation of the relativistically moving plasma, creating a radiation reaction adequate to overcome the magnetic force. The properties of such a relativistic magnetosphere have not yet been explored, and indeed our understanding of relativistic magnetohydrodynamics is very limited. In the present case the coupling to the electromagnetic radiation field would assume a major role in the bulk dynamical behavior of the magnetosphere.

The evidence so far shows that pulses occupy about $\frac{1}{30}$ of the time of each repetition period. This limits the region responsible to dimensions of the order of $\frac{1}{30}$ of the circumference of the "velocity of light circle." In the radial direction equally, dimensions must be small; one would suspect small enough to make the pulse rise-times comparable with or larger than the flight time of light across the region that is responsible. This would imply that the radiation emanates from the plasma that is moving within 1 per cent of the velocity of light. That is the region of velocity where radiation effects would in any case be expected to become important.

The axial asymmetry that is implied needs further comment. A magnetic field of a neutron star may well have a strength of 10^{12} gauss at the surface of the 10 km object. At the "velocity of light circle," the circumference of which for the observed periods would range from 4×10^{10} to 0.75×10^{10} cm, such a field will be down to values of the order of 10^3–10^4 gauss (decreasing with distance slower than the inverse cube law of an undisturbed dipole field. A field pulled out radially by the stress of the centrifugal force of a whirling plasma would decay as an inverse square law with radius). Asymmetries in the radiation could arise either through the field or the plasma content being non-axially symmetric. A skew and non-dipole field may well result from the explosive event that gave rise to the neutron star; and the access to plasma of certain tubes of force may be dependent on surface inhomogeneities of the star where sufficiently hot or energetic plasma can be produced to lift itself away from the intense gravitational field (10–100 MeV for protons; much less for space charge neutralized electron-positron beams).

The observed distribution of amplitudes of pulses makes it very unlikely that a modulation mechanism can be responsible for the variability . . . but rather the effect has to be understood in a variability of the emission mechanism. In that case the observed very sharp dependence of the instantaneous intensity on frequency (1 MHz change in the observation band gives a substantially different pulse amplitude) represents a very narrow-band emission mechanism, much narrower than synchrotron emission, for example. A coherent mechanism is then indicated, as is also necessary to account for the intensity of the emission per unit area that can be estimated from the lengths of the sub-pulses. Such a coherent mechanism would represent non-uniform static configurations of changes in the relativistically rotating region. Non-uniform distributions at rest in a magnetic field are more readily set up and maintained than in the case of high individual speeds of charges, and thus the configuration discussed here may be particularly favorable for the generation of a coherent radiation mechanism.

If this basic picture is the correct one it may be possible to find a slight, but steady, slowing down of the observed repetition frequencies. Also, one would then suspect that more sources exist with higher rather than lower repetition frequency, because the rotation rates of neutron stars are capable of going up to more than 100/s and the observed periods would seem to represent the slow end of the distribution.

65 / The Infrared Sky and the Galactic Center

A stronomers had been attempting to gather data in the infrared region of the electromagnetic spectrum since 1800, when the British astronomer William Herschel passed sunlight through a prism and first noted that there were invisible rays, just beyond the red end of the spectrum, that heated his thermometer. In the summer of 1856, while on an expedition to the peak of Teneriffe in the Canary archipelago to test the merits of observing at high altitudes, Scotland's astronomer royal, C. Piazzi Smyth, used a thermocouple and detected infrared radiation from the Moon.[45]

Toward the end of the nineteenth century, Thomas Edison invented an infrared sensor that he took out to Wyoming in 1878 to observe a solar eclipse. He apparently measured the infrared radiance of the solar corona, although "science seems to have forgotten both the detector and the measurement," reports solar physicist John Eddy.[46] Planets and stars were studied in the infrared in the first half of the twentieth century, but infrared astronomy did not gain real momentum until the 1960s, when astronomers were at last able to adopt and refine sensitive infrared instrumentation first developed by the military for wartime use.

A turning point for infrared astronomy arrived with a small two-page note in the *Astrophysical Journal* in 1965. Caltech astronomers Gerry Neugebauer, D. E. Martz, and Robert Leighton had begun the first overall survey of the infrared sky with a specially built telescope on nearby Mount Wilson and announced the detection of extremely cool stars—so cool that some were invisible to optical telescopes. "Since most of these 'superred' stars occur in the Milky Way," they reported, "interstellar reddening may be of some consequence; but in at least a few cases the stars seem to be intrinsically extremely red."[47] Many of the infrared sources turned out to be older stars immersed in cocoons of dust and gas ejected by the stars. The survey produced the first catalog of infrared stars—some fifty-five hundred sources.

The true power of the new technology was revealed when Eric Becklin and Neugebauer went on to examine the center of the Milky Way galaxy in 1966 with an infrared photometer mounted at separate times on the 24-, 60-, and 200-inch reflecting telescopes at the Mount Wilson and Palomar observatories, which allowed them to see through the obscuring curtain of interstellar dust that keeps the galactic nucleus hidden to optical telescopes. What they discovered was the first hint of the densest collection of stars in the central region of our galaxy, a site that radio astronomers had earlier labeled Sagittarius A (Sgr A). Their work verified that Sgr A was indeed the very center of our galaxy. In 1974 Bruce Balick and Robert Brown used a radio interferometer at the National Radio Astronomy Observatory in West Virginia to look more deeply into Sgr A and found within that region a bright and compact source of radio radiation, which came to be known as Sgr A* [pronounced "Sadge A-star"].[48] Over the succeeding decades radio, x-ray, and infrared telescopes have peered even closer in, gathering evidence that Sgr A* is a supermassive black hole containing the mass of around four million Suns in a region likely smaller than Mars's orbit around the Sun. Within a few light-years of its position are some ten million stars orbiting this central mass at high velocity. These were the stars that Becklin and Neugebauer first detected in 1966.

Also starting in the 1960s University of Arizona astronomer Frank Low developed a liquid-helium-cooled germanium bolometer that was hundreds of times more sensitive than previous detectors, allowing him and Douglas Kleinmann of Rice University to observe galaxies in the far infrared. To their surprise they discovered that galaxies could emit more radiation in the infrared than all other wavelengths combined.[49] This infrared light arises from bursts of massive star formation hidden behind clouds of dust and gas.

Cooled infrared detectors were eventually lofted to high altitudes aboard aircraft, balloons, and rockets. By 1983 the Infrared
Astronomical Satellite (IRAS), a joint project of the United States,
Great Britain, and the Netherlands, was launched into space and
during its ten months of operation increased the number of known
infrared sources by 70 percent. This included seventy-five thousand
"starburst galaxies" (galaxies extremely bright in the infrared due to
intense star formation) and dusty disks around other stars in our
galaxy (possible evidence of new planetary systems in the making;
see Chapter 74).

"Infrared Observations of the Galactic Center."
Astrophysical Journal, Volume 151 (1968)
by Eric E. Becklin and Gerry Neugebauer

Introduction

The dynamical center of the Galaxy lies 10 kpc* from the Sun in the
direction of the constellation Sagittarius. Some details about the galactic
center have been obtained through observations at radio wavelengths; for
example, data in the decimeter and centimeter range, recently summarized
by Downes and Maxwell, show a discrete 10-pc diameter non-thermal
source, Sagittarius A, which is believed to coincide with a small galactic
nucleus.[50]

Indications as to the nature of the galactic center have also been
obtained from visual and near-infrared observations in selected areas
where obscuration by interstellar dust is low. Further information has been
derived from optical observations of galaxies thought to be similar to the
Milky Way; specifically, the spiral galaxy M31 has a central starlike region
with a diameter less than 10 pc.

Observations of the galactic nucleus are not possible at visible wavelengths because of strong obscuration by intervening dust; it is well known,
however, that the amount of obscuration decreases at longer wavelengths.
Stebbins and Whitford using a photocell, scanned across the galactic equa-

* More recent measurements set the distance to the Milky Way's center at 8 kiloparsecs, around 26,000 light-years.

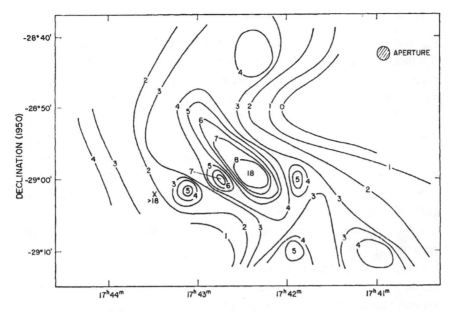

Figure 65.1: "A contour map of the galactic center region at 2.2 μ taken with an aperture of 1'.8 diameter . . . "

tor at an effective wavelength of 1.03 μ [microns] but detected no small discrete source.[51] Moroz made scans at an effective wavelength of 1.7 μ in the vicinity of Sagittarius A, but no radiation was detected.[52] In August, 1966, one of us (E. B.) scanned the region of Sagittarius A using the wavelength band from 2.0 to 2.4 μ. On these scans infrared radiation was discovered which agrees both in position and extent with the radio source Sagittarius A. . . .

Observations

Observations of the galactic center region were made in the wavelength bands 0.8–1.1 μ, 1.5–1.8 μ, 2.0–2.4 μ, and 3.1–3.8 μ; the most extensive and highest-quality observations are those made in the 2.0–2.4-μ band. . . . All of the data were obtained with the photometer mounted at the f/16 Cassegrain foci of the 24-, 60-, and 200-inch reflecting telescopes of the Mount Wilson and Palomar Observatories. This range of telescope apertures, when used with either a 6-, 4-, or 2-mm focal-plane diaphragm, resulted in resolutions ranging from 1'.8 to 0'.08 [arcminute] diameter. . . .

Figure [65.1] is a map of the brightness distribution in the galactic center region at 2.2 μ covering ¼ square degree with a resolution of 1'.8. . . .

Summary of Data

Radiation from the galactic center region at 2.2 μ can conveniently be discussed in four parts: (1) a dominant source which agrees in position and extent with the radio source Sagittarius A; (2) a pointlike source of radiation located within the dominant source (1) and near the position of its maximum brightness; (3) a general background radiation distributed predominantly along the galactic plane; (4) several smaller extended sources.

1. The dominant source, which is approximately in the center of Figure [65.1], has a full width at half-maximum of 3′–5′ when observed with 1′.8 resolution, and has a total extent of 5′–10′ diameter with a definite elongation along the galactic plane. . . .

2. . . . a pointlike source displaced 10″ from the centroid of the bright core of radiation.

3. The general background radiation is best seen in Figure [65.1]. The extent and brightness of the background is not well determined, although it appears that it may be an extension of the dominant source. . . .

4. Figure [65.1] shows seven additional localized sources of radiation. . . . The 2.2-μ flux density in each secondary source is about one fifth that measured in the dominant source. . . .

Comparison of Infrared and Radio Observations

Radio observations of the center of the Galaxy show (a) a bright discrete source 3′–4′ in diameter—Sagittarius A; (b) several weaker secondary sources a few arcmin[utes] in diameter, and (c) an extended background about 1° in diameter.

The dominant infrared source (1) agrees in position with Sagittarius A; the 1950 coordinates of the position of maximum infrared brightness with 0′.25 resolution and the mean position of Sagittarius A are:

	Right Ascension	Declination
Radio	$17^h42^m28^s \pm 2^s$	$-28° 58'.5 \pm 0'.5$
Infrared	$17^h42^m30^s \pm 1^s$	$-28° 59'.4 \pm 0'.1$

When observed with 1′.8 resolution, the dominant source has a full width at half-maximum of 3′–5′, which agrees with the width measured by Downes, Maxwell, and Meeks at 1.9 cm using 2′.2 resolution. . . .[53] The agreement in the positions and sizes of the radio source Sagittarius A and the dominant infrared source strongly suggests that the two sources are spatially coincident.

An extrapolation of the radio flux density into the infrared predicts a 3.4-μ flux density a factor of 10^3 less than that observed; therefore the mechanisms for generating the infrared and radio energy require further discussion. . . .

Discussion

We now consider the nature of the source of the infrared radiation from the galactic center. For the most part we shall assume that the infrared radiation is of stellar origin and that the observed 900° K black-body spectrum arises from the effects of interstellar absorption upon the spectra of ordinary stars. . . . Although the assumption that the observed infrared radiation arises from a stellar population similar to that present in the nuclear region of M31 [Andromeda galaxy] is consistent with the data, other models for the source of radiation which cannot be ruled out at present will be considered.[54] Infrared radiation could originate from a non-thermal source at the center of the Galaxy. . . .

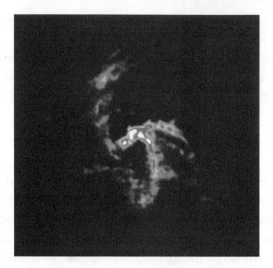

Figure 65.2: A mini-spiral in the center of the Milky Way, covering an area about 30 light-years wide. These are streamers of ionized gas falling into the supermassive black hole at the galaxy's center. This radio image was taken with the Very Large Array in New Mexico.[55]

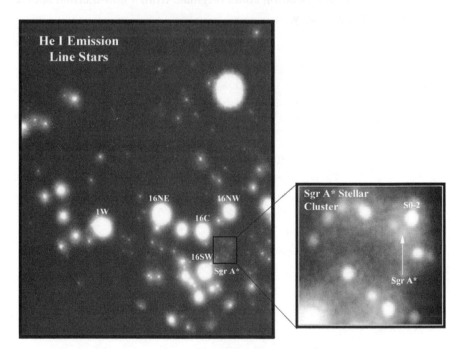

Figure 65.3: Infrared image of the central half parsec of our galaxy at 2.2 microns taken with Hawaii's Keck telescope in 1999. Becklin and Neugebauer saw all these sources as one object in 1968. The inset zooms in on an even smaller cluster of stars, believed to be orbiting around the supermassive black hole.[56]

66 / Neutrino Astronomy

I. Solar Neutrinos

In the 1960s Raymond Davis, a physicist at the Brookhaven National Laboratory on New York's Long Island, launched the field of "underground solar astronomy" when he went to the Black Hills of South Dakota and set up a detector nearly a mile below the surface to trap neutrinos emanating from the Sun. His results (or rather lack of them) originated a mystery about the Sun's workings that lasted three decades—a mystery that required new physics in order to be solved.

In 1930 the Viennese physicist Wolfgang Pauli had proposed that a new particle existed, invisible to ordinary instruments, to explain an energy that was mysteriously lost during radioactive decay. Others deduced that this particle had no mass and no charge. Indeed, it was nothing more than a phantom that carried off a bit of energy. This elusive mote was dubbed the neutrino, Italian for "little neutral one." Deep in the core of the Sun, where temperatures reach up to 15 million degrees Celsius, half a billion tons of hydrogen are converted into helium every second, releasing a flood of neutrinos in the process. In the 1960s solar physicist John Bahcall concluded from his theoretical models that the Sun should be emitting enough high-energy neutrinos to be detectable on the Earth— a phenomenon that offered the opportunity to look directly into the heart of the Sun.

Encouraged by Bahcall's prediction, Davis proceeded with his plans to construct the world's first underground solar telescope, based on an earlier neutrino detector that he had designed and operated to prove the existence of the neutrino.[57] This "telescope" was a huge tank filled with 100,000 gallons of tetrachloroethylene—dry-cleaning fluid—and placed deep inside South Dakota's Homestake gold mine to block out disruptive cosmic rays such as muons (heavy electrons). Neutrinos usually pass through matter as if it weren't there, but sporadically a neutrino bumped into the nucleus of a chlorine atom in the tank and turned it into argon. Thus the neutrinos were detected by extracting the resulting argon atoms from the fluid and counting them.

The first report by Davis and his colleagues in 1968 was both historic and disturbing: neutrinos were detected, which was the first direct evidence that the Sun shines by nuclear fusion, but they were capturing just one neutrino every other day—at most, one-third the number that solar physicists were expecting. Some blamed Davis's equipment, but others were worried that the shortfall indicated that physicists didn't fully understand how the Sun worked, and that would have forced astronomers to reexamine many other models in astronomy, from how the universe produced its first matter to the course of stellar evolution. Over three decades of collecting data, the Homestake results didn't vary, and with the construction of bigger and more sophisticated neutrino observatories, Davis's initial findings were sustained.

The problem of the solar-neutrino deficit that Davis first noticed was eventually resolved with new physics. By the 1970s particle physicists knew that neutrinos come in three "flavors" (as physicists put it): the electron neutrino, the muon neutrino, and the tau neutrino. And investigations in the 1990s by Japan's Super-Kamiokande neutrino detector and the Sudbury Neutrino Observatory in Canada indicated that some of the electron neutrinos racing out of the Sun were transforming themselves into the other flavors before they got to the Earth, an idea earlier proposed by theorists.[58] Both advanced underground detectors are huge vats of water surrounded by photomultiplier tubes, which register the bursts of light (Cherenkov radiation) released whenever a neutrino occasionally interacts with the water. Thus the Homestake detector, capable of seeing just electron neutrinos, had been measuring only a portion of the Sun's neutrino production. This discovery of neutrino "oscillations" was of great importance to the standard model of elementary particles. For a neutrino to change identity meant it had to have a bit of mass, which was a revolutionary finding. It was also interesting in that it was an astronomical observation that first revealed this new property in particle physics.

"Search for Neutrinos from the Sun." *Physical Review Letters*, Volume 20 (May 20, 1968) by Raymond Davis Jr., Don S. Harmer, and Kenneth C. Hoffman

Recent solar-model calculations have indicated that the sun is emitting a measurable flux of neutrinos from decay of B^8 [boron with five protons and three neutrons] in the interior. The possibility of observing these energetic neutrinos has stimulated the construction of four separate neutrino detectors. This paper will present the results of initial measurements with a detection system based upon the neutrino capture reaction Cl^{37} (v, e^-) Ar^{37} [chlorine transformed to argon]. It was pointed out by Bahcall that the energetic neutrinos from B^8 would feed the analog state of Ar^{37} (a superallowed transition) that lies 5.15 MeV above the ground state. . . .[59] On the basis of these predictions, the total solar-neutrino-capture rate in 520 metric tons of chlorine would be in the range of 2 to 7 per day.

The detector design—A detection system that contains 390,000 liters (520 tons chlorine) of liquid tetrachloroethylene, C_2Cl_4 in a horizontal cylindrical tank was built. . . . The system is located 4,850 feet underground in the Homestake gold mine at Lead, South Dakota. It is essential to place the detector underground to reduce the production of Ar^{37} from reactions by protons formed in cosmic-ray muon interactions. . . .

Neutrino detection depends upon removing the Ar^{37} from a large volume of liquid contained in a sealed tank, and observing the decay of Ar^{37} (35-day half-life) in a small proportional counter. . . . The Ar^{37} activity is removed by purging with helium gas. Liquid is pumped uniformly from the bottom of the tank and returned to the tank through a series of 40 eductors arranged along two horizontal header pipes inside the tank. . . .

Argon is extracted by circulating the helium from the tank through an

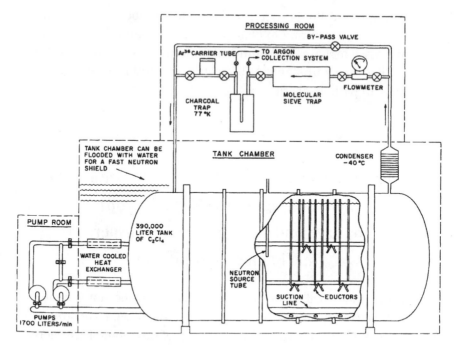

Figure 66.1: "Schematic arrangement of the Brookhaven solar neutrino detector."

argon extraction system. . . . The tetrachloroethylene vapor is removed by a condenser at −40° C followed by a bed of molecular sieve adsorber at room temperature. The helium then passes through a charcoal bed at 77° K to adsorb the argon, and is finally returned to the tank. This arrangement is shown schematically in Fig. [66.1].

[Omitted here is a technical description of how the argon is removed from the charcoal and isolated from other impurities.]

Counting—The argon sample is counted in a small proportional counter with an active volume 3 cm long and 0.5 cm in diameter. A small amount of methane is added to the argon to improve the counting characteristics of the gas. . . . The counter is shielded from external radiations by a cylindrical iron shield 30 cm thick lined with a ring of 5-cm-diam proportional counters for registering cosmic-ray muons. The argon counter is held in the well of a 12.5- by 12.5-cm sodium-iodide scintillation counter located inside the ring counters. Events in anticoincidence with both the ring counters and the scintillation counter are recorded on a 100-channel pulse-height analyzer. . . .

Figure 66.2: Detector in the Homestake mine.

Results and discussion—Two experimental runs have been performed. . . . The first exposure was 48 days. The tank was purged with 0.50 million liters of helium. A volume of 1.27 std cc of argon was recovered from the tank, and this volume contained 94% of the carrier Ar^{36} introduced at the start of the exposure. It was counted for 39 days and the total number of counts observed in the Ar^{37} peak position (full width at half-maximum) in the pulse-height spectrum was 22 counts. This rate is to be compared with a background rate of 31 $\pm$ 10 counts for this period. The neutrino-capture rate in the tank deduced from the exposure, counter efficiency, and argon recovery from this experiment was (-1.1 ± 1.4) per day.

A second exposure was made for 110 days from 23 June to 11 October 1967. The tank was purged with 0.53 million liters of helium yielding 0.62 cm^3 of argon with a 95% recovery of the added carrier Ar^{36}. The pulse-height spectra are shown in Fig. [66.3] for the first 35 days of counting and also for a total period of 71 days. This rate can be compared with the background rate for the counter filled with Ar^{36} purified in an identical manner (shown in Fig. [66.3]).

It may be seen from the pulse-height spectrum for the first 35 days of counting that 11 $\pm$ 3 counts were observed in the 14 channels where Ar^{37}

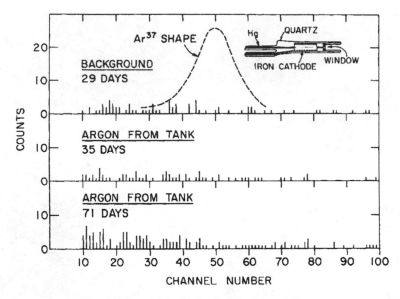

Figure 66.3: "Pulse-height spectra."

should appear. The counter background for this period of time corresponded to 12 ± 4 counts. Thus, there is no increase in counts from the sample over that expected from background counting rate of the counter. One would deduce from these rates that the neutrino-capture rate in 610 tons of tetrachloroethylene was equal to or less than 0.5 per day based upon one standard deviation. . . . It may be seen that this limit is approximately a factor of 7 below that expected from . . . solar-model calculations. . . .

It is possible to improve the sensitivity of the present experiment by reducing the background of the counter. However, background effects from cosmic-ray muons will eventually limit the detection sensitivity of the experiment at its present location. . . .

II. Supernova Neutrinos

In the early morning hours of February 24, 1987, Canadian astronomer Ian Shelton was taking survey photographs of the Large Magellanic Cloud from the Las Campanas Observatory in northern Chile when he noticed a new pinpoint of light in the cloud. What he had discovered was the first visible supernova in our galactic neighborhood since 1604 (before the invention of the astronomical telescope). The stellar explosion, around 167,000 light-years distant, was officially designated Supernova 1987A.

What Shelton and others did not realize at the time was that the explosion was unwittingly recorded nearly a day before SN1987A was glimpsed visibly. The supernova had been triggered when a blue supergiant star ran out of fuel and its iron core suddenly collapsed to form an ultradense neutron star. This created a firestorm of 10^{58} neutrinos that sped out of the star in all directions at near the speed of light. Of the ten thousand trillion trillion neutrinos that eventually made it to the Earth, some were caught by two underground detectors, originally built to observe protons decay.

Hearing of the supernova, researchers at Japan's Kamiokande II detector, which had just been upgraded the year before to detect neutrinos, pored through their computer records and discovered eleven neutrino events that likely emanated from the distant explosion over a thirteen-second span. The timing and energy of the events matched the expected signal, and the first two neutrinos seemed to arrive from the general direction of the Large Magellanic Cloud.

Physicists at the Irvine-Michigan-Brookhaven proton-decay detector set 2,000 feet beneath the shore of Lake Erie searched their data as well and found eight supernova-neutrino candidates over the same time period, about eighteen hours before the visible flaring.*[60] It was a momentous find: in capturing these neutrinos, the researchers at each facility were witnessing, for the first time, the very formation of a neutron star.

---

* The neutrinos power a shock wave that takes hours to work its way through the bulk of the star. The visible explosion occurs once the shock wave surfaces.

"Observation of a Neutrino Burst from Supernova SN1987A." *Physical Review Letters*, Volume 58 (April 6, 1987)
by K. Hirata, T. Kajita, M. Koshiba, M. Nakahata, Y. Oyama, N. Sato, A. Suzuki, M. Takita, Y. Totsuka, T. Kifune, T. Suda, K. Takahashi, T. Tanimori, K. Miyano, M. Yamada, E. W. Beier, L. R. Feldscher, S. B. Kim, A. K. Mann, F. M. Newcomer, R. Van Berg, W. Zhang, and B. G. Cortez

Following the optical sighting on 24 February 1987 of the supernova now called SN1987A, a search was made of the data taken in the detector Kamiokande II during the period from 16:09, 21 February 1987 to 07:31, 24 February 1987. We report here the results of that search.

The Kamiokande II detector, directed primarily at nucleon decay and solar-^{8}B-neutrino detection, has been operating since the beginning of 1986. It is described in detail elsewhere. . . . [61] The inner detector fiducial volume containing 2,140 tons of water is viewed by an array of [948] 20-in-diameter photomultiplier tubes (PMT's) on a 1×1-m^2 lattice on the surface. . . .

Neutrinos of different flavors are detected through the scattering reaction $\nu e \rightarrow \nu e$ [neutrino/electron scattering]. The kinematics of this reaction and the subsequent multiple scattering of the recoiling electron preserve knowledge of the incident neutrino direction within approximately 28° rms at electron energies in the vicinity of 10 MeV. . . . The Cherenkov light of a 10-MeV electron gives on average 26.3 hit PMT's (N_{hit}) at ⅓ photoelectron threshold. . . .

The search for a neutrino burst from SN1987A was carried out on the data of run 1892, which, except for a pedestal run of 105 sec duration every hour, continuously covered the period from 16:09, 21 February 1987 to 07:31, 24 February 1987, in Japanese Standard Time (JST), which is UT [universal time] plus 9 h. Events satisfying the following three criteria were selected: (1) The total number of photoelectrons per event in the inner detector had to be less than 170, corresponding to a 50-MeV electron; (2) the total number of photoelectrons in the outer detector had to be less than

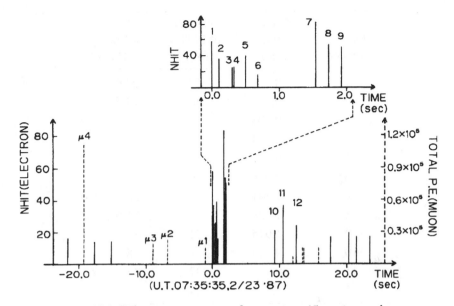

Figure 66.4: "The time sequence of events in a 45-sec interval centered on 07:35:35 UT, 23 February 1987. The vertical height of each line represents the relative energy of the event. Solid lines represent low-energy electron events in units of the number of hit PMT's, N_{hit} (left-hand scale). Dashed lines represent muon events in units of the number of photoelectrons (right-hand scale). Events $\mu1$–$\mu4$ are muon events which precede the electron burst at time zero. The upper right figure is the 0–2-sec time interval on an expanded scale."

30, ensuring event containment; and (3) the time interval from the preceding event had to be longer than 20 μsec [microseconds], to exclude electrons from muon decay.

The short-time correlation of these low-energy contained events was investigated and the event sequence as shown in Fig. [66.4] was observed at 16:35:35 JST (7:35:35 UT) of 23 February 1987. In Fig. [66.4] we show the time sequence of all low-energy events (solid lines) and all cosmic-ray muon events (dashed lines) in the given interval. The event sequence during 0 to 2 sec is shown expanded in the upper right corner. The properties of the events in the burst (numbered 1 to 12 in Fig. [66.4]) are summarized in Table [66.1]. Event number 6 has $N_{hit} < 20$ and has been excluded from the signal analysis. . . .

[Omitted here is a discussion of the statistical methods used to assess the data.]

Event Number	Event Time (sec)	Number of PMT's (N_{hit})	Electron Energy (MeV)
1	0	58	20.0 ± 2.9
2	0.107	36	13.5 ± 3.2
3	0.303	25	7.5 ± 2.0
4	0.324	26	9.2 ± 2.7
5	0.507	39	12.8 ± 2.9
6	0.686	16	6.3 ± 1.7
7	1.541	83	35.4 ± 8.0
8	1.728	54	21.0 ± 4.2
9	1.915	51	19.8 ± 3.2
10	9.219	21	8.6 ± 2.7
11	10.433	37	13.0 ± 2.6
12	12.439	24	8.9 ± 1.9

Table 66.1: Measured properties of the twelve electron events detected in the neutrino burst. Due to its low number of photomultiplier hits, event 6 is excluded from the analysis.

The only background process that might conceivably give rise to a burst of events in a short interval of time would be the production of an energetic nuclear cascade by an incident cosmic-ray muon. . . . [But] the overall probability that any of the muons, μ1 to μ4 [see Figure 66.4], was the progenitor of the event burst in Table [66.1] is extremely low. . . .

We conclude that the event burst at 7:35:35 UT, 23 February 1987, displayed in Fig. [66.4] and Table [66.1] is a genuine neutrino burst. This is the only such burst found by us during the period from 9 January to 25 February 1987. We therefore associate it with SN1987A. This association is supported by the time structure of the events in the burst, their energy distribution, and uniform volume distribution. Additional support is provided by the correlation in angle of the first two observed events with the direction to SN1987A. The event burst occurred roughly 18 h prior to the first optical sighting.

[Omitted here are calculations showing that the energy of SN1987A in neutrinos was around 8×10^{52} ergs.*]

---

* For comparison, that is nearly 100 times larger than the energy output of the Sun over its entire 10-billion-year lifetime.

This observation is the first direct observation in neutrino astronomy, and coincides remarkably well with the current model of supernova collapse and neutron-star formation. In that model an aged, massive star, having exhausted its nuclear fuel, undergoes a supernova explosion. In supernovae of Type II almost all of the gravitational binding energy of the resultant neutron star, $\sim 3 \times 10^{53}$ ergs, is radiated within a few seconds in the form of 10^{58} neutrinos of all flavors [types] with average energy in the vicinity of 10–15 MeV. . . .

67 / Gamma-Ray Bursts

The Vela ("watchman" in Spanish) series of U.S. spacecraft were designed to monitor worldwide compliance with the 1963 nuclear test ban treaty. Along with detecting covert nuclear-bomb tests in space and on the Earth, the satellites used their x-ray, neutron, and gamma-ray detectors to monitor both solar activity (providing radiation warnings for manned missions) and terrestrial lightning activity.

Some astronomers, aware of the spacecrafts' capabilities, encouraged Ray Klebesadel, Ian Strong, and Roy Olson at the Los Alamos National Laboratory in New Mexico to use the Vela network to look for bursts of energetic photons from distant space, possibly from exploding supernovae. Klebesadel and his team came across the first candidate by accident. When plowing through computer listings by hand to check on the Vela detectors' performance, they noticed one particular gamma-ray event from July 2, 1967, that could not be explained. It was a spike, a dip, a second spike, and then a long gradual trailing off. "One thing that was immediately apparent," said Klebesadel, "was that this was not a response to a clandestine nuclear test."[62]

Going on to search through records spanning three years—from July 1969 to July 1972—the Los Alamos team discovered sixteen gamma-ray bursts occurring over that period, originating from random points in the sky. Triangulation among the satellites suggested these bursts came from outside the solar system. It was a type of celestial event never before detected (and by satellites not designed for astronomy). The duration of the bursts ranged from less than a second to some thirty seconds—popping off like a cosmic flashbulb, flickering for a moment, then fading away.

At the time that Klebesadel and his colleagues published their historic findings in 1973, gamma-ray astronomy was well under way. The first gamma-ray telescope carried into orbit, aboard the Explorer 11 satellite in 1961, had detected fewer than a hundred cosmic gamma-ray photons, but these appeared to come from all directions, implying some sort of uniform gamma-ray background. By the 1970s, other U.S. and European space-borne detectors discovered a number of new gamma-ray sources. The greatest source of gamma rays emanated from the center of the Milky Way, but radiation was also detected from such objects as the Crab nebula (a supernova remnant), Cygnus X-1 (a black-hole candidate), quasars, and active galaxies.

The gamma-ray bursts, however, offered the field its greatest mystery and led to a decades-long debate on the cause and exact location of these powerful events. At first it was assumed that the bursts were largely occurring within the disk of the Milky Way, but the Burst and Transient Source Experiment (BATSE) detector aboard the Compton Gamma Ray Observatory launched in 1991 conclusively established that the bursts were uniformly distributed over the celestial sky. Some astronomers argued that they were still relatively nearby events, confined to a halo surrounding our galaxy. But by the late 1990s, evidence arrived proving the bursts were cosmological. Starting in 1997, alerted by gamma-ray astronomers to the latest bursts and getting more precise coordinates, optical astronomers were at last able to spot a burst's visible afterglow. Obtaining the redshifts of these glows, they confirmed the gamma-ray bursts were up to billions of light-years distant.[63]

By the end of the twentieth century, the exact cause of a gamma-ray burst was not fully determined, but astronomers were coming to believe that the longer and more common bursts (lasting more than a second) are generated by the collapse of a very massive star into a black hole. In this scenario, the bursts appear particularly powerful because the energy is beamed into narrow jets by the spinning hole. The shorter bursts, on the other hand, may originate from the collision of neutron stars or black holes. Bursts might also involve beams jetting out as a black hole draws in and destroys a companion star.

"Observations of Gamma-Ray Bursts of Cosmic Origin." *Astrophysical Journal*, Volume 182 (June 1, 1973) by Ray W. Klebesadel, Ian B. Strong, and Roy A. Olson

Abstract

Sixteen short bursts of photons in the energy range 0.2–1.5 MeV [million electron volts] have been observed between 1969 July and 1972 July using widely separated spacecraft. Burst durations ranged from less than 0.1 s to ~30 s. . . . Directional information eliminates the Earth and Sun as sources.

I. Introduction

On several occasions in the past we have searched the records of data from early *Vela* spacecraft for indications of gamma-ray fluxes near the times of appearance of supernovae. These searches proved uniformly fruitless. Specific predictions of gamma-ray emission during the initial stages of the development of supernovae have since been made by [Stirling] Colgate.[64] Also, more recent *Vela* spacecraft are equipped with much improved instrumentation. This encouraged a more general search, not restricted to specific time periods. The search covered data acquired with almost continuous coverage between 1969 July and 1972 July, yielding records of 16 gamma-ray bursts distributed throughout that period. . . .

II. Instrumentation

The observations were made by detectors on the four *Vela* spacecraft, *Vela 5A, 5B, 6A,* and *6B,* which are arranged almost equally spaced in a circular orbit with a geocentric radius of ~1.2×10^5 km.

On each spacecraft six 10 cm^3 CsI scintillation counters are so distributed as to achieve a nearly isotropic sensitivity. Individual detectors respond to energy depositions of 0.2–1.0 MeV for *Vela 5* spacecraft and 0.3–1.5 MeV for *Vela 6* spacecraft, with a detection efficiency ranging between 17 and 50 percent. The scintillators are shielded against direct penetration by electrons below ~0.75 MeV and protons below ~20 MeV. A high-Z shield attenuates photons with energy below that of the counting threshold. . . .

III. Observations

. . . A count-rate record is generated only in response to a rapid rise in count rate to a level significantly above background. . . . Present processing requires that at least two spacecraft record the burst with a deviation from simultaneity of 4 s or less. Sixteen events have been observed to meet these criteria, two of which were recorded by all four spacecraft. Absence of consistent response from all four spacecraft can be attributed in most cases to an inappropriate mode of operation or to marginal signal levels.

These bursts display a wide variety of characteristics. Time durations range from less than a second to about 30 s. Some count-rate records have a number of clearly resolved peaks while others do not appear to display any significant structure. The time-integrated flux density in the measured energy interval ranges from the minimum identifiable level of $~10^{-5}$ ergs cm^{-2} to more than 2×10^{-4} ergs cm^{-2}. Instantaneous flux densities have exceeded 4×10^{-4} ergs cm^{-2} s^{-1}. . . .

Allowing for differing energy thresholds and statistical fluctuations, the integrated flux for a particular event is independent of the recording spacecraft. Differences in the time of arrival of the signals at the various spacecraft imply that the spacecraft are not equidistant from any given source. Inverse-square law considerations thereby place the sources at a distance of at least 10 orbit diameters, or several million kilometers.

Arrival-time differences have been derived approximately in all cases, and fairly accurately ($\pm$ 0.05 s) for a number of cases. For a two-spacecraft coincidence the transit delay defines a circle on the celestial sphere on which the source position must lie. For three spacecraft we can define intersecting circles, whose points of intersection represent the source position and its mirror image in the orbital plane of the spacecraft, a presently unresolved ambiguity. Nevertheless, it has been possible by this technique to rule out the sun as a source. Also, in none of the 16 cases was there found any close correlation with any recorded indications of solar activity. . . .

A burst observed on 1970 August 22 is presented as an example. Figure [67.1] shows the count rate as a function of time. Each plot is presented in two parts. On the left, on a linear time scale, are plotted 10 measurements of count rate made at 4-minute intervals for the time immediately

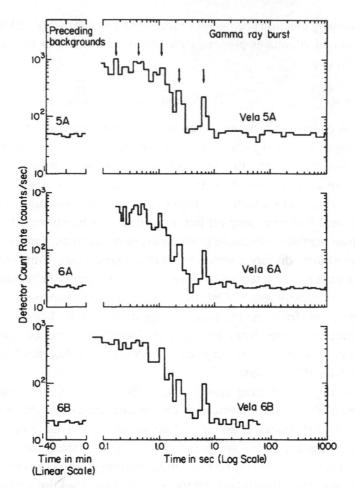

Figure 67.1: "Count rate as a function of time for the gamma-ray burst of 1970 August 22 as recorded at three *Vela* spacecraft. Arrows indicate some of the common structure. Background count rates immediately preceding the burst are also shown. *Vela* 5A count rates have been reduced by 100 counts per second (a major fraction of the background) to emphasize structure."

preceding the burst. These establish a background count rate. The record of the burst is plotted on the right on a logarithmic time scale. All the *Vela 5A* data have had a uniform 100 counts per second (a major fraction of the background) subtracted before plotting in order to facilitate comparison of time structure.

[In] the initial part of the burst (extending to ~4 s) . . . there appears structure common to the records of all three spacecraft. Although the exact statistical significance of this structure has not yet been firmly established, it has been used to adjust these three records in time, relative to the initiation of the recordings. . . .

In addition to the initial structure, all three records show a distinct peak centered around 6.5 s. For each record this peak is statistically significant to about 6 standard deviations. It represents integrated flux densities of 10^{-5} ergs cm^{-2} and 4×10^{-6} ergs cm^{-2} in the lower and higher energy ranges, respectively. The spectrum is clearly softer than that of the initial part of the burst.

IV. Discussion

A search was made for reports of a nova or supernova within a reasonable time (~ several weeks) of each gamma-ray burst. No reported novae were related in time or direction to any of the bursts. Only two reported supernovae reached maximum apparent magnitude within a few days of an observed burst. In both cases, however, reports of prediscovery observations were later made which preceded the gamma-ray burst by at least several days. In addition, the source positions derived from preliminary timing data are inconsistent with the locations of the supernovae.

The lack of correlation between gamma-ray bursts and reported supernovae does not conclusively argue against such an association, since it is possible that there are supernovae, not necessarily bright in the optical region ("theoreticians' supernovae"), whose rate of occurrence may exceed those which are optically visible. A source at a distance of 1 Mpc [megaparsec] would need to emit ~10^{46} ergs in the form of electromagnetic radiation between 0.2 and 1.5 MeV in order to produce the level of response observed here. Since this represents only a small fraction ($<10^{-3}$) of the energy usually associated with supernovae, the energy observed is not inconsistent with a supernova as a source.

68 / Binary Pulsar and Gravity Waves

In 1974 two radio astronomers discovered the first pair of neutron stars orbiting one another. What made this find historic was that the binary system provided a unique relativistic laboratory to prove one of the last predictions of general relativity yet to be tested.

In 1916, just a few months after he had introduced his theory of general relativity, Einstein wrote a paper on another possible outcome of his new vision of space and time—gravitational radiation.[65] He recognized that just as electromagnetic waves, such as radio waves, are generated when electrical charges travel up and down an antenna, waves of gravitational radiation are produced when masses move about, such as two stars orbiting one another. These waves of gravitational energy would flow outward from the source, much like starlight does. The gravity waves wouldn't be traveling *through* space, however; rather they would be vibrations in space-time itself. Moreover, these spacequakes would carry energy away from the system. In the case of a binary, the two stars would consequently draw closer together and their orbital period would decrease. Since such an effect was too small to detect in the binaries of ordinary stars, Einstein expected gravity waves to remain strictly theoretical, but that changed with the discovery of compact neutron star systems, more powerful sources of gravitational energy.

As with so many astronomical discoveries, the opportunity arose through serendipity. Seven years after the discovery of the first pulsar (see Chapter 64), Joseph Taylor, then at the University of Massachusetts at Amherst, initiated the most sensitive pulsar survey to date. The search was carried out at the Arecibo radio observatory in Puerto Rico, using a computerized search technique developed by Taylor and his graduate student Russell Hulse. By the end of his fourteen-month stay at Arecibo, Hulse had found forty pulsars, which made for a nice doctoral thesis, but one particular pulsar, PSR 1913+16 (right ascension 19 hours and 13 minutes, declination 16 degrees), eclipsed all the others. Hulse first detected it on July 2, 1974. Its pulsing was particularly fast (59 milliseconds, or 17 beeps per second), making it the second fastest pulsar then known. "Fantastic," wrote Hulse on his discovery sheet.[66] When confirming the pulse rate nearly two months later, though, he found that the pulse rate of PSR 1913+16 was mysteriously changing from day to day, unlike any other pulsar previously found. After noting the fourth new period in his records, he scratched them all out in frustration.

Eventually Hulse came to suspect that the pulsar was circling another star, its period regularly increasing and decreasing due to the orbital motion. The proof arrived on September 16 when he was at last able to verify the "turnaround," the moment when the pulsar swung around the end of its orbit, causing its apparent pulse period to stop decreasing and to start increasing. Hearing the news, Taylor immediately flew down to Puerto Rico with better pulsar timing equipment. Within days, they confirmed that the two objects—the pulsar and its unseen companion—were orbiting one another every 7 hours and 45 minutes. While one was assuredly a neutron star, due to the pulsing, they concluded the companion was likely a neutron star as well; the size of the binary orbit is not much bigger than the radius of our Sun (~430,000 miles or 2 light-seconds).

Immediately on its discovery, Hulse and Taylor recognized that this binary pulsar, located some 16,000 light-years away, was the perfect system on which to carry out tests of general relativity beyond the solar system. Taylor and several other colleagues continued to travel to Arecibo over the years to monitor PSR 1913+16 and make measurements through the changing ticks of the pulsar clock. By 1979 they were ready to file their first report. Along with improving the accuracy of the measured parameters (seen by comparing the tables in the 1975 and 1979 reports), they also noted some dramatic changes: as Taylor and Hulse originally predicted, it was found that the binary's orbit was precessing (pivoting around) at a whopping 4.2 degrees per year, thirty-five thousand times more than the annual change in the planet Mercury's orbit due to the same relativistic effect (see Chapters 20 and 36). But more importantly, Taylor's group found definitive evidence that the binary was emitting gravity waves.

After analyzing some five million pulses and correcting for a number of subtle factors (such as the Earth's motion, the perturbations of other planets, variations in the Earth's rotation, signal delays due to interstellar gas, and even the solar system's movement around the galaxy), they saw that the binary's orbital period was definitely decreasing—by about seventy-five millionths of a second each year—which meant the binary was losing energy and the two stars were drawing closer together by some 3 feet a year. This change exactly matched what would be expected if the system was losing energy in the form of gravity waves alone. This accomplishment was later described as "an example of . . . science at its best."[67] Einstein's prediction was at last confirmed sixty-three years after it had been made.

"Discovery of a Pulsar in a Binary System."
Astrophysical Journal, Volume 195 (January 15, 1975)
by Russell A. Hulse and Joseph H. Taylor

I. Introduction

We wish to report the detection of an unusual pulsar discovered during the course of a systematic survey for new pulsars being carried out at the Arecibo Observatory in Puerto Rico. The object has a pulsation period of about 59 ms [milliseconds]—shorter than that of any other known pulsar except the one in the Crab Nebula—and periodic changes in the observed pulsation rate indicate that the pulsar is a member of a binary system with an eccentric orbit of 0.3230 [day] period. Thus for the first time it is possible to observe the gravitational interactions of a pulsar and another massive object, and additional observations should make it possible to determine the masses of the two objects unambiguously.

Table 68.1: Some Parameters of the Binary Pulsar and Elements of the Orbit

Right Ascension (1950.0)	19 h 13 m 13 s ± 4 s
Declination (1950.0)	+ 16° 00′ 24″ ± 60″
Pulsar Period	0.059030 ± 0.000001 second
Radial Velocity Variation	199 ± 5 km/second
Period of Binary Orbit	27908 ± 7 seconds
Orbit Eccentricity	0.615 ± 0.010
Projected Semimajor Axis of Orbit	1.00 ± 0.02 R_{sun}

II. Discovery of the Binary Pulsar

. . . Forty pulsars have now been detected in this work, of which 32 were not previously known. . . . The 59-ms pulsar, PSR 1913+16, was first detected in 1974 July. Attempts to measure its period to an accuracy of ±1 μs [microsecond] were frustrated by apparent changes in period of up to ~80 μs from day to day, and sometimes by as much as 8 μs over 5 minutes. Such behavior is quite uncharacteristic of other pulsars: the largest known secular changes of period are of order 10 μs per *year,* and irregular changes of period are many orders of magnitude smaller. It soon became clear that Doppler shifts resulting from orbital motion of the pulsar could account for the observed period changes, and by the end of September an accurate velocity curve of this "single-line spectroscopic binary" had been obtained (see figure [68.1]). . . .

The orbital elements given in [Table 68.1] were obtained from direct measurements of the pulsar period over about 200 different 5-minute intervals distributed over 10 days. The 5-minute intervals are long enough that the period can be measured to an accuracy of about 1 μs, but short enough that the period does not change too drastically within the interval.

III. Physical Parameters of the Binary Pair

The mass of the pulsar is, of course, a quantity of great interest, as is the size and mass of the unseen companion. The observed mass function permits a wide range of values for M_1 and M_2. However, if we restrict attention to values of M_1 [the pulsar] thought to be reasonable for neutron stars [around 1.4 solar masses], the picture becomes clearer.

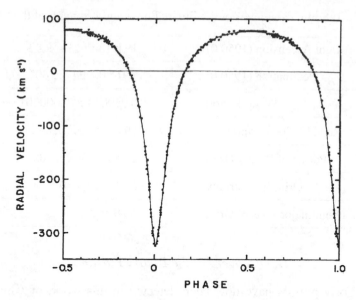

Figure 68.1: Velocity curve for the binary pulsar showing changes in its radial velocity as the pulsar periodically approaches us then recedes.

[Omitted here are the authors' calculations from the orbital parameters that the invisible companion is likely not a main-sequence star, supported by the fact that no eclipses are observed.]

We conclude that the companion must be a compact object, probably a neutron star or a black hole. A white dwarf companion cannot be ruled out, but seems unlikely for evolutionary reasons.

IV. Additional Observations

. . . Timing data much more accurate than that already available can in principle be obtained by recording the absolute time of arrival of the pulses. Observations of this sort done on other pulsars yield absolute arrival times accurate to ~10^{-4} s. Measurements of comparable quality are now being acquired for PSR 1913+16, and in due course the data will yield greatly improved accuracies for the celestial coordinates and for the orbital elements of the binary system. This in turn will allow a number of interesting gravitational and relativistic phenomena to be studied. The binary configuration provides a nearly ideal relativity laboratory including an accurate clock in a high-speed, eccentric orbit and a strong gravitational field. We note, for example, that the changes of both v^2/c^2 and GM/c^2r during the orbit are sufficient to cause changes in observed period of several parts in 10^6. Therefore, both the relativistic Doppler shift and the gravitational redshift will be easily measurable. Furthermore, the general relativistic advance of periastron [precession] should amount to about 4° per year, which will be detectable in a short time. The measurements of these effects, not usually observable in spectroscopic binaries, would allow the orbit inclination and the individual masses to be obtained.

The star field in the direction of the pulsar is crowded, and the observed dispersion measure suggests that PSR 1913+16 is about 5 kpc [kiloparsecs] distant. . . .

"Measurements of General Relativistic Effects in the
Binary Pulsar PSR 1913+16."
Nature, Volume 277 (February 8, 1979)
by Joseph H. Taylor, Lee A. Fowler, and Peter M. McCulloch

The earliest observations of binary pulsar PSR 1913+16 showed that, because its orbit involved large velocities [around a thousandth the speed of light], a high eccentricity, and relatively strong gravitational fields, several special and general relativistic effects should eventually be observable. The advance of periastron [at the rate of ~4.2° per year] was the first of these effects to be measured, and the rate of advance has now been determined to better than 0.1% accuracy. We report here the detection of four more effects of relativistic origin, including quantitative measurements of three of them. Together with the much larger effects already measured, the new parameters over-determine the system and provide: (1) the first determination of the mass of a radio frequency pulsar; (2) constraints on the nature of the companion star, and a measurement of its mass; (3) determination of the angle of inclination between the plane of the orbit and the plane of the sky; (4) quantitative confirmation of the existence of gravitational radiation at the level predicted by general relativity; and (5) qualitative observation of geodetic precession of the pulsar spin axis. The data are consistent with the general theory of relativity and provide some strong constraints on any alternative theory of gravitation.

Pulse Arrival-Time Data

Most of the new information comes as the result of pulse arrival-time measurements, which now comprise approximately 1,000 observations spanning 4.1 yr. The data were obtained at frequencies near 430 and 1,410 MHz, using the 305-m radio telescope at Arecibo, Puerto Rico. The normal observing procedure involves adding together approximately 5,000 pulses, using a pre-computed ephemeris to define the expected pulsation period. The resulting mean pulse profile is then fitted by the method of least squares to a long-term average "standard profile," to obtain the precise pulse arrival time. Because of improvements in equipment and techniques between 1974 and 1978, our timing accuracy has gradually improved from ~1 ms [millisecond] to ~50 μs [microsecond]. Data have been acquired at intervals not exceeding 7 months, and in spite of the short period of the pulsar ($P = 0.059$ s), there has been no problem in keeping track of the number of elapsed pulse periods.

Our analysis of the timing data . . . proceeds by the following steps. First, the pulse arrival times are corrected from the location of the observatory to the barycenter of the Solar System, including a relativistic clock correction to account for annual changes in gravitational potential at the Earth. A correction is then made for the dispersive delay in the interstellar medium, using the frequency of observation as Doppler-shifted by the Earth's motion. Finally, the proper time t_p in the pulsar's reference frame is

Table 68.2: "Parameters Derived from Timing Data"

Right Ascension (1950.0)	19 hr 13 min 12.474 sec ± 0.004 sec
Declination (1950.0)	16° 01′ 08.02″ ± 0.06″
Pulsar Period	0.059029995269
Change in Pulsar Period	$(8.64 \pm 0.02) \times 10^{-18}$ second per second
Projected Semimajor Axis	2.3424 ± 0.0007 light-second
Orbital Eccentricity	0.617155 ± 0.000007
Binary Orbit Period (P_b)	27906.98172 ± 0.00005 seconds
Rate of Advance of Periastron	4.226 ± 0.002 degrees per year
Change Rate in Orbit Period ($\dot{P}_b$)	$(-3.2 \pm 0.6) \times 10^{-12}$ second per second

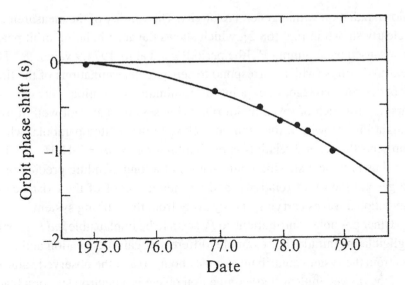

Figure 68.2: The decay of the binary pulsar's orbital period (dots) matches the plotted curve, which corresponds to the change in orbital period predicted by general relativity if $m_p = m_c = 1.41\ M_\odot$.

obtained by correcting for (1) the projection onto the line of sight of the pulsar's orbital position, (2) the integrated effects of gravitational redshift and transverse Doppler shift in the highly eccentric pulsar orbit, and (3) the gravitational propagation delay. . . .

[Omitted here is a discussion of the statistical methods used in determining the confidence levels of the results.]

Parameter values based on data acquired through October 1978 are listed in Table [68.2]. . . . [A]ll of them have now been determined to within ≤20% accuracy. . . . [The] parameters . . . provide strong constraints on the masses of the pulsar and its companion (m_p and m_c, respectively). . . . [A] probable solution for the masses [is] $m_p = 1.39 \pm 0.15\ M_\odot$ [mass of the sun], $m_c = 1.44 \pm 0.15\ M_\odot$, and that the observed rate of periastron advance is entirely caused by the general relativistic effect. In this case, the observed magnitude of $\dot{P}_b$ [the rate of change in orbital period] agrees with the value expected from gravitational radiation damping to within a factor 1.3 ± 0.3. There is no compelling evidence for significant tidal or rotationally-induced contributions. . . .

Undoubtedly the most important new result of this work is the measurement of $\dot{P}_b$ with the sign and magnitude expected on the basis of gravitational radiation within general relativity. The validity of the measurement is clearly shown in Fig. [68.2], which shows the accumulating orbit phase error caused by assuming P_b [the period] fixed at its 1974.9 value. . . . The measured points (which correspond to separate determinations of the time of periastron passage [the point of minimum separation between the stars] . . . for each of seven major observing sessions) are not well fit by a straight line. However, they fall very close to the plotted parabola, which represents the general relativistic prediction for $m_p = m_c = 1.41\,M_\odot$ . . . The data thus provide a striking confirmation of a long-standing prediction of the general theory of relativity, and an indirect proof of the existence of gravitational waves carrying energy away from the orbiting system.

Other possible contributions to $\dot{P}_b$ seem to be implausible, *ad hoc,* or of negligible magnitude. For example, differential galactic rotation and mass loss from the system contribute at most about ~1% of the observed value of $\dot{P}_b$. If the (as yet unidentified) companion object is a neutron star or a black hole, then tidal dissipation is also of no significance. Tidal effects could be important in a white dwarf companion, but the only white dwarfs consistent with [an advance in periastron of] 4.266 deg yr^{-1} have either rapid rotation or masses close to the Chandrasekhar limit. Such stars would produce either positive or negligible contributions to $\dot{P}_b$. We conclude that the only straightforward interpretation of the observed orbital decay is in terms of gravitational radiation damping. . . .

[In the last section of their paper, the authors proceed to argue that their measured parameters of the binary-pulsar system do not agree with alternative theories of gravity and are consistent with both members being neutron stars. They then claim that they are seeing evidence of the pulsar's spin axis precessing about 1 degree per year, as also predicted by general relativity.]

Our measurement of the rate of change of orbital period is an entirely new test of general relativity, and at present the only one that depends on the theory's structure beyond the first post-newtonian approximation. The only straightforward interpretation of the data is that gravitational waves exist and carry energy away from an orbiting system at the rate predicted by general relativity. The conclusions that $m_p \approx m_c \approx 1.4\,M_\odot$ . . . give strong support to the interpretation of the PSR1913+16 system as a pair of neutron stars, and provide an important test of the theory of late stages of stellar evolution. . . .

VIII

ACCELERATING OUTWARD

In the latter half of the twentieth century, the rate of astronomical discoveries quickened at an astounding clip as observatories and satellites were sent into space to gather a wide range of radiations and as ever bigger telescopes became available on the ground. The 200-inch Hale telescope atop Palomar mountain in southern California, which had dominated astronomy for many decades, was at last surpassed in size by telescopes with advanced designs situated in South America, Hawaii, and Arizona. At the same time, electronic detectors made small telescopes as powerful in gathering light as larger telescopes with photographic plates, enabling observers to image faint sources with greater clarity. Multi-object spectrographs allowed international collaborations to carry out surveys of stars and galaxies that dwarfed those of the past.

What resulted from these investigations was an unexpected revision of the celestial landscape. For one, astronomers began to gather evidence that extrasolar planets did indeed circle other stars like our Sun within the Milky Way galaxy, a first step in assessing whether life exists elsewhere in the universe. One of the more surprising discoveries was the realization that the luminous galaxies themselves are mere whitecaps immersed within a hidden cosmic sea of "dark matter," believed to be composed of exotic elementary particles that exert their gravitational pull on ordinary matter. Sensitive measurements of the cosmic microwave background suggest that this invisible material accounts for around 85 percent of the matter in the universe.

The cosmos, long thought to be incredibly uniform over very large scales, was also found to exhibit a wondrous texture. By mapping how galaxies congregate through vast reaches of space, astronomers saw that the spirals and ellipticals were not smoothly distributed but rather distinctly arranged as if they sit on the surfaces of huge, nested bubbles, each bubble spanning several hundreds of millions of light-years. Assembled at the dawn of creation, these structures serve as a road map for theorists as they carry out powerful computer simulations to trace how galaxies, clusters of galaxies, and superclusters originated. Meanwhile, using the enhanced vision provided by larger telescopes, observers were able to peer farther out into space to see how galaxies evolved over time. They came to recognize that galaxies did not emerge all at once in the early universe but continued to coalesce and interact with one another over many eons. Most galaxies were found to have constructed massive central black holes in the process.

By the 1980s particle physicists and cosmologists forged a fruitful partnership, which expanded understanding of the universe's birth. New insights into the ways in which the basic forces of nature are related to one another allowed theorists to explore the earliest moments of the Big Bang—since the first 10^{-35} second of cosmic time—which provided answers to mysteries that had long perplexed them. The standard cosmological model of the universe's birth was amended to include a brief moment of hyperexpansion known as inflation, which helped explain why the early universe was so hot, how it became so big and homogeneous, and why it keeps expanding.

The most astonishing find occurred right before the close of the twentieth century, when astronomers discovered that the expansion of the universe is not slowing down as the eons progress (as they long assumed) but rather accelerating. This extraordinary behavior suggests that a "dark energy" permeates the cosmos, serving as the engine for the escalating velocity of space-time's stretching.

69 / Dark Matter

Halfway through the twentieth century astronomers were confident that a final tally of the universe's material content was nearing completion. But by the 1970s substantial evidence was accumulating that the stars and galaxies constituted just a tiny fraction—merely the luminous component—of a more extensive and hidden ocean of matter distributed throughout the universe.

The first hint that something was awry in astronomy's bookkeeping actually arrived in the 1930s. The discoverer was Caltech astronomer Fritz Zwicky, a legend in the astronomical community for his strong opinions, wide-ranging interests, and resolute personality. Many stories of his life dwell on how this Swiss national trained in physics was frequently in conflict with his astronomy colleagues. Absent in these accounts was his compassionate and solicitous behavior toward students and those in need. After World War II Zwicky personally carried out a campaign to help restock the wartorn libraries of Europe; he and his wife packed the books by hand for shipment.[1] As an astronomer, Zwicky is noted for his copious theoretical insights: decades before the phenomena were directly observed, he predicted that supernovae would create tiny neutron stars and that galaxies could act as gravitational lenses (see Chapters 41 and 70). He also recognized that some of the matter in the universe seemed to be missing.

In 1933 Zwicky examined the velocity information then available in the literature on galaxies congregated within the famous Coma cluster, a rich group of hundreds of galaxies some 300 million light-years distant. His statistical analysis revealed that the Coma galaxies are moving around in the cluster at a fairly rapid pace—fast enough that the cluster by all rights should be breaking apart. But because the cluster is very much intact, Zwicky concluded that some kind of unseen matter—cold stars, gas, dust—pervaded the cluster to provide an additional gravitational glue. In his first report to the Swiss journal *Helvetica Physica Acta,* he referred to this hidden ingredient as *dunkle Materie,* or dark matter.[2] Zwicky extended his argument in the 1937 *Astrophysical Journal* paper excerpted below. Mount Wilson astronomer Sinclair Smith discerned a similar effect in the Virgo cluster of galaxies and noted that the missing matter would "remain unexplained until further information becomes available."[3]

Over the years there were occasional references to a missing-mass problem by others, but astronomers largely ignored the issue for three decades, believing the dilemma would disappear once the motions of galaxies in clusters were better understood.[4] But Zwicky's prescient observation came to the forefront of astronomical concerns once Vera Rubin, W. Kent Ford, and several colleagues at the Carnegie Institution of Washington gathered substantial evidence that dark matter was a major component of individual galaxies as well. Rubin had begun studying the rotations of spiral galaxies intending to learn why spirals vary—from ones with arms tightly wrapped to those with arms spread out widely. Soon she discovered that the galaxies were not rotating as expected. Astronomers had largely assumed that galaxies rotated like our solar system, where—following Newton's laws—the outer planets (far from the system's primary concentration of mass, the Sun) move more slowly than the inner planets. But Rubin found that outlying stars on the edges of spiral galaxies were traveling just as fast as the stars closer in. On a graph the rotation velocity remained "flat"—the same speed with distance outward. This meant that an extra, hidden mass—up to ten times more than seen visibly—had to be permeating the galaxy to provide the gravitational force to keep the outlying stars from flying off. "Astronomers can approach their tasks with some amusement," noted Rubin, "recognizing that they study only the 5 or 10 percent of the universe which is luminous."[5]

Radio astronomers were actually the first to see this effect in the early 1970s but only in a handful of galaxies.[6] By eventually studying hundreds of galaxies, Rubin and her team provided overwhelming proof. The report on their first substantial set of galaxies came out in 1978, followed by others over the next decade.[7] Theorists had already suggested a potential hiding place for the extra matter; in 1974 astrophysicists in both the United States and Estonia concluded from theoretical models that a flat spiral disk had to be embedded in a massive halo of material to remain stable.[8] Further evidence provided by x-ray measurements of hot gas around clusters and "weighing" clusters by their gravitational lensing effects (see Chapter 70) added observational support that galaxies and clusters are indeed immersed in dark halos.

By the end of the twentieth century, the exact nature of the dark matter still remained a mystery. Substellar remnants, black holes, dim white-dwarf stars, and gas likely make up a small part of the missing mass, but measurements of the helium and deuterium produced in the early universe suggest the Big Bang didn't make enough ordinary matter—baryons (the collective term for protons and neutrons)—to account for it all. Around 85 percent of the matter in the universe is in a form that makes its presence known only through gravity. Neutrinos, now known to harbor a bit of mass (see Chapter 66), certainly contribute to the total, but from the dynamics of the dark halos around galaxies and clusters cosmologists believe the bulk of the dark matter is composed of weakly interacting massive particles (WIMPs), which are yet to be discovered and which may point the way to new particle physics.

"On the Masses of Nebulae and of Clusters of Nebulae."
Astrophysical Journal, Volume 86 (October 1937)
by Fritz Zwicky

The determination of the masses of extragalactic nebulae [galaxies] constitutes at present one of the major problems in astrophysics. Masses of nebulae until recently were estimated either from the luminosities of nebulae or from their internal rotations. In this paper it will be shown that both these methods of determining nebular masses are unreliable. . . .

The observed absolute luminosity of any stellar system is an indication of the approximate amount of luminous matter in such a system. In order to derive trustworthy values of the masses of nebulae from their absolute luminosities, however, detailed information on the following three points is necessary.

1. According to the mass-luminosity relation, the conversion factor from absolute luminosity to mass is different for different types of stars. The same holds true for any kind of luminous matter. In order to determine the conversion factor for a nebula as a whole, we must know, therefore, in what proportions all the possible luminous components are represented in this nebula.

2. We must know how much dark matter is incorporated in nebulae in the form of cool and cold stars, macroscopic and microscopic solid bodies, and gases.

3. Finally, we must know to what extent the apparent luminosity of a given nebula is diminished by the internal absorption of radiation because of the presence of dark matter.

Data are meager on point 1. Accurate information on points 2 and 3 is

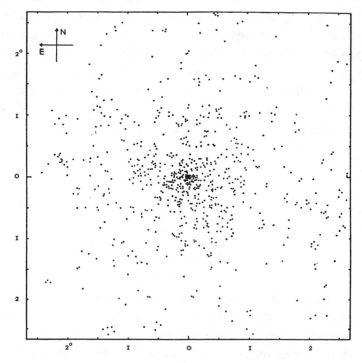

Figure 69.1: "The Coma cluster of nebulae."

almost entirely lacking. Estimates of the masses of nebulae from their observed luminosities are therefore incomplete and can at best furnish only the lowest limits for the values of these masses. . . .

If the total masses of clusters of nebulae were known, the average masses of cluster nebulae could immediately be determined from counts of nebulae in these clusters, provided internebular material is of the same density inside and outside of clusters.

As a first approximation, it is probably legitimate to assume that clusters of nebulae such as the Coma cluster (see Fig. [69.1]) are mechanically stationary systems. With this assumption, the virial theorem of classical mechanics gives the total mass of a cluster in terms of the average square of the velocities of the individual nebulae which constitute this cluster. But even if we drop the assumption that clusters represent stationary configurations, the virial theorem . . . allows us to draw important conclusions concerning the masses of nebulae. . . .

[Omitted here are four pages of calculations in which Zwicky first uses the virial theorem—a relationship between the kinetic and gravitational potential energy of a system—to arrive at a total mass for the Coma cluster.]

We find

$$M > 9 \times 10^{46} \text{ grams} \qquad [1]$$

The Coma cluster contains about one thousand nebulae. The average mass of one of these nebulae is therefore

$$\overline{M} > 9 \times 10^{43} \text{ grams} = 4.5 \times 10^{10} M_\odot \qquad [2]$$

Inasmuch as we have introduced at every step of our argument inequalities which tend to depress the final value of the mass M, the foregoing value [2] should be considered as the lowest estimate for the average mass of nebulae in the Coma cluster. This result is somewhat unexpected, in view of the fact that the luminosity of an average nebula is equal to that of about 8.5×10^7 suns. . . . This discrepancy is so great that a further analysis of the problem is in order. Parts of the following discussion were published several years ago, when the conclusion expressed in [2] was reached for the first time.[9]

We inquire first what happens if the cluster considered is not stationary. . . . [Then obtaining a lower mass to resolve the above discrepancy] means that the cluster is expanding . . . the cluster will ultimately just fly apart. . . . By assuming this, however, we run into . . . difficulties. In the first place, it is difficult to understand why under these circumstances there are any great clusters of nebulae remaining in existence at all, since the formation of great clusters by purely geometrical chance is vanishingly small. . . .

The distribution of nebulae in the Coma cluster, illustrated in Figure [69.1], rather suggests that stationary conditions prevail in this cluster. It is proposed, therefore, to study the Coma cluster in more detail. . . . Sufficiently large amounts of internebular matter in clusters might seriously change our estimate [2] of the average value of nebular masses as derived from the preceding application of the virial theorem to clusters of nebulae. . . .

"Extended Rotation Curves of High-Luminosity Spiral Galaxies. IV. Systematic Dynamical Properties, Sa → Sc." *Astrophysical Journal,* Volume 225 (November 1, 1978) by Vera C. Rubin, W. Kent Ford, Jr., and Norbert Thonnard

Introduction

In the 50 years since Hubble (1926) introduced his classification sequence for galaxies, few systematic observational programs have attempted to study dynamical properties of galaxies as a function of Hubble type (HT). The constraints have been principally instrumental. Optical rotation curves have furnished valuable dynamical information, but generally only for the inner regions of late-type spirals. Neutral hydrogen [radio] observations have revealed integral properties of gas-rich systems, but with limited spatial resolution.

Available optical instrumentation now permits the detection of emission across a very large portion of the disks of spirals, well beyond the "turnover point" in the rotation curves. We have initiated a program to obtain spectra of spiral galaxies at high velocity resolution and at a large spatial scale, in order to study their properties as a function of HT. We now have velocities for 10 spirals, Sa through Sc, of high intrinsic luminosity. . . .

The Rotation Curves

Optical spectra were obtained with the Kitt Peak and Cerro Tololo 4 m [four-meter] spectrographs plus Carnegie image tube. . . . [One striking feature] can be observed directly from the spectra. *All rotation curves are approximately flat,* with only a slight rise or fall following the initial steep gradient. . . .

Parameters for these galaxies discussed are listed in Table [69.1]. . . . Columns [4], [5], [6] give the galaxy radius, the radius of the last measured velocity, and the ratio of the two. In the mean, our velocities extend over 80% of the galaxy radius.

Rotation curves are plotted in Figures [69.2] and [69.3]. The general flatness of the curves . . . are notable. . . .

Table 69.1: Data for Program Galaxies

(1)	(2)	(3)	(4)	(5)	(6)	(7)
NGC	Hubble Class*	Distance (Mpc)	Radius r (kpc)	Nuclear Distance of Last Velocity (kpc)	Fraction r Observed	Rotation Curve at Large r
4378	Sa I	48.6	23.4	22.0	0.94	Falling
4594	Sa	18.2	23.6	15.	0.64	Rising
7217	Sb–Sab III	24.7	13.8	11.0	0.80	Falling
2590	Sb	95.9	34.8	17.4	0.50	Rising
1620	Sbc	68.5	30.8	21.9	0.71	Rising
3145	Sbc I	68.3	30.8	25.3	0.82	Flat
801	Sbc–Sc	119.	57.1	49.1	0.86	Flat or Rising
7541	Sbc–Sc III	57.5	28.6	23.2	0.81	Rising
7664	Sbc–Sc	74.2	35.6	28.1	0.79	Flat or Rising
2998	Sc I	95.6	39.1	34.0	0.87	Flat
3672	Sc I–II	33.1	19.8	17.6	0.89	Flat

* In Hubble's classification scheme, the capital letter S stands for spiral galaxy, while the lowercase *a, b,* and *c* denote the type of spiraling. An Sa-type galaxy has tightly wound spiral arms, while Sb and Sc are successively more open. The combination of letters, such as Sab and Sbc, was later introduced by astronomer Gerard de Vaucouleurs to accommodate transitional stages.

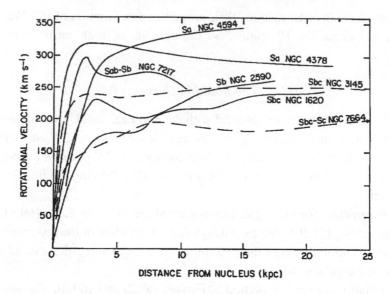

Figure 69.2: "Rotational velocities for seven galaxies, as a function of distance from nucleus. Curves have been smoothed to remove velocity undulations across arms and small differences between major-axis velocities on each side of nucleus. . . ."

Conclusions

The major result of this work is the observation that rotation curves of high-luminosity spiral galaxies are flat, at nuclear distances as great as $r = 50$ kpc. Roberts and his collaborators deserve credit for first calling attention to flat rotation curves.[10] Recent 21 cm observations by Krumm and

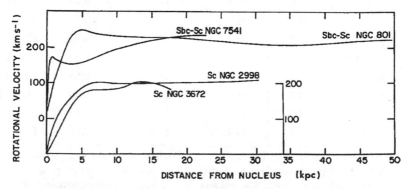

Figure 69.3: "Rotation curves for two pairs of galaxies."

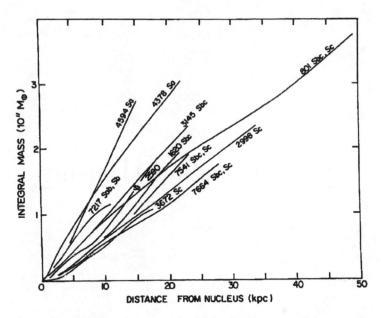

Figure 69.4: "Integral mass within disk of radius *r*, as a function of *r*, for 11 galaxies, Sa through Sc, out to last measured velocity. . . . Linear increase of mass with radius is a consequence of flat rotation curves. . . ."

Salpeter[11] have strengthened this conclusion. These results take on added importance in conjunction with the suggestion of Einasto, Kaasik, and Saar,[12] and Ostriker, Peebles, and Yahil[13] that galaxies contain massive halos extending to large *r*. Such models imply that the galaxy mass increases significantly with increasing *r* [see Figure 69.4] which in turn requires that rotational velocities remain high for large *r*. The observations presented here are thus a necessary but not sufficient condition for massive halos. . . .

70 / Gravitational Lensing

The first cosmic gravitational lens was serendipitously discovered in 1979, nearly seventy years after the phenomenon was first imagined. And within a decade of the discovery, an effect that Einstein once thought insignificant became one of astronomy's most valuable tools for exploring the universe.

When presenting his theory of general relativity in 1915, Einstein made the prediction that a beam of light would noticeably bend as it passed by a massive celestial body, such as the Sun—twice the bending predicted from Newton's laws alone.[14] And when astronomers, monitoring a 1919 solar eclipse, saw starlight grazing the darkened Sun get deflected by the amount forecast by general relativity, Einstein became world-famous overnight (see Chapter 36). The Sun, in this case, becomes the gravitational equivalent of an optical lens.

Others soon wondered whether such an effect might be seen farther out. In 1920 the British astronomer Arthur Eddington suggested the possibility of seeing multiple images of a star if the star was properly situated behind another stellar body acting as a lens.[15] Four years later, Orest Chwolson in Germany noted that if a distant star is aligned just right—lying precisely behind its gravitational lens—its light would be spread out as a ring that completely surrounds the lens.[16]

Einstein himself was aware of these effects. As early as the spring of 1912, three years before his theory of general relativity was complete, he carried out some calculations on gravitational lensing in his notebook and jotted down the possibility that a lens might not only create a double image of a star but also magnify the intensity of its light.[17] But he seems to have dropped the subject and did not publish his findings until prompted by a young Czech electrical engineer and amateur scientist in 1936 to once again consider the problem of cosmic lensing. "Some time ago, [Rudi] W. Mandl paid me a visit and asked me to publish the results of a little calculation, which I had made at his request," wrote Einstein in his paper to the journal *Science* entitled "Lens-Like Action of a Star by the Deviation of Light in the Gravitational Field." Einstein declared it "a most curious effect" but also concluded there was "no hope of observing this phenomenon directly," since it defied "the resolving power of our instruments."[18] Privately, Einstein told the editor of *Science* that his findings had "little value, but it makes the poor guy [Mandl] happy."[19]

The following year Caltech astronomer Fritz Zwicky noted that while the effect for stars is extremely small, "extragalactic *nebulae* [galaxies] offer a much better chance . . . for the observation of gravitational lens effects," enabling astronomers to "see nebulae at distances greater than those ordinarily reached by even the greatest telescopes."[20] It was a prescient notion, but one that was not confirmed for another forty-two years.

In 1979 British astronomer Dennis Walsh was closely perusing a photographic plate to locate the visible counterpart to a newly discovered radio source, 0957+561, when he noticed that the radio object's position coincided with *two* starlike bodies instead of just one. Additional telescopic observations at the Kitt Peak National Observatory confirmed that the cozy pair were quasars. More importantly, the spectra of these quasars were nearly identical, which hinted that they were not just the chance alignment of two separate objects (which often happens). A celestial object's spectrum is as distinctive and exclusive as a fingerprint. The spectral matchup strongly suggested that it was the *same* quasar—the brilliant core of a young galaxy some 9 billion light-years distant—seen in duplicate. In their report to the journal *Nature*, Walsh and his colleagues suspected a gravitational lens was at work, and further observations confirmed it.

At the time of the discovery the lens was unknown, but within months it was found to be a giant elliptical galaxy, the brightest member of a rich cluster located halfway between the quasar and the Earth.[21] Although the physical principle is not the same, one can think of the quasar light as a stream of water that comes upon a massive object (the lens) and gets diverted to either side of it. One stream becomes two or more streams. Thus our eyes detect multiple images of the quasar rather than just one. Because of the differing lengths of each light path, the astronomers noted in their report that any variation in the quasar's light intensity should be seen at different times in each image. This delay was measured many times over the succeeding years: any change in the intensity of image A of the quasar 0957+561 is duplicated by image B after the passage of around 420 days.[22]

Many other gravitational lens systems were later uncovered. When galaxies are lensed (often by intervening clusters of galaxies), their broader shapes are smeared into arcs or rings. Since the amount of deflection depends on the total mass of the gravitational lens, astronomers use this information to "weigh" clusters of galaxies (an idea first suggested by Zwicky in 1937).[23] Their results confirm that up to 90 percent of the mass in clusters is composed of an unknown dark matter (see Chapter 69).

And with improved technology, astronomers were eventually able to detect the gravitational lensing of individual stars as well, an enterprise that Einstein deemed hopeless. Background stars in our Milky Way and in the Magellanic clouds have been seen to briefly magnify—"microlense"—due to dark objects passing in front of them.

"0957 + 561 A, B: Twin Quasistellar Objects or Gravitational Lens?" *Nature*, Volume 279 (May 31, 1979) by Dennis Walsh, Robert F. Carswell, and Ray J. Weymann

Abstract

0957 + 561 A, B are two QSOs [quasistellar objects or quasars] of mag 17 with 5.7 arc s[econd] separation at redshift 1.405 [~9 billion light-years distant]. Their spectra leave little doubt that they are associated. Difficulties arise in describing them as two distinct objects and the possibility that they are two images of the same object formed by a gravitational lens is discussed.

Spectroscopic observations have been in progress for several years on QSO candidates using a survey of radio sources made at 966 MHz with the MKIA telescope at Jodrell Bank. Many of the identifications have been published [but some were] sources that were either too extended or too confused for accurate interferometric positions to be measured, and these were observed with the pencil-beam of the 300 ft telescope at NRAO, Green Bank at λ 6 cm and λ 11 cm. This gave positions with typical accuracy 5–10 arc s and the identifications are estimated as ~80% reliable.

The list . . . includes the source 0957 + 561 which has within its field a close pair of blue stellar objects, separated by ~6 arc s, which are suggested as candidate identifications. Their positions and red and blue magnitudes, m_R and m_B, estimated from the Palomar Observatory Sky Survey (POSS) are given in Table [70.1]. . . . Since the images on the POSS overlap, the magnitude estimates may be of lower accuracy than normal, but they are very nearly equal and object A is definitely bluer than object B. The mean position of the two objects is 17 arc s from the radio position, so the identification is necessarily tentative.

Observations

The two objects 0957 + 561 A, B were observed on 29 March 1979 at the 2.1 m telescope of the Kitt Peak National Observatory (KPNO) using the intensified image dissector scanner (IIDS). . . . After 20-minute integration on each object it was clear that both were QSOs with almost identical spectra and redshifts of ~1.40 on the basis of strong emission lines

Table 70.1: Positions and Magnitudes of 0957 + 561 A, B

Object	RA	Dec (1950.0)	M_R	M_B
0957 + 561A	09 57 57.3	+56 08 22.9	17.0	16.7
0957 + 561B	09 57 57.4	+56 08 16.9	17.0	17.0

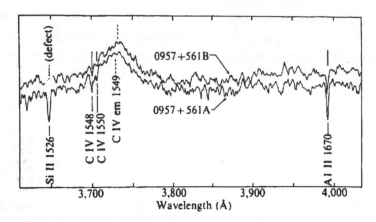

Figure 70.1: "Microdensitometer tracings of portions of the spectra of 0957 + 561 A and B. . . . The solid lines mark the position of absorption features in the two QSOs and the dashed lines mark the adopted centers of the C IV emission line."

identified as C IV λ 1549 and C III λ 1909.* Further observations were made on 29 March and on subsequent nights. . . . By offsetting to observe empty sky a few arc seconds from one object on both 29 and 30 March it was confirmed that any contamination of the spectrum of one object by light from the other was negligible. . . .

The data on the C IV λ 1549 and C III λ 1909 lines are much more accurate than those on the other lines. . . . Within the limits of observational error, the corresponding lines in each object are identical in observed wavelength and equivalent width [see Figure 70.1]. . . .

Although no attempt was made to carry out accurate spectrophotometry, some characteristics of the continua seem fairly well defined. Below about 5,300 Å they appear to have identical shapes, with QSO A brighter than B by 0.35 mag. Above 5,300 Å, however, the flux from B rises more steeply than that from A and they are equal at ~6,500 Å [see figure 70.2]. These results are consistent with the magnitude estimates of Table [70.1].

Discussion

The great similarity in the spectral characteristics of these two QSOs which have the same redshift and which are separated by only 6 arc s seems

---

* C III and C IV denote different ionized states of carbon.

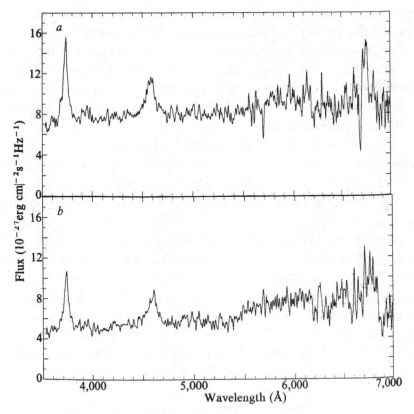

Figure 70.2: "IIDS scans of 0957 + 571 A (*a*) and B (*b*). The data are smoothed over 10 Å and the spectral resolution is 16 Å."

to constitute overwhelming evidence that the two are physically associated. . . . [W]e shall assume the QSO redshifts are cosmological. The same similarities further suggest that we may be dealing with a single source which has been split into two images by a gravitational lens. . . .

In the conventional interpretation of two adjacent QSOs we must either regard it as a coincidence that the emission spectra are so nearly the same, or assume that the initial conditions, age and environment influencing the development of the QSOs have been so similar that they have evolved nearly identically. . . . The conventional interpretation of the sources as two QSOs requires additional coincidences to explain the absorption line systems regardless of the mechanism invoked to explain the absorption. . . .

We now consider the possibility that a gravitational lens is operating. The theory of gravitational imaging in a cosmological context has been considered elsewhere[24] . . . and we simply quote the main results of applying this theory. The following are the relevant parameters involved in considering the gravitational lens hypothesis: the angular separation of the images, the shape of the images and their sizes, and the amplification of the two images. There is no evidence on the plate taken on 2 April or on the POSS for any departure of the images from stellar images. The magnitude difference between A and B (Table [70.1]) is ~0.3. mag and this is confirmed by our observations.

The 0.3 mag difference between the two components requires that the amplification of QSO light is ~4 for the brighter image, and thus implies a normal luminosity for the QSO. . . . The maximum angular size of the lens is only ~8 times that of the object, so we should not expect to resolve it on the sky.

An apparent objection arises from the difference in the shapes of the continua between the two QSOs. It is possible that differential reddening along the two light paths may be responsible. Note that the observed break at 5,300 Å corresponds to an emitted wavelength of 2,200 Å in the rest system of the QSOs. This is the wavelength of a well known resonance in interstellar extinction by dust in our Galaxy, and a model can be constructed to explain the observed continuum ratio incorporating the 2,200 Å feature at the redshift of the QSOs. This would imply that the intrinsic flux from B exceeds that from A.

Further observations would shed light on the gravitational lens hypothesis. If the flux from the object is variable, the light curves of the two images should be similar but with a relative time delay due to the difference in path lengths. The lag depends on the details of the geometry, but with the parameters discussed above would be expected to be of the order of months to years. Determination of the radio structure would also clearly be of great value.

Since submission of this article we have heard that on 19, 20, and 21 April the two QSOs were observed by N. Carleton, F. Chaffee and M. Davis (of the Smithsonian Astrophysical Observatory) and R.J.W. using the SAO photon-counting reticon spectrograph attached to the SAO-UA multiple mirror telescope. The observations covered the range 5,900–7,100 Å. . . . The main results are: (1) to within the measuring errors the Mg II emission lines have the same profiles and observed equivalent widths (85 and 76 $\pm$ 12 Å for A and B respectively) and the same redshift (1.4136 $\pm$ 0.0015 for both). (2) Absorption lines due to Fe II $\lambda\lambda$ 2586, 2599, Mg II $\lambda\lambda$ 2795, 2802 and Mg I $\lambda\lambda$ 2852 are present in both objects but are somewhat stronger in A. The mean heliocentric redshifts of the two absorption systems are 1.3915 for A and 1.3914 for B. A cross-correlation analysis confirms that the difference in the two absorption redshifts is remarkably small and corresponds to a velocity difference of 7 ± 10 km s^{-1}. These observations strengthen the case for a gravitational lens.

Figure 70.3: Since the twin quasars were discovered in 1979, many other gravitational lens systems have been found. Here is a Hubble space telescope image of the rich cluster of galaxies Abell 2218. This cluster is so massive and compact that light rays passing through it are deflected by its enormous gravitational field. This lensing phenomenon magnifies, brightens, and distorts the images of galaxies farther out into arcs and streaks.

71 / Inflation

In the late 1970s Alan Guth, then a young postdoc at Stanford University exploring how the latest theories in particle physics would affect the behavior of the early universe, came to see that our cosmos may have begun not only with a bang but with a cosmic *burp*—an exceedingly brief moment of superaccelerated expansion that not only helped solve a number of cosmological mysteries but also became the foundation for cosmological models in the decades to follow. Guth called this special event in our universe's history *inflation.*

Guth did not start out to do cosmology. Working with a Cornell University colleague, Henry Tye, he was trying to determine if the most current grand unified theories in physics—theories that attempt to unify the forces of nature—might give rise to magnetic monopoles (hypothetical particles of magnetic charge). The two particle physicists concluded that monopoles would be generated, and proceeded to see how many might be produced in the Big Bang. So many would have been created that "we began to wonder why the universe was here at all," said Guth. "Their tremendous weight would have closed the universe back up eons ago."[25] Guth and Tye eventually surmised that monopole production could be curtailed if the early universe "supercooled" as it expanded—the forces of nature in effect staying unified for a while as temperatures plunged, just as water can sometimes supercool and remain liquid below its freezing point. The notion of inflation was encountered when Tye casually reminded Guth to check how this supercooling might affect the expansion of the infant universe.

Guth carried out the initial calculation at his home office on the night of December 6, 1979. He started at about 10^{-35} second into the universe's birth, when the cosmos was smaller than a proton and starting to cool below 10^{27} degrees. His equations told him that the supercooling would endow the universe with a tremendous potential energy. A pressure contribution to gravity became so substantial that it reversed the effect of gravity: rather than being an attractive force, gravity became repulsive, causing the tiny universe to balloon outward at a superaccelerated rate for a minuscule moment ($\sim 10^{-35}$ second or so), stretching space-time by a factor of 10^{30} or more. When this supercooled state came to an end, its latent energy was released as the fireball that eventually cooled into the matter and radiation that surround us today.

Though not trained in cosmology, Guth came to realize within a few weeks that his scenario solved two puzzles that had long been troubling cosmologists: why the universe was so uniform (horizon problem) and why it was right at the brink between eternal expansion and eventual recollapse (its curvature geometrically "flat"). With general relativity, Einstein showed that matter causes space to warp and bend (see Chapter 36). With too little mass in the universe, the curvature of space-time will never curl back up on itself; instead, the universe remains "open," destined to expand forever. But a certain quantity of mass/energy — just beyond a critical density — can provide enough gravity to "close" the universe, slowing down the galaxies and eventually drawing them back inward. As Guth points out in his paper, cosmologists had no fundamental explanation for the universe to be at the very cusp between these two states, as our universe seems to be. Such a condition required extreme fine-tuning. It was far more probable for the universe to be *very* open or to have closed up within 10^{-44} second after the Big Bang. A burst of inflation, though, essentially jump-starts the universe and immediately flattens its curvature, driving it naturally to the very boundary between open and closed (a condition known in cosmology as $\Omega = 1$).*

* Ω is the simple ratio of the density of cosmic matter divided by the critical density necessary to bring the universe to the brink of closure. If both densities are the same, then Ω equals 1. If Ω is less than 1, the universe remains open; if Ω is greater than 1, the expanding universe will eventually halt and then contract at some distant time in the future.

Inflation also explained the universe's puzzling uniformity. Over scales of billions of light-years, matter and remnant radiation from the Big Bang are distributed fairly smoothly, which had been a mystery because there is not enough time in a standard Big Bang model to obtain the same temperature and density in all parts of the embryonic universe. But the preinflationary kernel that gave birth to our universe—a trillion times smaller than a proton—did have time to blend its contents well, with inflation then stepping in to maintain this uniform mixture throughout the growing bubble of space-time. Other theorists were independently arriving at similar scenarios, in particular Alexei Starobinsky in Moscow.[26] But Guth's more accessible and widely circulated model (with its catchy name) became the most influential.

Guth's initial approach had a fatal flaw, which he pointed out in his paper. At the end of the hyperaccelerated burst, he was left with a chaotic collection of tiny "bubble universes," none of which could evolve into the universe we see around us. But in the following years other theorists, including Andrei Linde, Andreas Albrecht, and Paul Steinhardt, developed other versions of inflation involving different mechanisms that allowed any one of Guth's many bubbles to balloon into a suitable cosmos.[27] Inflation soon became an essential feature in standard cosmological models.

Inflation also came to be important in explaining the universe's large-scale structure. It predicted that quantum fluctuations in the universe's preinflationary seed would have blown up to astronomical scales as the universe swiftly expanded. According to the theory, it was these perturbations that eventually pushed and squeezed primordial matter into galaxies and clusters. Inflationary models received valuable observational support in 1992 when NASA's Cosmic Background Explorer (COBE) satellite saw in the universe's microwave background (the residual heat from the Big Bang) the pattern of fluctuations predicted.[28] The match was even more precise with measurements taken by the Wilkinson Microwave Anisotropy Probe (WMAP) a decade later.

"Inflationary Universe: A Possible Solution to the Horizon and Flatness Problems."
Physical Review D, Volume 23 (January 15, 1981)
by Alan H. Guth

Abstract

The standard model of hot big-bang cosmology requires initial conditions which are problematic in two ways: (1) The early universe is assumed to be highly homogeneous, in spite of the fact that separated regions were causally disconnected (horizon problem); and (2) the initial value of the Hubble constant must be fine tuned to extraordinary accuracy to produce a universe as flat (i.e., near critical mass density) as the one we see today (flatness problem). These problems would disappear if, in its early history, the universe supercooled to temperatures 28 or more orders of magnitude below the critical temperature for some phase transition. A huge expansion factor would then result from a period of exponential growth, and the entropy of the universe would be multiplied by a huge factor when the latent heat is released. Such a scenario is completely natural in the context of grand unified models of elementary-particle interactions. In such models, the supercooling is also relevant to the problem of monopole suppression. Unfortunately, the scenario seems to lead to some unacceptable consequences, so modifications must be sought.

Introduction: The Horizon and Flatness Problems

The standard model of hot big-bang cosmology relies on the assumption of initial conditions which are very puzzling in two ways. . . . The purpose of this paper is to suggest a modified scenario which avoids both of these puzzles. . . .

In the standard model, the initial universe is taken to be homogeneous and isotropic, and filled with a gas of effectively massless particles in thermal equilibrium at temperature T_0. The initial value of the Hubble expansion "constant" H is taken to be H_0, and the model universe is then completely described.

Now I can explain the puzzles. The first is the well-known horizon problem.[29] The initial universe is assumed to be homogeneous, yet it consists of at least ~10^{83} separate regions which are causally disconnected (i.e., these regions have not yet had time to communicate with each other via light signals). . . .

The second puzzle is the flatness problem. This puzzle seems to be much less celebrated than the first, but it has been stressed by [Robert] Dicke and [P. James E.] Peebles.[30] I feel that it is of comparable importance to the first. It is known that the energy density ρ of the universe today is near the critical value ρ_{cr} (corresponding to the borderline between an open and closed universe). . . .

[Here Guth defines the term Ω, the ratio of the universe's energy density ρ to the critical density ρ_{cr} necessary to bring our universe to the border between being open (eternally expanding) or closed (the expansion eventually stopping and turning into a contraction). In other words, $\Omega = \rho/\rho_{cr}$.]

One can safely assume that

$$0.01 < \Omega_p < 10$$

. . . and the subscript p denotes the value at the present time. Although these bounds do not appear at first sight to be remarkably stringent, they, in fact, have powerful implications. The key point is that the condition $\Omega \approx 1$ [the universe's density equaling the critical density] is unstable. Furthermore, the only time scale which appears in the equations for a radiation-dominated universe is the Planck time . . . 5.4×10^{-44} sec. A typical closed universe will reach its maximum size on the order of this time scale, while a typical open universe will dwindle to a value of ρ much less than ρ_{cr}. A universe can survive ~10^{10} years only by extreme fine tuning of the initial values of ρ and H, so that ρ is very near ρ_{cr}. For the initial conditions taken at $T_0 = 10^{17}$ GeV, the value of H_0 must be fine tuned to an accuracy of one part in 10^{55}. In the standard model this incredibly precise initial relationship must be assumed without explanation. . . .

[Omitted here is Guth's summary of the basic equations of the standard model of the universe.]

The Inflationary Universe

. . . Suppose the equation of state for matter . . . exhibits a first-order phase transition at some critical temperature T_c. Then as the universe cools through the temperature T_c, one would expect bubbles of the low-temperature phase to nucleate and grow. However, suppose the nucleation rate for this phase transition is rather low. The universe will continue to cool as it expands, and it will then supercool in the high-temperature phase. Suppose that this supercooling continues down to some temperature T_s, many orders of magnitude below T_c. When the phase transition finally takes place at temperature T_s, the latent heat is released. . . .

Let us examine the properties of the supercooling universe in more detail. . . . As T → 0, the system is cooling not toward the true vacuum, but rather toward some metastable false vacuum with an energy density ρ_0 which is necessarily higher than that of the true vacuum. . . . The universe is expanding exponentially, in a false vacuum state of energy density ρ_0. . . . [T]he pressure is negative . . . [and] it can be seen that the negative pressure is also the driving force behind the exponential expansion. . . .

If the universe reaches a state of exponential growth, it is quite plausible for it to expand and supercool by a huge number of orders of magnitude before a significant fraction of the universe undergoes the phase transition. . . . Assuming that at least some region of the universe started at temperatures high compared to T_c, one would expect that, by the time the temperature in one of these regions falls to T_s, it will be *locally* homogeneous, isotropic, and in thermal equilibrium. . . . When the temperature of such a region falls below T_c, the inflationary scenario will take place. The end result will be a huge region of space which is homogeneous, isotropic, and of nearly critical mass density. . . . [T]his region can be bigger than (or much bigger than) our observed region of the universe.

[Omitted here is Guth's discussion of the inflationary model in the context of grand unified models of elementary-particle interactions.]

Problems of the Inflationary Scenario

. . . [T]he inflationary scenario seems to lead to some unacceptable consequences. It is hoped that some variation can be found which avoids these undesirable features but maintains the desirable ones. . . .

The central problem is the difficulty in finding a smooth ending to the period of exponential expansion. . . . The randomness of the bubble formation process . . . leads to gross inhomogeneities. . . . [This does] not quite prove that the scenario is impossible, but these consequences are at best very unattractive. Thus, it seems that the scenario will become viable only if some modification can be found which avoids these inhomogeneities. . . .

Conclusion

I have tried to convince the reader that the standard model of the very early universe requires the assumption of initial conditions which are very implausible for two reasons:

(i) *The horizon problem.* Causally disconnected regions are assumed to be nearly identical; in particular, they are simultaneously at the same temperature.

(ii) *The flatness problem.* For a fixed initial temperature, the initial value of the Hubble "constant" must be fine tuned to extraordinary accuracy to produce a universe which is as flat as the one we observe.

Both of these problems would disappear if the universe supercooled by 28 or more orders of magnitude below the critical temperature for some phase transition. (Under such circumstances, the universe would be growing exponentially in time.) However, the random formation of bubbles of the new phase seems to lead to a much too inhomogeneous universe. . . .

In conclusion, the inflationary scenario seems like a natural and simple way to eliminate both the horizon and the flatness problems. I am publishing this paper in the hope that it will highlight the existence of these problems and encourage others to find some way to avoid the undesirable features of the inflationary scenario.

72 / The Bubbly Universe

The long-established assumption that galaxies are generally spread uniformly over great distances throughout the universe was jolted in the 1980s when astronomers at the Harvard-Smithsonian Center for Astrophysics discovered a distinct pattern to their dispersal: the galaxies were found to form bubblelike structures, which surrounded voids deplete of bright galaxies. The distribution of galaxies, it was reported, resembled "suds in the kitchen sink."[31]

Almost as soon as galaxies were discovered in the 1920s (see Chapter 51), surveys were initiated to trace their positions across the celestial sky. By 1926 Edwin Hubble was concluding that galaxies are smoothly distributed.[32] Harlow Shapley at Harvard soon questioned this premise, noting that galaxies tended to huddle in groups and clusters. A major example was the prominent Virgo cluster of galaxies near the north galactic pole. By 1933 Shapley had cataloged twenty-five clusters. "The irregularities in distribution are large," he reported, "and . . . probably not to be attributed to chance."[33] But a deeper survey by Hubble published the following year seemed to confirm his initial impression that the universe was indeed homogeneous.[34] Whatever clustering there was, he countered, tended to smooth out as you viewed larger and larger swathes of space. "On the grand scale," he said, ". . . the tendency to cluster averages out."[35] This became the cosmological principle: that the universe will appear the same from any position within it, its contents having no distinct structure over an immense range.

Surveys in the 1950s, however, greatly increased the number of known clusters of galaxies and brought a further step in the cosmic hierarchy to the attention of astronomers. In 1953 Gérard de Vaucouleurs began describing a particular collection of clusters—which included our Local Group of galaxies, the Virgo cluster, and a cloud of galaxies in Ursa Major—as the "Local Supergalaxy" (later "supercluster"). Forming a large, flat system, the clusters together spanned some 160 million light-years from end to end.[36] Using the Palomar Observatory Sky Survey and other survey information, George Abell in 1958 produced the first comprehensive catalog of clusters of galaxies—2,712 in all—and saw further evidence that clusters assembled into larger systems.[37] C. Donald Shane and Carl Wirtanen, conducting a galaxy survey from Lick Observatory, also reported that the "association of clusters to form larger aggregations . . . seems to be a rather general feature."[38]

By the 1970s, with the introduction of new technologies to determine the distances to galaxies more efficiently, astronomers began to carry out extensive redshift surveys (the distances revealed by the amount a galaxy's light is reddened by the universe's expansion) to map the large-scale structure of the cosmos. A number of filamentary superclusters were traced across the celestial sky, such as the Coma supercluster, the Hercules supercluster, and the Perseus-Pisces supercluster.[39] While carrying out these surveys, observers were disconcerted to find that there were also vast regions that appeared to be devoid of galaxies. In 1981 Robert Kirshner and his colleagues reported on a spherical region some 300 million light-years wide that contained virtually no bright galaxies. It came to be known as the Boötes void for its location in that constellation.[40] Theorists began speculating that the superclusters might all be connected to form a cell-like structure through the universe.[41]

At the Harvard-Smithsonian Center for Astrophysics, Margaret Geller and John Huchra, who suspected that many of the large-scale structures being sighted might be optical illusions, decided in 1985 to survey a deep and narrow slice of the sky—a wedge 6 degrees thick and 117 degrees wide. Distances to all the galaxies in this swath were measured out to some 650 million light-years, using telescopes on Mount Hopkins in Arizona. Geller and Huchra initially assumed that clusters and superclusters would just show up as random groupings amid a more diffuse distribution of galaxies. But when they instructed their graduate student, Valérie de Lapparent, to construct a map of the survey's redshifts, they were surprised by the results. Instead of being randomly scattered over the chart, the galaxies assembled to form a unique architecture; they congregated as if they resided on the surfaces of gigantic, nested bubbles (see Figure 72.1). Inside the bubbles were equally huge voids, where matter is less dense. All the previous hints of a universal frothiness came into dramatic focus. Over the succeeding years, other redshift surveys probed farther into space and saw the same foamlike pattern repeated outward (see Figure 72.2).

Initially the Harvard-Smithsonian team wondered whether the voids were chiefly generated by the shock waves of early and powerful supernova explosions, sweeping the primordial gases into thin, spherical shells. But computer simulations, as well as evidence from the cosmic microwave background, currently suggest that the seeds of the bubblelike structures were largely forged when pressure waves moved through the early universe's hot primordial plasma, creating regions of compressed and rarefied matter. Gravity then amplified these irregularities, leading to galaxies and clusters forming predominantly in the areas of compression and the less dense voids enlarging over time and remaining relatively empty.

"A Slice of the Universe." *Astrophysical Journal*, Volume 302 (March 1, 1986) by Valérie de Lapparent, Margaret J. Geller, and John P. Huchra

Abstract

We describe recent results obtained as part of the extension of the [Harvard-Smithsonian] Center for Astrophysics redshift survey to m_B [magnitude] = 15.5. The new sample contains 1,100 galaxies (we measured 584 new redshifts) in a 6° × 117° strip going through the Coma cluster. Several features of the data are striking. The galaxies appear to be on the surfaces of bubble-like structures. The bubbles have a typical diameter of ~25 h^{-1} Mpc [megaparsecs].* The largest bubble in the survey has a diameter of ~50 h^{-1} Mpc, comparable with the most recent estimates of the diameter of the void in Boötes. . . . The edge of the largest void in the survey is remarkably sharp.

All of these features pose serious challenges for current models for the formation of large-scale structure. . . . These new data might be the basis for a new picture of the galaxy and cluster distributions.

I. Introduction

The behavior of the distribution of galaxies on scales $\gtrsim$ 10 h^{-1} Mpc ($H_0 = 100h$ km s^{-1} Mpc^{-1}) is a critical constraint on models for the formation of large-scale structure. . . . On the observational side there have been a number of discoveries of structures which extend for tens of Mpc. Zeldovich, Einasto, and Shandarin emphasize a general cell-like structure in the galaxy distribution on these and larger scales.[42] Two striking examples of individual large structures are the Perseus-Pisces chain, an apparently filamentary structure extending for 40 h^{-1} Mpc, and the void in Boötes which has a volume of order 10^5 h^{-3} Mpc^3.[43] Taken at face value, the break in the correlation function at a scale of 15 h^{-1} Mpc indicates that structures as large as these must not be common. However, the determination of the behavior of the correlation function on large scales may be suspect because of biases in the galaxy catalogs.

This *Letter* is a preliminary discussion of recent results obtained as part of the extension of the Center for Astrophysics redshift survey. Several features of the results are striking. The distribution of galaxies in the redshift survey slice looks like a slice through the suds in the kitchen sink; it appears that the galaxies are on the surfaces of bubble-like structures with diameter 25–50 h^{-1} Mpc. This topology poses serious challenges for current models for the formation of large-scale structure.

II. The Data

. . . On the sky, this survey covers a strip of ~117° × 6°, centered near the north Galactic pole. We assigned magnitudes to the individual objects

---

* h is a widely used factor in astronomy ($h = H_0/100$), which allows distances to be given without specifying a particular Hubble constant. For a Hubble constant (H_0) of 100, $h = 1$; for a H_0 of 50, $h = \frac{1}{2}$ and the distance of 25 Mpc would be doubled to 50 Mpc.

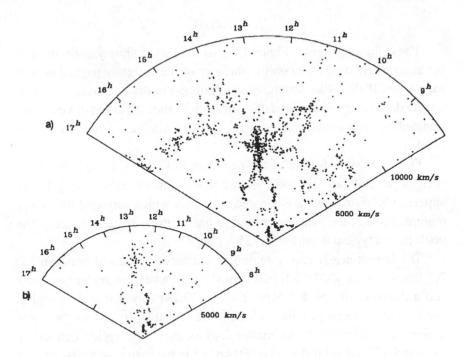

Figure 72.1: "(*a*) [top] Map of the observed velocity plotted vs. right ascension in the declination wedge 26°.5 ≤ δ ≤ 32°.5. The 1061 objects plotted have m_B ≤ 15.5 and V ≤ 15,000 km s^{-1}. (*b*) [bottom] Same as [Figure 72.1*a*] for m_B ≤ 14.5 and V ≤ 10,000 km s^{-1}. The plot contains 182 galaxies. . . ."

in multiple systems: galaxies are included in the sample only if they satisfy m_B ≤ 15.5. The resulting catalog contains 1099 galaxies. . . .

Figure [72.1*a*] shows a plot of the observed velocity versus right ascension for the galaxies brighter than m_B = 15.5 in the full 6° thick wedge; we only plot the 1061 objects with velocities less than 15,000 km s^{-1}. The effective depth of the sample is ~100h^{-1} Mpc. For contrast, we plot in Figure [72.1*b*] the 182 objects with m_B ≤ 14.5 and V ≤ 10,000 km s^{-1}.

Figures [72.1*a*] and [72.1*b*] differ significantly in appearance. In Figure [72.1*a*] nearly every galaxy with V ≤ 10,000 km s^{-1} appears to be in a large structure. We argue below that the size of the largest of these bubble-like structures is comparable with the depth of the earlier survey. Thus undersampling explains the difference in the appearance of the surveys. . . .

III. Analysis

The cellular pattern of Figure [72.1a] . . . can be simply understood if the galaxies are distributed on the surfaces of shells tightly packed next to each other. If shell-like structures are common in the universe, any sufficiently deep wedge-shaped redshift survey will show a pattern of voids surrounded by connected filaments of galaxies similar to that in Figure [72.1a].

One impressive feature of the new data in Figure [72.1a] is the presence of several large regions almost devoid of galaxies. The galaxies appear to be distributed in elongated structures which surround the empty regions. Most of the galaxies belong to one of these large structures. The voids have a typical diameter of ~25 h^{-1} Mpc. . . .

The largest nearly empty region of the survey—located between 13^h 20^m and 17^h, with $4000 \leq V \leq 9000$ km s^{-1}—has a nearly circular boundary and a diameter of ~50 h^{-1} Mpc. Figure [72.1b] shows that this void is *not* caused by large peculiar velocities of galaxies located in the same region. . . . Analogously, the smaller void located in the region defined by $11^h \leq \alpha \leq 12^h$ and $7000 \leq V \leq 10,000$ km s^{-1} is likely to have a diameter of at least $14\,h^{-1}$ Mpc.

Another striking feature in Figure [72.1a] is the sharpness of the boundaries of the high density regions which surround the voids. The edge of the 50 h^{-1} Mpc void is remarkably sharp. In the incomplete regions of the survey we find other voids surrounded by similarly sharp edges. . . .

The galaxy luminosity function could of course be different in the voids. They could be full of low-luminosity galaxies. The smaller void in the survey, centered at 3500 km s^{-1} and 13^h20^m, does not seem to support this hypothesis. The diameter of this structure is ~20 h^{-1} Mpc. In Figure [72.1b] the void is already visible, and in Figure [72.1a] no galaxies have been added inside the void by going 1 mag fainter. . . .

Other less complete surveys lend some support to the picture suggested by the new survey data. The Local Supercluster might sit on the surface of a shell. Shells with diameters as large as 50 h^{-1} Mpc can explain the presence of a filament-like supercluster in Hercules extending for ~100–150 h^{-1} Mpc in declination at $V \approx 10,000$ km s^{-1}.[44] The southern part of this supercluster is separated by a void from another supercluster at ~5000 km s^{-1}. We suggest that here, as in Figure [72.1a], the apparent filament is a cut through boundaries of several bubble-like structures. In this picture the filaments are *not* gravitationally bound.

The clustering in the survey is as remarkable as the appearance of the voids. The Coma cluster is the richest system and appears to lie at the intersection of two large shells and several smaller shells. . . .

IV. Implications

The best available model for generating the bubble-like structures observed in the survey is the explosive galaxy formation theory of Ostriker and Cowie (hereafter OC) in which galaxies form on the surfaces of expanding shock waves.[45] Most of the current models ideally assume that the structures form directly from the action of gravity on the matter perturbations, but the sharpness of the transition between the high-density regions and the voids in the survey indicates that hydrodynamic processes must be important in the formation of galaxies. In the OC galaxy formation theory, energetic explosions sweep the protogalactic gas and lead to the formation of dense cooled shells which can then fragment under gravitational instability and form new stellar systems. However, the current versions of the OC explosive scenario cannot account for the observed bubbles with diameter larger than 20 h^{-1} Mpc: the large thermal energy

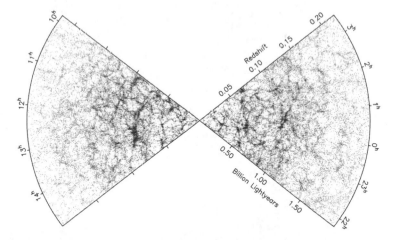

Figure 72.2: Large-scale structure in the Two-Degree Field Galaxy Redshift Survey, carried out in the 1990s by a team of British, Australian, and U.S. astronomers led by Matthew Colless of the Australian National University, Steve Maddox of the University of Nottingham, and John Peacock of the University of Edinburgh. Each of the points in the image represents a galaxy—tens of thousands in two slices of the sky. (© Two-Degree Galaxy Redshift Survey and Anglo-Australian Observatory)

input required to create larger bubbles would introduce small-scale fluctuations in the microwave background larger than the observational upper limits. In the OC model, the size of the bubbles (and their surface mass density) places limits on the energy input required to create them, and thus on the nature of the explosions. The size and the density contrast of the largest observed bubbles thus restrict the range of models capable of producing these structures. . . .

Finally we caution that the largest structures in this survey (and in other surveys) are comparable with the sample depth. Deeper surveys could, of course, lead to the discovery of even larger structures.

73 / Galaxy Evolution and the Hubble Deep Field

In the first half of the twentieth century, astronomers largely used distant galaxies as simple markers in their cosmological studies. By tracking the shapes, luminosities, and redshifts of galaxies outward (and back in time), they hoped to determine whether the universe was "open"—destined to expand forever—or "closed"— the expansion eventually stopping and reversing someday toward a Big Crunch. In carrying out this program, though, the astronomers had to assume that galaxies, following a tumultuous birth shortly after the Big Bang, drifted onward in tranquil isolation and experienced little change up to the present day. But starting in the 1950s, hints began arriving that the universe has substantially evolved over the eons. In 1958 the British radio astronomer Martin Ryle reported that he counted more far-off radio sources than expected, which implied that those remote objects were more luminous and active than galaxies in our present-day universe.[46] "It is apparent," he said, "that if most of the radio stars are extragalactic, then their luminosity is considerably in excess of that of the ordinary galaxies."[47]

A decade later Beatrice Tinsley at the University of Texas began what has been described as "one of the boldest graduate thesis projects ever undertaken."[48] It was the first serious theoretical attempt to figure out how galaxies transformed over time, so that the light from far-away galaxies could be correctly interpreted. She created complex computer models to show how galaxies changed in color and luminosity as the stars within them aged. Her results created a stir and brought galaxy evolution to the forefront of astronomical concerns when they showed that galaxies were likely evolving far more than previously expected.[49] In her models, galaxies started out very bright and blue, when gaseous resources were at their peak, then gently faded and reddened as the eons went by. Her ideas, considered highly controversial by more senior authorities at the time, inspired observers to start pushing outward with their telescopes to discern this evolution.

By 1978 Harvey Butcher, then at the Kitt Peak National Observatory, and Yale astronomer Augustus Oemler directly observed that the light from clusters of galaxies some 5 billion light-years distant (and hence five billion years back in time) did radiate more blue light than the more reddish clusters near us today. They surmised that blue, star-forming disk galaxies were more common at that time and that some kind of activity within the cluster exhausted the galaxies' gas as the clusters evolved to the present day.[50] By the 1990s, as larger ground-based telescopes were built, the Hubble Space Telescope launched, and more sensitive detectors introduced to see farther out into the universe, this "Butcher-Oemler effect" was confirmed and extended. Astronomers gradually came to see that galaxies did not emerge fully formed in one great burst of fireworks, as once thought. Rather, they seem to have arisen more like an ever-growing fire. While some elliptical and spiral galaxies were in place early on, other galaxies continued to coalesce, interact, merge with one another, and evolve over several more eons.

The superb resolution of the Hubble Space Telescope in particular allowed astronomers to see directly what had only been suspected before then: the bluer objects in the far clusters were indeed spiral galaxies, existing in greater numbers than found in today's clusters. Moreover, these spirals were particularly ragged and asymmetric. Looking in the field, away from the clusters, the Hubble spied many faint, blue, irregular galaxies—all in all, a universe far different from the more settled one today.

This new vision of an evolving universe was captured most vividly when Robert Williams, then director of the Space Telescope Science Institute, decided to train Hubble's mirror on one tiny spot of the sky—a dark, starless region near the handle of the Big Dipper in the constellation Ursa Major—for ten consecutive days in December 1995. Over that period, the telescope took a series of 342 time-exposure photographs in ultraviolet, blue, red, and infrared light with the Wide-Field Planetary Camera 2, images that were combined and computer-enhanced to produce the most deeply penetrating astronomical picture ever taken at the time—the Hubble Deep Field. This stunning portrait allowed astronomers to see simultaneously some 2,000 galaxies in different stages of development—in the local, intermediate, and distant universe—and verified that the "universe at high redshift looks rather different than it does at the current epoch."[51] Three years later a similar deep field, a 12-billion-light-year-long corridor through space and time, was obtained by the Hubble Space Telescope in the southern celestial hemisphere. Both fields provided a rich source of data that was deeply mined by astronomers in follow-up studies on galaxy evolution using ground-based telescopes.

"The Hubble Deep Field: Observations, Data Reduction, and Galaxy Photometry." *Astronomical Journal*, Volume 112 (October 1996)

by Robert E. Williams, Brett Blacker, Mark Dickinson,
W. Van Dyke Dixon, Henry C. Ferguson,
Andrew S. Fruchter, Mauro Giavalisco, Ronald L. Gilliland,
Inge Heyer, Rocio Katsanis, Zolt Levay, Ray A. Lucas,
Douglas B. McElroy, Larry Petro, Marc Postman,
Hans-Martin Adorf, and Richard N. Hook

Abstract

The Hubble Deep Field (HDF) is a Director's Discretionary program on *HST* [Hubble Space Telescope] in Cycle 5* to image an undistinguished field at high Galactic latitude in four passbands as deeply as reasonably

---

* A year-long period where proposals accepted from the astronomical community are executed by the Hubble Space Telescope. It's approximately 3,000 orbits of observing time. Cycle 5, in this case, was the fifth such cycle for the HST.

possible. These images provide the most detailed view to date of distant field galaxies and are likely to be important for a wide range of studies in galaxy evolution and cosmology. In order to optimize observing in the time available, a field in the northern continuous viewing zone was selected and images were taken for ten consecutive days, or approximately 150 orbits. Shorter 1–2 orbit images were obtained of the fields immediately adjacent to the primary HDF in order to facilitate spectroscopic follow-up by ground-based telescopes. The observations were made from 1995 December 18–30, and both raw and reduced data have been put in the public domain as a community service. . . .

Introduction

The HDF program is an outgrowth of previous, highly successful *Hubble Space Telescope* imaging projects which have elucidated the evolution of galaxies at high redshift. During Cycles 1 through 5, a variety of *HST* General Observer and Guaranteed Time Observer programs, as well as the Medium Deep Survey (MDS) key project, imaged distant galaxies in both cluster and field environments, providing (for the first time) kiloparsec-scale morphological data at all redshifts. The MDS used the WFPC-1 [Wide-Field Planetary Camera] and WFPC-2 cameras in parallel mode to image random galaxies near the fields of targeted objects. Analyzing 144 field galaxies having $I < 22$* [magnitude] from six fields, Driver *et al.* found from visual classification that early-type spirals, ellipticals, and late-type spirals/irregulars were observed in roughly equal proportions, with the Sd/Irr's† having much higher surface density than their counterparts at the current epoch.[52] Driver *et al.* extended this analysis with a similar study of one very deep field for which they showed that galaxy counts beyond $I = 22$ continue to be increasingly dominated by Sd/Irr galaxies. Combining ground-based redshift information with *HST* imaging, Lilly and collaborators[53] obtained B and I images for 32 galaxies from their CFHT [Canada-France-Hawaii Telescope] survey ($17.5 < I < 22.5$) with known redshifts in the range $0.5 < z < 1.2$.‡ They found that the observed galaxy morphologies were similar to those seen locally, but that the B images (rest frame UV) looked far less regular than observed at longer wavelengths. In addition, they determined that the central surface brightnesses of the disks in their

---

* The I band is a filter centered around 8,140 Angstroms.
† An Sd galaxy is a spiral whose arms are loosely wound.
‡ Distances between approximately 5 billion and 8 billion light-years.

sample of late-type spirals were more than 1.2 magnitude brighter than found locally. Also, they found that many of the bluer galaxies were nucleated, and they concluded that both of these effects must be responsible for much of the observed evolution of the luminosity function of blue galaxies.

Other *HST* programs targeted galaxies with known redshifts based upon their membership in clusters that had been studied from the ground, e.g., 0939+4713 (Dressler *et al.*)[54] and the cluster(s) associated with the radio galaxy 3C 324 at $z = 1.21$ (Dickinson).[55] Both of these programs demonstrated the ability of the refurbished *HST* to resolve galaxy structure at moderate to high redshift in a way that made morphological classification and a quantitative study of various parameters possible. Cluster 0939+4713 does not look entirely unlike nearby clusters insofar as it is populated largely by spiral and elliptical galaxies. However, the disk systems are bluer and more numerous than spiral galaxies in the cores of clusters today, and often show signs of disturbance and tidal interactions. Evidently, these spirals are responsible for the rapidly evolving blue galaxy population first noted in distant clusters by Butcher & Oemler.[56] Looking back to $z = 1.21$, the cluster associated with 3C 324 includes apparently normal, mature E/S0s, but readily recognizable spiral galaxies appear to be rare, and a large number of irregular, amorphous objects are present (Dickinson)[57]. . . .

While much of the information available in these images remains to be interpreted, two things have become clear. First, *HST* can indeed resolve galaxy-sized systems out to high redshift. Second, the Universe at high redshift looks rather different than it does at the current epoch. The fact that *HST* can image galaxies back at epochs when they were apparently forming and evolving rapidly is of fundamental importance to our understanding of galaxy evolution, and it is imperative that this capability be fully exploited. Based on the current excellent performance of the telescope, a decision was made to devote a substantial fraction of the Director's Discretionary time in Cycle 5 to the study of distant galaxies. A special Institute Advisory Committee was convened which recommended to the Director that deep imaging of one "typical" field at high galactic latitude be done with the Wide-Field Planetary Camera 2 (WFPC-2) in several filters, and that the data be made available immediately to the astronomical community for study. Following this recommendation a working group was formed to develop and carry out the project.

It is not our purpose here to interpret the data, but rather to present the images and source catalogs, along with the necessary background to facilitate the use of the HDF in studies of galaxy evolution. . . .

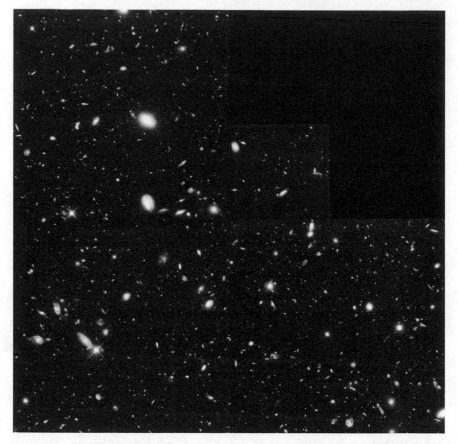

Figure 73.1: "A . . . composite image of the full HDF field."

[From here the article spends more than fifty pages describing the criteria for selecting the deep field, the scientific rationale for the selection of filters used, technical aspects of planning the observations, and details of data reduction and calibration. It also presents the images and a complete source catalog.]

The Images

Figure [73.1] shows a . . . composite of the HDF full field. . . . This image was produced from the initial version 1 data reduction and . . . reveals the striking variety of colors and morphologies of the distant galaxies visible in the field. . . .

. . . At brightness levels well above the detection limit there are relatively few stars compared to the number of galaxies. The galaxies have a wide distribution of brightnesses, sizes, and shapes, and the brightest galaxies appear to correspond to the normal Hubble types. Most of the fainter sources also appear to be galaxies, although this must be established carefully because many of them are only marginally spatially resolved, but their morphologies are frequently chaotic and asymmetric. Not surprisingly, the fainter sources have a more compact appearance. . . . Some of the fainter objects are undoubtedly bright H II regions and massive star complexes in galaxies, which appear above the background threshold while the remaining lower surface brightness regions of the galaxy do not. A cursory study of the images does not reveal any obvious heretofore unobserved class of objects compared to earlier *HST* images of moderately distant clusters such as 0939+4713 and those associated with 3C 324. There do appear to be a number of the linear structures having the sizes and luminosities of galaxies that have been noted by others (e.g., Cowie *et al.*). . . . [58] Color differences among the various galaxies are notable in that some of them appear much more prominent relative to neighboring galaxies in one of the filters than in others. . . .

The Hubble Deep Field observations were taken with the expectation that they will contribute to the resolution of some of the outstanding questions in studies of galaxy formation. . . .

74 / Extrasolar Planets

Speculation that planetary systems like ours circle other stars has a long tradition. In the fourth century B.C. the Greek philosopher Epicurus, in a letter to his student Herodotus, surmised that there are "infinite worlds both like and unlike this world of ours."[59] The noted eighteenth-century astronomer William Herschel, too, conjectured that every star might be accompanied by its own band of planets and comets but figured they could "never be perceived by us on account of the faintness of light."[60] A planet, visible only by reflected light, would be lost in the glare of its sun when viewed from afar.

In the twentieth century, astronomers recognized that a planet might be detected by its gravitational pull on the star, causing the star to systematically wobble. At first they attempted to observe this wobble by tracking a star's proper motion over the years as it journeyed across the heavens (with no success). By the 1980s they began focusing on how the wobble affects the star's light. When the star is tugged radially toward the Earth, its light gets slightly dopplershifted toward the blue; pulled away by the gravitational tug of a companion, the star's light is shifted the other way, toward the red end of the spectrum. Over time these periodic changes in radial velocity can become discernible (see Chapter 26). In 1979 Canadian astronomers Bruce Campbell and Gordon Walker single-handedly pioneered a technique to detect velocity changes as small as a dozen meters a second, sensitive enough for extrasolar-planet hunting, and a number of groups in both Europe and the United States soon refined the method.[61] Searchers were energized in 1984 when the Infrared Astronomical Satellite (IRAS) began seeing circumstellar material surrounding several stars in our galaxy, and optical astronomers, taking a special image of the dwarf star Beta Pictoris, revealed a dusty disk (seen edge-on) that extends from the star four hundred times the distance from the Earth to the Sun (400 AU).[62] It was the first striking evidence of planetary systems in the making, suggesting that such systems might be common (see Figure 74.1).

The first indication of a planet orbiting another star arrived unexpectedly and within an unusual environment. In 1991 radio astronomers Alex Wolszczan and Dale Frail, while searching for millisecond pulsars at the Arecibo Observatory in Puerto Rico, saw systematic variations in the beeping of pulsar PSR 1257+12, which suggested that three bodies were orbiting the neutron star. Fast, millisecond pulsars are spun up by accreting matter from a stellar companion. So this system, reported Wolszczan and Frail, "probably consists of 'second generation' planets created at or after the end of the pulsar's binary history."[63]

But the principal goal for extrasolar planet hunters was finding evidence for "first generation" planets around stars like our Sun. That long anticipated event at last occurred in 1994 when Geneva Observatory astronomers Michel Mayor and Didier Queloz, working from the Haute-Provence Observatory in southern France, discerned the presence of an object similar to Jupiter orbiting 51 Pegasi, a Sun-like star 45 light-years distant in the constellation Pegasus. They first revealed their discovery at a conference in Florence, Italy, and their fellow astronomers declared it a "spectacular detection."[64] Unlike our own solar system, this extrasolar planet is located a mere 0.05 AU (4.6 million miles) from its star, far closer in than Mercury is to our Sun, and completes one orbit every four days. Planet hunters had assumed it would take years of collecting data before detecting the subtle and gradual stellar wobbles caused by a planet orbiting its parent star, but the small orbit of 51 Peg B enabled them to spot its variations quickly.

Other discoveries followed swiftly. Geoffrey Marcy and R. Paul Butler, then both at San Francisco State University and friendly

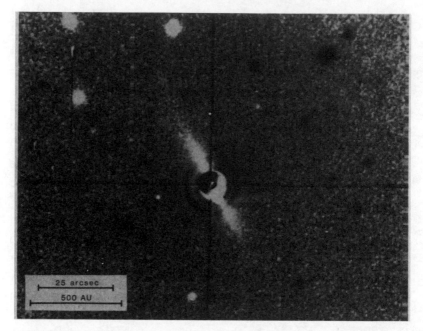

Figure 74.1: Image of Beta Pictoris showing the edge-on circum-stellar disk extending 25 arcsec (400 AU) to the northeast and southwest of the star, which is situated behind an obscuring mask. North is at the top.

competitors of the Geneva observers, had been gathering radial velocity data at Lick Observatory since 1987. Searching through their records, they found evidence for a planet similar to 51 Peg B, a body at least seven times the mass of Jupiter closely circling within 0.43 AU (40 million miles) of the star 70 Virginis.[65] These finds challenged theorists, who had not imagined planets with eccentric orbits so close to their sun. These unusual planets, though, were quickly overshadowed by a simultaneous discovery by Marcy and Butler—a large planet orbiting 47 Ursae Majoris at a more distant 2.1 AU (~200 million miles). This companion of 47 Ursae Majoris thus gained special distinction for being more "reminiscent of solar system planets."[66] In 1999 Butler, Marcy, and several colleagues found the first multiple planetary system, a trio of planets circling the star Upsilon Andromedae.[67] By the end of the twentieth century, more than forty extrasolar planets had been detected, with astronomers continuing to add candidates at the rate of more than one a month. Much like the history of asteroid hunting, it was far easier to spot such hidden objects once astronomers knew where and how to look.

"A Search for Substellar Companions to Solar-Type Stars Via Precise Doppler Measurements: A First Jupiter Mass Companion Detected." In *Cool Stars, Stellar Systems, and the Sun: Ninth Cambridge Workshop, Astronomical Society of the Pacific Conference Series*, Volume 109 (1996) by Michel Mayor and Didier Queloz

Abstract

Since 1994, we have initiated with ELODIE, the new echelle spectrograph of the Observatoire de Haute-Provence (France), a high-precision radial velocity survey of a sample of 140 nearby G and K stars in order to detect Jovian planets and brown dwarfs. We present here, after 18 months of measurements, a first analysis of this survey. Our most important result is the detection of the first Jovian mass companion to a solar-type star.

Program

The range of masses for companions to solar-type stars, from 0.001 to 0.080 $M_\odot$ [mass of the sun], is still largely unexplored. The gap between low-mass stars and planets . . . illustrates our poor knowledge of formation processes of substellar objects. Nevertheless, . . . a clear distinction can be made between planets formed in disks (quasi-circular orbits) and more massive objects formed by fragmentation of collapsing clouds (eccentric orbits). . . .

We have initiated in April 1994 with ELODIE a high-precision radial velocity survey on a sample of about 140 nearby G and K stars in order to find very low mass companions in the transition region from stars to giant planets. The stars sample is composed of constant radial velocity stars already observed with CORAVEL during the last 15 years.[68]

The aim of this survey is twofold: We want to determine the minimum mass of stellar companions supposed to be formed by fragmentation of collapsing clouds and the maximum mass for heavy planets. Then, we want to

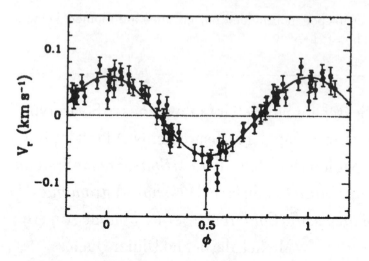

Figure 74.2: "Radial velocity measurement of 51 Peg from April 1994 to December 1995 corrected for the long-term variation of the γ-velocity. The solid line is a circular orbit with a 4.2293 day period and a 59 m s^{-1} amplitude."

have a first estimate of the rate and the orbital characteristics of heavy planets and brown dwarfs.

Accuracy of the Survey

The accuracy of the radial velocity measurements made with ELODIE is about 15 m s^{-1} [meters per second] up to 9th magnitude stars. No spectrograph drift has been detected.

The First Extrasolar Planet

After only 16 months of measurements a few stars of our sample already appear to be good candidates for having a very low mass companion. But the real surprise was the discovery of 51 Peg, a solar-type star with an extremely short period velocity variation of 4.2293 days. This velocity variation is due to the presence of a Jupiter-mass companion of 0.47 M_J / sin i in circular orbit (see Figure [74.2]). A conservative upper limit of 2 M_J [mass of Jupiter] for the companion can be set. . . . Using the most recent [measurements], the probable range of mass for the planet is between half and less than one Jupiter mass.

Alternative explanations such as pulsation or rotation of spots can be rejected because 51 Peg is nearly a twin of the Sun (a little older) and extremely stable photometrically (better than 0.1% at the given period). The Sun is known to pulsate with a period of 5 minutes and gravity modes as long as four days are not seen. Such high order modes would not be stable anyway and 51 Peg exhibits a perfect sinusoidal velocity variation since more than one and a half years. A G dwarf with a rotational period of 4.2 days would be extremely active. On the contrary 51 Peg has a very quiet chromospherical activity. Its low activity index indicates a rotation period of 30 days.

The short distance between the planet and the star (0.05 AU) is in obvious contradiction with the core-collapse scenario for the formation of heavy planets. Today we are far from understanding the formation mechanism of this planet.

A gaseous heavy planet, even at such a short distance from a solar-type star, is not significantly affected by evaporation processes. A brown dwarf at the same distance is *a fortiori* still more stable and cannot be the progenitor of this planet after a strong evaporation. If the companion of 51 Peg was formed at a distance larger than 5 AU, we have to find a mechanism efficient enough to induce a strong orbital decay (a factor of a hundred). If the tidal interaction with the protoplanetary disk can be suspected, then we have to explain why the solar system has not followed the same path. Clearly, we need a lot of new detections of heavy planets to establish the distribution of masses and periods as a guide for our understanding of formation scenarios.

"A Planet Orbiting 47 Ursae Majoris."
Astrophysical Journal, Volume 464 (June 20, 1996)
by R. Paul Butler and Geoffrey W. Marcy

Abstract

The G0 V star 47 UMa [Ursae Majoris] exhibits very low amplitude radial velocity variations having a period of 2.98 yr, a velocity amplitude of $K = 45.5$ m s^{-1}, and small eccentricity. The residuals scatter by 11 m s^{-1} from a Keplerian fit to the 34 velocity measurements obtained during 8 yr. The minimum mass of the unseen companion is $M_2 \sin i = 2.39$ M_J, and for likely orbital inclinations of 30°–90°, its mass is less than 4.8 M_J. This mass resides in a regime associated with extrasolar giant planets. Unlike the planet candidates 70 Vir B and 51 Peg B, this companion has an orbital radius (2.1 AU) and eccentricity ($e = 0.03$) reminiscent of giant planets in our solar system. Its effective temperature will be at least 180 K due simply to absorbed stellar radiation, and probably slightly higher due to intrinsic heating from gravitational contraction. For 47 UMa B to be, instead, an orbiting brown dwarf of mass $M > 40 M_J$, the inclination would have to be $i < 3°.4$, which occurs for only 0.18% of randomly oriented orbits. In any case, this companion is separated from the primary star by ~0″.2, which portends follow-up work by astrometric and direct IR techniques.

Introduction

The observed disks of gas and dust around young solar-type stars have characteristics similar to those required for formation of the planets in our solar system. The recent detections of planet-like companions to 51 Peg and to 70 Vir provide the first evidence that planet formation may indeed occur commonly around solar-type stars.

However, these first two "planets" exhibit characteristics that are not represented among the nine in our solar system. The planet around 51 Peg has an orbital radius of only 0.051 AU, which is 7 times smaller than the semimajor axis of Mercury. The companion to 70 Vir has a minimum mass of 6.6 M_J and an eccentricity of 0.40, both of which exceed the range of values found among solar system planets. The simple term "planet" may not adequately represent the formation process of 70 Vir B. This planet-like companion may belong to a new class characterized by eccentric orbits $e > 0.2$ and masses 5–15 M_J. . . . This class apparently does not extend to higher masses (between 15 and 40 M_J), as they would be easily detected, but have not been found (within 5 AU). At higher masses, "brown dwarf" companions have apparently been found within 5 AU and all exhibit masses greater than 40 M_J.

Theories of the formation of gas giants predict that the final orbits will be circular, having radii of at least several AU.[69] These expectations follow from the dissipation that occurs in eccentric orbits within a gaseous disk and from the survival of ice grains beyond 3 AU, where low-equilibrium temperatures are found (T < 200 K). The predictions of both circular orbits and large orbital radii are subject to caveats. First, protostellar disks exhibit a range of masses (up to 0.1 $M_\odot$) with some masses exceeding that of the "minimum-mass solar nebula." Such massive disks may have sufficiently high densities of the refractory grains, inward of several AU, to permit rapid growth of large rocky cores, leading to gas giants. Even planetary cores that form outside 5 AU may migrate inward, resulting in gas giants at small orbital radii. Second, significant orbital eccentricities may arise from nonaxisymmetric disk instabilities, which may in fact be driven by a massive protoplanet itself. Thus, the "eccentric planets" around 70 Vir and HD 114762 may stem naturally from massive disks by mechanisms yet to be explored.

Neither 51 Peg B nor 70 Vir B appears to be an obvious analog of planets in our solar system. Here we describe observations of 47 UMa, which exhibits Doppler variations consistent with a planetary companion that has properties reminiscent of solar system planets.

The Doppler Technique and Stellar Sample

In 1987 June we began precise Doppler monitoring of the solar-like star 47 UMa. . . . 47 UMa and the Sun have similar properties. The effective temperature, absolute visual magnitude, and surface gravity of 47 UMa are all consistent with its being a normal, old disk, G0 V main-sequence star. The relatively low chromospheric [activity] and the modest rotation period of 16 days suggest that its age is 4–8 Gyr [billion years]. . . . The metallicity of 47 UMa is solar. . . .

. . . In brief, we use the Lick Observatory 0.6 m CAT and 3 m Shane telescopes to feed the "Hamilton" coudé echelle spectrograph. Wavelength calibration is accomplished by placing an iodine gas absorption cell in the star beam. The superimposed absorption lines of iodine serve as indelible wavelength markers. . . . Relative Doppler errors were 10 m s^{-1} until 1994 November, when improvements in the spectrometer brought the errors to 3 m s^{-1}. For comparison, the reflex motion of the Sun due to Jupiter is 12.5 m s^{-1}, thus rendering gas giants detectable at 5 AU. . . .

Velocities of 47 Ursae Majoris and an Orbital Solution

A total of 34 observations of 47 UMa have been obtained, spanning 8.7 yr from 1987.5 through 1996.2. The measured Doppler velocities from these observations are shown in Figure [74.3]. The error bars represent the internal error of these observations, ~10 m s^{-1}. The rms [root mean square] of the velocity variations of the 47 UMa observations is 35 m s^{-1}, much greater than the errors. A periodogram analysis finds an extremely strong peak for a period of 3.0 yr, which agrees with an eyeball inspection of Figure [74.3].

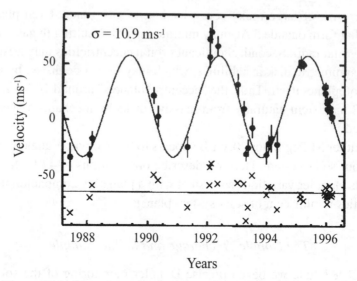

Figure 74.3: "Doppler velocities for 47 Uma (G0 V). A total of 34 observations have been made over 8.7 yr, shown as the filled circles. A Keplerian orbit with a period of 1090 days fits the observed velocities, implying a companion mass, $M \sin i = 2.4\ M_J$. . . ."

We employ a nonlinear least-squares fitting routine to determine the best-fit Keplerian orbit. . . . The period of 2.98 yr and low amplitude of $K = 45.5\ \mathrm{m\ s^{-1}}$ are consistent with an extremely low mass companion orbiting 47 UMa. In this interpretation, the semimajor axis of the orbit is 2.11 AU. We assume that the primary star has a mass of 1.05 $M_\odot$, based on its spectral classification as G0 V. The derived minimum mass of the companion is

$$M_{\mathrm{comp}} \sin i = 2.39\ M_J \ldots$$

Alternative Explanations

There are two alternative explanations that could in principle explain the observed radial velocity signal, namely, spots and pulsation. A large dark stellar spot would cause a net apparent Doppler redshift if it were on the approaching limb of the star, and a blueshift if it were on the receding limb. The period of such a Doppler signal would be the rotation period of the star. However, the estimated rotation period of 47 UMa is 16 days, which is inconsistent with the observed 1090 day periodicity of the velocity variations. Radial pulsation can be ruled out because of the long period. . . .

We also considered multiple companions to explain the velocity variations. The periodogram analysis reveals only one strong peak at $P = 1090$ days, with no other peaks having height even 50% as strong. This suggests that only one companion is predominantly responsible for perturbing the primary star. Nonetheless, additional companions that induce perturbations less than 20 m s^{-1} could be responsible for the 2 σ departures in the velocities seen around 1992. Further Doppler measurements are in progress.

Discussion

The most viable explanation for the observed velocity variations is that a companion orbits 47 UMa, having a minimum mass of 2.39 M_J. The actual mass remains unknown pending determination of the orbital inclination. . . .

Theoretical models of gas giant planets have been computed. . . .[70] The models contain the relevant interior physics, approximate atmospheric boundary conditions, evolutionary effects (including gravitational contraction), and the absorbed radiation from the host star. The radius of 47 UMa B is estimated from the models to be $R = 1.1$ R_J slightly larger than 1 R_J due to radiation from the star. . . .

One wonders how the companion to 47 UMa should be classified. Since deuterium burning does not take place in objects of 2–3 M_J, 47 UMa B is qualitatively different from the brown dwarfs. Indeed, deuterium may portend a method of distinction between planets and brown dwarfs. In addition, [Alan] Boss has argued that masses lower than 20 M_J cannot easily form by standard star formation processes. Assistance from grain-grain coagulation may be necessary. The nearly circular orbit suggests formation in a dissipative environment, presumably a disk. For the reasons above, it is tempting to classify 47 UMa B as a giant "planet." We caution that the term "planet" is loaded with implications stemming from the nature and supposed formation of the planets in our solar system. Thus, the firm adoption of the term "planet" for 47 UMa B must await its empirical placement in the context of other low-mass objects orbiting FGK stars. Nonetheless, the orbit and mass of 47 UMa B offer little compelling argument that it differs qualitatively from the gas giants in our solar system. . . .

75 / The Accelerating Universe

As telescopic instrumentation improved and astronomers pushed outward into the universe (and further back in time) over the twentieth century, they hoped to measure how the expansion of the universe since its explosive birth has been slowing down—decelerating—as the eons pass due to the gravitational pull of matter. Thus it came as a great surprise in 1998 when two international teams, analyzing distant supernovae in painstaking detail, discovered that the universe behaves in the completely opposite manner: the universe is now ballooning outward at a pace that is *accelerating* over time.

In the 1930s Walter Baade first suggested that supernovae—exploding stars—would be convenient distance markers for measuring cosmic expansion, but it took many years for astronomers to learn how to use them.[71] Massive stars blow up as brilliant supernovae when they run out of fuel. But these Type II explosions, as they are called, are hardly standard candles; they can vary greatly in brightness depending on the original size and mass of the exploding star. Fortunately, astronomers discovered that there's another way for a star to blow up: white dwarf stars, tiny as the Earth, can be in binary star systems and steal gas from their companions. If the white dwarf pilfers enough mass, it collapses under the weight of gravity, and the extra material ignites, consuming the entire star in a second. Most exciting to cosmologists, such explosions (called Type Ia) were assumed to have similar brightnesses at their peak, making them excellent standard candles.[72]

The Supernova Cosmology Project, headed by physicist Saul Perlmutter of the Lawrence Berkeley National Laboratory in California, was set up in 1988 to advance the methods for spotting supernovae in distant galaxies reliably and quickly (before this it could take years to detect just one distant supernova).[73] Crucial was developing software that would allow computers to find the supernovae automatically in the digital images taken at the telescope. By the mid-1990s Perlmutter and his colleagues from Europe, South America, and Australia were finding batches of supernovae in any one search, which convinced major observatories that it was at last worthwhile to schedule telescope time to examine them as distance markers.

Meanwhile, another international group of astronomers, known as the High-z Supernova Search Team, led by Brian Schmidt of the Mount Stromlo and Siding Spring Observatories in Australia, formed in 1994 to join the search. Since many on the High-z team were world experts on supernovae, they also focused on how to interpret the explosions. Astronomers had come to learn that Type Ia supernovae (SNe Ia) can vary in brightness—they weren't perfect standard candles. But by 1993 Mark Phillips of the Cerro Tololo Inter-American Observatory in Chile had demonstrated that the brighter Type Ia explosions faded more slowly than the dimmer ones and in a predictable way.[74] Mario Hamuy of the University of Arizona, Adam Riess, then with Harvard University, and others proceeded to show how to use such light-curve information to calibrate a supernova— that is, correct for dust obscuration and determine the supernova's peak intrinsic brilliance, both crucial for distance measurements.[75]

The two teams separately observed many fields on the celestial sky once or twice a year, right after a new moon when the sky was darkest. Three weeks later, they examined the same fields to detect changes. Type Ia supernovae are rare, but with hundreds of thousands of galaxies being examined, a dozen or so supernovae were usually sighted in any one session. They then followed up with a variety of ground-based telescopes to examine the candidates in more spectral detail.

In 1997, once they had detected some appreciably distant supernovae, both teams were puzzled by their results and reluctant to believe them. By the January 1998 meeting of the American Astronomical Society in Washington, D.C., Perlmutter and his group tentatively discussed that there might be a cosmological constant—signified by the Greek letter lambda (Λ)—at work in the universe, a repulsive force once proposed by Einstein, then abandoned. But they stressed their results were still uncertain. Six weeks later at an international symposium in California the High-z Supernova Search Team, able to correct for a major source of systematic uncertainty—dust—and spurred by their competitors' earlier remarks, reported that the cosmological constant Λ was indeed present, causing the universe to accelerate outward. They saw that the total density of the universe was such that it was caught between being open or closed—a condition denoted by astronomers as $\Omega = 1$ (see Chapter 71). This had been long suspected, but now they realized it was due to a combination of both mass *and* vacuum energy ($\Omega_M + \Omega_\Lambda = 1$).

Both groups, to their shock and amazement, had discovered that supernovae a few billion light-years out and farther were fainter than they were expecting. Maybe supernovae in the past were less luminous because of differing chemical compositions? But since the spectra of the distant supernovae do not differ from stellar explosions nearby, the supernova hunters concluded that it's more likely the universe had expanded more quickly—leaving the supernovae farther behind—over the billions of years since the stars exploded. The results were convincing because two independent teams had reached the same conclusion using different supernovae and different analytic techniques.

The cosmological constant was chosen to explain the extra push on space-time because it was familiar. Einstein had first introduced the Λ term into his equations of general relativity in 1917 for another reason: to keep the universe, which he thought was static, from collapsing due to the gravitational attraction of all its matter (see Chapter 37). But once he learned that the universe is coasting outward, a movement that could be accounted for with his original equations, he dropped the term, calling it "theoretically unsatisfying anyway."[76] Now Λ was being added back as a possible motor for the acceleration.

An accelerating universe settled some old disputes. Previous cosmological measurements had suggested that the universe was fairly young, ten billion years or less, yet certain stars in our galaxy were known to be far more ancient. With an added energy to gradually boost expansion over time, though, the universe suddenly became older, eliminating the conflict. It also clarified why astronomers were finding just a fraction of the matter required for a flat universe (as ours seems to be).

The cosmological constant is but one explanation; even before the discovery, physicists had been contemplating other sources of vacuum energy based on new and speculative theories of high-energy physics. Since the origin of the acceleration was not yet pinpointed, the new accelerative force was soon referred to in general as "dark energy."

These findings were sizably enhanced in 2001 when a particularly distant stellar explosion, farther out than the previously sampled supernovae, was found to be comparatively *brighter* for its redshift, pinpointing the time when the universe was still slowing down before speeding up, as cosmological models predicted.[77] There was further backing in 2003 with NASA's Wilkinson Microwave Anisotropy Probe (WMAP), which made the most sensitive map to date of the universe's microwave background, the afterglow of the Big Bang. The spectral patterns seen in this mapping suggested that 4 percent of the universe's mass-energy contents resides in ordinary matter, 23 percent (give or take a few percent) in an unknown dark matter, and the rest—the most prominent contribution—as dark energy, similar to the portions supported by the supernova data.[78]

"Observational Evidence from Supernovae for an
Accelerating Universe and a Cosmological Constant."
Astronomical Journal, Volume 116 (September 1998)
by Adam G. Riess, Alexei V. Filippenko, Peter Challis,
Alejandro Clocchiatti, Alan Diercks, Peter M. Garnavich,
Ron L. Gilliland, Craig J. Hogan, Saurabh Jha,
Robert P. Kirshner, B. Leibundgut, M. M. Phillips,
David Reiss, Brian P. Schmidt, Robert A. Schommer,
R. Chris Smith, J. Spyromilio, Christopher Stubbs,
Nicholas B. Suntzeff, and John Tonry

Abstract

We present spectral and photometric observations of 10 Type Ia supernovae (SNe Ia) in the redshift range $0.16 \leq z \leq 0.62$.* The luminosity distances of these objects are determined by methods that employ relations between SN Ia luminosity and light curve shape. Combined with previous data from our High-z Supernova Search Team and recent results by Riess et al., this expanded set of 16 high-redshift supernovae and a set of 34 nearby supernovae are used to place constraints on the following cosmological parameters: the Hubble constant (H_0), the mass density (Ω_M), the cosmological constant (i.e., the vacuum energy density, Ω_Λ), the deceleration parameter (q_0), and the dynamical age of the universe (t_0). The distances of the high-redshift SNe Ia are, on average, 10%–15% farther than expected in a low mass density ($\Omega_M = 0.2$) universe without a cosmological constant. Different light curve fitting methods, SNe Ia subsamples, and prior constraints unanimously favor eternally expanding models with positive cosmological constant (i.e., $\Omega_\Lambda > 0$) and a current acceleration of the expansion (i.e., $q_0 < 0$). ... A universe closed by ordinary matter (i.e., $\Omega_M = 1$) is formally ruled out at [a high] confidence level for the two differ-

---

* Such redshifts represent distances from roughly 2 billion to 5 billion light-years from Earth; z is the astronomical measure of an object's reddening due to its light waves being stretched by cosmic expansion.

ent fitting methods. We estimate the dynamical age of the universe to be 14.2 ± 1.7 Gyr [billion years]. . . .

Introduction

This paper reports observations of 10 new high-redshift Type Ia supernovae (SNe Ia) and the values of the cosmological parameters derived from them. Together with the four high-redshift supernovae previously reported by our High-z Supernova Search Team[79] and two others,[80] the sample of 16 is now large enough to yield interesting cosmological results of high statistical significance. Confidence in these results depends not on increasing the sample size but on improving our understanding of systematic uncertainties.

The time evolution of the cosmic scale factor depends on the composition of mass-energy in the universe. While the universe is known to contain a significant amount of ordinary matter, Ω_M, which decelerates the expansion, its dynamics may also be significantly affected by more exotic forms of energy. Preeminent among these is a possible energy of the vacuum (Ω_Λ), Einstein's "cosmological constant," whose negative pressure would do work to accelerate the expansion. Measurements of the redshift and apparent brightness of SNe Ia of known intrinsic brightness can constrain these cosmological parameters.

The High-z Program

Measurement of the elusive cosmic parameters Ω_M and Ω_Λ through the redshift-distance relation depends on comparing the apparent magnitudes of low-redshift SNe Ia with those of their high-redshift cousins. This requires great care to assure uniform treatment of both the nearby and distant samples.

The High-z Supernova Search Team has embarked on a program to measure supernovae at high redshift and to develop the comprehensive understanding of their properties required for their reliable use in cosmological work. Our team pioneered the use of supernova light curve shapes to reduce the scatter about the Hubble line. . . . This dramatic improvement in the precision of SNe Ia as distance indicators increases the power of statistical inference for each object by an order of magnitude and sharply reduces their susceptibility to selection bias. Our team has also pioneered methods for using multicolor observations to estimate the reddening to each individual supernova, near and far, with the aim of minimizing the confusion between effects of cosmology and dust. . . . As the use of SNe Ia for measuring Ω_M and Ω_Λ progresses from its infancy into childhood, we can expect a . . . shift in the discussion from results limited principally by statistical errors to those limited by our depth of understanding of SNe Ia.

This Paper

Our own High-z Supernova Search Team has been assiduously discovering high-redshift supernovae, obtaining their spectra, and measuring their light curves since 1995. The goal is to provide an independent set of measurements that uses our own techniques and compares our data at high and low redshifts to constrain the cosmological parameters. Early results from four SNe Ia (three observed with *HST* [Hubble Space Telescope]) hinted at a non-negligible cosmological constant and "low" Ω_M but were limited by statistical errors. . . .[81] Our aim in this paper is to move the discussion forward by increasing the data set from four high-redshift SNe to 16. . . .

[Omitted here are comprehensive descriptions of the authors' supernova observations, spectral identifications, calibration techniques, and light-curve fittings. Also discussed are the various approaches used in correcting the data and determining the supernovae's luminosities to judge their distances. They include the Multicolor Light Curve Shape Method (MLCS), template fitting, and snapshot method.]

Cosmological Implications of Type Ia Supernovae

. . . From the nine spectroscopic high-redshift SNe Ia with well-observed light and color curves, a non-negligible positive cosmological constant is strongly preferred at the 99.6% and greater than 99.9% confidence levels for the MLCS and template-fitting methods, respectively. . . .

We can include external constraints on Ω_M, Ω_Λ, or their sum to further refine our determination of the cosmological parameters. For a spatially flat universe (i.e. $\Omega_M + \Omega_\Lambda \equiv \Omega_{tot} \equiv 1$), we find $\Omega_\Lambda = 0.68 \pm 0.10$ ($\Omega_M = 0.32 \pm 0.10$) and $\Omega_\Lambda = 0.84 \pm 0.09$ ($\Omega_M = 0.16 \pm 0.09$) for MLCS and template fitting, respectively. The hypothesis that matter provides the closure density (i.e. $\Omega_M = 1$) is ruled out . . . by either method. Again, $\Omega_\Lambda > 0$ [a positive cosmological constant] and an eternally expanding universe are strongly preferred. . . .

Other measurements based on the mass, light, x-ray emission, num-

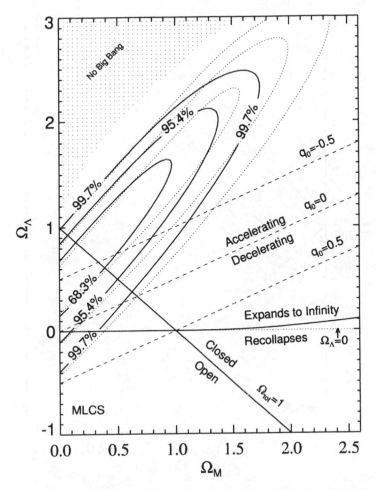

Figure 75.1: "Joint confidence levels for (Ω_M, Ω_Λ) from SNe Ia. The solid contours are results from the MLCS method applied to well-observed SNe Ia light curves together with the snapshot method applied to incomplete SNe Ia light curves.... Regions representing specific cosmological scenarios are illustrated...."

bers, and motions of clusters of galaxies provide constraints on the mass density that have yielded typical values of $\Omega_M \approx$ 0.2–0.3.... Using the constraint that $\Omega_M \equiv$ 0.2 provides a significant indication for a cosmological constant: $\Omega_\Lambda =$ 0.65 $\pm$ 0.22 and $\Omega_\Lambda =$ 0.88 $\pm$ 0.19 for the MLCS and template-fitting methods, respectively....

If we instead demand that $\Omega_\Lambda \equiv$ 0 [no cosmological constant] ... doing so yields an unphysical value of $\Omega_M =$ −0.38 $\pm$ 0.22

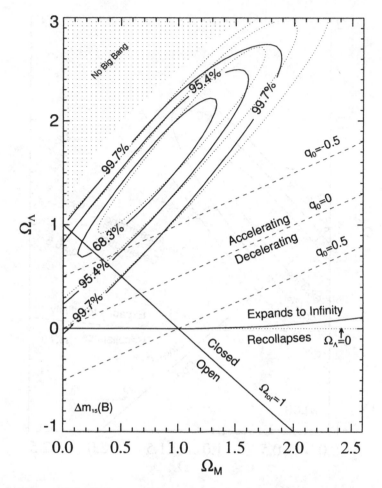

Figure 75.2: "Joint confidence levels for $(\Omega_M, \Omega_\Lambda)$ from SNe Ia. The solid contours are results from the template-fitting method applied to well-observed SNe Ia light curves together with the snapshot method applied to incomplete SNe Ia light curves.... Regions representing specific cosmological scenarios are illustrated...."

and $\Omega_M = -0.52 \pm 0.20$ [negative mass density] for the MLCS and template-fitting approaches, respectively. This result emphasizes the need for a positive cosmological constant for a plausible fit.

... Contours of [probability] from the MLCS method and the template-fitting method, each combined with the snapshot [probability], are shown in Figures [75.1] and [75.2]. ...

Discussion

The results . . . suggest an eternally expanding universe that is accelerated by energy in the vacuum. Although these data do not provide independent constraints on Ω_M and Ω_Λ to high precision without ancillary assumptions or inclusion of a supernova with uncertain classification, specific cosmological scenarios can still be tested without these requirements.

High-redshift SNe Ia are observed to be dimmer than expected in an empty universe (i.e., $\Omega_M = 0$) *with no cosmological constant.* A cosmological explanation for this observation is that a positive vacuum energy density accelerates the expansion. Mass density in the universe exacerbates this problem, requiring even more vacuum energy. For a universe with $\Omega_M = 0.2$, the MLCS and template-fitting distances to the well-observed SNe are 0.25 and 0.28 mag farther on average than the prediction from $\Omega_\Lambda = 0$. The average MLCS and template-fitting distances are still 0.18 and 0.23 mag farther than required for a 68.3% consistency for a universe with $\Omega_M = 0.2$ and without a cosmological constant.

Depending on the method used to measure all the spectroscopically confirmed SNe Ia distances, we find Ω_Λ to be inconsistent with zero at confidence levels from 99.7% to more than 99.9%. Current acceleration of the expansion is preferred at the 99.5% to greater than 99.9% confidence level. The ultimate fate of the universe is sealed by a positive cosmological constant. Without a restoring force provided by a surprisingly large mass density (i.e., $\Omega_M > 1$) the universe will continue to expand forever. . . .

[Omitted here is the authors' detailed exploration of the systematic uncertainties that might lead to overestimates of the SNe Ia distances.]

The systematic uncertainties presented by gray extinction, sample selection bias, evolution, a local void, weak gravitational lensing, and sample contamination currently do not provide a convincing substitute for a positive cosmological constant. Further studies are needed to determine the possible influence of any remaining systematic uncertainties.

"Measurements of Ω and Λ from 42 High-Redshift
Supernovae." *Astrophysical Journal*, Volume 517
(June 1, 1999)
by Saul Perlmutter, Gregory Aldering, Gerson Goldhaber,
Robert A. Knop, Peter Nugent, P. G. Castro,
Susana Deustua, Sebastien Fabbro, Ariel Goobar,
Donald E. Groom, Isobel M. Hook, Alex G. Kim, M. Y. Kim,
J. C. Lee, N. J. Nunes, R. Pain, Carlton R. Pennypacker,
Robert Quimby, Christopher Lidman, Richard S. Ellis,
Michael Irwin, Richard G. McMahon, Pilar Ruiz-Lapuente,
Nic Walton, Brad Schaefer, Brian J. Boyle,
Alex V. Filippenko, T. Matheson, Andrew S. Fruchter,
Nino Panagia, Heidi J. M. Newberg,
and Warrick J. Couch
(The Supernova Cosmology Project)

Abstract

We report measurements of the mass density, Ω_M, and cosmological-constant energy density, Ω_Λ, of the universe based on the analysis of 42 type Ia supernovae discovered by the Supernova Cosmology Project. The magnitude-redshift data for these supernovae, at redshifts between 0.18 and 0.83, are fitted jointly with a set of supernovae from the Calán/Tololo Supernova Survey, at redshifts below 0.1, to yield values for the cosmological parameters. All supernova peak magnitudes are standardized using a SN Ia light-curve width-luminosity relation. . . . For a flat ($\Omega_M + \Omega_\Lambda = 1$) cosmology we find $\Omega_M^{\text{flat}} = 0.28^{+0.09}_{-0.08}$. . . . The data are strongly inconsistent with a $\Lambda = 0$ [no cosmological constant] flat cosmology, the simplest inflationary model. An open, $\Lambda = 0$ cosmology also does not fit the data well: the data indicate that the cosmological constant is nonzero and positive, with a [probability] confidence of . . . 99%, including the identified systematic uncertainties. . . .

Introduction

Since the earliest studies of supernovae, it has been suggested that these luminous events might be used as standard candles for cosmological measurements.[82] At closer distances they could be used to measure the Hubble constant if an absolute distance scale or magnitude scale could be established, while at higher redshifts they could determine the deceleration parameter. The Hubble constant measurement became a realistic possibility in the 1980s, when the more homogeneous subclass of type Ia supernovae (SNe Ia) was identified. Attempts to measure the deceleration parameter, however, were stymied for lack of high-redshift [very distant] supernovae. Even after an impressive multiyear effort by Nørgaard-Nielsen et al., it was only possible to follow one SN Ia, at $z = 0.31$, discovered 18 days past its peak brightness.[83]

The Supernova Cosmology Project was started in 1988 to address this problem. The primary goal of the project is the determination of the cosmological parameters of the universe using the magnitude-redshift relation of type Ia supernovae. In particular, Goobar & Perlmutter showed [in 1995] the possibility of separating the relative contributions of the mass density, Ω_M, and the cosmological constant, Λ, to changes in the expansion rate by studying supernovae at a range of redshifts. . . .[84]

A first presentation of analysis techniques, identification of possible sources of statistical and systematic errors, and first results based on seven of these supernovae at redshifts $z \sim 0.4$ were given in Perlmutter *et al.* (1997).[85] These first results yielded a confidence region that was suggestive of a flat, $\Lambda = 0$ universe but with a large range of uncertainty. Perlmutter *et al.* (1998) added a $z = 0.83$ SN Ia to this sample, with observations from the *Hubble Space Telescope* (*HST*) and Keck 10 m telescope, providing the first demonstration of the method of separating Ω_M and Λ contributions.[86] This analysis offered preliminary evidence for a low-mass-density universe with a best-fit value of $\Omega_M = 0.2 \pm 0.4$, assuming $\Lambda = 0$. Independent work by Garnavich *et al.* (1998), based on three supernovae at $z \sim 0.5$ and one at $z = 0.97$, also suggested a low mass density. . . .[87]

Perlmutter *et al.* (1997) presented a preliminary analysis of 33 additional high-redshift supernovae, which gave a confidence region indicating an accelerating universe and barely including a low-mass $\Lambda = 0$ cosmology.[88] Recent independent work of Riess *et al.,* based on 10 high-redshift supernovae added to the Garnavich *et al.* set, reached the same conclusion.[89] Here we report on the complete analysis of 42 supernovae from the

Supernova Cosmology Project, including the reanalysis of our previously reported supernovae with improved calibration data and improved photometric and spectroscopic SN Ia templates.

Basic Data and Procedures

The new supernovae in this sample of 42 were all discovered while still brightening, using the Cerro Tololo Inter-American Observatory (CTIO) 4 m telescope with the 2048^2 pixel prime-focus CCD camera or the 4×2048^2 pixel Big Throughput Camera. The supernovae were followed with photometry over the peak of their light curves and approximately 2–3 months further . . . using the CTIO 4 m, Wisconsin-Indiana-Yale-NOAO (WIYN) 3.6 m, ESO 3.6 m, Isaac Newton Telescope (INT) 2.5 m, and the William Herschel Telescope (WHT) 4.2 m telescopes. (SN 1997ap and other 1998 supernovae have also been followed with *HST* photometry.) The supernova redshifts and spectral identifications were obtained using the Keck I and II 10 m telescopes with the Low-Resolution Imaging Spectrograph and the ESO 3.6 m telescope. . . .

[Omitted here are the authors' extensive discussions of their correction and fitting procedures on the data, as well as their many checks on systematic uncertainties.]

Results

. . . [From our primary fit], we find $\Omega_M^{\text{flat}} = 0.28^{+0.09}_{-0.08}$ in a flat universe [that is, cosmic matter in total provides 28% of the critical density, give or take 8–9%, necessary for a flat universe. Ω_Λ makes up the remaining 72%]. Cosmologies with $\Omega_\Lambda = 0$ [no cosmological constant] are a poor fit to the data at the 99.8% confidence level. The contours of Figure [75.3] more fully characterize the best-fit confidence regions. . . . The $(\Omega_M, \Omega_\Lambda) = (1,0)$ line on Figure [75.4b] shows that 38 out of 42 high-redshift supernovae are fainter than predicted for this model. These supernovae would have to be over 0.4 mag brighter than measured (or the low-redshift supernovae 0.4 mag fainter) for this model to fit the data.

The $(\Omega_M, \Omega_\Lambda) = (0, 0)$ upper solid line on Figure [75.4a] shows that the data are still not a good fit to an "empty universe," with zero mass density and cosmological constant. The high-redshift supernovae are as a group fainter than predicted for this cosmology; in this case, these supernovae would have to be almost 0.15 mag brighter for this empty cosmology to fit

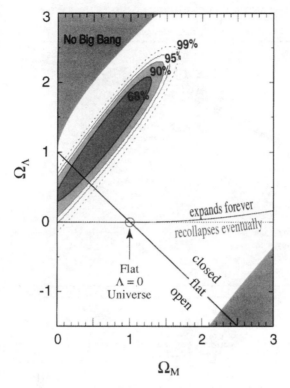

Figure 75.3: "Best-fit confidence regions in the Ω_M-Ω_Λ plane for our primary analysis. . . . In cosmologies above [the] near-horizontal line the universe will expand forever, while below this line the expansion of the universe will eventually come to a halt and recollapse. . . ."

the data, and the discrepancy is even larger for $\Omega_M > 0$. This is reflected in the high probability (99.8%) of $\Omega_\Lambda > 0$. . . .

Conclusions and Discussion

The confidence regions of Figure [75.3] and the residual plot of Figure [75.4b] lead to several striking implications. First, the data are strongly inconsistent with the $\Lambda = 0$, flat universe model (indicated with a circle) that has been the theoretically favored cosmology. If the simplest inflationary theories are correct and the universe is spatially flat, then the supernova data imply that there is a significant, positive cosmological constant. Thus the universe may be flat *or* there may be little or no cosmological constant, but the data are not consistent with both possibilities simultaneously. This is the most unambiguous result of the current data set.

Ω AND Λ FROM 42 HIGH-REDSHIFT SUPERNOVAE

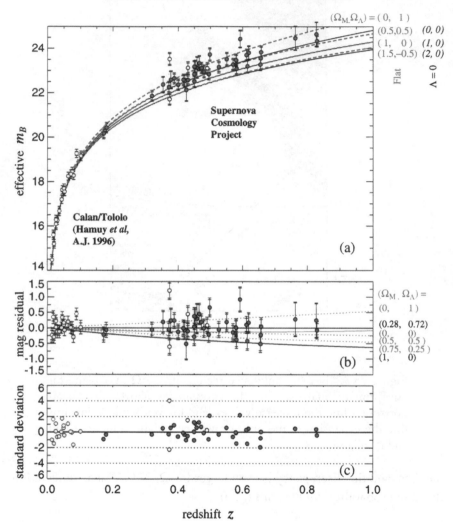

Figure 75.4: "(a) [top] Hubble diagram for 42 high-redshift type Ia supernovae from the Supernova Cosmology Project and 18 low-redshift type Ia supernovae from the Calán/Tololo Supernova Survey, plotted on a linear redshift scale to display details at high redshift. . . . (b) [middle] Magnitude residuals from the best-fit cosmology for the [primary] fit supernova subset, $(\Omega_M, \Omega_\Lambda) = (0.28, 0.72)$. The dashed curves are for a range of flat cosmological models: $(\Omega_M, \Omega_\Lambda) = (0, 1)$ on top, $(0.5, 0.5)$ third from bottom, $(0.75, 0.25)$ second from bottom, and $(1, 0)$ is the solid curve on bottom. The middle solid curve is for $(\Omega_M, \Omega_\Lambda) = (0, 0) \ldots$ (c) [bottom] Uncertainty-normalized residuals from the best-fit cosmology for the [primary] fit supernova subset, $(\Omega_M, \Omega_\Lambda) = (0.28, 0.72)$."

Second, this data set directly addresses the age of the universe relative to the Hubble time. . . . For any value of the Hubble constant less than $H_0 = 70$ km s^{-1} Mpc^{-1}, the implied age of the universe is greater than 13 Gyr [billion years], allowing enough time for the oldest stars in globular clusters to evolve. . . .

Third, even if the universe is not flat, the confidence regions of Figure [75.3] suggest that the cosmological constant is a significant constituent of the energy density of the universe. The best-fit model (the center of the shaded contours) indicates that the energy density in the cosmological constant is ~0.5 more than that in the form of mass energy density. . . .

Given the potentially revolutionary nature of this third conclusion, it is important to reexamine the evidence carefully to find possible loopholes. None of the identified sources of statistical and systematic uncertainty . . . could account for the data in a $\Lambda = 0$ universe. If the universe does in fact have zero cosmological constant, then some additional physical effect or "conspiracy" of statistical effects must be operative—and must make the high-redshift supernovae appear almost 0.15 mag (~15% in flux) fainter than the low-redshift supernovae. At this stage in the study of SNe Ia, we consider this unlikely but not impossible. For example . . . some carefully constructed smooth distribution of large-grain-sized gray dust that evolves similarly for elliptical and spiral galaxies could evade our current tests. Also, the full data set of well-studied SNe Ia is still relatively small, particularly at low redshifts, and we would like to see a more extensive study of SNe Ia in many different host-galaxy environments before we consider all plausible loopholes . . . to be closed.

Notes

Citations in the notes are given in the author/year format when the source is also listed in the bibliography of materials consulted through the course of my research. Full citations are given here for those additional sources referred to only in relation to the text, either for further reading or for clarification, or as a citation that is part of one of the original papers excerpted in this anthology.

I. The Ancient Sky

1. Evans, 1998, p. 4.
2. For a more detailed explanation of the Mayan calendar, see Thurston, 1994.
3. Heath, 1991, p. 6.
4. Evans, 1998, p. 67.
5. Abetti, 1952, p. 37.
6. Toomer, 1998, p. 45.

II. Revolutions

1. Swerdlow, 1973, p. 434.
2. Ibid., p. 436.
3. Copernicus, 1995, p. 26.
4. Ibid., p. 1.
5. Swerdlow, p. 436.
6. Copernicus, 1995, p. 14.
7. Copernican expert Owen Gingerich has noted that Wallis incorrectly translated this bracketed sentence. I substituted the translation from *De revolutionibus*, translated by John F. Dobson and assisted by Selig Brodetsky, from *Occasional Notes of the Royal Astronomical Society* 10 (May 1947), p. 20.
8. Shapley and Howarth, 1929, p. 14.
9. Dreyer, 1953, p. 366.
10. Voelkel, 1999, p. 57.
11. Kepler, 1995, p. 124.
12. Current scholarship on Tycho Brahe's aristocratic status suggests that the post of imperial mathematician was actually beneath him. Tycho called himself "imperial counsellor." Still, it's conventional to say that Kepler inherited the position of imperial mathematician. (My thanks to James Voelkel for bringing this to my attention.)
13. Kepler, 1992, p. 256.
14. Ibid., p. 495.

15. Ibid., p. 508.
16. Voelkel, p. 66.
17. Abetti, 1952, p. 94.
18. Gingerich and Van Helden, 2003.
19. Abetti, p. 108.
20. Ibid., p. 110.
21. Ibid., p. 134.
22. Communication with Newton expert George Smith, July 15, 2003.
23. Cohen, I. B. "Newton." *Dictionary of Scientific Biography*, volume 10. New York: Scribner's, 1974, p. 64.
24. Newton, 1999, p. 794.
25. Newton, 1974, p. 78.
26. Newton, 1999, p. 943.
27. Rule 3 appears for the first time in the second (1713) edition of the *Principia*.
28. Rule 4 appears for the first time in the third (1726) edition of the *Principia*.
29. Ibid., p. 270.
30. Ibid., p. 895.
31. Halley, 1705, p. 24.
32. Herschel discusses binary star systems in "Catalogue of 500 New Nebulae, Nebulous Stars, Planetary Nebulae, and Clusters of Stars; with Remarks on the Construction of the Heavens," *Philosophical Transactions of the Royal Society of London* 92 (1802): 477–528.

III. Taking Measure

1. Bradley, James. "A Letter from the Reverend Mr. James Bradley Savilian Professor of Astronomy at Oxford, and F.R.S. to Dr. Edmond Halley Astronom. Reg. &c. Giving an Account of a New Discovered Motion of the Fix'd Stars." *Philosophical Transactions* 35 (1727–28): 637–61.
2. The full statement is found in the introduction of Laplace's *Théorie analytique des probabilités* [1812–20].
3. Laplace, 1809, p. 366.
4. Halley, 1717–19, p. 737.
5. Hoskin, 1999, p. 160.
6. Abetti, Giorgio. "Giuseppe Piazzi." *Dictionary of Scientific Biography*, Volume 10. New York: Scribner's, 1974, p. 592.
7. I would like to thank Michael Hoskin of Cambridge University for making me aware of this English translation archived in his university library.
8. *Memoirs of the Royal Astronomical Society* 12 (1841), p. 453.
9. Shapley and Howarth, 1929, p. 252.
10. *Astronomische Nachrichten* 89 (1877): 349–52.
11. Wright, 1971, p. x.
12. Wright, 1750, p. 62.
13. Ibid., p. 84.
14. Kant, 1900, p. 190
15. Ibid., p. 64.
16. Herschel, 1784b, p. 220.

17. Halley, 1714–16, p. 390.

18. Herschel, William. "On Nebulon Stars, Properly So Called." *Philosophical Transactions of the Royal Society of London* 81 (1791): 73.

19. Hoskin, 1999, p. 213.

IV. Touching the Heavens

1. Huggins, Sir William, and Lady Huggins (eds.). *The Scientific Papers of Sir William Huggins.* London: William Wesley and Son, 1909.

2. Comte, Auguste. *The Positive Philosophy.* Translated by Harriet Martineau. New York: AMS Press, 1974.

3. DeVorkin, 1999, p. 28.

4. Newton, Isaac. "A Letter of Mr. Isaac Newton, Professor of the Mathematicks in the University of Cambridge; Containing His New Theory about Light and Colors." *Philosophical Transactions* 6 (1671): 3075–87.

5. Wollaston, William Hyde. "A Method of Examining Refractive and Dispersive Powers, by Prismatic Reflection." *Philosophical Transactions of the Royal Society of London* 92 (1802): 365–80.

6. Fraunhofer, 1898, p. 9.

7. Kirchhoff, Gustav, and Robert Bunsen. "Chemical Analysis by Observation of Spectra." *Annalen der Physik und der Chemie* 110 (1860): 161–89.

8. Huggins, William, and William Allen Miller. "Note on the Lines in the Spectra of Some of the Fixed Stars." *Proceedings of the Royal Society of London* 12 (1863): 444–45.

9. Rutherford, L. M. *American Journal of Science* 35 (1863).

10. Secchi, A. *Astronomische Nachrichten,* No. 1405 (March 3, 1863).

11. Huggins, 1897, p. 911.

12. See note 8 above.

13. Huggins, 1897, p. 915.

14. Ibid., p. 916.

15. Ibid., pp. 916–17.

16. Bowen, I. S. "The Origin of the Nebular Lines and the Structure of the Planetary Nebulae." *Astrophysical Journal* 67 (1928): 1–15.

17. Fizeau, H. L. *Annales de chimie et de physique* 19 (1870): 217.

18. Keeler, 1890.

19. Pickering, E. C. "A New Class of Binary Stars." *Monthly Notices of the Royal Astronomical Society* 50 (March 1890): 296–98.

20. Hoskin, 1999, p. 255.

21. Rutherfurd, L. *American Journal of Science and Arts* 35 (1863).

22. Cannon, Annie Jump. "The Henry Draper Memorial." *Journal of the Royal Astronomical Society of Canada* 9 (1915): 203–15.

23. DeVorkin, 2000, p. 85.

24. Michelson, A. A., and F. G. Pease. "Measurement of the Diameter of α Orionis with the Interferometer." *Astrophysical Journal* 53 (1921): 249–59.

25. DeVorkin, 2000, p. 86.

26. Saha, Meghnad. "Ionization in the Solar Chromosphere." *Philosophical Magazine* 40 (1920): 479–88.

27. Lang and Gingerich, 1979, p. 243.

28. Letter from H. N. Russell to C. Payne, January 14, 1925. H. N. Russell papers, Princeton University Library.

29. Payne, 1925, pp. 186, 188.

30. Unsöld, A. *Zeitschrift für Physik* 46 (1928): 765.

31. McCrea, W. H. *Monthly Notices of the Royal Astronomical Society* 89 (1929): 483.

32. Russell, Henry Norris. "On the Composition of the Sun's Atmosphere." *Astrophysical Journal* 70 (1929): 78–79.

33. Ibid., p. 65.

34. Strömgren, B. *Zeitschrift für Astrophysik* 4 (1932): 118; also 7 (1933): 222.

35. Lang and Gingerich, 1979, p. 291.

36. *Monthly Notices of the Royal Astronomical Society* 17 (February 1857): 129.

37. Hale, G. E. "On the Probable Existence of a Magnetic Field in Sun-Spots." *Astrophysical Journal* 28 (1908): 315–43.

38. Larmor, Joseph. "How Could a Rotating Body Such as the Sun Become a Magnet?" *Report for the British Association for the Advancement of Science* (1919): 159–60.

39. Birkeland, K. *Comptes rendus de l'Académie des sciences* (Paris) 157 (1913): 275; Störmer, C. *Archives des sciences physiques et naturelles* (Geneva) 24 (1907): 317.

40. Biermann, L. "Solar Corpuscular Radiation and the Interplanetary Gas." *Observatory* 77 (1957): 109–10.

41. Leighton, Robert B., Robert W. Noyes, and George W. Simon. "Velocity Fields in the Solar Atmosphere. I. Preliminary Report." *Astrophysical Journal* 135 (1962): 474–99.

42. Janssen, P. *Comptes rendus de l'Académie des sciences* 67 (1868).

43. Ramsay, William. "On a Gas Showing the Spectrum of Helium." *Proceedings of the Royal Society of London* 58 (1895): 65–67.

44. Chladni, Ernst F. F. *Über den Ursprung der von Pallas gefundenen und anderer ihr änlicher Eisenmassen und über einige damit in Verbindung stehende Naturerscheinungen.* Riga: J. F. Hartknoch, 1794.

45. Howard, Edward, John Lloyd Williams, and Count de Bournon. "Experiments and Observations on Certain Stony and Metalline Substances, Which at Different Times Are Said to Have Fallen on the Earth; Also on Various Kinds of Native Iron." *Philosophical Transactions of the Royal Society of London* 92 (1802): 168–212.

46. Clarke, 1836.

47. Newton, H. A. "The original accounts of the displays in former times of the November Star-Shower; together with a determination of the length of its cycle, its annual period, and the probable orbit of the group of bodies around the sun." *The American Journal of Science and Arts* 37 (1864): 377–89.

48. Schiaparelli, M. J. V. "On the Relation Which Exists Between Comets and Shooting Stars." *Astronomische Nachrichten* 68 (1867): 331; also, *Monthly Notices of the Royal Astronomical Society* 27 (1867): 246–47.

49. De Maria and Russo, 1989, p. 216.

50. Compton, Arthur H. "A Geographic Study of Cosmic Rays." *Physical Review* 43 (March 1933): 387–403.

51. Baade and Zwicky, 1934b.

52. Fermi, Enrico. "On the Origin of the Cosmic Radiation." *Physical Review* 75 (April 1949): 1169–74.

53. "The Discovery of the Satellites of Mars." *Monthly Notices of the Royal Astronomical Society* 38 (November 9, 1877): 208.

54. Maxwell, James Clerk. *On the Stability of the Motion of Saturn's Rings.* Cambridge: Macmillan, 1859.

55. Keeler, James E. "A Spectroscopic Proof of the Meteoric Constitution of Saturn's Rings." *Astrophysical Journal* 1 (1895): 416–27.

56. MacPherson, 1906, p. 122.

57. Standage, 2000, p. 179.

58. Jewitt, D., and J. Luu. "Discovery of the Candidate Kuiper Belt Object 1992 QB1." *Nature* 362 (1993): 730–32.

V. Einsteinian Cosmos

1. Fölsing, 1997, p. 196.

2. Einstein, A. "Über die vom Relativitätsprinzip geforderte Trägheit der Energie [On the Inertia of Energy Demanded by the Relativity Principle]." *Annalen der Physik* 23 (1907): 371–84. In this paper the relationship is given as $\mu = E_0 \, / \, V^2$, where μ is the mass, E_0 the energy of the system, and V the speed of light in a vacuum. See also *Jahrbuch der Radioaktivität und Elektronik* 4 (1907): 433–62.

3. Lorentz et al., 1923, p. 48.

4. Ibid.

5. Ibid., pp. 49–50.

6. Einstein, A. *Sitzungsberichte der Preussischen Akademie der Wissenschaften zu Berlin* 11 (1915): 831–39.

7. Hoffmann, 1972, p. 125.

8. Several scientists used Newton's laws to contemplate gravity acting upon light. They include the British scientist John Michell in 1783 (see Chapter 42), the French astronomer Pierre-Simon Laplace in 1799, and the German mathematician and astronomer Johann Soldner (see "On the Deflection of a Light Ray from Its Straight Motion Due to the Attraction of a World Body Which it Passes Closely." *Berliner Astronomisches Jahrbuch auf das Jahr 1804*). Newton himself briefly considered the possibility as early as 1704.

9. Eddington, 1920, p. 115.

10. Shapiro, Irwin I., et al. "Fourth Test of General Relativity: Preliminary Results." *Physical Review Letters* 20 (May 27, 1968): 1265–69.

11. Lorentz et al., 1923, p. 188.

12. De Sitter, 1917, p. 26.

13. Friedmann, A. "On the Curvature of Space." *Zeitschrift für Physik* 10 (1922): 377–86.

14. Lemaître, 1931a, p. 489.

15. Lemaître, Canon Georges. *The Primeval Atom.* New York: Van Nostrand, 1950, p. 78.

16. Eddington, 1931, p. 450.

17. Hoyle, F. "A New Model for the Expanding Universe." *Monthly Notices of the Royal Astronomical Society* 108 (1948): 372–82.

18. Hoyle, Fred. *The Nature of the Universe.* Oxford: Basil Blackwell, 1950. As early as 1932, British theorist Arthur Eddington foreshadowed the expression. He was disturbed that an expanding universe implied a "peculiar beginning of things" and wrote in his book *The Nature of the Physical World* that "as a scientist I simply do not believe that the present order of things started off with a bang; unscientifically I feel equally unwilling to accept the implied discontinuity in the divine nature."

19. Philip, A. G. Davis, and D. H. DeVorkin (eds.). "In Memory of Henry Norris Russell." *Dudley Observatory Report* 13 (1977): 90.

20. Eddington, Arthur. *Stars and Atoms.* Oxford: Clarendon Press, 1927, p. 50.

21. Adams, W. S. "The Relativity Displacement of the Spectral Lines in the Companion of Sirius." *Proceedings of the National Academy of Sciences* 11 (July 15, 1925): 382–87.

22. Greenstein, J. L., J. B. Oke, and H. L. Shipman. "Effective Temperature, Radius, and Gravitational Redshift of Sirius B." *Astrophysical Journal* 169 (1971): 563–66.

23. Chandrasekhar, 1931, p. 81.

24. Landau, L. "On the Theory of Stars." *Physicalische Zeitschrift der Sowjetunion* 1 (1932): 285–88.

25. Chandrasekhar, 1932, p. 67.

26. Chandrasekhar, 1934, p. 79.

27. Chandrasekhar, 1935, p. 207.

28. "Discussion," 1935, p. 38.

29. Wali, 1984, p. 30

30. See note 24.

31. Gamow, G. *Atomic Nuclei and Nuclear Transformations.* Oxford: Oxford University Press, 1937.

32. Oppenheimer, J. R., and Robert Serber. "On the Stability of Stellar Neutron Cores." *Physical Review* 54 (1938): 540; Oppenheimer, J. R., and G. M. Volkoff. "On Massive Neutron Cores." *Physical Review* 55 (1939): 374–81.

33. Michell, John. "On the Means of Discovering the Distance, Magnitude, &c. of the Fixed Stars, in Consequence of the Diminution of the Velocity of Their Light, in Case Such a Diminution Should Be Found to Take Place in Any of Them, and Such Other Data Should Be Procured from Observations, as Would Be Farther Necessary for That Purpose." *Philosophical Transactions of the Royal Society of London* 74 (1784): 35–57.

34. Laplace, 1809.

35. Schwarzschild, Karl. "On the Gravitational Field of a Point Mass According to the Einsteinian Theory." *Sitzungberichte der K. Preussischen Akademie der Wissenschaften zu Berlin* 1 (1916): 189–96.

36. See note 32; Oppenheimer and Volkoff, p. 381.

37. Oppenheimer and Snyder, 1939, p. 456.

38. Einstein, A. "Stationary Systems with Spherical Symmetry Consisting of Many Gravitating Masses." *Annals of Mathematics* 40 (1939): 922–36.

39. Webster, B. Louise, and Paul Murdin. "Cygnus X-1: A Spectroscopic Binary with a Heavy Companion?" *Nature* 235 (1972): 37–38; Bolton, Charles Thomas. "Dimensions of the Binary System HDE 226868: Cygnus X-1." *Nature* 240 (1972): 124–27.

40. Hawking, S. W. "Particle Creation by Black Holes." *Communications in Mathematical Physics* 43 (1975): 199–220. This paper was based on his report to the Oxford Symposium on quantum gravity held in Harwell, England, February 15–16, 1974.

41. Lockyer, J. Norman. "The Bakerian Lecture: Researches in Spectrum-Analysis in Connexion with the Spectrum of the Sun. No. III." *Philosophical Transactions of the Royal Society of London* 164 (1874), p. 493.

42. Nye, Mary Jo. *Molecular Reality: A Perspective on the Scientific Work of Jean Perrin.* London: MacDonald, 1972, p. 54.

43. Gamow, G. *Zeitschrift für Physik* 52 (1928): 510. The same conclusion was simultaneously arrived at by Princeton physicists Edward Condon and Ronald Gurney; see *Nature* 122 (1928): 439.

44. Atkinson, R. d'E., and F. G. Houtermans. *Zeitschrift für Physik* 54 (1929): 656 (Their working title for this article, ultimately rejected by the editor, was "How Can One Cook a Helium Nucleus in a Potential Pot?"); Atkinson, Robert d'E. "Atomic Synthesis and Stellar Energy I, II." *Astrophysical Journal* 73 (1931): 250–95, 308–47.

45. Bethe, H. A., and C. L. Critchfield. "The Formation of Deuterons by Proton Combination." *Physical Review* 54 (1938): 248–54.

46. Weizsäcker, C. F. von. "Element Transformation Inside Stars. II." *Physikalische Zeitschrift* 39 (1938): 633–46.

47. Eddington, A. S. "The Internal Constitution of Stars." *Nature* 106 (1920): 19.

48. Bethe, 1939, p. 434.

49. The word *ylem* is widely associated with Gamow, but Alpher was the first to use the term in a 1948 *Physical Review* paper entitled "A Neutron-Capture Theory of the Formation and Relative Abundance of the Elements." *Physical Review* 74: 1577–89.

50. "Remarks on the Evolution of the Expanding Universe." *Physical Review* 75 (1949): 1089–95; "Neutron-Capture Theory of Element Formation in an Expanding Universe." *Physical Review* 84 (1951): 60–68; "Physical Conditions in the Initial Stages of the Expanding Universe." *Physical Review* 92 (1953): 1347–61.

51. Weinberg, Steven. *The First Three Minutes.* New York: Basic Books, 1977, p. 124.

52. Salpeter, Edwin E. *Astrophysical Journal* 115 (1952): 326–28; Öpik, E. J. *Proceedings of the Royal Irish Academy* 54 (1951): 49.

53. Hoyle, F. "On Nuclear Reactions Occurring in Very Hot Stars. I. The Synthesis of Elements from Carbon to Nickel." *Astrophysical Journal Supplement* 1 (1954): 121–46.

54. Hoyle, F. "The Synthesis of the Elements from Hydrogen." *Monthly Notices of the Royal Astronomical Society* 106 (1946): 343–83.

55. Merrill, Paul W. "Technetium in the Stars." *Science* 115 (May 2, 1952): 484.

56. Cameron, A. G. W. "Nuclear Reactions in Stars and Nucleogenesis." *Publications of the Astronomical Society of the Pacific* 69 (June 1957): 201–22.

57. Gamow, G. "Nuclear Energy Sources and Stellar Evolution." *Physical Review* 53 (1938): 595–604.

58. See, for example, Schönberg, M., and S. Chandrasekhar. "On the Evolution of the Main-Sequence Stars." *Astrophysical Journal* 96 (1942): 161–72; Sandage,

Allan R., and Martin Schwarzschild. "Inhomogeneous Stellar Models II: Models with Exhausted Cores in Gravitational Contraction." *Astrophysical Journal* 116 (1952): 463–76; Hoyle, F., and M. Schwarzschild. "On the Evolution of Type II Stars." *Astrophysical Journal Supplement* 2 (1955): 1–40; Walker, Merle F. "Studies of Extremely Young Clusters I: NGC 2264." *Astrophysical Journal Supplement* 2 (1956): 365–87.

VI. The Milky Way and Beyond

1. Kant, 1900.
2. Huggins, W. and Mrs. Huggins. "On the Spectrum, Visible and Photographic, of the Great Nebula in Orion." *Proceedings of the Royal Society of London* 46 (1889): 60.
3. Clerke, A. M. *The System of the Stars.* 2nd edition. London: A. & C. Black, 1905, p. 349.
4. Berendzen, Richard, ed. *Education in and History of Modern Astronomy.* New York: New York Academy of Sciences, 1972, p. 203.
5. Leavitt, Henrietta S. "1777 Variables in the Magellanic Clouds." *Annals of the Astronomical Observatory of Harvard College*, volume 60, number 4. Cambridge: Harvard Observatory, 1908 (volume actually distributed in 1907), p. 107.
6. Leavitt, 1912, p. 3.
7. Plummer, H. G. "Note on the Velocity of Light and Doppler's Principle." *Monthly Notices of the Royal Astronomical Society* 74 (1914): 660–63; Shapley, Harlow. "On the Nature and Cause of Cepheid Variation." *Astrophysical Journal* 40 (1914): 448–65.
8. Bohlin, K. *Kungliga Svenska Vetenskapsakademiens handlingar* 43:10 (1909).
9. Gingerich, "Harlow Shapley." *Dictionary of Scientific Biography*, volume 12. New York: Scribner's, 1975, p. 346.
10. Kapteyn, J. C. "First Attempt at a Theory of the Arrangement and Motion of the Sidereal System." *Astrophysical Journal* 55 (1922): 302–27.
11. Kapteyn, J. C. *Report of the British Association for the Advancement of Science* (1905): 237–65.
12. Lindblad, B. "Star-Streaming and the Structure of the Stellar System." *Arkiv för matematik, astronomi och fysik* 19A:21 (1925): 1–8.
13. Oort, Jan H. "Observational Evidence Confirming Lindblad's Hypothesis of a Rotation of the Galactic System." *Bulletin of the Astronomical Institutes of the Netherlands* 3 (1927): 275–82.
14. Baade, 1963, p. 9.
15. Houghton, H. E. "Sir William Herschel's 'Hole in the Sky.' " *Monthly Notes of the Astronomical Society of South Africa* 1 (February 1942): 107.
16. Abetti, 1952, p. 293; see also Secchi, Angelo. *L'astronomia in Roma nel pontificato di Pio IX.* Rome: Tipografia della pace, 1877.
17. Barnard, E. E. "On the Dark Markings of the Sky, with a Catalogue of 182 Such Objects." *Astrophysical Journal* 49 (1919): 1–24.
18. Ibid., p. 1.
19. Wolf, M. "On the Dark Nebula NGC 6960." *Astronomische Nachrichten* 219 (1923): 109–16.
20. Eddington, A. S. "Diffuse Matter in Interstellar Space." *Proceedings of the Royal Society of London. Series A.* 111 (July 1, 1926): 447.

21. Berry, 1961, p. 406.

22. Clerke, 1902, p. 403.

23. MacPherson, 1906, p. 34.

24. Ritchey, G. W. "Novae in Spiral Nebulae." *Publications of the Astronomical Society of the Pacific* 29 (1917): 210–12.

25. Curtis, H. D. "New Stars in Spiral Nebulae." *Publications of the Astronomical Society of the Pacific* 29 (1917): 180–82.

26. Hoskin, 1976, p. 49.

27. Öpik, E. "An Estimate of the Distance of the Andromeda Nebula." *Astrophysical Journal* 55 (1922): 406–10.

28. *Publications of the American Astronomical Society* 5 (1925), p. 246.

29. Hubble, 1926.

30. Duncan, John C. "Three Variable Stars and a Suspected Nova in the Spiral Nebula M 33 Trianguli." *Publications of the Astronomical Society of the Pacific* 34 (October 1922): 290–91.

31. Shapley, Harlow. "Studies Based on the Colors and Magnitudes in Stellar Clusters. Sixth Paper: On the Determination of the Distances of Globular Clusters." *Astrophysical Journal* 48 (1918): 89–124.

32. Slipher, V. M. "The Radial Velocity of the Andromeda Nebula." *Lowell Observatory Bulletin* 2 (1913): 56–57.

33. Slipher, V. M. "Spectrographic Observations of Nebulae." *Popular Astronomy* 23 (1915): 21–24; see also Hetherington, Norriss S. "The Measurement of Radial Velocities of Spiral Nebulae." *Isis* 62 (September 1971): 309–13.

34. Wirtz, Carl. "De Sitter's Cosmology and the Motion of the Spiral Nebulae." *Astronomische Nachrichten* (October 1924): 21.

35. Humason, Milton L. "The Large Apparent Velocities of Extra-Galactic Nebulae." *Astronomical Society of the Pacific Leaflet* 1 (1931): 152.

36. Freedman, W. L., et al. "Final Results from the Hubble Space Telescope Key Project to Measure the Hubble Constant." *Astrophysical Journal* 553 (2001): 47–72.

37. Astronomer Otto Struve in Evans, 1959, p. 177.

38. Baade, 1963, p. 38.

39. Eggen, O. J., D. Lynden-Bell, and A. R. Sandage. "Evidence from the Motions of Old Stars that the Galaxy Collapsed." *Astrophysical Journal* 136 (1962): 748–66.

40. Alexander, Stephen. "On the Origin of the Forms and the Present Condition of Some of the Clusters of Stars and Several of the Nebulae." *Astronomical Journal* 2 (April 17, 1852): 101.

41. Proctor, R. A. "A New Theory of the Milky Way." *Monthly Notices of the Royal Astronomical Society* 30 (1869): 50–56.

42. Easton, C. "A New Theory of the Milky Way." *Astrophysical Journal* 12 (1900): 157.

43. Westerhout, 1972, p. 216.

44. Lin, C.-C., and F. H. Shu. "On the Spiral Structure of Disk Galaxies." *Astrophysical Journal* 140 (1964): 646–55.

45. Öpik, E. J. *Proceedings of the American Academy of Arts and Sciences* 67 (1932): 169.

VII. New Eyes, New Universe

1. *Astronomy and Astrophysics for the 1980s: Report of the Astronomy Survey Committee.* Washington, D.C.: National Academy Press, 1982.

2. Hey, 1973, p. 7.

3. Reber, Grote. "Cosmic Static." *Astrophysical Journal* 91 (1940): 621–24; followed by "Cosmic Static," *Proceedings of the IRE* 30 (August 1942): 367–78.

4. Hey, James S., S. J. Parsons, and J. W. Phillips. "Fluctuations in Cosmic Radiation at Radio Frequencies." *Nature* 158 (1946): 234.

5. Baade, Walter, and Rudolph Minkowski. "Identification of the Radio Sources in Cassiopeia, Cygnus A, and Puppis A." *Astrophysical Journal* 119 (1954): 206–14.

6. Bolton, John G., Gordon J. Stanley, and O. B. Slee. "Positions of Three Discrete Sources of Galactic Radio-Frequency Radiation." *Nature* 164 (1949): 101–2.

7. Townes, C. H. *International Astronomical Union Symposium Number 4.* Edited by H. C. van de Hulst. Cambridge: Cambridge University Press, 1957, p. 92.

8. Shklovsky, I. S. *Doklady akademii nauk SSSR* 92 (1953): 25.

9. Cheung, A. C., D. M. Rank, C. H. Townes, D. D. Thornton, and W. J. Welch. "Detection of NH_3 Molecules in the Interstellar Medium by Their Microwave Emission." *Physical Review Letters* 21 (December 16, 1968): 1701–5; "Detection of Water in Interstellar Regions by Its Microwave Radiation." *Nature* 221 (February 15, 1969): 626–28.

10. Snyder, Lewis E., David Buhl, B. Zuckerman, and Patrick Palmer. "Microwave Detection of Interstellar Formaldehyde." *Physical Review Letters* 22 (1969): 679–81.

11. Weaver, Harold, David R. W. Williams, N. H. Dieter, and W. T. Lum. "Observations of a Strong Unidentified Microwave Line and of Emission from the OH Molecule." *Nature* 208 (1965): 29–31.

12. Perkins, F., T. Gold, and E. E. Salpeter. *Astrophysical Journal* 145 (1966): 361–66.

13. Herrnstein, J. R., J. M. Moran, L. J. Greenhill, P. J. Diamond, M. Inoue, N. Nakai, M. Miyoshi, C. Henkel, and A. Riess. *Nature* 400 (1999): 539–41.

14. Birkeland, K. *Comptes rendus de l'Académie des sciences* (Paris) 157 (1913): 275; and Störmer, C. *Archives des sciences physiques et naturelles* (Geneva) 24 (1907): 317.

15. Barabashov, N. P., and Yu. N. Lipskii. "The First Results Obtained from Photographs of the Invisible Side of the Moon." *Doklady academii nauk SSSR* 129 (1959): 288–89.

16. Van Helden, Albert. *Measuring the Universe.* Chicago: University of Chicago Press, 1985, p. 1.

17. Herschel, William. "On the Remarkable Appearances at the Polar Regions of the Planet Mars; the Inclination of Its Axis, the Position of Its Poles, and Its Spheroidical Figure; With a Few Hints Relating to Its Real Diameter and Atmosphere." *Philosophical Transactions of the Royal Society of London* 74 (1784): 273.

18. Lowell, Percival. *Mars.* Boston: Houghton Mifflin and Company, 1895.

19. Murray, B.C., L. A. Soderblom, J. A. Cutts, R. Sharp, D. Milton, and R. Leighton. "Geological Framework of the South Polar Region of Mars." *Icarus* 17 (1972): 328.

20. Leighton, R. B., and B. C. Murray. "Behavior of carbon dioxide and other volatiles on Mars." *Science* 153 (1966): 136–44.

21. Sagan, C. "The Long Winter Model of Martian Biology." *Icarus* 15 (1971): 511–14.

22. Friedman, H., S. W. Lichtman, and E. T. Byram. "Photon Counter Measurements of Solar X-Rays and Extreme Ultraviolet Light." *Physical Review* 83 (September 1, 1951): 1025–30.

23. Friedman, 1972, p. 272.

24. Sandage, A., et al. "On the Optical Identification of Sco X-1." *Astrophysical Journal* 146 (1966): 316–21; Johnson, Hugh M., and C. B. Stephenson. "A Possible Old Nova Near Sco X-1." *Astrophysical Journal* 146 (1966): 602–4.

25. Giacconi, R., H. Gursky, E. Kellogg, E. Schreier, and H. Tananbaum. "Discovery of Periodic X-Ray Pulsations in Centaurus X-3 from *Uhuru*." *Astrophysical Journal* 167 (July 15, 1971): L67–L73.

26. Schreier, E., R. Levinson, H. Gursky, E. Kellogg, H. Tananbaum, and R. Giacconi. "Evidence for the Binary Nature of Centaurus X-3 from *Uhuru* X-Ray Observations." *Astrophysical Journal* 172 (March 15, 1972): L79–L89.

27. Bartusiak, 1986, pp. 53–54.

28. Gold, Thomas. "The Origin of Cosmic Radio Noise." *Proceedings of the Conference on Dynamics of Ionized Media*. London: University College, 1951.

29. Bartusiak, 1986, p. 151.

30. Hazard, C., M. B. Mackey, and A. J. Shimmins. "Investigation of the Radio Source 3C 273 by the Method of Lunar Occultations." *Nature* 197 (March 16, 1963): 1037–39.

31. Oke, J. B. "Absolute Energy Distribution in the Optical Spectrum of 3C 273." *Nature* 197 (March 16, 1963): 1040–41.

32. Greenstein, Jesse L., and Thomas A. Matthews. "Red-Shift of the Unusual Radio Source: 3C 48." *Nature* 197 (March 16, 1963): 1041–42.

33. Dicke, R. H., P. J. E. Peebles, P. G. Roll, and D. T. Wilkinson. "Cosmic Black-Body Radiation." *Astrophysical Journal* 142 (1965): 416.

34. McKellar, A. *Publications of the Dominion Astrophysical Observatory* (Victoria) 7 (1941): 251.

35. Bennett, C. L., et al. "Scientific Results from the Cosmic Background Explorer." *Proceedings of the National Academy of Sciences* 90 (1993): 4766–73.

36. Dicke, R. H., P. J. E. Peebles, P. G. Roll, and D. T. Wilkinson. "Cosmic Black-Body Radiation." *Astrophysical Journal* 142 (1965): 414–19.

37. Penzias, A. A., and R. W. Wilson. "Measurement of the Flux Density of CAS A at 4080 Mc/s." *Astrophysical Journal* 142 (1965): 1149–55.

38. DeGrasse, R. W., D. C. Hogg, E. A. Ohm, and H. E. D. Scovil. "Ultra-low Noise Receiving System for Satellite or Space Communication." *Proceedings of the National Electronics Conference* 15 (1959): 370.

39. Ohm, E. A. *Bell System Technical Journal* 40 (1961): 1065.

40. Hey, 1973, p. 139.

41. Wheeler, John Archibald. "Superdense Stars." *Annual Review of Astronomy and Astrophysics* 4 (1966): 393–432.

42. Kellermann, 1983, pp. 164–65.

43. Large, M. I., A. E. Vaughn, and B. Y. Mills. *Nature* 220 (1968): 340;

Lovelace, R. B. E., J. M. Sutton, and H. D. Craft. *International Astronomical Union Circular 2113* (1968).

44. Richards, D. W., and J. M. Comella. *Nature* 222 (1969): 551.

45. Smyth, C. Piazzi. "Astronomical Experiment on the Peak of Teneriffe, Carried Out Under the Sanction of the Lords Commissioners of the Admiralty." *Philosophical Transactions of the Royal Society of London* 148 (1858): 465–533.

46. Eddy, 1972, p. 165.

47. Neugebauer, G., D. E. Martz, and R. B. Leighton. "Observations of Extremely Cool Stars." *Astrophysical Journal* 142 (1965): 399.

48. Balick, Bruce, and Robert L. Brown. "Intense Sub-Arcsecond Structure in the Galactic Center." *Astrophysical Journal* 194 (December 1, 1974): 265–70.

49. Kleinmann, D. E., and F. J. Low. "Observations of Infrared Galaxies." *Astrophysical Journal* 159 (March 1970): L165–L172.

50. Downes, D., and A. Maxwell. "Radio Observations of the Galactic Center Region." *Astrophysical Journal* 146 (1966): 653–65.

51. Stebbins, Joel, and A. E. Whitford. "Six-Color Photometry of Stars. V. Infrared Radiation from the Region of the Galactic Center." *Astrophysical Journal* 106 (1947): 235–42.

52. Moroz, V. I. *Astronomicheskiĭ zhurnal* 38 (1961): 487.

53. Downes, D., A. Maxwell, and M. L. Meeks. *Nature* 208 (1965): 1189.

54. Becklin today says that most of the infrared radiation they detected at that time is now known to arise from the dense concentration of stars in the galactic center.

55. Yusef-Zadeh, F., M. Morris, and R. D. Ekers. *Nature* 348 (November 1, 1990): 45–47; Yusef-Zadeh, F., D. A. Roberts, and J. Biretta. "Proper Motions of Ionized Gas at the Galactic Center: Evidence for Unbound Orbiting Gas." *Astrophysical Journal Letters* 499 (1998): L159–L192.

56. Ghez, A. M., B. L. Klein, M. Morris, and E. E. Becklin. "High Proper-Motion Stars in the Vicinity of Sagittarius A*." *Astrophysical Journal* 509 (1998): 678–86.

57. Davis was in friendly competition with physicists Frederick Reines and Clyde Cowan; each group had set up their detectors, each using different methods, at the Savannah River nuclear plant in South Carolina for their source of neutrinos. Reines and Cowan detected the neutrino first in 1956.

58. Gribov, Vladmir, and Bruno Pontecorvo. "Neutrino Astronomy and Lepton Charge." *Physics Letters* B28 (1969): 493–96; Wolfenstein, Lincoln. "Neutrino Oscillations in Matter." *Physical Review* D17 (1978): 2369; Mikheyev, Stanislav, and Alexei Smirnov. *Soviet Journal of Nuclear Physics* 42 (1985): 913.

59. Bahcall, John N. "Solar Neutrino Cross Sections and Nuclear Beta Decay." *Physical Review* 135 (July 13, 1964): B137–B146.

60. Bionta, R. M., et al. "Observation of a Neutrino Burst in Coincidence with Supernova 1987A in the Large Magellanic Cloud." *Physical Review Letters* 58 (April 6, 1987): 1494–96.

61. Beier, E. W., in *Seventh Workshop on Grand Unification/ICOBAN '86: Toyama University, Japan, April 16–18, 1986.* Jiro Arafune (editor). Singapore: World Scientific, 1987.

62. Meegan, Charles A., Robert D. Preece, and Thomas M. Koshut. *Gamma-*

Ray Bursts: 4th Huntsville Symposium, Huntsville, Alabama, September 1997. Woodbury, N.Y.: American Institute of Physics, 1998.

63. Galama, T. J., et al. "Optical Follow-Up of GRB 970508." *Astrophysical Journal* 497 (April 10, 1998): L13–L16.

64. Colgate, S. A. *Canadian Journal of Physics* 46 (1968): S476.

65. Einstein, A. *Sitzungsberichte der Königlich Preussischen Akademie der Wissenschaften* (1916): 688–96.

66. Bartusiak, 2000, p. 79.

67. Kleppner, Daniel. "The Gem of General Relativity." *Physics Today* 46 (April 1993): 11.

VIII. Accelerating Outward

1. From discussions with Zwicky's daughter, Barbarina Zwicky.

2. Zwicky, F. "The Redshift of Extragalactic Nebulae." *Helvetica Physica Acta* 6 (1933): 110. In 1922, the Dutch astronomer Jacobus Kapteyn used the term "dark matter" to describe the nonluminous contributions to the total mass of the galaxy. He concluded, though, that "it appears at once that this mass cannot be excessive." (See note 10 in Part VI, "The Milky Way and Beyond," above.)

3. Smith, Sinclair. "The Mass of the Virgo Cluster." *Astrophysical Journal* 83 (January 1936): 23–30.

4. Kahn, F. D., and L. Woltjer. "Intergalactic Matter and the Galaxy." *Astrophysical Journal* 130 (November 1959): 705–17.

5. Rubin, Vera C. "The Rotation of Spiral Galaxies." *Science* 220 (June 24, 1983): 1339–44.

6. Roberts, M. S., and A. H. Rots. "Comparison of Rotation Curves of Different Galaxy Types." *Astronomy and Astrophysics* 26 (1973): 483–85.

7. For example, see Rubin, V. C., and W. K. Ford Jr. *Astrophysical Journal* 238 (1980): 471–87; Burstein, D., V. C. Rubin, N. Thonnard, and W. K. Ford Jr. *Astrophysical Journal* 253 (1982): 70–85; Rubin, V. C., W. K. Ford Jr., and N. Thonnard. *Astrophysical Journal* 261 (1982): 439–56.

8. Einasto, Jaan, Ants Kaasik, and Enn Saar. "Dynamic Evidence on Massive Coronas of Galaxies." *Nature* 250 (1974): 309–10; Ostriker, J. P., and P. J. E. Peebles. "A Numerical Study of the Stability of Flattened Galaxies: Or, Can Cold Galaxies Survive?" *Astrophysical Journal* 186 (December 1, 1973): 467–80.

9. See Zwicky note 2 above.

10. Roberts, M.S. "The Rotation Curves of Galaxies." *Comments on Astrophysics* 6 (1976): 105.

11. Krumm, N., and E. E. Salpeter. *Astrophysical Journal* 208 (1976): L7; and *Astronomy and Astrophysics* 56 (1977): 465.

12. Einasto, Kaasik, and Saar, cited in note 8 above.

13. Ostriker, J. P., P. J. E. Peebles, and A. Yahil. "The Size and Mass of Galaxies, and the Mass of the Universe." *Astrophysical Journal* 193 (October 1, 1974): L1–L4.

14. See note 8 in Part V, "Einsteinian Cosmos," above.

15. Eddington, A. S. *Space, Time, and Gravitation.* Cambridge: Cambridge University Press, 1920, pp. 133–35, 208–9.

16. Chwolson, O. "Über eine mögliche Form fiktiver Doppelsterne [Regarding a Possible Form of Fictitious Double Stars]." *Astronomische Nachrichten* 221 (1924): 329.

17. Renn, Jürgen, Tilman Sauer, and John Stachel. "The Origin of Gravitational Lensing: A Postscript to Einstein's 1936 *Science* Paper." *Science* 275 (January 10, 1997): 184–86.

18. Einstein, Albert. "Lens-Like Action of a Star by the Deviation of Light in the Gravitational Field." *Science* 84 (December 4, 1936): 506–7.

19. See Renn, Sauer, and Stachel, 186, cited in note 17 above.

20. Zwicky, F. "Nebulae as Gravitational Lenses." *Physical Review* 51 (February 15, 1937): 290.

21. Young, Peter, James E. Gunn, Jerome Kristian, J. B. Oke, James A. Westphal. "The Double Quasar Q0957 + 561 A, B: A Gravitational Lens Image Formed by a Galaxy at z = 0.39." *Astrophysical Journal* 241 (October 15, 1980): 507–20.

22. Oscoz, A., et al. "Time Delay in QSO 0957 + 561 from 1984–1999 Optical Data." *Astrophysical Journal* 552 (2001): 81–90.

23. Zwicky, F. "On the Masses of Nebulae and of Clusters of Nebulae." *Astrophysical Journal* 86 (October 1937): 237–38.

24. Sanitt, N. *Nature* 234 (1971): 199.

25. Bartusiak, 1986, pp. 241–42.

26. Starobinsky, A. A. *JETP Letters* 30 (1979): 682; "A New Type of Isotropic Cosmological Models Without Singularity." *Physics Letters B* 91 (1980): 99–102.

27. Albrecht, Andreas, and Paul J. Steinhardt. "Cosmology for Grand Unified Theories with Radiatively Induced Symmetry Breaking." *Physical Review Letters* 48 (1982): 1229–23; Linde, A. D. "A New Inflationary Universe Scenario: A Possible Solution of the Horizon, Flatness, Homogeneity, Isotropy and Primordial Monopole Problems." *Physics Letters B* 108 (1982): 389–93; Linde, A. D. "Chaotic Inflation." *Physics Letters B* 129 (1983): 177–81.

28. Smoot, G. F., et al. "Structure in the COBE Differential Microwave Radiometer First-Year Maps." *Astrophysical Journal* 396 (September 1, 1992): L1–L5.

29. See Rindler, W. *Monthly Notices of the Royal Astronomical Society* 116 (1956): 663; Weinberg, S. *Gravitation and Cosmology.* New York: Wiley, 1972, pp. 489–90; Misner, C. W., K. S. Thorne, and J. A. Wheeler. *Gravitation.* San Francisco: Freeman, 1973, pp. 740, 815.

30. See Dicke, R. H., and P. J. E. Peebles. *General Relativity: An Einstein Centenary Survey.* Edited by S. W. Hawking and W. Israel. Cambridge: Cambridge University Press, 1979.

31. De Lapparent, Geller, and Huchra, 1986, p. L1.

32. Hubble, Edwin. "Extra-Galactic Nebulae." *Astrophysical Journal* 64 (1926): 321–69.

33. Shapley, Harlow. "Luminosity Distribution and Average Density of Matter in Twenty-Five Groups of Galaxies." *Proceedings of the National Academy of Sciences* 19 (1933): 591–96.

34. Hubble, Edwin. "The Distribution of Extra-Galactic Nebulae." *Astrophysical Journal* 79 (1934): 8–76.

35. Ibid., p. 62.

36. De Vaucouleurs, Gérard. "Evidence for a Local Supergalaxy." *Astronomical Journal* 58 (1953): 30–32.

37. Abell, George O. "The Distribution of Rich Clusters of Galaxies." *Astrophysical Journal Supplement* 3 (1958): 211–88.

38. Shane, C. D., and C. A. Wirtanen. "The Distribution of Extragalactic Nebulae." *Astrophysical Journal* 59 (1954): 285.

39. Gregory, S. A., L. A. Thompson, and W. G. Tifft. "The Perseus Supercluster." *Astrophysical Journal* 243 (1981): 411–26; Chincarini, Guido, Herbert J. Rood, and Laird A. Thompson. "Supercluster Bridge between Groups of Galaxy Clusters." *Astrophysical Journal* 249 (1981): L47–L50.

40. Kirshner, Robert P., Augustus Oemler Jr., and Paul L. Schechter. "A Million Cubic Megaparsec Void in Boötes?" Astrophysical Journal 248 (September 1, 1981): L57–L60.

41. Zeldovich, Ya. B., J. Einasto, and S. F. Shandarin. "Giant Voids in the Universe." *Nature* 300 (1982): 407–13.

42. Ibid.

43. See Gregory, Thompson, and Tifft, cited in note 39 above.

44. See Chincarini, Guido, Rood, and Thompson, cited in note 39 above.

45. Ostriker, Jeremiah P., and Lennox L. Cowie. "Galaxy Formation in an Intergalactic Medium Dominated by Explosions." *Astrophysical Journal* 243 (1981): L127–L131.

46. Ryle, M. "The Nature of the Cosmic Radio Sources." *Proceedings of the Royal Society of London. Series A.* 248 (November 25, 1958): 289–308.

47. Ibid., p. 298.

48. Kennicutt, Robert C. Jr. "Evolution of the Stars and Gas in Galaxies" in *The Astrophysical Journal: American Astronomical Society Centennial Issue.* Edited by Helmut A. Abt. Chicago: University of Chicago Press, 1999, p. 1165.

49. Tinsley, Beatrice M. "Evolution of the Stars and Gas in Galaxies." *Astrophysical Journal* 151 (February 1968): 547–65.

50. Butcher, Harvey, and Augustus Oemler Jr. "The Evolution of Galaxies in Clusters. I. ISIT Photometry of CL 0024 + 1654 and 3C 295." *Astrophysical Journal* 219 (January 1, 1978): 18–30; Butcher, Harvey, and Augustus Oemler Jr. "Nature of Blue Galaxies in the Cluster CL 1447 + 2619." *Nature* (July 5, 1984): 31–33.

51. Williams, 1996, p. 1336.

52. Driver, Simon P., Rogier A. Windhorst, and Richard E. Griffiths. "The Contribution of Late-Type/Irregulars to the Faint Galaxy Counts from Hubble Space Telescope Medium-Deep Survey Images." *Astrophysical Journal* 453 (November 1, 1995): 48–64; Driver, Simon P., Rogier A. Windhorst, Eric J. Ostrander, William C. Keel, Richard E. Griffiths, and Kavan U. Ratnatunga. "The Morphological Mix of Field Galaxies to m_I = 24.25 Magnitudes (b_s ~ 26 Magnitudes) from a Deep *Hubble Space Telescope* WFPC2 Image." *Astrophysical Journal* 449 (August 10, 1995): L23–L27.

53. Schade, David, S. J. Lilly, David Crampton, F. Hammer, O. Le Fèvre, and L. Tresse. "Canada-France Redshift Survey: *Hubble Space Telescope* Imaging of High-Redshift Field Galaxies." *Astrophysical Journal* 451 (September 20, 1995): L1–L4.

54. Dressler, Alan, Augustus Oemler Jr., Harvey R. Butcher, and James E. Gunn. "The Morphology of Distant Cluster Galaxies. I. *HST* Observations of CL 0939+4713." *Astrophysical Journal* 430 (July 20, 1994): 107–20; Dressler, Alan, Augustus Oemler, Jr., W. B. Sparks, and R. A. Lucas. "New Images of the Distant, Rich Cluster CL 0939+4713 with WFPC2." *Astrophysical Journal* 435 (November 1, 1994): L23–L26.

55. Dickinson, Mark, in *Galaxies in the Young Universe*. Edited by H. Hippelein, H.-J. Meisenheimer, and H.-J. Röser. Berlin: Springer, 1995, p. 144.

56. See note 50.

57. See note 55 and Dickinson, Mark, in *Fresh Views on Elliptical Galaxies* (ASP Conference Series). Edited by A. Buzzoni, A. Renzini, and A. Serrano. San Francisco: Astronomical Society of the Pacific, 1995, p. 283.

58. Cowie, Lenox L., Esther M. Hu, and Antoinette Songaila. "Faintest Galaxy Morphologies from *HST* WFPC2 Imaging of the Hawaii Survey Fields." *Astronomical Journal* 110 (October 1995): 1576–83.

59. Oates, Whitney Jennings (ed.). *The Stoic and Epicurean Philosophers: The Complete Extant Writings of Epicurus, Epictetus, Lucretius, Marcus Aurelius*. New York: Modern Library, 1957.

60. Herschel, William. "On Nebulous Stars, Properly So Called." *Philosophical Transactions of the Royal Society of London* 81 (1791): 74.

61. Campbell, B., and G. A. H. Walker. "Precision Radial Velocities with an Absorption Cell." *Publications of the Astronomical Society of the Pacific* 91 (August 1979): 540–45.

62. Aumann, H. H., et al. "Discovery of a Shell Around Alpha Lyrae." *Astrophysical Journal* 278 (March 1, 1984): L23–L27; Smith, Bradford A., and Richard J. Terrile. "A Circumstellar Disk Around β Pictoris." *Science* 226 (December 21, 1984): 1421–24.

63. Wolszczan, A., and D. A. Frail. "A Planetary System Around the Millisecond Pulsar PSR1257+12." *Nature* 355 (January 9, 1992): 147; six months earlier a British team had announced a planet orbiting a neutron star but later realized it was fictitious due to an error in their calculations (see Bailes, M., A. G. Lyne, and S. L. Shemar. "A Planet Orbiting the Neutron Star PSR1829-10." *Nature* [July 25, 1991]: 311–13).

64. Marcy, Geoffrey W., and R. Paul Butler. "A Planetary Companion to 70 Virginis." *Astrophysical Journal* 464 (June 20, 1996): L147.

65. Ibid., L147–L151.

66. Butler and Marcy, 1996, p. L153.

67. Butler, R. Paul, et al. "Evidence for Multiple Companions to υ Andromedae." *Astrophysical Journal* 526 (December 1, 1999): 916–27.

68. Baranne, A., M. Mayor, and J.-L. Poncet. "CORAVEL: A New Tool for Radial Velocity Measurements." *Vistas in Astronomy* 23 (1979): 279–316; Duquennoy, A., and M. Mayor. "Multiplicity Among Solar-Type Stars in the Solar Neighborhood." *Astronomy and Astrophysics* 248 (1991): 485–524.

69. Boss, Alan P. "Proximity of Jupiter-like Planets to Low-Mass Stars." *Science* 267 (January 20, 1995): 360–62.

70. Burrows, A., et al. "Prospects for Detection of Extra-Solar Giant Planets by Next-Generation Telescopes." *Nature* 375 (1995): 299; Saumon, D., et al. "A Theory of Extrasolar Giant Planets." *Astrophysical Journal* 460 (1996): 993–1018; Guillot, T., et al. "Giant Planets at Small Orbital Distances." *Astrophysical Journal* 459 (1996): L35–L38.

71. Baade, W. "The Absolute Photographic Magnitude of Supernovae." *Astrophysical Journal* 88 (1938): 285–304.

72. Colgate, S. "Supernovae as a Standard Candle for Cosmology." *Astrophysical Journal* 232 (1979): 404–8; Tammann, G. A., in *ESA/ESO Workshop on Astronomical Uses of the Space Telescope*. F. Macchetto, F. Pacini, and M. Tarenghi (eds.). Geneva: ESO, 1979, p. 329.

73. In the late 1980s Danish astronomers Hans Ulrik Nørgaard-Nielsen, Leif Hansen, and Henning Jørgensen pioneered many of the methods for finding distant supernova, but with their cruder technology it took a couple of years to find just one good candidate. See Nørgaard-Nielsen, H. U., et al. *Nature* 339 (1989): 523.

74. Phillips, M. M. "The Absolute Magnitudes of Type Ia Supernovae." *Astrophysical Journal* 413 (1993): L105–L108.

75. Hamuy, Mario, et al. "A Hubble Diagram of Distant Type Ia Supernovae." *Astronomical Journal* 109 (January 1995): 1–13; Riess, Adam G., William H. Press, and Robert P. Kirshner. "Using Type Ia Supernova Light Curve Shapes to Measure the Hubble Constant." *Astrophysical Journal* 438 (1995): L17–L20.

76. Pais, 1982, p. 288. That Einstein called the cosmological constant his "biggest blunder" does not appear in any of his writings; the anecdote first appeared in George Gamow's autobiography (*My World Line*, New York: Viking Press, 1970): ". . . when I was discussing cosmological problems with Einstein, he remarked that the introduction of the cosmological term was the biggest blunder he ever made in his life."

77. Riess, Adam G., et al. "The Farthest Known Supernova: Support for an Accelerating Universe and a Glimpse of the Epoch of Deceleration." *Astrophysical Journal* 560 (October 10, 2001): 49–71.

78. Bennett, C. L., et al. "First-year Wilkinson Microwave Anisotropy Probe (WMAP) Observations: Preliminary Maps and Basic Results." *Astrophysical Journal Supplement* 148 (2003): 1–27.

79. Garnavich, P., et al. "Constraints on Cosmological Models from Hubble Space Telescope Observations of High-z Supernovae." *Astrophysical Journal* 493 (February 1, 1998): L53–L57; Schmidt, Brian P., et al. "The High Z Supernova Search: Measuring Cosmic Deceleration and Global Curvature of the Universe Using Type Ia Supernovae." *Astrophysical Journal* 507 (November 1, 1998): 46–63.

80. Riess, Adam G., et al. "Snapshot Distances to Type Ia Supernovae: All in 'One' Night's Work." *Astrophysical Journal* 504 (September 10, 1998): 935–44.

81. Garnavich, P., et al, cited in note 79.

82. See note 71 above.

83. See note 73 above.

84. Goobar, A., and S. Perlmutter. "Feasibility of Measuring the Cosmological Constant Λ and Mass Density Ω Using Type IA Supernovae." *Astrophysical Journal* 450 (September 1, 1995): 14–18.

85. Perlmutter, S., et al. "Measurements of the Cosmological Parameters Ω and Λ from the First Seven Supernovae at $z \geq 0.35$." *Astrophysical Journal* 483 (July 10, 1997): 565–81.

86. Perlmutter, S., et al. "Discovery of a Supernova Explosion at Half the Age of the Universe." *Nature* 391 (January 1, 1998): 51–54.

87. Garnavich, P., et al. "Constraints on Cosmological Models from *Hubble Space Telescope* Observations of High-z Supernovae." *Astrophysical Journal* 493 (February 1, 1998): L53–L57.

88. Perlmutter, S., et al. "Cosmology from Type Ia Supernovae." *Bulletin of the American Astronomical Society* 29 (1997): 1351.

89. Riess, Adam G., et al. "Observational Evidence from Supernovae for an Accelerating Universe and a Cosmological Constant." *Astronomical Journal* 116 (September 1998): 1009–38.

Bibliography

Abell, George O. "Evidence Regarding Second-Order Clustering of Galaxies and Interactions Between Clusters of Galaxies." *Astronomical Journal* 66 (December 1961): 607–13.

———. "Clustering of Galaxies." *Annual Review of Astronomy and Astrophysics* 3 (1965): 1–22.

Abetti, Giorgio. *The History of Astronomy.* New York: H. Schuman, 1952.

Abt, Helmut A. (ed.). *The Astrophysical Journal: American Astronomical Society Centennial Issue.* Chicago: University of Chicago Press, 1999.

Adams, Walter S. "An A-Type Star of Very Low Luminosity." *Publications of the Astronomical Society of the Pacific* 26 (October 1914): 198

———. "The Spectrum of the Companion of Sirius." *Publications of the Astronomical Society of the Pacific* 27 (December 1915): 236–37.

Adler, Mortimer J. (ed.). "Ptolemy, Copernicus, Kepler." *Great Books of the Western World,* volume 15. Chicago: Encyclopedia Britannica, Inc., 1993.

Alexander, Stephen. "On the Origin of the Forms and the Present Condition of Some of the Clusters of Stars, and Several of the Nebulae." *Astronomical Journal* 2 (1852): 95–103.

Allen, David A. "Infrared Astronomy: An Assessment." *Quarterly Journal of the Royal Astronomical Society* 18 (1977): 188–98.

Alpher, R. A., H. Bethe, and G. Gamow. "The Origin of Chemical Elements." *Physical Review* 73 (April 1, 1948): 803–4.

Alpher, Ralph A., and Robert Herman. "Evolution of the Universe." *Nature* 162 (November 13, 1948): 774–75.

———. *Genesis of the Big Bang.* New York: Oxford University Press, 2001.

Aristarchus of Samos. "On the Sizes and Distances of the Sun and Moon." Translated by Sir Thomas L. Heath in *Aristarchus of Samos.* Oxford: Clarendon Press, 1913.

Aristotle. "De Caelo." Translated by J. L. Stocks in *The Works of Aristotle, Volume 2,* edited by W. D. Ross. Oxford: Clarendon Press, 1930.

Armitage, Angus. *William Herschel.* London: Thomas Nelson and Sons, 1962.

Baade, W. "The Resolution of Messier 32, NGC 205, and the Central Region of the Andromeda Nebula." *Astrophysical Journal* 100 (September 1944): 137–46.

———. "A Revision of the Extra-Galactic Distance Scale." *Transactions of the International Astronomical Union* 8 (1952): 397–98.

———. *Evolution of Stars and Galaxies.* Cambridge: Harvard University Press, 1963.

Baade, W., and F. Zwicky. "On Super-Novae." [1934a] *Proceedings of the National Academy of Sciences* 20 (May 15, 1934): 254–59.

———. "Cosmic Rays From Super-Novae." [1934b]. *Proceedings of the National Academy of Sciences* 20 (May 15, 1934): 259–63.

Bailes, M., A. G. Lyne, and S. L. Shemar. "A Planet Orbiting the Neutron Star PSR1829-10." *Nature* 352 (July 25, 1991): 311–13.

Barnard, E. E. "On a Nebulous Groundwork in the Constellation Taurus." *Astrophysical Journal* 25 (1907): 218–25.

——. "On a Great Nebulous Region and the Question of Absorbing Matter in Space and the Transparency of the Nebulae." *Astrophysical Journal* 31 (1910): 8–14.

Bartusiak, Marcia. *Thursday's Universe.* New York: Times Books, 1986.

——. *Einstein's Unfinished Symphony.* Washington, D.C.: Joseph Henry Press, 2000.

Becklin, Eric E., and Gerry Neugebauer. "Infrared Observations of the Galactic Center." *Astrophysical Journal* 151 (1968): 145–61.

Berendzen, Richard (ed.). *Education in and History of Modern Astronomy.* New York: New York Academy of Sciences, 1972.

Berendzen, Richard, Richard Hart, and Daniel Seeley. *Man Discovers the Galaxies.* New York: Columbia University Press, 1984.

Berry, Arthur. *A Short History of Astronomy.* New York: Dover Publications, 1961.

Bessel, F. W. "On the Parallax of 61 Cygni." *Monthly Notices of the Royal Astronomical Society* 4 (November 1838): 152–61.

Bethe, H. A. "Energy Production in Stars." *Physical Review* 55 (March 1, 1939): 434–56.

Bionta, R. M., et al. "Observation of a Neutrino Burst in Coincidence with Supernova 1987A in the Large Magellanic Cloud." *Physical Review Letters* 58 (April 6, 1987): 1494–96.

Bobrovnikoff, N. T. "The Discovery of Variable Stars." *Isis* 33 (June 1942): 687–89.

Bondi, Hermann, and Thomas Gold. "The Steady-State Theory of the Expanding Universe." *Monthly Notices of the Royal Astronomical Society* 108 (1948): 252–70.

Brush, Stephen G. "Why Was Relativity Accepted?" *Physics in Perspective* 1 (1999): 184–214.

Burbidge, E. Margaret, G. R. Burbidge, William A. Fowler, and F. Hoyle. "Synthesis of the Elements in Stars." *Reviews of Modern Physics* 29 (October 1957): 547–650.

Butler, R. Paul, and Geoffrey W. Marcy. "A Planet Orbiting 47 Ursae Majoris." *Astrophysical Journal* 464 (June 20, 1996): L153–L156.

Butcher, Harvey, and Augustus Oemler Jr. "The Evolution of Galaxies in Clusters. I. ISIT Photometry of CI 0024 + 1654 and 3C 295." *Astrophysical Journal* 219 (January 1, 1978): 18–30.

Cannon, Annie J., and Edward C. Pickering. "Spectra of Bright Southern Stars." *Annals of the Astronomical Observatory of Harvard College*, volume 28, part II. Cambridge: Harvard Observatory, 1901.

——. "The Henry Draper Catalogue." *Annals of the Astronomical Observatory of Harvard College*, volume 91. Cambridge: Harvard Observatory, 1918.

Chandrasekhar, S. "The Maximum Mass of Ideal White Dwarfs." *Astrophysical Journal* 74 (1931): 81–82.

——. "Some Remarks on the State of Matter in the Interior of Stars." *Zeitschrift für Astrophysik* 5 (1932): 321–27.

———. "Stellar Configurations with Degenerate Cores." *The Observatory* 57 (1934): 373–77.

———. "The Highly Collapsed Configurations of a Stellar Mass (Second Paper)." *Monthly Notices of the Royal Astronomical Society* 95 (1935): 207–25.

Chapman, Allan. "William Herschel and the Measurement of Space." *Quarterly Journal of the Royal Astronomical Society* 30 (1989): 399–418.

Cheung, A. C., D. M. Rank, C. H. Townes, D. D. Thornton, and W. J. Welch. "Detection of NH_3 Molecules in the Interstellar Medium by Their Microwave Emission." *Physical Review Letters* 21 (December 16, 1968): 1701–5.

———. "Detection of Water in Interstellar Regions by Its Microwave Radiation." *Nature* 221 (February 15, 1969): 626–28.

Clarke, W. A. "On the Origin of Shooting Stars." *The American Journal of Science and Arts* 30 (1836): 369–70.

Cleomedes. "On the Orbits of the Heavenly Bodies." Translated by T. L. Heath in *Greek Astronomy.* London: J. M. Dent and Sons, 1932.

Clerke, Agnes M. *A Popular History of Astronomy During the Nineteenth Century.* London: Adam and Charles Black, 1902.

Cohen, Morris R., and I. E. Drabkin (eds.). *A Source Book in Greek Science.* Cambridge: Harvard University Press, 1958.

Copernicus, Nicolaus. *On the Revolutions of the Heavenly Spheres.* Translation by Charles Glenn Wallis. Amherst, N.Y.: Prometheus Books, 1995.

Curtis, Heber D. "New Stars in Spiral Nebulae." *Publications of the Astronomical Society of the Pacific* 29 (1917): 180–82.

———. "Novae in Spiral Nebulae and the Island Universe Theory." *Publications of the Astronomical Society of the Pacific* 29 (1917): 206–7.

Davis, Raymond, Don S. Harmer, and Kenneth C. Hoffman. "Search for Neutrinos from the Sun." *Physical Review Letters* 20 (May 20, 1968): 1205–9.

De Lapparent, Valérie, Margaret J. Geller, and John P. Huchra. "A Slice of the Universe." *Astrophysical Journal* 302 (March 1, 1986): L1–L5.

De Maria, M., M. G. Ianniello, and A. Russo. "The Discovery of Cosmic Rays: Rivalries and Controversies Between Europe and the United States." *Historical Studies in the Physical and Biological Sciences* 22 (1991): 165–92.

De Maria, M., and A. Russo. "Cosmic Ray Romancing: The Discovery of the Latitude Effect and the Compton-Millikan Controversy." *Historical Studies in the Physical and Biological Sciences* 19 (1989): 211–66.

"A Demonstration Concerning the Motion of Light, Communicated from Paris, in the *Journal des Scavans,* and Here Made English." *Philosophical Transactions* 12 (1677): 893–94.

De Sitter, Willem. "On Einstein's Theory of Gravitation, and Its Astronomical Consequences. Third Paper." *Monthly Notices of the Royal Astronomical Society* 78 (1917): 3–28.

De Vaucouleurs, Gérard. "The Distribution of Bright Galaxies and the Local Supergalaxy." *Vistas in Astronomy* 2 (1956): 1584–1606.

DeVorkin, David H. "Steps Toward the Hertzsprung-Russell Diagram." *Physics Today* 31 (March 1978): 32–39.

———. *The History of Modern Astronomy and Astrophysics: A Selected, Annotated Bibliography.* New York: Garland Publishing, 1982.

——. *Henry Norris Russell: Dean of American Astronomers.* Princeton: Princeton University Press, 2000.

—— (ed.). *The American Astronomical Society's First Century.* Washington, D.C.: American Astronomical Society, 1999.

Dicke, Robert H., Robert Beringer, Robert L. Kyhl, and A. B. Vane. "Atmospheric Absorption Measurements with a Microwave Radiometer." *Physical Review* 70 (September 1946): 340–48.

Dicke, R. H., P. J. E. Peebles, P. G. Roll, and D. T. Wilkinson. "Cosmic Black-Body Radiation." *Astrophysical Journal* 142 (1965): 414–19.

"Discussion of Papers by A. S. Eddington and E. A. Milne." *The Observatory* 58 (1935): 37–39.

Dreyer, J. L. E. *A History of Astronomy from Thales to Kepler.* New York: Dover Publications, 1953.

Dyson, F. W., A. S. Eddington, and C. Davidson. "A Determination of the Deflection of Light by the Sun's Gravitational Field, from Observations Made at the Total Eclipse of May 29, 1919." *Philosophical Transactions of the Royal Society of London, Series A* 220 (1920): 291–333.

Earman, John, and Clark Glymour. "Relativity and Eclipses: The British Eclipse Expeditions of 1919 and Their Predecessors." *Historical Studies in the Physical Sciences* 11 (1980): 49–85.

Easton, C. "A New Theory of the Milky Way." *Astrophysical Journal* 12 (1900): 136–58.

Eddington, Arthur S. "On the Radiative Equilibrium of the Stars." *Monthly Notices of the Royal Astronomical Society* 77 (November 10, 1916): 16–35.

——. "Further Notes on the Radiative Equilibrium of the Stars." *Monthly Notices of the Royal Astronomical Society* 77 (June 1917): 596–612.

——. *Space, Time, and Gravitation.* Cambridge: Cambridge University Press, 1920.

——. "On the Relation Between the Masses and Luminosities of the Stars." *Monthly Notices of the Royal Astronomical Society* 84 (1924): 308–32.

——. "The End of the World: From the Standpoint of Mathematical Physics." *Nature* 127 (March 21, 1931): 447–53.

Eddy, John A. "Thomas A. Edison and Infra-Red Astronomy." *Journal for the History of Astronomy* 3 (October 1972): 165–87.

Einasto, Jaan, Mihkel Jôeveer, and Enn Saar. "Structure of Superclusters and Supercluster Formation." *Monthly Notices of the Royal Astronomical Society* 193 (1980): 353–75.

Einstein, A. "Kosmologische Betrachtungen zur allgemeinen Relativitätstheorie," *Sitzungsberichte der Königlich Preussischen Akademie der Wissenschaften* (1917): 142–52.

Eisberg, Joann. "Making a Science of Observational Cosmology." *Journal for the History of Astronomy* 32 (August 2001): 263–78.

Ekers, R. D., J. H. Van Gorkom, U. J. Schwarz, and W. M. Goss. "The Radio Structure of Sgr A." *Astronomy and Astrophysics* 122 (1983): 143–50.

Evans, James. *The History and Practice of Ancient Astronomy.* New York: Oxford University Press, 1998.

Evans, Herbert McLean. *Men and Moments in the History of Science.* Seattle: University of Washington Press, 1959.

Ewen, H. I., and E. M. Purcell. "Radiation from Galactic Hydrogen at 1,420 Mc./sec." *Nature* 168 (September 1, 1951): 356.

Fernie, J. D. "The Period-Luminosity Relation: A Historical Review." *Publications of the Astronomical Society of the Pacific* 81 (December 1969): 707–31.

——. "The Historical Quest for the Nature of the Spiral Nebulae." *Publications of the Astronomical Society of the Pacific* 82 (December 1970): 1189–1230.

Festou, Michel C., Hans Rickman, and Richard M. West. *Astronomy and Astrophysics Review* 4 (1993): 363–447.

Filippenko, Alexei. "Einstein's Biggest Blunder? High-Redshift Supernovae and the Accelerating Universe." *Publications of the Astronomical Society of the Pacific* 113 (December 2001): 1441–48.

Flam, Faye. "COBE Finds the Bumps in the Big Bang." *Science* 256 (May 1, 1992): 612.

Fölsing, Albrecht. *Albert Einstein: A Biography*. New York: Viking, 1997.

Forbes, Eric G. "Gauss and the Discovery of Ceres." *Journal for the History of Astronomy* 2 (October 1971): 195–99.

Fowler, R. H. "On Dense Matter." *Monthly Notices of the Royal Astronomical Society* 87 (December 1926): 114–22.

Fraunhofer, Joseph. *Prismatic and Diffraction Spectra*. Translated and edited by J. S. Ames. New York: Harper & Brothers, 1898.

Friedman, Herbert. "Ultraviolet and X Rays from the Sun." *Annual Review of Astronomy and Astrophysics* 1 (1963): 59–96.

——. "Rocket Astronomy." *Education in and History of Modern Astronomy*. New York: New York Academy of Sciences, 1972.

——. *Sun and Earth*. New York: Scientific American Books, 1986.

Friedmann, Aleksandr. "On the Curvature of Space." *Zeitschrift für Physik* 10 (1922): 377–86. Translated by Brian Doyle in *A Source Book in Astronomy and Astrophysics, 1900–1975*. Edited by Kenneth R. Lang and Owen Gingerich. Cambridge: Harvard University Press, 1979.

Galilei, Galileo. *Sidereus Nuncius or The Sidereal Messenger*. Translated with introduction, conclusion, and notes by Albert Van Helden. Chicago: University of Chicago Press, 1989.

Gamow, G. "Expanding Universe and the Origin of Elements." *Physical Review* 70 (1946): 572–73.

——. "The Evolution of the Universe." *Nature* 162 (October 30, 1948): 680–82.

Giacconi, Riccardo, Herbert Gursky, and Frank R. Paolini. "Evidence for X Rays from Sources Outside the Solar System." *Physical Review Letters* 9 (December 1, 1962): 439–43.

Gillispie, Charles Coulston (ed.). *Dictionary of Scientific Biography*. New York: Scribner's, 1970–78.

Gingerich, Owen. "Johannes Kepler and the New Astronomy." *Quarterly Journal of the Royal Astronomical Society* 13 (1972): 346–73.

——. "Session I: 300 Years of Astronomy." *Vistas in Astronomy* 20 (1976): 1–9.

——. *The Eye of Heaven: Ptolemy, Copernicus, Kepler*. New York: American Institute of Physics, 1993.

——. *The Great Copernicus Chase and Other Adventures in Astronomical History*. Cambridge, Mass.: Sky Publishing Corp.; Cambridge: Cambridge University Press, 1992.

Gingerich, Owen, and Albert Van Helden. "From Occhiale to Printed Page: The Making of Galileo's *Sidereus nuncius.*" *Journal for the History of Astronomy* 34 (August 2003): 251–67.

Glyn Jones, Kenneth. *The Search for the Nebulae.* Chalfont St. Giles, U.K.: Alpha Academic, 1975.

Gold, Thomas. "Rotating Neutron Stars as the Origin of the Pulsating Radio Sources." *Nature* 218 (1968): 731–32.

Goldsmith, Donald. *Worlds Unnumbered: The Search for Extrasolar Planets.* Sausalito, Calif.: University Science Books, 1997.

Goodricke, John. "A Series of Observations on, and a Discovery of, the Period of the Variation of the Light of the Bright Star in the Head of Medusa, Called Algol." *Philosophical Transactions of the Royal Society of London* 73 (1783): 474–82.

Gordon, Kurtiss. "History of Our Understanding of a Spiral Galaxy: Messier 33." *Quarterly Journal of the Royal Astronomical Society* 10 (1969): 293–307.

Grant, Robert. *History of Physical Astronomy.* London: R. Baldwin, 1852.

Greenstein, Jesse L. "Quasi-Stellar Radio Sources." *Scientific American* 209 (December 1963): 54–62.

Greenstein, Jesse L., and Thomas A. Matthews. "Red-Shift of the Unusual Radio Source: 3C 48." *Nature* 197 (March 16, 1963): 1041–42.

Gregory, David. *The Elements of Astronomy, Physical and Geometrical.* London: J. Nicholson, 1715.

Guth, Alan H. "Inflationary Universe: A Possible Solution to the Horizon and Flatness Problems." *Physical Review D* 23 (January 15, 1981): 347–56.

Hall, Asaph. "The Discovery of the Satellites of Mars." *Monthly Notices of the Royal Astronomical Society* 38 (November 9, 1877): 205–9.

Halley, Edmund [or Edmond]. *A Synopsis of the Astronomy of Comets.* London: Printed for John Senex, 1705.

———. "An Account of Several Nebulae or Lucid Spots Like Clouds, Lately Discovered Among the Fixt Stars by Help of the Telescope." *Philosophical Transactions* 29 (1714–16): 390–92.

———. "Considerations on the Change of the Latitudes of Some of the Principal Fixt Stars." *Philosophical Transactions* 30 (1717–19): 736–38.

Hardin, Clyde L. "The Scientific Work of the Reverend John Michell." *Annals of Science* 22 (1966): 27–47.

Harper, Eamon. "George Gamow: Scientific Amateur and Polymath." *Physics in Perspective* 3 (2001): 335–72.

Harwit, Martin. *Cosmic Discovery.* Cambridge, Mass.: MIT Press, 1984.

Hazard, C., M. B. Mackey, and A. J. Shimmins. "Investigation of the Radio Source 3C 273 by the Method of Lunar Occultations." *Nature* 197 (March 16, 1963): 1037–39.

Heath, Thomas L. *Aristarchus of Samos.* Oxford: Clarendon Press, 1913.

———. *Greek Astronomy.* New York: Dover, 1991.

Herschel, John F. W. *Outlines of Astronomy.* New York: P. F. Collier, 1902.

Herschel, William. "Account of a Comet." *Philosophical Transactions of the Royal Society of London* 71 (1781): 492–501.

———. "On the Proper Motion of the Sun and Solar System; With an Account of Several Changes That Have Appeared Among the Fixed Stars Since the Time of

Mr. Flamstead." *Philosophical Transactions of the Royal Society of London* 73 (1783): 247–83.

———. "On the Remarkable Appearances at the Polar Regions of the Planet Mars, the Inclination of Its Axis, the Position of Its Poles, and Its Spheroidical Figure; With a Few Hints Relating to Its Real Diameter and Atmosphere." [1784a]. *Philosophical Transactions of the Royal Society of London* 74 (1784): 233–73.

———. "On the Construction of the Heavens." [1784b] *Philosophical Transactions of the Royal Society of London* 75 (1784): 213–66.

———. "Astronomical Observations Relating to the Construction of the Heavens, Arranged for the Purpose of a Critical Examination, the Result of Which Appears to Throw Some New Light upon the Organization of the Celestial Bodies." *Philosophical Transactions of the Royal Society of London* 101 (1811): 269–336.

Hetherington, Norriss S. "The Measurement of Radial Velocities of Spiral Nebulae." *Isis* 62 (September 1971): 309–13.

———. "The First Measurements of Stellar Parallax." *Annals of Science* 28 (May 1972): 319–25.

———. "Philosophical Values and Observation in Edwin Hubble's Choice of a Model of the Universe." *Historical Studies in the Physical Sciences* 13 (1982): 41–67.

Hewish, A., S. J. Bell, J. D. H. Pilkington, P. F. Scott, and R. A. Collins. "Observation of a Rapidly Pulsating Radio Source." *Nature* 217 (February 24, 1968): 709–13.

Hey, J. S. *The Evolution of Radio Astronomy*. New York: Science History Publications, 1973.

Hey, J. S., J. W. Phillips, and S. J. Parsons. "Cosmic Radiations at 5 Metres Wave-Length." *Nature* 157 (March 9, 1946): 296–97.

Hirata, K., et al. "Observation of a Neutrino Burst from the Supernova SN 1987A." *Physical Review Letters* 58 (April 6, 1987): 1490–93.

Hirshfeld, Alan W. *Parallax: The Race to Measure the Cosmos*. New York: W. H. Freeman and Company, 2001.

Hoffmann, Banesh. *Albert Einstein: Creator and Rebel*. New York: Viking, 1972.

Hoskin, Michael A. "Ritchey, Curtis and the Discovery of Novae in Spiral Nebulae." *Journal for the History of Astronomy* 7 (February 1976): 47–53.

———. "Goodricke, Pigott and the Quest for Variable Stars." *Journal for the History of Astronomy* 10 (1979): 23–41.

———. "William Herschel's Early Investigations of Nebulae: A Reassessment." *Journal for the History of Astronomy* 10 (1979): 165–75.

———. "William Herschel and the Construction of the Heavens." *Proceedings of the American Philosophical Society* 133 (1989): 427–32.

———. (ed.). *The Cambridge Concise History of Astronomy*. Cambridge: Cambridge University Press, 1999.

Hubble, Edwin P. "Cepheids in Spiral Nebulae." *Publications of the American Astronomical Society* 5 (1925): 261–64.

———. "Extragalactic Nebulae." *Astrophysical Journal* 64 (1926): 321–69.

———. "A Relation Between Distance and Radial Velocity Among Extra-Galactic Nebulae." *Proceedings of the National Academy of Sciences* 15 (March 15, 1929): 168–73.

Huggins, William. "The New Astronomy." *The Nineteenth Century* 41 (June 1897): 907–29.

——. *The Scientific Papers of Sir William Huggins*. London: W. Wesley and Son, 1909.

Huggins, William, and W. A. Miller. "On the Spectra of Some of the Nebulae." *Philosophical Transactions of the Royal Society of London* 154 (1864): 437–44.

Hulse, R. A., and J. H. Taylor. "Discovery of a Pulsar in a Binary System." *Astrophysical Journal* 195 (January 15, 1975): L51–L53.

Humason, Milton. "The Large Radial Velocity of N.G.C. 7619." *Proceedings of the National Academy of Sciences* 15 (March 15, 1929): 167–68.

Humboldt, Alexander von. *Cosmos: A Sketch of a Physical Description of the Universe*. Translated by E. C. Otté and B. H. Paul. London: Henry G. Bohn, 1849.

Jaki, Stanley L. *The Milky Way*. New York: Science History Publications, 1972.

Jansky, Karl G. "Directional Studies of Atmospherics at High Frequencies." *Proceedings of the Institute of Radio Engineers* 20 (December 1932): 1920–32.

——. "Electrical Disturbances Apparently of Extraterrestrial Origin." *Proceedings of the Institute of Radio Engineers* 21 (October 1933): 1387–98.

——. "A Note on the Source of Interstellar Interference." *Proceedings of the Institute of Radio Engineers* 23 (October 1935): 1158–63.

Jôeveer, Mihkel, and Jaan Einasto. "Has the Universe the Cell Structure?" In *The Large Scale Structure of the Universe*, pp. 241–51. Edited by M. S. Longair and J. Einasto. Dordrecht: D. Reidel Publishing Company, 1978.

Kaler, James. *Stars*. New York: Scientific American Library, 1998.

Kant, Immanuel. *Kant's Cosmogony as in His Essay on the Retardation of the Rotation of the Earth and His Natural History and Theory of the Heavens*. Edited and translated by W. Hastie. Glasgow: James Maclehose and Sons, 1900.

Keeler, James E. "On the Motions of the Planetary Nebulae in the Line of Sight." *Publications of the Astronomical Society of the Pacific* 2 (November 29, 1890): 265–80.

Kellermann, K., and B. Sheets (eds.). *Serendipitous Discoveries in Radio Astronomy: Proceedings of a Workshop Held at the National Radio Astronomy Observatory, Green Bank, West Virginia on May 4, 5, 6, 1983*. Green Bank, W.Va.: National Radio Astronomy Observatory, 1983.

Kepler, Johannes. *Mysterium Cosmographicum*. Translated by A. M. Duncan. New York: Abaris Books, 1981.

——. *New Astronomy*. Translated by William H. Donahue. Cambridge: Cambridge University Press, 1992.

——. *Epitome of Copernican Astronomy and Harmonies of the World*. Translated by Charles Glenn Wallis. Amherst, N.Y.: Prometheus Books, 1995.

Kirchhoff, Gustav Robert. *Untersuchungen über das Sonnenspectrum und die Spectrum der chemischen Elemente* [Researches on the Solar Spectrum and the Spectra of the Chemical Elements]. Translated by Henry E. Roscoe. Cambridge: Macmillan, 1862–63.

Kirshner, Robert P. *The Extravagant Universe*. Princeton: Princeton University Press, 2002.

Kirshner, Robert P., Augustus Oemler Jr., Paul L. Schechter, and Stephen A. Shectman. "A Million Cubic Megaparsec Void in Boötes?" *Astrophysical Journal* 248 (September 1, 1981): L57–L60.

Klebesadel, Ray W., Ian B. Strong, and Roy A. Olson. "Observations of Gamma-

Ray Bursts of Cosmic Origin." *Astrophysical Journal* 182 (June 1, 1973): L85–L88.

Kleinmann, D. E., and F. J. Low. "Observations of Infrared Galaxies." *Astrophysical Journal* 159 (March 1970): L165–L172.

Kragh, Helge. "The Beginning of the World: Georges Lemaître and the Expanding Universe." *Centaurus* 32 (1987): 114–39.

Lang, Kenneth R., and Owen Gingerich (ed.). *A Source Book in Astronomy and Astrophysics, 1900–1975.* Cambridge, Mass.: Harvard University Press, 1979.

Laplace, P. S. *The System of the World.* Translated by J. Pond. London: 1809.

Leavitt, Henrietta. "1777 Variables in the Magellanic Clouds." *Annals of the Astronomical Observatory of Harvard College,* volume 60. Cambridge: Harvard Observatory, 1908 (volume actually distributed in 1907), pp. 104–7.

———. "Periods of 25 Variable Stars in the Small Magellanic Cloud." *Harvard College Observatory Circular No. 173* (1912): 1–3.

Lemaître, Abbé G. "A Homogeneous Universe of Constant Mass and Increasing Radius Accounting for the Radial Velocity of Extra-galactic Nebulae." [1931a]. *Monthly Notices of the Royal Astronomical Society* 91 (March 1931): 483–90.

———. "The Beginning of the World from the Point of View of Quantum Theory." [1931b]. *Nature* 127 (May 9, 1931): 706.

———. *Nature* 128 [1931c] (October 24, 1931): 704–6.

Lin, C. C., and Frank H. Shu. "On the Spiral Structure of Disk Galaxies." *Astrophysical Journal* 140 (August 1964): 646–55.

Littmann, Mark. *Planets Beyond: Discovering the Outer Solar System.* New York: Wiley, 1988.

Lo, K. Y. "The Galactic Center: Is It a Massive Black Hole?" *Science* 233 (September 26, 1986): 1394–1403.

Lo, K. Y., and M. J. Claussen. "High-Resolution Observations of Ionized Gas in Central 3 Parsecs of the Galaxy: Possible Evidence for Infall." *Nature* 306 (December 15, 1983): 647–51.

Lockyer, J. Norman. "Spectroscopic Observations of the Sun—No. II." *Philosophical Transactions of the Royal Society of London* 159 (1869): 425–44.

———. *The Chemistry of the Sun.* London: Macmillan and Company, 1887.

Lorentz, H. A., A. Einstein, H. Minkowski, and H. Weyl. *The Principle of Relativity.* Translated by W. Perrett and G. B. Jeffery. London: Methuen and Company, 1923.

Low, F. J., and D. E. Kleinmann. "Infrared Observations of Seyfert Galaxies, Quasistellar Sources, and Planetary Nebulae." *Astronomical Journal* 73 (1968): 868–69.

Lowell, Percival. *Mars.* Boston: Houghton, Mifflin and Company, 1895.

MacPherson, Hector. *A Century's Progress in Astronomy.* Edinburgh: William Blackwood and Sons, 1906.

Magie, William Francis (ed.). *A Source Book in Physics.* New York: McGraw-Hill, 1935.

Martens, Rhonda. *Kepler's Philosophy and the New Astronomy.* Princeton: Princeton University Press, 2000.

Marvin, Ursula B. "Ernst Florens Friedrich Chladni (1756–1827) and the Origins of Modern Meteorite Research." *Meteoritics and Planetary Sciences* 31 (1996): 545–88.

Mayor, M., et al. "51 Pegasi." *International Astronomical Union Circular 6251* (October 25, 1995).

Mayor, Michel, and Didier Queloz. "A Jupiter-Mass Companion to a Solar-Type Star." *Nature* 378 (November 23, 1995): 355–59.

———. "A Search for Substellar Companions to Solar-Type Stars Via Precise Doppler Measurements: A First Jupiter Mass Companion Detected." In *Cool Stars, Stellar Systems, and the Sun. 9th Cambridge Workshop.* Edited by Roberto Pallavicini and Andrea K. Dupree. San Francisco: Astronomical Society of the Pacific, 1996.

McCarthy, Martin F. "Fr. Secchi and Stellar Spectra." *Popular Astronomy* 58 (April 1950): 153–69.

McCauley, J. F., M. H. Carr, J. A. Cutts, W. K. Hartmann, Harold Masursky, D. J. Milton, R. P. Sharp, and D. E. Wilhelms. "Preliminary Mariner 9 Report on the Geology of Mars." *Icarus* 17 (1972): 289–327.

McGrayne, Sharon Bertsch. *Nobel Prize Women in Science.* Washington, D.C.: Joseph Henry Press, 1998.

Meadows, A. J. *Science and Controversy: A Biography of Sir Norman Lockyer.* Cambridge, Mass.: MIT Press, 1972.

Meadows, A. J., and J. E. Kennedy. "The Origin of Solar-Terrestrial Studies." *Vistas in Astronomy* 25 (1982): 419–26.

Memoirs of the Royal Astronomical Society 12 (1841): 453.

Michell, John. "An Inquiry into the Probable Parallax, and Magnitude of the Fixed Stars, from the Quantity of Light Which They Afford Us, and the Particular Circumstances of Their Situation." *Philosophical Transactions* 57 (1767): 234–64.

Moore, Patrick. *William Herschel: Astronomer and Musician of 19 New King Street, Bath.* Bath, U.K.: William Herschel Society, 1983.

Morgan, W. W., Stewart Sharpless, and Donald Osterbrock. "Some Features of Galactic Structure in the Neighborhood of the Sun." *Astronomical Journal* 57 (1952): 3.

Neugebauer, G., D. E. Martz, and R. B. Leighton. "Observations of Extremely Cool Stars." *Astrophysical Journal* 142 (1965): 399–401.

Newton, Isaac. "De Motu Corporum in Gyrum," in *The Mathematical Papers of Isaac Newton,* volume VI. Edited by D. T. Whiteside. Cambridge: Cambridge University Press, 1974.

———. *The Principia.* Translated by I. Bernard Cohen and Anne Whitman. Berkeley: University of California Press, 1999.

Oke, J. B. "Absolute Energy Distribution in the Optical Spectrum of 3C 273." *Nature* 197 (March 16, 1963): 1040–41.

Olmsted, Denison. "Observations on the Meteors of November 13th, 1833." *The American Journal of Science and Arts* 25 (January 1834): 363–411.

———. "Observations on the Meteors of November 13th, 1833 (Continued)." *The American Journal of Science and Arts* 26 (July 1834): 132–74.

Oort, J. H. "The Structure of the Cloud of Comets Surrounding the Solar System, and a Hypothesis Concerning Its Origin." *Bulletin of the Astronomical Institutes of the Netherlands* 11 (January 13, 1950): 91–110.

Oort, J. H., F. T. Kerr, and G. Westerhout. "The Galactic System as a Spiral Nebula." *Monthly Notices of the Royal Astronomical Society* 118 (1958): 379–89.

Öpik, Ernst. "An Estimate of the Distance of the Andromeda Nebula." *Astrophysical Journal* 55 (1922): 406–10.

———. "Stellar Structure, Source of Energy, and Evolution." *Publications de l'Observatoire astronomique de l'Université de Tartu* 30:3 (1938): 1–115. Translated in *A Source Book in Astronomy and Astrophysics, 1900–1975*. Edited by Kenneth R. Lang and Owen Gingerich. Cambridge: Harvard University Press, 1979.

Oppenheimer, J. R., and H. Snyder. "On Continued Gravitational Contraction." *Physical Review* 56 (September 1, 1939): 455–59.

Osterbrock, Donald E., John R. Gustafson, and W. J. Shiloh Unruh. *Eye on the Sky: Lick Observatory's First Century.* Berkeley: University of California Press, 1988.

Ostriker, J. P., and P. J. E. Peebles. "A Numerical Study of the Stability of Flattened Galaxies: Or, Can Cold Galaxies Survive?" *Astrophysical Journal* 186 (December 1, 1973): 467–80.

Ostriker, J. P., P. J. E. Peebles, and A. Yahil. "The Size and Mass of Galaxies, and the Mass of the Universe." *Astrophysical Journal* 193 (October 1, 1974): L1–L4.

Pais, Abraham. *"Subtle Is the Lord . . .": The Science and the Life of Albert Einstein.* Oxford: Oxford University Press, 1982.

Partridge, R. B. "The Seeds of Cosmic Structure." *Science* 257 (July 10, 1992): 178–79.

Payne, Cecilia H. *Stellar Atmospheres.* Cambridge, Mass.: Harvard Observatory, 1925.

Peebles, P. J. E. *Principles of Physical Cosmology.* Princeton: Princeton University Press, 1993.

Penzias, A. A., and R. W. Wilson. "A Measurement of Excess Antenna Temperature at 4080 Mc/s." *Astrophysical Journal* 142 (1965): 419–21.

Perlmutter, S., et al. "Cosmology from Type Ia Supernovae." *Bulletin of the American Astronomical Society* 29 (1997): 1351.

———. "Measurements of Ω and Λ from 42 High-Redshift Supernovae." *Astrophysical Journal* 517 (June 1, 1999): 565–86.

Plotkin, Howard. "Edward C. Pickering, the Henry Draper Memorial, and the Beginnings of Astrophysics in America." *Annals of Science* 35 (1978): 365–77.

Proctor, R. A. "A New Theory of the Milky Way." *Monthly Notices of the Royal Astronomical Society* 30 (1869): 50–56.

Reber, Grote. "Cosmic Static." *Astrophysical Journal* 91 (1940): 621–24.

———. "Cosmic Static." *Proceedings of the Institute of Radio Engineers* 28 (February 1940): 68–70.

———. "Cosmic Static." *Proceedings of the Institute of Radio Engineers* 30 (August 1942): 367–78.

———. "Cosmic Static." *Astrophysical Journal* 100 (November 1944): 279–87.

Riess, Adam G. "An Accelerating Universe and Other Cosmological Implications from SNe Ia." *Bulletin of the American Astronomical Society* 30:2 (1998): 843.

Riess, Adam G., et al. "Observational Evidence from Supernovae for an Accelerating Universe and a Cosmological Constant." *Astronomical Journal* 116 (September 1998): 1009–38.

Romer, Alfred. "The Welcoming of Copernicus's *De revolutionibus*: The *Commentariolus* and Its Reception." *Physics in Perspective* 1 (1999): 157–83.

Ronan, Colin A. *The Astronomers.* London: Evans Brothers, 1964.

Rubin, Vera C. "The Rotation of Spiral Galaxies." *Science* 220 (June 24, 1983): 1339–44.

Rubin, Vera C., W. Kent Ford Jr., and Norbert Thonnard. "Extended Rotation Curves of High-Luminosity Spiral Galaxies. IV. Systematic Dynamical Properties, Sa→Sc." *Astrophysical Journal* 225 (November 1, 1978): L107–L111.

Russell, Henry Norris. "Relations Between the Spectra and Other Characteristics of the Stars." *Popular Astronomy* 22 (1914): 275–94.

Sabine, Edward. "On Periodical Laws Discoverable in the Mean Effects of the Larger Magnetic Disturbances — No. II." *Philosophical Transactions of the Royal Society of London* 142 (1852): 103–24.

Sandage, Allan. "Edwin Hubble 1889–1953." *Journal of the Royal Astronomical Society of Canada* 83 (December 1989): 351–62.

Schmidt, M. "3C 273: A Star-Like Object with Large Red-Shift." *Nature* 197 (March 16, 1963): 1040.

Shane, C. D., and C. A. Wirtanen. "The Distribution of Extragalactic Nebulae." *Astronomical Journal* 59 (September 1954): 285–304.

Shapley, Harlow. "Note on the Magnitudes of Novae in Spiral Nebulae." *Publications of the Astronomical Society of the Pacific* 29 (1917): 213–17.

———. "Studies Based on the Colors and Magnitudes in Stellar Clusters. Seventh Paper: The Distances, Distribution in Space, and Dimensions of 69 Globular Clusters." *Astrophysical Journal* 48 (October 1918): 154–81.

———. "Studies Based on the Colors and Magnitudes in Stellar Clusters. Twelfth Paper: Remarks on the Arrangement of the Sidereal Universe." *Astrophysical Journal* 49 (June 1919): 311–36.

———. "Luminosity Distribution and Average Density of Matter in Twenty-Five Groups of Galaxies." *Proceedings of the National Academy of Sciences* 19 (June 15, 1933): 591–96.

——— (ed.). *Source Book in Astronomy 1900–1950*. Cambridge, Mass.: Harvard University Press, 1960.

Shapley, Harlow, and Helen E. Howarth (eds.). *A Source Book in Astronomy*. New York: McGraw-Hill, 1929.

Slipher, V. M. "Nebulae." *Proceedings of the American Philosophical Society* 56 (1917): 403–9.

———. "The Discovery of a Solar System Body Apparently Trans-Neptunian." *Lowell Observatory Observation Circular*, March 13, 1930.

Smith, George E. "How Did Newton Discover Universal Gravity?" *The St. John's Review* 45:2 (1999): 32–63.

———. "Comments on Ernan McMullin's 'The Impact of Newton's Principia on the Philosophy of Science.' " *Philosophy of Science* 68 (September 2001): 327–38.

Smith, R. W. "The Origins of the Velocity-Distance Relation." *Journal for the History of Astronomy* 10 (October 1979): 133–65.

Smoot, G. F., et al. "Structure in the COBE Differential Microwave Radiometer First-Year Maps." *Astrophysical Journal* 396 (September 1, 1992): L1–L5.

"Spiral Arms of the Galaxy." *Sky and Telescope* 11 (April 1952): 138–39.

Stachel, John (ed.). *Einstein's Miraculous Year*. Princeton: Princeton University Press, 1998.

Standage, Tom. *The Neptune File*. New York: Walker and Company, 2000.

Stocks, J. L. "De Caelo." In *The Works of Aristotle*, volume 2. Edited by W. D. Ross. Oxford: Clarendon Press, 1930.

Sullivan, Woodruff T. III. *Classics in Radio Astronomy.* Dordrecht: D. Reidel Publishing, 1982.

——— (ed.). *The Early Years of Radio Astronomy.* Cambridge: Cambridge University Press, 1984.

Swerdlow, Noel. "The Derivation and First Draft of Copernicus's Planetary Theory: A Translation of the *Commentariolus* with Commentary." *Proceedings of the American Philosophical Society* 117 (1973): 423–512.

Taylor, J. H., L. A. Fowler, and P. M. McCulloch. "Measurements of General Relativistic Effects in the Binary Pulsar PSR1913 + 16." *Nature* 277 (February 8, 1979): 437–40.

Thompson, J. Eric S. *A Commentary on the Dresden Codex.* Philadelphia: American Philosophical Society, 1972.

Thurston, Hugh. *Early Astronomy.* New York: Springer, 1994.

Tinsley, Beatrice M. "Possibility of a Large Evolutionary Correction to the Magnitude-Redshift Relation." *Astrophysics and Space Science* 6 (1970): 344–51.

———. "Evolution of the Stars and Gas in Galaxies." *Astrophysical Journal* 151 (February 1968): 547–65.

Toomer, G. J. *Ptolemy's Almagest.* Princeton: Princeton University Press, 1998.

Trumpler, Robert J. "Preliminary Results on the Distances, Dimensions and Space Distribution of Open Star Clusters." *Lick Observatory Bulletin* 14:420 (1930): 154–88.

Van Allen, J. A., G. H. Ludwig, E. C. Ray, and C. E. McIlwain. "Observation of High Intensity Radiation by Satellites 1958 Alpha and Gamma." *Jet Propulsion* 28 (September 1958): 588–92.

Van de Hulst, Hendrik C. "Radio Waves from Space: Origin of Radiowaves." *Nederlands tijdschrift voor natuurkunde* 11 (1945): 210–21.

Van de Hulst, H. C., C. A. Muller, and J. H. Oort. "The Spiral Structure of the Outer Part of the Galactic System Derived from the Hydrogen Emission at 21 cm Wave Length." *Bulletin of the Astronomical Institutes of the Netherlands* 12 (May 14, 1954): 117–49.

Van den Bergh, Sidney. "The Early History of Dark Matter." *Publications of the Astronomical Society of the Pacific* 111 (June 1999): 657–60.

Van Helden, Albert. *Measuring the Universe.* Chicago: University of Chicago Press, 1985.

Voelkel, James R. *Johannes Kepler and the New Astronomy.* New York: Oxford University Press, 1999.

Vogel, H. C. "On the Spectrographic Method of Determining the Velocity of Stars in the Line of Sight." *Monthly Notices of the Royal Astronomical Society* 52 (1892): 87–96.

Vogel, H. C. "List of the Proper Motions in the Line of Sight of Fifty-One Stars." *Monthly Notices of the Royal Astronomical Society* 52 (1892): 541–43.

Vogel, H. C., and Scheiner, J. "Orbit and Mass of the Variable Star Algol (β Persei)." *Publications of the Astronomical Society of the Pacific* 2 (1890): 27.

Wali, Kameshwar C. *Chandra: A Biography of S. Chandrasekhar.* Chicago: University of Chicago Press, 1984.

Walsh, D., R. F. Carswell, and R. J. Weymann. "0957+561 A, B: Twin Quasistellar Objects or Gravitational Lens?" *Nature* 279 (May 31, 1979): 381–84.

Weinrab, S., A. H. Barrett, M. L. Meeks, and J. C. Henry. "Radio Observations of OH in the Interstellar Medium." *Nature* 200 (November 30, 1963): 829–31.

Westerhout, Gart. "The Early History of Radio Astronomy." In *Education in and History of Modern Astronomy*. Edited by Richard Berendzen. New York: New York Academy of Sciences, 1972.

Wheeler, John Archibald. "Superdense Stars." *Annual Review of Astronomy and Astrophysics* 4 (1966): 393–432.

Whipple, Fred L. "A Comet Model. I. The Acceleration of Comet Encke." *Astrophysical Journal* 111 (March 1950): 375–94.

Whitrow, G. J. "Kant and the Extragalactic Nebulae." *Quarterly Journal of the Royal Astronomical Society* 8 (1967): 48–56.

Wilkins, G. A. "Sir Norman Lockyer's Contributions to Science." *Quarterly Journal of the Royal Astronomical Society* 35 (1994): 51–57.

Williams, Henry Smith. *The Great Astronomers*. New York: Newton, 1932.

Williams, Robert E., et al. "The Hubble Deep Field: Observations, Data Reduction, and Galaxy Photometry." *Astronomical Journal* 112 (October 1996): 1335–89.

Wolszczan, A. "Two Planets Around a 6.2-ms Pulsar 1257+12?" *Bulletin of the American Astronomical Society* 23 (1991): 1347.

Wolszczan, A., and D. A. Frail. "A Planetary System Around the Millisecond Pulsar PSR1257+12." *Nature* 355 (January 9, 1992): 145–47.

Wright, Thomas. *An Original Theory or New Hypothesis of the Universe* (facsimile reprint). Introduction and transcription by Michael A. Hoskin. New York: American Elsevier, 1971.

——. *An Original Theory or New Hypothesis of the Universe*. London: Chapelle, 1750.

Zwicky, Fritz. "On the Red Shift of Spectral Lines Through Interstellar Space." *Physical Review* 33 (1929): 1077.

——. "On the Masses of Nebulae and of Clusters of Nebulae." *Astrophysical Journal* 86 (October 1937): 217–46.

Acknowledgments

My journey into the annals of astronomy began at the Burndy Library of the Dibner Institute on the campus of the Massachusetts Institute of Technology. My deepest gratitude is extended to the archivists and librarians there, who treated me as a colleague during the months that I spent with them. For their assistance, patience, and many kindnesses, I thank Ben Weiss, Judith Nelson, David McGee, Anne Battis, Howard Kennett, and Larry Leier. Barbara Palmer at the Wolbach Library of the Harvard-Smithsonian Center for Astrophysics was similarly helpful.

With centuries of history to sift through, deciding which astronomical events made the cut became easier after discussions with Virginia Trimble, Donald Osterbrock, David DeVorkin, Peter Saulson, Robert Rosenstein, and Ken Croswell. Virginia, Don, and David helped even further by later reviewing various sections of this work.

My deep appreciation is extended to the other scientists and historians who graciously agreed to critique segments of the book that dealt with their areas of expertise, offering advice and constructive criticism. They include Ralph Alpher, Eric Becklin, Geoffrey Burbidge, Paul Butler, Michael Carr, Margaret Geller, Norriss Hetherington, Michael Hoskin, John Huchra, Russell Hulse, Ray Klebesadel, Alan Lightman, Adam Reiss, Vera Rubin, Maarten Schmidt, Harvey Tananbaum, Joseph Taylor, Ed Turner, James Voelkel, Craig Waff, D. L. Wark, and Robert Williams. Special thanks must go to both Owen Gingerich and George Smith, who provided me with invaluable lessons on the Copernican and Newtonian eras, respectively, as well as to Woodruff ("Woody") T. Sullivan III, who offered important insights on the history of radio astronomy. Barbarina Zwicky kindly furnished background information on her father, Caltech astronomer Fritz Zwicky.

Friends who provided both support and restful interludes during the course of this project include Elizabeth and Goetz Eaton, Tara and Paul McCabe, Suzanne Szescila and Jed Roberts, Sarah Saulson, and Linda and Steve Wohler. I thank Russell Galen, my esteemed agent, for once again shepherding my proposal to its publishing home, and my husband, Steve Lowe, for his love, support, and editorial assistance throughout the course of this endeavor.

Lastly, heartfelt thanks goes to my editor, J. Edward Kastenmeier, for initiating this project. As a new and enthusiastic recruit to amateur astronomy, Edward went looking for a sourcebook that would serve both scientist and general reader alike; failing to find one, he decided to get one published. I was both honored and pleased by his invitation to sign on as the anthology's editor and commentator. It turned out to be one of the most enjoyable experiences in my science-writing career, a splendid opportunity to spend two years perusing more than two thousand years of astronomical history.

Index

Page numbers in *italics* refer to figures and tables.

PERMISSIONS ACKNOWLEDGMENTS

1. Illustrations from Thompson, J. Eric S. *A Commentary on the Dresden Codex*. Philadelphia: American Philosophical Society, 1972. Copyright © 1972 by the American Philosophical Society. Thompson's original source was a copy of the 1892 Förstemann edition of the Codex loaned by the Yale University Library.

2. Aristotle. "De Caelo" translated by J. L. Stocks in *The Works of Aristotle, Volume 2*, edited by W. D. Ross. Oxford: Clarendon Press, 1930. Selection from Book II, Section 4: 286b, 291b, 294a, 297a, 297a, 297b, 298a. Reprinted by permission of Oxford University Press.

3. Aristarchus of Samos. "On The Sizes and Distances of the Sun and Moon," translated by Thomas L. Heath in *Aristarchus of Samos*. Oxford: Clarendon Press, 1913, pages 353, 355, 377, 379, 381. Reprinted by permission of Oxford University Press.

4. Cleomedes. "On the Orbits of the Heavenly Bodies," translated by Thomas L. Heath in *Greek Astronomy*. London: J. M. Dent and Sons, 1932.

5. Ptolemy. *Ptolemy's Almagest*. Translated and annotated by G. J. Toomer. Princeton: Princeton University Press, 1998. Copyright © 1998 by Princeton University Press. Reprinted by permission of Princeton University Press.

6. Ptolemy. *Ptolemy's Almagest*. Translated and annotated by G. J. Toomer. Princeton: Princeton University Press, 1998. Copyright © 1998 by Princeton University Press. Reprinted by permission of Princeton University Press.

7. Copernicus, Nicolaus. *On the Revolutions of the Heavenly Spheres*. Translated by Charles Glenn Wallis. Amherst, New York: Prometheus Books, 1995. Reprinted from Annapolis: The St. John's Bookstore, 1939.

8. Brahe, Tycho. "On a New Star, Not Previously Seen Within the Memory of Any Age Since the Beginning of the World," in *A Source Book in Astronomy*. Edited by Harlow Shapley and Helen E. Howarth. New York: McGraw-Hill, 1929. Reprinted from Brahe, Tycho. "De Nova Stella" in *Opera Omnia, Tomus I*. Edidit J. L. E. Dreyer (1913). Translated by John H. Walden (1928).

9. Kepler, Johannes. *Epitome of Copernican Astronomy & Harmonies of the World*. Translated by Charles Glenn Wallis. Amherst, New York: Prometheus Books, 1995. Reprinted from Annapolis: The St. John's Bookstore, 1939.

10. Galilei, Galileo. *Sidereus Nuncius or The Sidereal Messenger*. Translated with introduction, conclusion, and notes by Albert Van Helden. Chicago: The University of Chicago Press, 1989. Copyright © 1989 by The University of Chicago. Illustration of Huygens' Saturn from *Eustachius de Divinis Septempedanus pro sua annotatione in systema Saturnium: Christiani Hugenii adversus eiusdem assertionem*. Romae: Typis Dragondellianis, 1661.

11. Newton, Isaac. *The Principia: Mathematical Principles of Natural Philosophy*. Translated/edited by I. Bernard Cohen and Anne Whitman. Berkeley: University

of California Press, 1999. Copyright © 1999 The Regents of the University of California.

12. Halley, Edmund. *A Synopsis of the Astronomy of Comets*. London: Printed for John Senex, 1705.

13. Michell, John. "An Inquiry into the Probable Parallax, and Magnitude of the Fixed Stars, From the Quantity of Light Which They Afford Us, and the Particular Circumstances of Their Situation." *Philosophical Transactions of the Royal Society of London*, 57 (1767): 234–64.

14. "A Demonstration Concerning the Motion of Light, Communicated from Paris, in the *Journal des Scavans*, and Here Made English." *Philosophical Transactions*, Volume 12 (1677), a report on the work of Ole Römer.

15. Laplace, P. S. *The System of the World. Volume II.* Translated by J. Pond. London: 1809.

16. Herschel, William. "Account of a Comet." *Philosophical Transactions of the Royal Society of London* 71 (1781): 492–501.

Herschel, William. "A Letter from William Herschel." *Philosophical Transactions of the Royal Society of London* 73 (1783): 1–3.

17. Halley, Edmund. "Considerations on the Change of the Latitudes of Some of the Principal Fixt Stars." *Philosophical Transactions of the Royal Society of London* 30 (1717–1719): 736–38.

Herschel, William. "On the Proper Motion of the Sun and Solar System; With an Account of Several Changes That Have Appeared Among the Fixed Stars Since the Time of Mr. Flamstead." *Philosophical Transactions of the Royal Society of London* 73 (1783): 247–83.

Goodricke, John. "A Series of Observations on, and a Discovery of, the Period of the Variation of the Light of the Bright Star in the Head of Medusa, Called Algol." *Philosophical Transactions of the Royal Society of London* 73 (1783): 474–82.

18. "Bode and Piazzi" in *A Source Book in Astronomy*. Edited by Harlow Shapley and Helen E. Howarth. New York: McGraw-Hill, 1929. Translated from Bode, J. E. *Von Dem Neuen, Zwischen Mars und Jupiter Entdeckten Achten Haupt Planeten des Sonnensystems*. Berlin: 1802.

Piazzi, G. *Risultati delle Osservazioni della Nuova Stella*. Palermo: 1801. Translated by Antonio Parachinatti for Nevil Maskelyne of the Royal Greenwich Observatory. RGO manuscript 4/221 in Cambridge University Library. Reprinted by permission of the Syndics of the Cambridge University Library and the Particle Physics and Astronomy Research Council.

19. Bessel, F. W. "On the Parallax of 61 Cygni." *Monthly Notices of the Royal Astronomical Society* 4 (November 1838): 152–61.

20. Le Verrier, Urbain Jean Joseph. "Prediction of the Position of Neptune" in *A Source Book in Astronomy*. Edited by Harlow Shapley and Helen E. Howarth. New York: McGraw-Hill, 1929. Translated from *Astronomische Nachrichten* 25:580 (October 12, 1846).

21. Wright, Thomas. *An Original Theory or New Hypothesis of the Universe*. London: Chapelle, 1750.

Kant, Immanuel. *Kant's Cosmogony as in His Essay on the Retardation of the Rotation of the Earth and his Natural History and Theory of the Heavens*. Edited and Translated by W. Hastie. Glasgow: James Maclehose and Sons, 1900.

Herschel, William. "On the Construction of the Heavens." *Philosophical Transactions of the Royal Society of London* 75 (1785): 213–66.

22. Rosse, The Earl of. "Observations on the Nebulae." *Philosophical Transactions of the Royal Society of London* 140 (1850): 499–514.

23. Fraunhofer, Joseph. *Prismatic and Diffraction Spectra.* Translated and edited by J. S. Ames. New York: Harper & Brothers, 1898.

24. Kirchhoff, Gustav Robert. *Untersuchungen über das Sonnenspectrum und die Spectrum der chemischen Elemente* [Researches on the solar spectrum and the spectra of the chemical elements]. Translated by Henry E. Roscoe. Cambridge: Macmillan, 1862–1863.

25. Huggins, William and W. A. Miller. "On the Spectra of Some of the Nebulae." *Philosophical Transactions of the Royal Society of London* 154 (1864): 437–44.

26. Vogel, H. C. "On the Spectrographic Method of Determining the Velocity of Stars in the Line of Sight." *Monthly Notices of the Royal Astronomical Society* 52 (1892): 87–96.

Vogel, H. C. "List of the Proper Motions in the Line of Sight of Fifty-One Stars." *Monthly Notices of the Royal Astronomical Society* 52 (1892): 541–43.

Vogel, H. C. and J. Scheiner. "Orbit and Mass of the Variable Star *Algol* (β *Persei*)." *Publications of the Astronomical Society of the Pacific* 2 (1890): 27. Reprinted with permission of the Editors of the *Publications of the Astronomical Society of the Pacific.*

27. Cannon, Annie J. and Edward C. Pickering. "Spectra of Bright Southern Stars." *Annals of the Astronomical Observatory of Harvard College, Volume 28, Part II.* Cambridge, Mass.: Harvard Observatory, 1901. Reprinted with permission by John G. Wolbach Library, Harvard College Observatory, Cambridge, Massachusetts, USA.

28. Hertzsprung, Ejnar. "Giants and Dwarfs" in *A Source Book in Astronomy 1900–1950.* Edited by Harlow Shapley. Cambridge, Mass.: Harvard University Press, 1960. (Translation by Harlow Shapley of "Zur Strahlung der Sterne." *Zeitschrift für wissenshaftliche Photographie* 3 [1905]: 429–42.) Reprinted by permission of the Harvard University Press. Copyright © 1960 by the President and Fellows of Harvard College. Copyright © renewed 1988 by Harlow Shapley.

Russell, Henry Norris. "Relations Between the Spectra and Other Characteristics of the Stars." *Popular Astronomy* 22 (1914): 275–94.

29. Payne, Cecilia H. *Stellar Atmospheres.* Cambridge, Mass.: Harvard Observatory, 1925. Reprinted with permission by John G. Wolbach Library, Harvard College Observatory, Cambridge, Massachusetts, USA.

30. Eddington, A. S. "On the Radiative Equilibrium of the Stars." *Monthly Notices of the Royal Astronomical Society* 77 (November 10, 1916): 16–35. With permission of the *Monthly Notices of the Royal Astronomical Society* and Blackwell Publishing Ltd.

Eddington, A. S. "On the Relation Between Masses and Luminosities of the Stars." *Monthly Notices of the Royal Astronomical Society* 84 (1924): 308–32. With permission of the *Monthly Notices of the Royal Astronomical Society* and Blackwell Publishing Ltd.

31. Schwabe, Samuel. "The Periodicity of Sun Spots" in *A Source Book in Astronomy.* Edited by Harlow Shapley and Helen Howarth. New York: McGraw-Hill, 1929.

Humboldt, Alexander Von. *Cosmos: A Sketch of a Physical Description of the Universe.* Translated by E. C. Otté and B. H. Paul. London: Henry G. Bohn, 1849.

Sabine, Edward. "On Periodical Laws Discoverable in the Mean Effects of the Larger Magnetic Disturbances—No. II." *Philosophical Transactions of the Royal Society of London* 142 (1852): 103–24.

Lockyer, J. Norman. "Spectroscopic Observations of the Sun—No. II." *Philosophical Transactions of the Royal Society of London* 159 (1869): 425–44.

32. Olmsted, Denison. "Observations on the Meteors of November 13th, 1833." *The American Journal of Science and Arts* 25 (January 1834): 363–411.

Olmsted, Denison. "Observations on the Meteors of November 13th, 1833." *The American Journal of Science and Arts* 26 (July 1834): 132–74.

33. Hess, Victor Franz. "Concerning Observations of Penetrating Radiation on Seven Free Balloon Flights." *Physikalishe Zeitschrift* 13 (1912): 1084–91. Translated by Brian Doyle in *A Source Book in Astronomy and Astrophysics, 1900–1975.* Edited by Kenneth R. Lang and Owen Gingerich. Cambridge, Mass.: Harvard University Press, 1979. Reprinted by permission of the publisher. Copyright © 1979 by the President and Fellows of Harvard College.

34. Slipher, V. M. "The Discovery of a Solar System Body Apparently Trans-Neptunian." *Lowell Observatory Observation Circular* (March 13, 1930). Courtesy of Lowell Observatory. Used by Permission.

35. Einstein, A. "Zur Elektrodynamik beweger Körper [On the Electrodynamics of Moving Bodies]." *Annalen der Physik* 17 (1905): 891–921. From Lorentz, H. A., A. Einstein, H. Minkowski, and H. Weyl. *The Principle of Relativity.* Translated by W. Perrett and G. B. Jeffery. London: Methuen and Company, 1923. With permission of Taylor & Francis Books Ltd. and the Albert Einstein Archives at the Jewish National and University Library of the Hebrew University of Jerusalem.

Einstein, A. "Ist die Trägheit eines Körpers von seinem Energiegehalt abhängig? [Does the Inertia of a Body Depend Upon Its Energy-Content?]" *Annalen der Physik* 18 (1905): 639–41. From Lorentz, H. A., A. Einstein, H. Minkowski, and H. Weyl. *The Principle of Relativity.* Translated by W. Perrett and G. B. Jeffery. London: Methuen and Company, 1923. With permission of Taylor & Francis Books Ltd. and the Albert Einstein Archives at the Jewish National and University Library of the Hebrew University of Jerusalem.

36. Dyson, F. W., A. S. Eddington, and C. Davidson. "A Determination of the Deflection of Light by the Sun's Gravitational Field, from Observations Made at the Total Eclipse of May 29, 1919." *Philosophical Transactions of the Royal Society of London. Series A.* 220 (1920): 291–333.

37. Einstein, A. "Kosmologische Betrachtungen zur allgemeinen Relativitätstheorie [Cosmological Considerations of the General Theory of Relativity]." *Sitzungsberichte der Königlich Preussischen Akademie der Wissenschaften* (1917): 142–52. From Lorentz, H. A., A. Einstein, H. Minkowski, and H. Weyl. *The Principle of Relativity.* Translated by W. Perrett and G. B. Jeffery. London: Methuen and Company, 1923. With permission of Taylor & Francis Books Ltd. and the Albert Einstein Archives at the Jewish National and University Library of the Hebrew University of Jerusalem.

De Sitter, W. "On Einstein's Theory of Gravitation, and Its Astronomical Consequences. Third Paper." *Monthly Notices of the Royal Astronomical Society* 78

(November, 1917): 3–28. With permission of the *Monthly Notices of the Royal Astronomical Society* and Blackwell Publishing Ltd.

Lemaître, Abbé G. "A Homogeneous Universe of Constant Mass and Increasing Radius Accounting for the Radial Velocity of Extra-galactic Nebulae." *Monthly Notices of the Royal Astronomical Society* 91 (March, 1931): 483–90. With permission of the *Monthly Notices of the Royal Astronomical Society* and Blackwell Publishing Ltd.

38. Lemaître, G. "The Beginning of the World from the Point of View of Quantum Theory." *Nature* 127 (May 9, 1931): 706. Copyright © 1931 by Nature Publishing Group. Reproduced with permission of Nature Publishing Group in the format Trade Book via Copyright Clearance Center.

Lemaître, Abbé G. *Nature* 128 (October 24, 1931): 704–6. Copyright © 1931 by Nature Publishing Group. Reproduced with permission of Nature Publishing Group in the format Trade Book via Copyright Clearance Center.

Bondi, H. and T. Gold. "The Steady-State Theory of the Expanding Universe." *Monthly Notices of the Royal Astronomical Society* 108 (1948): 252–70. With permission of the *Monthly Notices of the Royal Astronomical Society* and Blackwell Publishing Ltd.

39. Adams, Walter S. "An A-Type Star of Very Low Luminosity." *Publications of the Astronomical Society of the Pacific* 26 (October 1914): 198. Reprinted with permission of the Editors of the *Publications of the Astronomical Society of the Pacific*.

Adams, Walter S. "The Spectrum of the Companion of Sirius." *Publications of the Astronomical Society of the Pacific* 27 (December 1915): 236–37. Reprinted with permission of the Editors of the *Publications of the Astronomical Society of the Pacific*.

Fowler, R. H. "On Dense Matter." *Monthly Notices of the Royal Astronomical Society* 87 (December 1926): 114–22. With permission of the *Monthly Notices of the Royal Astronomical Society* and Blackwell Publishing Ltd.

40. Chandrasekhar, S. "The Highly Collapsed Configurations of a Stellar Mass (Second Paper)." *Monthly Notices of the Royal Astronomical Society* 95 (1935): 207–25. With permission of the *Monthly Notices of the Royal Astronomical Society* and Blackwell Publishing Ltd.

"Discussion of Papers 11 and 12 by A. S. Eddington and E. A. Milne." *The Observatory* 58 (1935): 37–39. Reprinted with permission of the Editors of *The Observatory*.

41. Baade, W. and F. Zwicky. "On Super-Novae." *Proceedings of the National Academy of Sciences* 20 (May 15, 1934): 254–59. With the kind permission of Anna Zwicky.

Baade, W. and F. Zwicky. "Cosmic Rays From Super-Novae." *Proceedings of the National Academy of Sciences* 20 (May 15, 1934): 259–63. With the kind permission of Anna Zwicky.

42. Oppenheimer, J. R. and H. Snyder. "On Continued Gravitational Contraction." *Physical Review* 56 (September 1, 1939): 455–59. Copyright © 1939 by the American Physical Society.

43. Bethe, H. A. "Energy Production in Stars." *Physical Review* 55 (March 1, 1939): 434–56. Copyright © 1939 by the American Physical Society.

44. Alpher, R. A., H. Bethe, and G. Gamow. "The Origin of Chemical Elements." *Physical Review* 73 (April 1, 1948): 803–4. Copyright © 1948 by the American Physical Society.

45. Alpher, Ralph A. and Robert Herman. "Evolution of the Universe." *Nature* 162 (November 13, 1948): 774–75. Copyright © 1948 by Nature Publishing Group. Reproduced with permission of Nature Publishing Group in the format Trade Book via Copyright Clearance Center.

46. Burbidge, E. Margaret, G. R. Burbidge, William A. Fowler, and F. Hoyle. "Synthesis of the Elements in Stars." *Reviews of Modern Physics* 29 (October 1957): 547–650. Copyright © 1957 by the American Physical Society.

47. Öpik, Ernst. "Stellar Structure, Source of Energy, and Evolution." *Publications de l'Observatoire astronomique de l'Université de Tartu* 30:3 (1938): 1–115. Translated in *A Source Book in Astronomy and Astrophysics, 1900–1975.* Edited by Kenneth R. Lang and Owen Gingerich. Cambridge, Mass.: Harvard University Press, 1979.

48. Leavitt, Henrietta. "Periods of 25 Variable Stars in the Small Magellanic Cloud." *Harvard College Observatory Circular No. 173* (1912): 1–3. Reprinted with permission by John G. Wolbach Library, Harvard College Observatory, Cambridge, Massachusetts, USA.

49. Shapley, Harlow. "Studies Based on the Colors and Magnitudes in Stellar Clusters. Twelfth Paper: Remarks on the Arrangement of the Sidereal Universe." *Astrophysical Journal* 49 (June 1919): 311–36. Reproduced by permission of the American Astronomical Society.

50. Barnard, E. E. "On A Nebulous Groundwork in the Constellation *Taurus.*" *Astrophysical Journal* 25 (1907): 218–25. Reproduced by permission of the American Astronomical Society.

Trumpler, Robert J. "Preliminary Results on the Distances, Dimensions and Space Distribution of Open Star Clusters." *Lick Observatory Bulletin* 14, Number 420 (1930): 154–88. Reprinted with permission of UCO/Lick Observatory.

51. Hubble, Edwin P. "Cepheids in Spiral Nebulae." *Publications of the American Astronomical Society* 5 (1925): 261–64. Reproduced by permission of the American Astronomical Society.

52. Slipher, Vesto M. "A Spectrographic Investigation of Spiral Nebulae." *Proceedings of the American Philosophical Society* 56 (1917): 403–9. Permission granted by the American Philosophical Society.

Hubble, Edwin. "A Relation Between Distance and Radial Velocity Among Extra-Galactic Nebulae." *Proceedings of the National Academy of Sciences* 15 (March 15, 1929): 168–73. This item from the Edwin Hubble Papers is reproduced by permission of *The Huntington Library, San Marino, California.*

53. Baade, W. "The Resolution of Messier 32, NGC 205, and the Central Region of the Andromeda Nebula." *Astrophysical Journal* 100 (September 1944): 137–46. Reproduced by permission of the American Astronomical Society.

Baade, W. "A Revision of the Extra-Galactic Distance Scale." *Transactions of the International Astronomical Union* 8 (1952): 397–98. Published with permission of the International Astronomical Union.

54. Morgan, W. W., Stewart Sharpless, and Donald Osterbrock. "Some Features of Galactic Structure in the Neighborhood of the Sun." *The Astronomical Journal* 57 (1952): 3. Reproduced by permission of the American Astronomical Society.

Illustration from "Spiral Arms of the Galaxy." *Sky & Telescope* 11 (April 1952): 138–39. With permission of Yerkes Observatory and *Sky & Telescope.*

Van de Hulst, Hendrik C., C. Alex Muller, and Jan H. Oort. "The Spiral Structure of the Outer Part of the Galactic System Derived from the Hydrogen Emission at 21 cm Wavelength." *Bulletin of the Astronomical Institutes of the Netherlands* 12 (May 14, 1954): 117–49. Reproduced with permission of *Astronomy and Astrophysics*.

Illustration of Milky Way's arms from Oort, Jan. H., Frank J. Kerr, and Gart Westerhout. "The Galactic System as a Spiral Nebula." *Monthly Notices of the Royal Astronomical Society* 118 (1958): 379–89. With permission of the *Monthly Notices of the Royal Astronomical Society* and Blackwell Publishing Ltd.

55. Oort, J. H. "The Structure of the Cloud of Comets Surrounding the Solar System, and a Hypothesis Concerning Its Origin." *Bulletin of the Astronomical Institutes of the Netherlands* 11 (January 13, 1950): 91–110. Reproduced with permission of *Astronomy and Astrophysics*.

Whipple, Fred L. "A Comet Model. I. The Acceleration of Comet Encke." *Astrophysical Journal* 111 (March 1950): 375–94. Reproduced by permission of the American Astronomical Society.

56. Jansky, Karl G. "Directional Studies of Atmospherics at High Frequencies." *Proceedings of the Institute of Radio Engineers* 20 (December 1932): 1920–32. Portions reprinted with permission. Copyright © 1932 IRE (now IEEE).

Jansky, Karl G. "Electrical Disturbances Apparently of Extraterrestrial Origin." *Proceedings of the Institute of Radio Engineers* 21 (October 1933): 1387–98. Portions reprinted with permission. Copyright © 1933 IRE (now IEEE).

Jansky, Karl G. "A Note on the Source of Interstellar Interference." *Proceedings of the Institute of Radio Engineers* 23 (October 1935): 1158–63. Portions reprinted with permission. Copyright © 1935 IRE (now IEEE).

Reber, Grote. "Cosmic Static." *Astrophysical Journal* 100 (November 1944): 279–87. Reproduced by permission of the American Astronomical Society.

57. Van de Hulst, Hendrik C. "Origin of the Radio Waves from Space." *Nederlands tijdschrift voor natuurkunde* 11 (1945): 210–21. Translated by Woodruff T. Sullivan, III in *Classics in Radio Astronomy* (Dordrecht, Holland: D. Reidel Publishing, 1982) and reprinted with his kind permission.

Ewen, H. I. and E. M. Purcell. "Radiation from Galactic Hydrogen at 1,420 Mc./sec." *Nature* 168 (September 1, 1951): 356. Copyright © 1951 by Nature Publishing Group. Reproduced with permission of Nature Publishing Group in the format Trade Book via Copyright Clearance Center.

58. Weinrab, S., A. H. Barrett, M. L. Meeks, and J. C. Henry. "Radio Observations of OH in the Interstellar Medium." *Nature* 200 (November 30, 1963): 829–31. Copyright © 1963 by Nature Publishing Group. Reproduced with permission of Nature Publishing Group in the format Trade Book via Copyright Clearance Center.

59. Van Allen, James A., George H. Ludwig, Ernest C. Ray, and Carl E. McIlwain. "Observation of High Intensity Radiation by Satellites 1958 Alpha and Gamma." *Jet Propulsion* 28 (September 1958): 588–92. Copyright © 1958 by the American Institute of Aeronautics and Astronautics, Inc. Reprinted with permission.

60. McCauley, J. F., M. H. Carr, J. A. Cutts, W. K. Hartmann, Harold Masursky, D. J. Milton, R. P. Sharp, and D. E. Wilhelms. "Preliminary Mariner 9 Report on the Geology of Mars." *Icarus* 17 (1972): 289–327. Copyright © 1972 with permission from Elsevier.

61. Giacconi, Riccardo, Herbert Gursky, and Frank R. Paolini. "Evidence for X Rays from Sources Outside the Solar System." *Physical Review Letters* 9 (December 1, 1962): 439–43. Copyright © 1962 by the American Physical Society.

62. Schmidt, M. "3C 273: A Star-Like Object With Large Red-Shift." *Nature* 197 (March 16, 1963): 1040. Copyright © 1963 by Nature Publishing Group. Reproduced with permission of Nature Publishing Group in the format Trade Book via Copyright Clearance Center.

63. Penzias, A. A. and R. W. Wilson. "A Measurement of Excess Antenna Temperature at 4080 Mc/s." *Astrophysical Journal* 142 (1965): 419–21. Reproduced by permission of the American Astronomical Society.

64. Hewish, A., S. J. Bell, J. D. H. Pilkington, P. F. Scott, and R. A. Collins. "Observation of a Rapidly Pulsating Radio Source." *Nature* 217 (February 24, 1968): 709–13. Copyright © 1968 by Nature Publishing Group. Reproduced with permission of Nature Publishing Group in the format Trade Book via Copyright Clearance Center.

Gold, Thomas. "Rotating Neutron Stars as the Origin of the Pulsating Radio Sources." *Nature* 218 (1968): 731–32. Copyright © 1968 by Nature Publishing Group. Reproduced with permission of Nature Publishing Group in the format Trade Book via Copyright Clearance Center.

65. Becklin, Eric E. and Gerry Neugebauer. "Infrared Observations of the Galactic Center." *Astrophysical Journal* 151 (1968): 145–61. Reproduced by permission of the American Astronomical Society.

66. Davis, Raymond, Don S. Harmer, and Kenneth C. Hoffman. "Search for Neutrinos from the Sun." *Physical Review Letters* 20 (May 20,1968): 1205–9. Copyright © 1968 by the American Physical Society.

Hirata, K., et al. "Observation of a Neutrino Burst from the Supernova SN 1987A." *Physical Review Letters* 58 (April 6, 1987): 1490–93. Copyright © 1987 by the American Physical Society.

67. Klebesadel, Ray W., Ian B. Strong, and Roy A. Olson. "Observations of Gamma-Ray Bursts of Cosmic Origin." *Astrophysical Journal* 182 (June 1, 1973): L85–L88. Reproduced by permission of the American Astronomical Society.

68. Hulse, R. A. and J. H. Taylor. "Discovery of a Pulsar in a Binary System." *Astrophysical Journal* 195 (January 15, 1975): L51–L53. Reproduced by permission of the American Astronomical Society.

Taylor, J. H., L. A. Fowler, and P. M. McCulloch. "Measurements of General Relativistic Effects in the Binary Pulsar PSR1913+16." *Nature* 277 (February 8, 1979): 437–40. Copyright © 1979 by Nature Publishing Group. Reproduced with permission of Nature Publishing Group in the format Trade Book via Copyright Clearance Center.

69. Zwicky, F. "On the Masses of Nebulae and of Clusters of Nebulae." *Astrophysical Journal* 86 (October 1937): 217–46. Reproduced by permission of the American Astronomical Society.

Rubin, Vera C., W. Kent Ford, Jr., and Norbert Thonnard. "Extended Rotation Curves of High-Luminosity Spiral Galaxies. IV. Systematic Dynamical Properties, Sa→Sc." *Astrophysical Journal* 225 (November 1, 1978): L107–L111. Reproduced by permission of the American Astronomical Society.

70. Walsh, D., R. F. Carswell, and R. J. Weymann. "0957+561 A, B: Twin Qua-

sistellar Objects or Gravitational Lens?" *Nature* 279 (May 31, 1979): 381–84. Copyright © 1979 by Nature Publishing Group. Reproduced with permission of Nature Publishing Group in the format Trade Book via Copyright Clearance Center.

71. Guth, Alan H. "Inflationary Universe: A Possible Solution to the Horizon and Flatness Problems." *Physical Review D* 23 (January 15, 1981): 347–56. Copyright © 1981 by the American Physical Society.

72. De Lapparent, Valérie, Margaret J. Geller, and John P. Huchra. "A Slice of the Universe." *Astrophysical Journal* 302 (March 1, 1986): L1–L5. Reproduced by permission of the American Astronomical Society.

73. Williams, Robert E., et al. "The Hubble Deep Field: Observations, Data Reduction, and Galaxy Photometry." *Astronomical Journal* 112 (October 1996): 1335–89. Reproduced by permission of the American Astronomical Society.

74. Mayor, Michel and Didier Queloz. "A Search for Substellar Companions to Solar-Type Stars Via Precise Doppler Measurements: A First Jupiter Mass Companion Detected." In *Cool Stars, Stellar Systems, and the Sun. 9th Cambridge Workshop.* Roberto Pallavicini and Andrea K. Dupree (eds.). San Francisco: Astronomical Society of the Pacific, 1996. By the kind permission of the Astronomical Society of the Pacific.

Butler, R. Paul and Geoffrey W. Marcy. "A Planet Orbiting 47 Ursae Majoris." *Astrophysical Journal* 464 (June 20, 1996): L153–L156. Reproduced by permission of the American Astronomical Society.

75. Riess, Adam G., et al. "Observational Evidence from Supernovae for an Accelerating Universe and a Cosmological Constant." *Astronomical Journal* 116 (September 1998): 1009–38. Reproduced by permission of the American Astronomical Society.

Perlmutter, S., et al. "Measurements of Ω and Λ from 42 High-Redshift Supernovae." *Astrophysical Journal* 517 (June 1, 1999): 565–86. Reproduced by permission of the American Astronomical Society.

PHOTO CREDITS

Figure 54.1: Yerkes Observatory/*Sky & Telescope*

Figure 56.1: Lucent Technologies Inc./Bell Labs

Figure 56.3: Grote Reber

Figure 65.2: National Radio Astronomy Observatory/Associated Universities, Inc.

Figure 65.3: Andrea Ghez/UCLA

Figure 66.2: Brookhaven National Laboratory

Figure 70.3: NASA, Andrew Fruchter and the ERO Team, Sylvia Baggett (STScI), Richard Hook (ST-ECF), and Zoltan Levay (STScI)

Figure 73.1: R. Williams (STScI), the Hubble Deep Field Team, and NASA

Figure 74.1: Bradford Smith and Richard Terrile, Las Campanas Observatory

Front endpaper art (a drawing of the constellation Leo) and back endpaper art (a drawing of the constellation Aquila) are plates taken from *Uranometria* by Johann Bayer (1603). Courtesy of Jay M. Pasachoff.

WITHDRAWN
ERAU-PRESCOTT LIBRARY